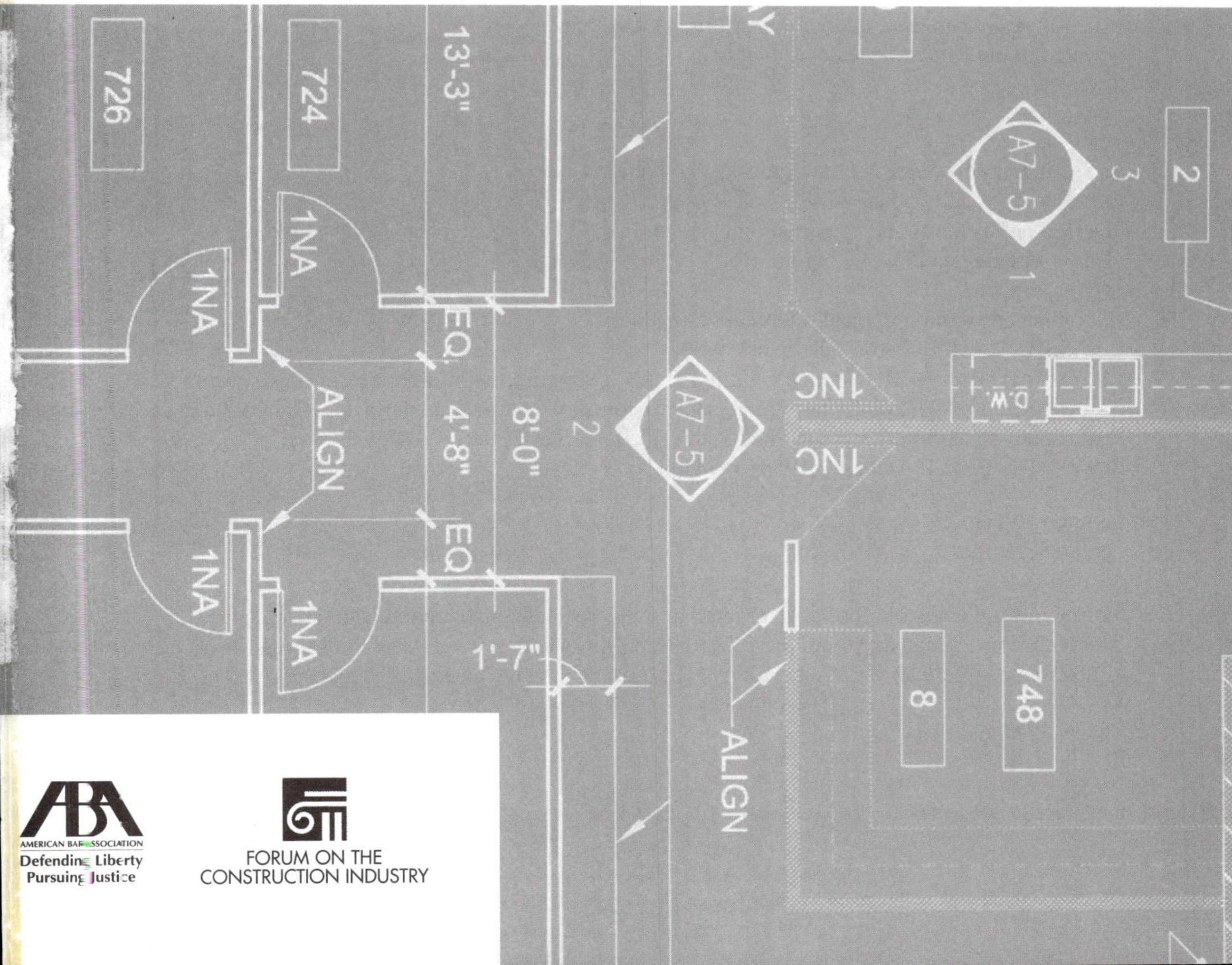

AMERICAN BAR ASSOCIATION FORUM ON THE CONSTRUCTION INDUSTRY

DESIGN PROFESSIONAL AND CONSTRUCTION MANAGER LAW

EDITORS:
STEPHEN A. HESS, JEROME V. BALES,
P. DOUGLAS FOLK AND L. TYRONE HOLT

Cover design by ABA Publishing

The materials contained herein represent the opinions of the authors and editors and should not be construed to be the action of either the American Bar Association or the Forum on the Construction Industry unless adopted pursuant to the bylaws of the Association.

Nothing contained in this book is to be considered as the rendering of legal advice for specific cases, and readers are responsible for obtaining such advice from their own legal counsel. This book and any forms and agreements herein are intended for educational and informational purposes only.

© 2007 American Bar Association. All rights reserved.
Printed in the United States of America.

10 09 08 07 06   5 4 3 2 1

Library of Congress Cataloging-in-Publication Data

Design professional and construction manager law / edited by Stephen A. Hess, Jerome V. Bales, P. Douglas Folk, L. Tyrone Holt.
    p. cm.
    Includes bibliographical references and index.
    ISBN 1-59031-786-6 (alk. paper)
    1. Architects—Malpractice—United States. 2. Engineers—Malpractice—United States. 3. Architecture—United States—State supervision. 4. Construction contracts—United States. 5. Dispute resolution (Law)—United States. I. Hess, Stephen A. II. Bales, Jerome V., 1949– III. Folk, P. Douglas, 1955– IV. Holt, L. Tyrone.

KF2925.3.D47 2007
343.73′07872—dc22   2006037349

Discounts are available for books ordered in bulk. Special consideration is given to state bars, CLE programs, and other bar-related organizations. Inquire at ABA Publishing, Book Publishing, American Bar Association, 321 N. Clark Street, Chicago, IL 60610.

www.ababooks.org

FROM THE ABA FORUM ON THE CONSTRUCTION INDUSTRY

For more information on these books, or to order copies, please visit the ABA Web Store at www.ababooks.org, or call the ABA Service Center at 1-800-285-2221.

The A201 Deskbook (PC 5570059)

Construction Accident Law (PC 5570063)

THE Construction Contracts Book (PC 5570203)

Construction Damages and Remedies (PC 5570201)

The Construction Lawyer's Guide to Labor & Employment Law (PC 5570073)

Design Professional and Construction Manager Law (PC 5570206)

The Design/Build Deskbook, Third Edition (PC 5570088)

Discovery Deskbook for Construction Disputes (PC 5570094)

Federal Government Construction Contracts (PC 5570083)

Forms & Substance: Specialized Agreements for the Construction Project (PC 5570098)

Fundamentals of Construction Law (PC 5570078)

Partnering in Construction (PC 5570064)

Sticks & Bricks: A Practical Guide to Construction Systems and Technology (PC 5570075)

Sweet on Construction Law (PC 5570053)

CONTENTS

Preface xxv

Introduction xxvii

About the Authors and Editors xxix

CHAPTER 1
State Regulation of the Design Professional 1
G. WILLIAM MYERS, P.E., ESQ.
    I. Introduction 1
   II. Individual Licensure 2
  III. Firm Licensure 5
       A. Overview 5
       B. Noncompliance Has Its Price 9
  IV. Firm Practice 11
       A. Who Is to Seal? 11
       B. What Is to Be Sealed and When? 14
       C. How Are Documents to Be Sealed? 14
       D. Other Requirements Imposed by Some States 15
   V. Conclusion 16
  VI. Practice Aids 17
       A. Checklist for Practice in a New Jurisdiction 17
       B. NCARB Certificates and NCEES Records 19
       C. Sample Interstate Commerce Letter 20

CHAPTER 2
Ethics and Rules of Conduct Governing Design Professionals 23
MOHAMMAD IQBAL, P.E., ESQ.
    I. Introduction 23
       A. Ethical Issues in the Design Process 24
       B. Rules of Professional Conduct 26
   II. Code of Ethics 27
       A. Basic Concepts and Fundamental Principles 27
       B. Historical Development 28
       C. Discussion of Professional Codes of Ethics 29
  III. AIA Ethical Standards and Rules of Professional Conduct 30
       A. Overview 30
       B. Enforcement 31
  IV. ASCE Code of Ethics and Rules of Professional Conduct 32

V.	NSPE Code of Ethics.	34
	A. Overview	34
	B. Indemnification, Limitation of Liability, and Ethics	34
VI.	Law, Ethics, and Professional Conduct.	36
VII.	Ethical Dilemma	38
VIII.	Licensing Boards Governing Design Professionals.	39
	A. Licensure Laws.	39
	B. Licensing Board Powers.	40
	C. Enforcement Powers.	41
	D. Investigative and Disciplinary Powers	42
	E. Revocation and Suspension of Licenses.	43
	F. Disciplinary Reciprocity.	44
	G. Judicial Review	45
	H. Standard of Review	46
IX.	Hyatt Regency Sky Walkway Collapse.	47
X.	Conclusion.	51
XI.	Appendices.	51
	A. Professional Organizations.	51
	B. Licensure Boards Web Site Addresses.	52

CHAPTER 3
Design Professional Contract Responsibilities—Pre-Design, Design, and Pre-Construction Phases 55

THEODORE D. LEVIN, P.E., ESQ.
MEHRDAD FARIVAR, ESQ., FAIA

I.	Introduction	55
II.	Conceptual/Pre-Design Phase: Contractual and Extra-Contractual Roles of the Design Professional	56
	A. The Process and the Product—Unique Aspects of the Design and Construction Process.	56
	1. The Design Professional as Advisor: To Be or Not to Be	57
	2. Understanding the Owner's Initial Concept.	59
	a. Project Use.	59
	b. Size of the Project	59
	3. Initial Formation of the Project Team	59
	a. Design Consultants.	59
	b. Project Consultants.	60
	4. Issues and Obligations in Dealing with Governmental Agencies/Approvals.	61
	a. Entitlements	61
	b. Building Permits and Approvals/Inspections.	62
	5. Implied Obligations of the Design Professional in the Conceptual/Pre-Design Phase: Dealing with Governmental Agencies and Approvals.	62
	B. Potential Liabilities of the Design Professional in the Pre-Design Phase	63
	1. Liability Arising from Initial Proposals	63

		2. Setting the Project Parameters 64
		a. Generally 64
		b. The Initial Program 65
		3. Initial Budget Estimates 66
		4. Initial Schedule Estimates 67
	III.	Contract Negotiation and Formation 68
		A. The Roles, Responsibilities, and Obligations of the Parties in Entering into the Contract 69
		B. The Contract 71
		1. Standard Form Contracts—the Test of Time? 71
		2. The Use of Custom Contracts 73
		3. Specific Contract Provisions—the Risks Defined 74
		4. Indemnity 75
		5. Basic vs. Additional Services.................................. 76
		6. Payments to the Design Professional 78
		7. Termination Issues 78
		8. Budget Estimates 80
		9. Schedule Estimates 81
		10. Copyright Issues 81
		11. Dispute Resolution 82
		12. Insurance 83
		13. Consequential Damages—Waiver and Related Issues 84
		14. Limitations of Liability 85
		15. Modifications and Changes to the Contract: Changes in Services 87
	IV.	The Design Services Phase: Roles and Responsibilities 88
		A. Contractual Roles and Liabilities 88
		1. Schematic Design 88
		2. Design Development 89
		3. Construction Documents 89
		B. Implied Obligations of the Design Professional during the Design Services Phases 90

CHAPTER 4
Design Professional Legal Responsibilities—Construction and Completion Phases 93
DAVID HATEM, ESQ.

I.	Introduction 93
II.	Construction Phase: Roles and Responsibilities 93
	A. Roles and Responsibilities of Project Participants in Design-Bid-Build 94
	1. Traditional Roles and Responsibilities 94
	2. Relationship of Roles and Responsibilities to Risk Allocation 94
	3. Variations on Traditional Roles and Responsibilities in Design-Bid-Build 95
	B. Roles and Responsibilities of the Design Professional during the Construction Phase 96
	1. Roles and Responsibilities Defined by Contract and Conduct 96
	2. Modification of Roles and Responsibilities during the Construction Phase 97

- 3. Standard Contract Terms Regarding Design Professional Roles and Responsibilities during the Construction Phase 98
  - a. Site Visits 98
  - b. Submittal Review 100
  - c. Construction Means/Methods 101
  - d. Safety 102
  - e. Quality Assurance 102
  - f. Rejection of Work 103
  - g. Contractor Claims, Payment Issues, and Interpretation of the Contract Documents 104
  - h. Termination of the Contractor 106
- 4. Commonly Recognized Implied Obligations of the Design Professional in the Performance of Construction Phase Services 107
- 5. Areas of Potential Professional Liability Exposure 108
  - a. Owner Claims 108
  - b. Third Party Claims 109
  - c. Contractor/Subcontractor Claims 110
  - d. Personal Injury Claims 111
  - e. OSHA 113
  - f. False Claims Act 113
- III. Completion and Commissioning Phase: Roles and Responsibilities 114
  - A. Substantial Completion 115
  - B. Punchlists 118
  - C. Final Completion 119
  - D. Warranties 120
  - E. Project Close-Out Documentation 121
  - F. Support to Owner in Evaluation/Defense of Contractor Claims 121

CHAPTER 5
Design Professionals and the Design-Build Project 123
PETER HALLS, ESQ.
BERNARD (B.J.) NODZON, ESQ.
NATHAN A. BRENNAMAN, ESQ.
- I. Introduction 123
- II. Differing Roles for Design Professionals in Design-Build Projects 124
  - A. Design-Build Led by the Design Professional 124
  - B. Design Professional as a Subcontractor to Design-Builder 125
  - C. Design Professional as a Member of a Joint Venture or Other Business Entity 125
  - D. Bridging 126
- III. General Considerations 126
  - A. Fast-Track Construction 126
  - B. Less Complete Design 127
- IV. Design-Build Led by Design Professional 128
  - A. Reduction of Claims by Practice 128
  - B. Warranty of Design 128
  - C. Strict Liability for the Design-Builder 131
  - D. Contractor-Type Liability 131

		E.	Insurance Coverage and Bonding. 132
		F.	Increased ADA Liability 132
		G.	Economic Loss Doctrine Eroded 134
		H.	Contractor Licensing 134
	V.	Contractor-Led Design-Builder. 135
		A.	Different Dynamics 135
		B.	Murder Clauses. 135
			1. Pay-If-Paid or Pay-When-Paid Clauses in Design-Build Contracts. 136
			2. Warranty Clauses 136
			3. Indemnification Clauses. 137
			4. Flow-Down Clauses. 137
		C.	Licensing Issues—Illegality of Design-Builder Not Having a Design Professional License. 138
		D.	Bond Claim Limitations 138
	VI.	Member of a Joint Venture or Other Business Entity 139
		A.	Choosing a Business Form 139
			1. Joint Ventures. 139
			2. Limited Liability Companies 139
			3. Partnerships 140
			4. Limited Liability Partnerships 140
			5. Corporations. 140
		B.	Insurance and Bonding. 141
		C.	Non-Compete and Exclusivity 141
		D.	Teaming Agreements 142
	VII.	Bridging. 142

CHAPTER 6
The Design Professional and Tort Liability **145**
TERENCE J. SCANLAN, ESQ.
LINDSEY E. MALONE, ESQ.

	I.	Claims against Design Professionals. 145
		A.	Strict Liability. 145
		B.	Negligence 146
			1. Standard of Care for Design Professionals 147
			2. Scope of Design Professional's Duty. 147
			3. Necessity of Expert Witness Testimony 148
			4. Causation 149
		C.	Gross Negligence 149
			1. Definition 149
			2. Effect on Contractual Limitations of Liability. 151
		D.	Breach of Fiduciary Duty. 151
		E.	Fraud—Elements. 153
		F.	Negligent Misrepresentation 154
	II.	Issues Affecting the Tort Liability of Design Professionals 155
		A.	Assumption of a Higher Degree of Care 155
		B.	Warranty Disclaimers. 156
		C.	Contract Modifications. 157

		1.	Use of a Master Agreement 157
		2.	Amendments and Modifications............................... 157
		3.	Scope Creep 157
	D.	Preparation of Plans and Specifications 158
	E.	Environmental Liability 159
III.	Who Can Sue Design Professionals: Economic Loss Rule 160
	A.	Application to Parties in Privity of Contract 160
	B.	Application to Third Parties Not in Privity 161
		1.	Two Tests 162
			a. Damages-Driven Test................................. 162
			b. Foreseeable Duty Test 163
	C.	Claims against Individual Design Professionals 164
	D.	Exceptions to Application/Enforcement of the Doctrine 165
		1.	Negligent Misrepresentation 165
		2.	Sudden/Calamitous Event 166
IV.	Site Safety 167
	A.	Background.. 167
	B.	Responsibility of the Contractor and Owner 168
	C.	Indemnity for the Design Professional............................. 171
	D.	Theories of Liability against the Design Professional 172
	E.	The Design Professional's Role on Site............................. 173
	F.	Design Professional Immunity Statutes 176
	G.	Control over the Work and OSHA Fines 177
	H.	Contract Provisions 179
	I.	Duty to Keep Own Employees Safe................................ 179
	J.	Hazardous Waste Sites 180
	K.	The Safety Dilemma... 181

CHAPTER 7
Design Professional Work Product—Ownership and Protection 183
BEN PATRICK, ESQ.

I.	Introduction 183
II.	What Types of Works Are Subject to Copyright Protection?............... 184
	A. Generally 184
	B. Copyrights Distinguished from Patents 184
III.	Who Owns the Copyright? 185
	A. The Default Rule: Plans Are Owned by the Designer 185
	1. The Work Made for Hire Doctrine............................ 185
	2. Co-Authorship 186
	B. Conveying Copyright Interests.................................... 186
	1. Ownership by the Designer with a Non-Exclusive License for the Owner 187
	2. A Reverse License 187
	3. Ownership of the Work by the Owner......................... 188
	4. Conveying Copyright Interests at a Later Date 189
	a. Licenses 189
	b. Transfer of Ownership................................. 190
	c. Transfer of Copyrights................................. 190

		C.	Common-Law Copyrights and Registered Copyrights 191
			1. Copyright Notice 191
			2. Formal Registration 191
IV.	What Protection Does the Copyright Afford? 192
	A.	A Brief History of Copyright Protection for Architectural Plans......... 192
	B.	The Useful Article Exception..................................... 192
	C.	The Architectural Works Copyright Protection Act................... 193
	D.	Rights Protected by Copyright 194
		1. The Right to Reproduce the Plans Themselves 194
		2. The Right to Prepare Derivative Works Based on the Plans 194
		3. The Right to Build the Structure Depicted in the Plans............. 194
	E.	Scope of Protection for Derivative Works 194
	F.	Scope of Protection for Works Consisting of Generic Features.......... 195
	G.	The Idea/Expression Dichotomy 195
V.	What Is Copyright Infringement? 197
	A.	Ownership of a Valid Copyright 197
	B.	Access to the Copyrighted Work by the Alleged Infringer 198
	C.	Substantial Similarity between the Copyrighted Work and the Allegedly Infringing Work 198
		1. The Extrinsic Test 199
		2. The Intrinsic Test 199
	D.	Defenses 199
		1. Independent Creation 199
		2. Fair Use 200
		3. Statute of Limitations... 201
	E.	Damages.. 201
		1. Plaintiff's Actual Damages 201
		2. Defendant's Profits Resulting from the Infringement 201
		3. Statutory Damages 202
		4. Injunctive Relief.. 202
		5. Attorneys' Fees and Costs 202
		6. What Information Is Protected by Trade Secret Laws?............ 203
VI.	Conclusion.. 203

CHAPTER 8
State Regulation of the Construction Manager 205
KEVIN L. KOLTON, ESQ.
VIRGIL R. MONTGOMERY, ESQ.

I.	Introduction 205
II.	State Licensing and Project Delivery Requirements 205
	A. States That Specifically Regulate Construction Managers............. 206
	1. Idaho... 206
	2. Oklahoma... 207
	B. Jurisdictions with Little or No Regulation 210
	C. States Regulating Construction Managers as Contractors or Design Professionals 210

		1.	Regulation as Contractor 211
			a. Express Regulation of Construction Managers as Contractors. 211
			b. Implicit Regulation as Contractors. 216
		2.	Regulation as Design Professional. 219
			a. Express Regulation as Design Professional 219
			b. Regulation under State Procurement Laws 221
	D.	Other Issues of Concern in Architect, Engineering, and Contractor Licensing 225	
		1.	Design Professional Licensing. 226
		2.	Contractor Licensing 227
III.	Ethical Considerations in the Relationship between Construction Manager and Owner. 230		
	A.	Public Projects 231	
		1.	Contingent Fees 232
			a. California 232
			b. Florida 232
		2.	Separate Contracts/Entities for Design, Management, Construction. 233
			a. Indiana 233
			b. Louisiana 233
			c. Michigan 233
			d. Missouri 234
			e. South Carolina 235
		3.	Exceptions to Prohibitions against Single Entities. 235
			a. Massachusetts. 235
			b. Minnesota. 235
			c. North Carolina 236
			d. Oklahoma. 236
			e. Texas 237
	B.	Private Projects. 237	
		1.	Certification 238
		2.	Accreditation 239
		3.	Codes of Ethics. 240
IV.	Practice Aids 242		
	A.	Checklist. 242	
	B.	Research and Resource Links. 243	

CHAPTER 9
Construction Management Overview 245
RICHARD D. CONNER, ESQ.
SHIV GHUMAN O'NEILL, ESQ.

I.	Introduction 245
II.	Historical Evolution of Construction Management. 248
III.	Recognition of Construction Management as a Profession. 253
IV.	Construction Management and the Professional Standard of Care 255
V.	Differences from Design and Construction Professions 256

VI. Construction Management Services Overlap with Other Professions. 257
VII. Major Categories of Service 258
    A. Agency CM 259
    B. CM At-Risk 260
    C. Program Management Services 261
    D. Value Engineering Services 262
    E. Owner's Representative Services. 262
VIII. Standard Form Agreements 263
    A. CMAA 263
        1. Pre-Construction Phase 263
        2. Construction Phase. 265
    B. AIA. 265
        1. Pre-Construction Phase 265
        2. Construction Phase. 266
    C. AGC 267
        1. Pre-Construction Phase 267
        2. Construction Phase. 267
    D. Comparison of the Standard Form Agreements 268
IX. Compensation 270
    A. Time-Based Fee 270
    B. Lump Sum/Fixed Fee 272
    C. Cost of Work Plus a Fixed Fee. 273
    D. Fee as a Percentage of Construction Cost 274
    E. Cost of Work Plus a Fee with a GMP 275
    F. Comparison of Compensation Arrangements 278
X. Practice Aids 279
    A. Treatises 279
    B. Web Sites 279

CHAPTER 10
Construction Manager's Responsibilities: Pre-Design, Design, and Pre-Construction Phase **281**
PAUL H. MARTINEZ, ESQ.
RASHIDA Y.V. MacMURRAY, ESQ.
I. Introduction 281
II. Relationship with Owner 282
III. Pre-Design 283
    A. Project Scope 283
    B. Environmental Impact Statements 284
        1. Phase I Environmental Assessment 285
        2. Phase II Environmental Assessment 285
    C. Site and Existing Building Survey Hazardous and
        Contaminated Substances. 286
    D. Geotechnical Investigation. 286
    E. Funding. 287
    F. Land Acquisition 287
    G. Engaging Design Professional 288

		H. Adjacent Site Investigation. 289

        H. Adjacent Site Investigation. 289
        I. Utility Service. 290
IV. Design 290
        A. Phases 290
            1. Schematic Design Drawings 292
            2. Design Development Drawings 293
            3. Construction Drawings and Specifications
               (Construction Documents). 293
        B. Estimate 294
        C. Project Progress Scheduling. 297
        D. Value Engineering 298
        E. Constructability 299
        F. Site Logistics 300
        G. Green Building and Related Compliance 301
V. Pre-Construction 303
        A. Long-Lead Materials and Equipment. 303
        B. Permits and Approvals 303
        C. Early Work 304
        D. GMP Estimate 304
VI. Compensation 304
        A. Consulting. 304
            1. Time and Reimbursable Expenses 305
            2. Not-to-Exceed 305
            3. Fixed Price 305
        B. Early Work 305
VII. Changes 306
        A. Consulting Additional Services 306
        B. Change Orders for Early Work. 306
VIII. Insurance 307
        A. Construction Manager's Professional Liability Insurance 308
        B. Contractor's Pollution Liability Insurance 308
        C. Builder's Risk Insurance 309
IX. Suspension or Termination 310
X. Claims 311
        A. *Force Majeure* and Extension of Time. 311
        B. Notice 311
XI. Contractual Risk 311
        A. Scope of Services 312
        B. Exclusion of Design Responsibility 312
        C. Waiver of Consequential Damages 313
        D. Limitation of Liability 314
        E. Guarantee of Estimates. 314
        F. Indemnification. 315
XII. Conclusion 316
XIII. Practice Aids 316
        A. Stand Alone Pre-Construction Agreement 316

  B. Part A—Pre-Construction Agreement 337
  C. Pre-Construction Agreement Checklist 348

CHAPTER 11
Construction Management Contract Responsibilities: Construction and Completion Phases. 349
STANLEY A. MARTIN, ESQ.
 I. Introduction 349
  A. Construction Manager's Responsibility during Construction 350
  B. Construction Manager's Responsibility during Completion 350
 II. The Construction Manager's Role, Risks, and Responsibilities during Construction 351
  A. Procurement of Trade Contractors 351
   1. Work Scope Allocation 352
   2. Trade Contractor Qualification. 352
   3. Trade Contractor Bidding and Selection 353
  B. Management of On-Site Work 354
   1. Coordination of Trades. 354
   2. Coordination of Changes and Modifications 354
  C. Cost Control 355
   1. Use of Contingency during Construction. 355
   2. Management of Trade Contractor Changes 357
   3. Continued Consulting on Value Engineering and Substitutions 358
  D. Schedule Control 358
   1. Establishing or Maintaining Baseline Schedule. 358
   2. Developing Schedule Updates 359
   3. Control of Trade Contractor Schedules and Coordination. 360
   4. Identification of Potential Schedule Impacts and Developing Alternatives. 360
   5. Float 361
  E. Quality Control and Quality Assurance 361
   1. Responsibilities of the Construction Manager as Design Is Finalized. 362
   2. Responsibility of the Construction Manager Vis-à-Vis Designer 363
  F. Risk Management. 363
   1. Contracts and Delegation of Risk. 364
   2. Insurable Risks. 364
   3. Surety Bonds and Performance Risk 365
   4. Management of Risk That Is Not Transferred or Insured 366
  G. Safety 366
  H. Dispute Resolution 366
  I. Customer Service and Customer Relations 367
 III. The Construction Manager's Role, Risks, and Responsibilities during Completion 367
  A. Management of Punch List. 368
  B. Commissioning and Turnover 369

		1.	Training 370
		2.	Start-Up of Equipment 370
	C.	Procuring and Handling Warranties/Guarantees 371
		1.	Responsibility to Obtain Warranty 371
		2.	Responsibility for Performance of Warranty Obligations by Vendors/Subcontractors 371
	D.	Audit or Cost Verification 372
IV.	Coordination with Others—an Independent Duty, or an Adjunct to All Other Tasks? 373
V.	Construction Manager as Advisor or Agent 373
	A.	Agency Role as Shield from Third Party Claims 373
	B.	Construction Manager's Duty to Owner as Agent 374
	C.	Imputation of Conduct to Owner 374
VI.	Conclusion: Use of Standard Form Contracts and Documents 375
	A.	CMAA 375
	B.	AGC 375
	C.	AIA 376
VII.	Treatises and Resources 376

CHAPTER 12
The Construction Manager and Tort Liability 377
WENDY KENNEDY VENOIT, ESQ.
DAVID A. DELMAN, ESQ.

I.	Introduction 377
II.	Tort Liability of the Construction Manager to the Owner 378
	A.	Theories of Liability 378
		1.	Professional Negligence 378
			a. Duty/Standard of Care 379
			b. Failure to Comply with Licensing Requirements 381
			c. Failure to Comply with Procurement Codes 384
			d. Failure to Comply with Applicable Building Codes 384
			e. Failure to Comply with Industry Standards 385
		2.	Breach of Fiduciary Duty 385
		3.	Intentional/Negligent Misrepresentation 386
		4.	Unfair Trade Practices 387
	B.	Bases of Liability 389
		1.	Defective Design/Design Errors 389
		2.	Faulty Workmanship and Improper Inspection/Supervision 391
		3.	Negligent Cost Estimates/The Over-Budget Project 393
		4.	Improper Certification of Progress Payments 395
		5.	Improper Schedule Preparation, Monitoring, and Coordination 396
		6.	Improper Processing of Submittals 397
		7.	Self-Dealing/Conflict of Interest/Lack of Independence 398
		8.	Collusion with Contractors 398
III.	Construction Manager's Liability to Third Parties 400
	A.	Theories of Liability 400

		1. Negligence 400
		2. Negligent Misrepresentation 402
		3. Interference with Contract 404
		4. Defamation 405
	B.	Bases for Liability 406
IV.	Owner's Liability to Third Parties Based on Imputation of Construction Manager's Misconduct 409
V.	Construction Manager Defenses 410
	A.	Contract Claims Precluding Tort Claims 410
	B.	Contributory/Comparative Negligence 411
	C.	Indemnity and Contribution 411
	D.	Superior Knowledge 412
	E.	Limitations of Authority 412
	F.	Exclusivity of Remedies/Limitation of Liability Provisions 412

CHAPTER 13
The Construction Manager and the Design-Build Project 415
JEAN C. ARNOLD, ESQ.
CHRIS SQUADRA

I.	Introduction 415
II.	The Construction Manager as Design-Build Team Leader 416
	A.	Conflicts of Interest 417
	B.	Vicarious Liability for Design 417
	C.	State Licensing Requirements 418
III.	Managing Total Project Costs on the Design-Build Project as a Design-Builder 418
	A.	Owner's Budget vs. Raw Construction Cost 419
	B.	Escalation: *Force Majeure* or Reasonably Inferable in Today's Economy 420
	C.	Design Team Fees and Reimbursables vs. Additional Services 421
		1. Redesign Requirements for the Design Team as Basic Services 421
		2. The Approvals Process: Unknown Hourly Billings or Predictable Lump Sum 423
		3. Controlling Owner Changes during Design as a Design-Builder 424
		4. Managing a Design Team Schedule: Contractual Responsibility 425
	D.	Predicting Permits, Fees, and Other Governmental Approval Costs 426
	E.	Owner and Design-Builder Contingencies 426
		1. Owner Contingency 426
		2. How Much Is Enough? 427
		3. Owner vs. Design-Builder Contingency 429
	F.	General Conditions vs. the Weatherman: Controlling Weather Protection 429
	G.	Value Engineering vs. Value Elimination 431
		1. Value Engineering—Owner's Life-Cycle Costs Defined and Categories Identified 431

	2. Managing the Implicit Design-Builder Incentive to Lower Costs at the Expense of Life-Cycle Cost 432
IV.	Ownership and Licensing of the Contract Documents 433
	A. Owner Retains the Right to Own Documents If the Design-Build Contract Is Terminated 435
	B. Documents in Electronic Format 436
	C. Use of Documents by Another Designer 436
V.	The Owner's Voice: Managing Input, Buy-In, and Approval during Pre-Construction. 437
VI.	The Construction Cost Estimate 440
VII.	Constructability Review 441
VIII.	The Construction Schedule 443
IX.	Application of Payment Review 443
X.	Use of a Commissioning Agent. 446
XI.	Friendly Contract Clauses for the Construction Manager as Design-Builder. 447
	A. Correcting Nonconforming Work 447
	B. Indemnity 449
XII.	Appendices. 451
	A. Sample Responsibility Matrix 451
	B. Instructions for Using the Fee and Reimbursables Worksheet. 461

CHAPTER 14
Design Professional/Construction Manager Indemnification and Professional Liability Insurance 463
KATHERINE DAVITT ENOS, ESQ.

I.	Introduction 463
II.	Standard Indemnification Obligations. 464
III.	Broad Form (Type I) Indemnification 464
IV.	Intermediate Form 465
V.	Comparative (Limited) Form. 465
VI.	Insurance Impacts 468
VII.	Restrictions on Indemnifications 469
VIII.	Understanding Standard Insurance Policies 469
	A. Professional Liability Coverage. 471
	B. Prior Acts Coverage 471
	C. Tail Coverage—Extended Reporting Periods 472
	D. General Liability Coverage 472
	E. Property Damage 473
	F. Bodily Injury 473
	G. Premises and Operations Liability 474
	H. Contractual Liability Coverage 475
	I. Personal and Advertising Injury. 475
	J. Libel and Slander 476
	K. Medical Payments 476
	L. Additional Insured Status on the CGL Policy 476
	M. Construction Management: Agency vs. At-Risk 478

	N.	Standard Exclusions 479
	O.	Express Warranties and Guarantees 480
	P.	Limits of Liability... 480
	Q.	Deductibles... 482
	R.	Who's Covered.. 482
	S.	Wrap-Up Project Policies.. 484
	T.	Considerations with Wrap-Ups 485
	U.	Attorneys' Fees and Litigation Costs 486
	V.	Pre-Claims Assistance 486
IX.	Evidence of Insurance 487
	A.	Binders 487
	B.	Certificates of Insurance... 487
X.	Recent Claim Trends 488
	A.	Meritless Claims... 488
	B.	Design–Build 489
	C.	Construction Managers... 489
XI.	Further Resources for Understanding Risk Transfer and Sharing............. 489
XII.	Conclusion.. 490

CHAPTER 15
Considerations for Subconsultant and Subcontract Agreements for Design Professionals and Construction Managers 491
L. TYRONE HOLT, ESQ.
CARRIE LYNN H. OKIZAKI, ESQ.

I.	Introduction 491
II.	The Process of Choosing a Subcontractor................................. 492
III.	Unique Legal Issues Applicable to the Prime Contractor/ Subcontractor Relationship 495
	A.	Vicarious Liability, Indemnification, and Contribution 496
	B.	Exculpatory, Risk Allocation, and Limitation of Liability Clauses....... 499
	C.	Payment Clauses... 499
	D.	Insurance Issues 501
	E.	Dispute Resolution 503
	F.	Flow-Down Clauses 504
	G.	Applicability of Economic Loss Rule 505
IV.	Drafting Subcontract Agreements 506
	A.	Rule No. 1: Get It in Writing, and Get It Signed! 507
	B.	Rule No. 2: Develop a "Form" Contract for All Subcontracts on the Project 507
		1. Form or Custom Subcontract Agreements....................... 508
		2. Developing a Standard or Template Base Subcontract Agreement.... 508
	C.	Rule No. 3: Identify Key Subcontract Terms and Conditions 511
V.	Subcontract Negotiations.. 511
VI.	Subcontract Issues Unique to Offshore Subcontracting Relationships 517
VII.	Conclusion.. 519
VIII.	Practice Aids 519
	A.	Standard Construction Manager and Subcontractor Contract Forms Available 519

IX. Appendix 522
    A. Sample Subcontract Agreement Consultant's Agreement to Furnish Services to PRIME CONTRACTOR 522

CHAPTER 16
Damages and Remedies 529
STEPHEN MILLER, ESQ.
STEPHEN A. HESS, ESQ.
  I. Introduction 529
 II. Causes of Action 530
    A. Breach of Contract and Related Claims 530
       1. Breach of Contract 530
       2. Promissory Estoppel 531
       3. Unjust Enrichment 532
    B. Tort Claims 532
III. General Measures of Damages 534
    A. Breach of Contract and Related Claims 534
       1. Breach of Contract 534
          a. Design Professional and Construction Manager Compensation ... 535
          b. Owner's Compensation 535
          c. Contractor's Compensation 537
       2. Promissory Estoppel 538
       3. Unjust Enrichment 538
    B. Tort Claims 539
IV. Calculation and Proof of Compensatory Damages 540
    A. General Damages 540
    B. Direct Costs 540
       1. In General 540
       2. Price Increases for Labor and Material 541
       3. Idle Equipment and Related Costs 541
       4. Home Office Overhead 542
       5. Lost Profits 544
       6. Delay/Disruption Damages 544
       7. Loss of Efficiency 545
       8. Liquidated Damages 545
       9. Interest 547
      10. Attorneys' Fees 547
    C. Alternative Methods for Calculating Damages 548
       1. Total Cost 548
       2. Modified Total Cost 549
       3. Measured Mile 549
       4. Should-Cost Estimates 550
       5. Industry Standards and Studies 550
       6. Time-and-Motion Studies 550
       7. Jury Verdict Method 550
    D. Consequential Damages 551
    E. Punitive Damages 551

V. Defenses to and Mitigation of Damage Claims..........................551
   A. Economic Waste..551
   B. Failure to Mitigate Damages 552
   C. Betterment 552
VI. Protection for Performance and Payment Rights.......................554
   A. In General..554
   B. Retention and Limitations Thereon 554
   C. Labor and Material Payment Bonds................................555
   D. Performance Bonds 555
   E. Mechanics' Liens 555
   F. Miller Act..556
   G. State Public Works Claims (Little Miller Acts)...................557
   H. Prompt Payment Statutes 557
VII. Contractual and Other Limitations on Liability......................558
   A. Exculpatory Clauses..558
   B. Waivers of Consequential/Incidental Damages 559
   C. Indemnification Clauses..559
   D. Disclaimer of Warranties 560
   E. Waiver of Lien Rights 560
   F. Limitations of Liability...560
   G. No Damages for Delay 561
   H. Pay-if-Paid/Pay-When-Paid Clauses 561
   I. Immunity under Workers' Compensation Laws 562
   J. Immunity from Site Safety Claims.................................563
   K. One-Call Statutes and Damage to Underground Structures............563

CHAPTER 17
Alternative Dispute Resolution and the Construction Dispute................565
DAVID W. RUDY, ESQ.
   I. Introduction: Litigation and Alternative Dispute Resolution 565
   II. Overview of Types of ADR..566
      A. Arbitration..566
      B. Mediation..567
      C. Partnering...567
      D. Judicial or Quasi-Judicial Procedures........................568
         1. Early Neutral Evaluation 569
         2. Mini-trial 569
   III. Strategies in Arbitration.......................................570
      A. Overview 570
      B. Law Governing..571
         1. The Contract Provision...................................572
         2. Contractual Preconditions 574
      C. Multiple Party Disputes and Fashioning Adequate Relief..............574
      D. Selection of the Forum and Procedural Considerations.................577
      E. Selection of the Arbitrator(s): Additional Considerations 579
      F. Pre-Hearing Issues 581
         1. Discovery..581

		2.	Expert Retention and Preparation 581
		3.	Pre-Arbitration Hearings 582
		4.	Briefing.. 583
	G.	The Arbitration Hearing 584
		1.	Scheduling 584
		2.	Special Issues in the Arbitration of Construction Cases 585
			a. Historical Revisionism..................................... 585
			b. Delay Claims 586
			c. Formality of the Process................................... 586
			d. Discontinuity in the Hearing 586
			e. Use of Technology in Proceedings 587
			f. Nature of the Award.. 588
	H.	Post-Hearing Procedures 588
		1.	Requesting Correction or Modification of the Award 588
		2.	Confirming the Award 588
		3.	Vacating the Award 589
IV.	Strategies in Mediation 590
	A.	Selection of the Process 590
		1.	Overview of the Mediation Process.............................. 590
		2.	Selection of a Specific Process................................... 591
	B.	Selection of the Mediator 595
	C.	Planning for and Attending the Mediation........................... 597
		1.	How Much Expert Work before Mediation?...................... 597
		2.	Use of Technology, Live Experts at the Mediation 598
		3.	Developing a Mediation Strategy................................ 599
			a. Pre-Mediation Negotiation or Positioning................... 599
			b. Proactive Participation in Defining the Process.............. 600
			c. Making Offers vs. Discussing Positions 601
			d. Briefing... 602
			e. Client and Advocate Exceptions of Process and Result........ 603
		4.	Wrapping Up the Successful Mediation......................... 605

CHAPTER 18
Special Topics in Litigation of Disputes Involving Design Professionals or Construction Managers 607
KENNETH STRONG, ESQ.
ERNEST ISOLA, ESQ.

I.	Introduction 607
II.	Time Limits for Filing Suit 607
	A.	Statutes of Limitation and Related Laws 607
	B.	Statutes of Repose 608
	C.	Using Tolling or Standstill Agreements to Defer Litigation 610
III.	Certificates of Merit.. 611
	A.	Who Is Covered? 612
	B.	What Are the Requirements?... 612
IV.	Management of Litigation through the Use of Special Masters 614

V. Use of Experts 616
    A. The Trial Judge Has a Gatekeeping Obligation Pursuant to
       Daubert, *Joiner*, and *Kumho Tire*. 616
    B. *Daubert/Kumho Tire* Extends to Both an Expert's Testimony
       and the Conclusions Reached by the Expert 617
    C. *Daubert/Kumho Tire* Applies in Construction or Closely Related Cases:
       It Must Be Reasonable for an Expert to Rely on That Which Forms
       the Basis of His or Her Opinion 618
VI. Discovery, Management, and Use of Evidence in Complex Construction
    Cases 619
    A. Document Management and Use of Databases 619
    B. Electronic Discovery 621
    C. Demonstrative Evidence. 625
VII. Using Joint Defense Agreements to Avoid Finger-Pointing and
    Manage the Dispute 628
    A. Practical Considerations for a Joint Defense 631
    B. The Necessity of a Writing. 633

Index **635**

PREFACE

The idea for this book came from the members of the Forum Governing and Publications Committees. In October 2004, Adrian Bastianelli of the Publications Committee asked the Steering Committee for the Forum's Division 3–Design to take responsibility for the development and publication of new book project. Division 3 agreed to his request with the understanding that its responsibilities included the coordination and management of the project, utilizing editors and authors from the Forum as a whole, and with the express understanding that the book would be a Forum-wide effort. In his first communication on this Project, Adrian asked us to consider whether the book should address both design professional and construction management liability.

At the outset, Division 3 concluded that the current needs of the industry required the book to address both design professional liability and the developing concepts and principles applicable to the liability of construction managers. This decision was made in recognition of the importance of construction managers and the emerging and expanding role of construction management as a professional discipline in the construction industry.

At the time that Division 3 started to work on this project, Ted Levin, Ken Strong, Michael Hanahan, Richard Tyler, Margaret Lineberry, and Jerry Bales served as the members of the Division 3 Steering Committee. I had the privilege of serving as the Chair of the Steering Committee for Division 3–Design.

Ted Levin, Ken Strong, Michael Hanahan, Mark Heley, and I developed the outline for the book content. This outline was approved by the Publications Committee in the Spring of 2005. Ty Holt and Mark Heley developed the organization structure for the production of the book and shepherded the outline and organizational structure through the Forum approval processes.

Early in 2005, the Governing Committee was very fortunate to get Stephen A. Hess to come on board as the Executive Editor of the project. He was very ably joined by Jerry Bales and P. Doug Folk to form the three-person editorial team. Jerry Bales assumed responsibility as the Editor for the design professional side of the book, and P. Doug Folk became the Editor in charge of the construction management subjects. Mark Heley and I assumed responsibility for overall project management and interface with the Governing Committee and Publications Committees.

The Editorial Team has worked day and night for many months to make this book a reality. Stephen Hess brought his significant prior experience as the author and editor of several legal publications for West Publications to the project. Jerry Bales and Doug Folk have each written extensively in their respective subject matter areas and are also well-known and respected lecturers on a variety of construction industry liability and contracting issues.

The members and Division Chairs of Division 1 Dispute Avoidance & Resolution, Division 2 Contract Documents, Division 4 Project Delivery Systems, Division 5 Contract Negotiations, Performance & Administration, Division 6 Employment Law & Labor Relations, Division 7 Bonds, Liens & Insurance, Division 8 International Contracting, Division 9 Specialty Trade Contractors & Suppliers, Division 10 Legislation & Environment, Division 11 Corporate Counsel, and Division 12 Owners & Lenders also provided key support and resources for this book. They provided authors, ideas, and enthusiastic support for the book from its inception. Without this support and the dedication of the members of Division 3 and the Editorial Team, this project would not have been completed.

Of course, the authors' role in the development of this book is fundamental to its existence. The authors for the chapters in this book come from across the Forum. They were selected based upon their respective depth of knowledge and experience on the subject matter for each chapter. They are committed members of the Forum on the Construction Industry. Each author volunteered his or her time and resources without any charge. Moreover, in a remarkable effort, the authors produced their outlines, draft chapters, and final chapter manuscripts for the book on the schedule established by the project management team. We cannot thank the authors enough for their dedication to this project.

We believe that this book is an important new resource for our members and the industry. It is intended both for practitioners that handle an errors and omissions case once every couple of years and a refresher and resource for the highly experienced construction lawyer who regularly handles professional liability claims. It will provide practical knowledge concerning the problems, needs, risks, rewards, duties, and liabilities of design professionals and construction managers.

We want to express our appreciation for everyone in the Forum on the Construction Industry, because without their support and assistance, this book would not have been possible.

L. Tyrone Holt, Project Manager
January 2007

INTRODUCTION

There may seem little in common between the work that a design professional performs and the tasks of a construction manager. At the same time, there are many facets of the design professional practice and the function of a construction manager that overlap, and trying to assign tasks as inherently identifiable to a design professional rather than a construction manager or vice versa betrays a simplistic understanding of the scope of each professional's work.

Moreover, the field of construction management is growing and changing. When design-build started finding more favor as a project delivery mechanism in recent decades, the legal field was forced to reckon with the changes, and the evolution of design-build was followed by evolution of the law necessary to govern design-build relationships. In that sense, construction management is very much the design-build of the new millennium; it is now far enough past its infancy for the construction industry to have a clear understanding of its contours, but it is not so mature that the legal rules governing construction management can be recited in black letter format. Chapter 9 provides an overview of the development of construction management as a specialized practice and a snapshot of the current forms of construction management.

These two factors—the intersection of design professional practice with construction management, and the fact that construction management is widely used but continues to evolve—suggest that this is a good time to take stock of one of the most important facets of the practice of design professionals and construction managers: the legal responsibilities, risk, and rights inherent in the respective practices. We have structured the book in a manner that reflects both the common ground and the divergence between the two areas of professional practice.

We started by allocating to each field a core set of chapters describing the fundamental legal responsibilities of design professionals (DP) and construction managers (CM), and because design professionals and construction managers have different responsibilities in these fundamental areas, the book separates the disciplines and treats them in separate but parallel chapters. The first topic this book addresses is the state regulation of professionals (DP Chapters 1 and 2, CM Chapter 8). The second general area encompasses the contractual responsibilities of these professionals, including those encountered in the pre-design, design, and construction phases (DP Chapter 3, CM Chapter 10) and those encountered in the construction and completion phases (DP Chapter 4, CM Chapter 11). The third area addressed is the potential liability in tort of DPs and CMs (DP Chapter 6, CM Chapter 12), and we have included special chapters dealing with each professional's role in design-build projects (DP Chapter 5, CM Chapter 13).

The remainder of the book is devoted to topics that are of common interest to design professionals and construction managers, and the analysis of which does not require discrete treatment for different professionals nearly as much as the other topics. These include chapters on professional liability insurance (Chapter 14), subcontractor arrangements (Chapter 15), damages and remedies (Chapter 16), alternative dispute resolution (Chapter 17), and special litigation topics (Chapter 18). One chapter (Chapter 7) is primarily of interest to design professionals, as it covers the intellectual property rights of design professionals in their work products.

One of our goals in producing this project was to make each chapter self-sufficient. Accordingly, the reader may notice some overlap in the treatment of particular topics throughout the book. We chose to err on the side of over-inclusion (with the result of a somewhat larger book) rather than burden the reader with constant cross-references to other portions of the book. In addition, our timing and editorial constraints made it impossible, as a practical matter, to try to address all changes in the forthcoming 2007 AIA documents.

Stephen A. Hess
Jerome V. Bales
P. Douglas Folk
L. Tyrone Holt
January 2007

ABOUT THE AUTHORS AND EDITORS

Jean C. Arnold (Chapter 13) is a graduate of the University of Denver with three degrees, B.S.B.A. (*summa cum laude*, 1980), M.B.A. (*beta gamma sigma*, 1983), and J.D. (1983). She was admitted to the practice of law in Colorado in 1983, is a member of the Denver and First Judicial District Bar Associations, and a founding member of the Arraj Inn of Court. She is also admitted to practice in the United States District Court for the District of Colorado, Bankruptcy Court and Tenth Circuit. Before pursuing a career in the law, Ms. Arnold worked for nine years for an electrical subcontractor with her last positions being in estimating and project engineering. In addition to a practical knowledge in the field of construction, she is a frequent lecturer, instructor, and speaker for various trade associations and organizations on a range of legal and construction topics, including Construction Law and Contracts, Mechanics' Liens, Negotiation Strategy, Arbitration and Mediation, Creditor's Rights in Bankruptcy and Commercial Collections.

Jerome V. Bales (Editor) is a member of the firm of Lathrop & Gage L.C. in Kansas City, Missouri, where he practices in the area of construction claims, litigation, arbitration, and contracts. He was graduated with distinction from the University of Missouri at Kansas City Law School in 1973, and earned his undergraduate degree from Rockhurst University, Kansas City, Missouri, in 1971. He is a member of the bar associations of Missouri and Kansas, and of the American Bar Association. He serves on the Steering Committee of Division 3 of the ABA Construction Forum, and is also a member of the Defense Research Institute Construction Law Committee and the American Institute of Architects (National Allied Member and Kansas City Chapter, Professional Affiliate Member). He is the former chair of the Kansas City, Missouri Metropolitan Bar Association Construction Law Committee. He is admitted to practice before all state and federal courts in Missouri and Kansas and serves as an arbitrator and mediator for the American Arbitration Association and the International Institute for Conflict Prevention and Resolution.

Nathan A. Brennaman (Chapter 5) practices with Faegre & Benson LLP, where his work focuses primarily on litigation of construction disputes. He holds degrees from Grinnell College (B.A. 1995) and the University of Minnesota Law School (J.D. 2003). Following his graduation from law school, Mr. Brennaman served as a law clerk to the Honorable Donald P. Lay of the United States Court of Appeals for the Eighth Circuit.

Richard D. Conner (Chapter 9) is a principal in the law firm of Conner Gwyn Schenck PLLC, in Greensboro and Raleigh, North Carolina. His practice is concentrated in construction litigation, arbitration, and mediation. He is certified as a mediator in the state and federal courts of North Carolina. Mr. Conner is a Fellow in the Construction Management Association of America, and he is a Fellow in the American College of Construction Lawyers. He is a member of the governing board of the Construction Law Section of the North Carolina Bar Association. He is a member of the LCC Panel and the Carolinas Construction Mediation Panel for the American Arbitration Association. Mr. Conner received his B.A. from Guilford College and is a 1972 graduate of University of North Carolina law school.

David A. Delman (Chapter 12) is a partner in the Hartford office of Pepe & Hazard, where he specializes in domestic and international professional liability disputes and delay/disruption claims in the engineering and construction industry. He was previously in the in-house legal department of Fluor Corporation, where he held the position of Assistant General Counsel, responsible for construction claims and disputes worldwide. Mr. Delman has lectured widely on issues of professional liability and maximization of financial returns in lump sum turnkey contracts. He earned a B.A., *summa cum laude*, from The American University School of International Service and a J.D., *magna cum laude*, from the Georgetown University Law Center.

Katherine Davitt Enos (Chapter 14) is Senior Vice President and manager of an A/E professional liability insurance underwriting unit for Victor O. Schinnerer & Company. Prior to her association with Schinnerer, Ms. Enos completed a term as a judicial law clerk in the United States District Court and worked for an employment law firm. She holds a law degree from West Virginia University College of Law and is a member of the Maryland and West Virginia bar associations. She also holds a Bachelor of Architecture from The Catholic University of America, is an Associate Member of The American Institute of Architects, and has worked for a number of design firms. Ms. Enos has presented various A/E risk management seminars throughout the United States and serves as Schinnerer's liaison to the American Institute of Architect's Risk Management Committee and ACEC's Legal Counsel Forum. In addition, she has served as an Adjunct Professor in the School of Architecture at The Catholic University of America in Washington, D.C., teaching Architectural Practice Management.

Mehrdad Farivar (Chapter 3) is a partner with Morris Polich & Purdy LLP in its Los Angeles office, where he practices construction law and litigation. Mr. Farivar is also a licensed architect in California and in Britain with seventeen years of international experience as an architect. In 2004, he was elected as a member of the College of Fellows, the American Institute of Architects, in recognition of his contributions to the architectural profession. As an architect, he has designed and has been involved in the design of many building types, including multi- and single-family housing, institutional buildings, hospitals, retail stores, and interiors. Mr. Farivar was educated at the Southwestern University School of Law, the Bartlette School of Architecture of London University, and the University of Manchester. He is a frequent lecturer and author on topics related to law and the practice of architecture.

P. Douglas Folk (Editor) has represented members of the design professions and construction industry since receiving his law degree with honors from the University of Iowa College of Law in 1980. Mr. Folk's primary practice areas include professional liability defense, construction law, architectural copyright matters, and ethics. From 1995 to 2001, he served as the public member of the Arizona Board of Technical Registration by appointment of the governor. Mr. Folk continues to serve on Enforcement Advisory Committees for the Board.

Mr. Folk is a past Chairman of the Construction Law Section of the State Bar of Arizona. He is also a member of the Forum on the Construction Industry, the Public Contract Law Section, and the Tort and Insurance Practice Section of the American Bar Association. Mr. Folk has written many articles for publication on construction law and risk management issues. He is also a co-author of the *Arizona Construction Law Practice Manual* published by the State Bar of Arizona. Mr. Folk's *Design Professionals Survival Course*™ and *Contractors Survival Course*™ provide specialized training to help contractors and design professionals improve their business practices and avoid or minimize the risk of claims and litigation.

Peter Halls (Chapter 5) is a partner in the Construction Law Practice Group at Faegre & Benson LLP, where he represents a wide variety of participants in the construction industry, including owners, architects, engineers, contractors, design-builders, subcontractors, and sureties. Mr. Halls has handled a full range of construction issues, including negotiating contract documents, identifying potential claims, negotiating claims, conducting mediations, and handling arbitrations and trials. He has been involved with construction projects in most U.S. states and in several foreign countries, including Russia, Germany, China, India, and Canada. He has served as Chair of Faegre & Benson's Construction Law Group, the International Construction Law Division 8 of the ABA Forum on the Construction Industry, the Minnesota State Bar Association's Construction Law Section, and as President of the Minnesota Construction Association and the Minnesota Chapter of CMAA. He received a B.A. *magna cum laude* with a double major in Government and Russian from Dartmouth College and his J.D. *magna cum laude* from the University of Minnesota Law School, where he was on the Minnesota Law Review and was Order of the Coif.

David Hatem (Chapter 4) is a founding partner of the Boston-based law firm, Donovan Hatem LLP, where he leads the firm's Professional Practices Group in its representation of engineers, architects, and construction management professionals. Mr. Hatem is nationally recognized for his expertise in the law related to the design and construction industry. He has provided risk management strategies, and advice for many of the major public construction projects that have taken place across the country. Mr. Hatem serves as legal counsel to the American Consulting Engineers of Massachusetts, American Council of Engineering Companies of Massachusetts, and General Counsel for Design Build Institute of New England, Inc. He is a member of the Massachusetts and American Bar Associations and the American Society of Civil Engineers. He is a prolific writer, having authored over 200 papers published in a wide spectrum of technical, business, and legal journals as well as two books. Mr. Hatem is also an instructor at Tufts University, where he teaches a course entitled "Legal Aspects of Engineering Process."

Stephen A. Hess (Executive Editor and Chapter 16) is a shareholder of Sparks Willson Borges Brandt and Johnson, P.C. in Colorado Springs, Colorado. His writings include numerous national articles on construction law topics, he serves as the Editor of *Construction Briefings* and is on the Board of Editors of the *Real Estate Review*. Mr. Hess is the author of several books published by West, including *The Colorado Handbook on Courtroom Evidence*, *The Colorado Handbook on Civil Litigation*, the two-volume *Colorado Civil Rules Annotated*, and seven chapters on litigation for *Colorado Methods of Practice*. He is also the author of two chapters of CLE in Colorado, Inc.'s *Practitioners Guide to Colorado Construction*

Law and has written the "Contracts" article for the past eleven years for its *Annual Survey of Colorado Law*. He is an adjunct faculty member of the University of Denver College of Law, where he created and teaches the school's Construction Law Seminar. He received degrees from Franklin and Marshall College (B.A. 1982), Rice University (M.A. 1985) and University of Colorado School of Law (J.D. 1987).

L. Tyrone Holt (Editor and Chapter 15) received his B.A., 1971, *summa cum laude*, Morehouse College; J.D., 1974, Stanford University; and M. Theology, 1997, Peace Theological Seminary and College of Philosophy. Mr. Holt has over thirty years of experience as a construction transaction and trial attorney, with more than twenty years of experience as a construction mediator and arbitrator. He was admitted to practice law in the State of Colorado, the United States District Court for the District of Colorado, the United States Court of Appeals for the Ninth and Tenth Circuits, and the United States Supreme Court. He was a law clerk to the Honorable Preston Devine of the California Court of Appeals in 1974.

Mohammad Iqbal (Chapter 2) is General Counsel for Walker Parking Consultants and a licensed P.E. and S.E. in several states nationwide. He has received the following degrees: B.S. from the University of Engineering and Technology, Lahore, Pakistan; M.S. from the Middle East Technical University, Ankara, Turkey; D.S. in structural engineering from Washington University in St. Louis; MBA from the University of Chicago; and J.D. from the Northern Illinois University. He is admitted to practice law in the State of Illinois, the United States Court of Appeals for the Seventh Circuit, and the United States Supreme Court.

Ernest Isola (Chapter 18) is a partner in the Construction and Commercial Litigation practice groups of Gordon Rees, LLP and practices in the firm's San Francisco office. He specializes in counseling and defending professionals and businesses involved in a variety of claims, including construction disputes, professional liability claims, litigation involving corporate officers and directors, and other business litigation matters. Mr. Isola has represented developers, contractors, and design professionals in every type of construction claim, assisted them in formulating contracts and handling transactional matters, and has also represented businesses in class action lawsuits and claims for unfair business practices under California Business & Professions Code §17200. Mr. Isola obtained his undergraduate degree in Quantitative Economics and Decision Science from the University of California, San Diego in 1992. He attended law school at Santa Clara Law School, where he graduated *cum laude* and earned his Juris Doctor degree in 1997. While at Santa Clara Law School, Mr. Isola was a recipient of the Emery Law Scholarship and an editor of the Santa Clara Law Review.

Kevin L. Kolton (Chapter 8) is a partner in Schiff Hardin LLP's Construction Law Group. His practice is concentrated in energy and commercial construction, procurement, and dispute resolution through mediation, arbitration, and litigation. His experience includes projects involving various industrial and energy construction projects, including selective catalytic reduction and other environmental controls, boilers and heat recovery steam generators, alternative energy projects, including nuclear and wind generation facilities, steam and gas turbines, and other new and emerging technologies. He draws from a broad understanding of all facets of the construction process, including his background in engineering that preceded his practicing law. Mr. Kolton received his undergraduate degree (B.S., Engineering, 1980) from the University of Illinois at Urbana-Champaign and his law degree (J.D., *magna cum laude*, 1989) from the South Texas College of Law. He is admitted to practice in Illinois.

Theodore D. Levin (Chapter 3) is a partner with Morris Polich & Purdy LLP in its Los Angeles office. He is a member of the California Bar, and is both a practicing lawyer and registered civil engineer. Mr. Levin represents design professionals, contractors, construction managers, and owners in all aspects of construction law, including matters involving breach of contract and defective design and construction, contract drafting and project risk management, project delay and disruption claims, and payment disputes. He also lectures and writes extensively on a broad range of legal topics related to the construction industry. Mr. Levin received his degree in Architectural Engineering from the University of Colorado, and after practicing as an engineer received his J.D. from the University of San Francisco School of Law, where he was a Law Review Editor.

Rashida Y.V. MacMurray (Chapter 10) is an associate in Holland and Knight LLP's Washington, D.C., office, where she practices in the area of construction and intellectual property law. Prior to entering private practice, Ms. MacMurray worked in the construction industry as a construction manager. Specifically, she was employed as a Project Engineer for Bovis Lend Lease (formally Bovis Construction Corp.) and as a Project Manager for Lucent Technologies. Ms. MacMurray is author of "Trademarks and Copyrights: Which Intellectual Property Right Affords its Owner the Greatest Protection," published in the May 2005 issue of Northwestern University School of Law's *Journal of Technology and Intellectual Property*. Ms. MacMurray received her Bachelors degree in Architecture in 1995 and a Master of Engineering in Civil Engineering in 1997, both from the University of Virginia. She earned her J.D. degree in January 2002 from Rutgers University-Newark Law School.

Lindsey E. Malone (Chapter 6) is an Associate with the law firm of Skellenger Bender P.S. in Seattle, Washington. Ms. Malone is admitted to practice in the federal and state courts of Washington. Her practice focuses on design professional defense, construction law, condominium defect litigation, and commercial litigation. Ms. Malone received her B.A. in English from the University of Iowa, and her J.D. from the University of Iowa School of Law.

Stanley A. Martin (Chapter 11) is a partner in the Boston office of the national law firm of Holland & Knight, LLP. His practice focuses on construction industry and public contract law. He is an author of several manuals, articles, and text chapters on construction law issues, has taught construction law courses at M.I.T. and Northeastern University, and has lectured extensively on industry topics. Mr. Martin has served on the Corporate Advisory Council of the Boston Society of Architects, and as a board member of the Associated General Contractors of Massachusetts, and is a member of the Construction Industry Panel of the American Arbitration Association. He has also served as a board member and Secretary of the Boston Architectural Center. Mr. Martin received a degree in architecture from M.I.T. and a J.D. degree from Boston College Law School.

Paul H. Martinez (Chapter 10) is in-house counsel for Turner Construction Company, which is ranked by *Engineering News Record* as the number one at-risk construction manager in the

United States. Prior to joining Turner, his practice was limited to construction and real estate development matters. He is a member of the ABA Forum Committee for the Construction Industry; was appointed to Construction Litigation committee by the New York Bar Association, was appointed to the Contract Documents Committee by the Associated General Contractors, and serves on American Arbitration Association's National Construction Dispute Resolution Committee. He graduated from the United States Merchant Marine Academy with a B.S.M.E. and from Pace Law School with a J.D. He is a member of the New York, New Jersey, and Connecticut bars. Before admission to the bar, he served as Group Manager for Construction Services at Nestle Enterprises, Inc.

Stephen Miller (Chapter 16) is a principal of the Miller Law Firm in Kansas City, Kansas, where he maintains a nationwide practice in commercial and construction law and litigation. He received three degrees from the University of Notre Dame (B.A. 1980; J.D. 1983; and M.A. 1989). He is a member of the Kansas City Metropolitan, Missouri, and American Bar Associations. Mr. Miller is admitted to practice in Missouri, the U.S. Court of Federal Claims, the U.S. Court of Appeals (Fifth, Eighth, and Tenth Circuits), and the U.S. Supreme Court.

Virgil R. Montgomery (Chapter 8) is a partner in Schiff Hardin LLP's Construction Law Group. His practice is concentrated in procurement and dispute resolution through mediation, arbitration, and litigation. Mr. Montgomery's practice is also heavily concentrated on project controls by assisting project owners and contractors with the development of appropriate corporate project oversight mechanisms, such as cost analysis, schedule analysis, and other project reporting systems. His experience includes projects involving various industrial and energy construction projects, including selective catalytic reduction and other environmental controls, boilers and heat recovery steam generators, alternative energy projects, including hydroelectric, nuclear, and wind generation facilities, sewage treatment facilities, steam and gas turbines, and other new and emerging technologies. Mr. Montgomery received his undergraduate degree (B.A., History and Political Science, 1989) from the University of Colorado and his law degree (J.D. 1992) from Harvard Law School. He is admitted to practice in Illinois and Ohio, as well as before the U.S. District Courts for the Northern and Southern Districts of Ohio.

G. William Myers (Chapter 1) is chief legal officer at L. Robert Kimball & Associates, Inc., an architectural and engineering firm headquartered in Ebensburg, Pennsylvania, and ranked among the top design firms by *Engineering News Record* and *World Architecture* magazines. Mr. Myers has provided Kimball counsel in the areas of regulatory compliance, risk and project management, contract negotiation, and dispute resolution for twenty-six years. Mr. Myers is admitted to the Supreme Court of Pennsylvania, the U.S. District Court for the Western and Middle Districts of Pennsylvania, and the United States Court of Federal Claims. He received a B.S. in Industrial Engineering from The Pennsylvania State University in 1972 and a J.D. from the Duquesne University School of Law in 1980, following four years of military service in the U.S. Navy Civil Engineer Corps.

Bernard (B.J.) Nodzon (Chapter 5) practices with Faegre & Benson LLP, where his work focuses on litigation and alternative dispute resolution of commercial disputes. Before joining Faegre & Benson, B.J. served as a law clerk to the Honorable Donald P. Lay of the

United States Court of Appeals for the Eighth Circuit. Mr. Nodzon earned degrees from the William Mitchell College of Law (J.D., *summa cum laude*, 2002) and University of St. Thomas (B.A., *summa cum laude*, 1999), and he is the author of a number of articles on diverse legal topics.

Carrie Lynn H. Okizaki (Chapter 15) is an attorney in Schiff Hardin LLP's Construction Law Group. Ms. Okizaki is a member of the Colorado, Illinois, and American Bar Associations. Ms. Okizaki currently concentrates her practice in advising owners, developers, general contractors, architects, engineers, and other design professionals in variety of commercial and litigation matters, including multifaceted, complex disputes involving delays, disruption and losses of efficiency, breaches of contracts for performance, scope of work and payment; and complex multiparty insurance coverage issues. Ms. Okizaki received a B.S. Degree in Foreign Service from Georgetown University in 1996, and her J.D. from the University of Colorado School of Law in 2000, where she was a casenote and comment editor for the University of Colorado Law Review.

Shiv Ghuman O'Neill (Chapter 9) is an attorney with Baker & Daniels LLP in Indianapolis, Indiana. Ms. O'Neill practices construction law, and construction and real estate litigation as a member of Baker & Daniels' Construction, Environmental, and Real Property Liability Group. She graduated *magna cum laude* from Indiana University Law School in 2001, where she was elected to Order of the Coif and served on the *Indiana Law Journal*. She earned her bachelor's degree from Loyola University of Chicago in 1998, and is admitted to practice in both Indiana and Ohio.

Ben Patrick (Chapter 7) is an attorney at Watt, Tieder, Hoffar & Fitzgerald in San Francisco. Mr. Patrick represents companies in the construction industry, including general contractors, architects, engineers, and sureties. He has represented companies varying in size from one-man shops to ENR Top 50 Firms in mediations, litigations, and arbitrations with claims ranging from under $100,000 to over $300,000,000. Mr. Patrick also counsels clients on contract drafting, intra-office policies, risk management, and intellectual property management. He is a frequent speaker at trade association seminars and has written articles for national construction trade journals. He is a member of the American Bar Association's Forum on the Construction Industry and the ABA's Construction Law Subsection of the Litigation Section.

David W. Rudy (Chapter 17) is a private mediator with a nationwide practice and the owner of Just Accord, Inc., with a principal office in Black Forest, Colorado. After graduating *magna cum laude* from the University of San Francisco School of Law in 1974, he became a trial practitioner. Over the next two decades, he completed approximately twenty court and jury trials to judgment. He was also an adjunct faculty member for seventeen years, teaching insurance law and occasionally trial practice and appellate practice. In 1991, he began the transition from full-time trial lawyer to full-time neutral. In the last fifteen years, he has successfully completed approximately 2000 mediations, most involving multiple parties and complex issues. Mr. Rudy also has degrees from Tufts University (B.A., *magna cum laude,* 1968) and the University of Chicago (M.A. 1969).

Terence J. Scanlan (Chapter 6) is a principal in the law firm of Skellenger Bender P.S., in Seattle, Washington. Mr. Scanlan is admitted to practice in the federal and state courts in

Washington, state courts in Oregon, and the United States Court of Appeals for the Ninth Circuit. His practice emphasizes representation of design professional firms on all types of claims including construction management, delay and cost issues, structural failures, landslides, pipeline design defects, and other civil and geotechnical engineering issues. Clients include many of ENR's Top 50 international design firms, as well as regional and local firms in the Pacific Northwest and in Alaska. Mr. Scanlan received a B.A. in Economics from Seattle University, and his J.D. from Seattle University School of Law.

Chris Squadra (Chapter 13) has worked in many capacities in the construction industry over the last thirty years, including filling roles with the construction contracting group, working for design teams, and most recently representing owner's interests on major projects throughout North and Central America. As the founding principal of ARC, one of the nation's proven leaders in providing project controls, his career has focused on "mission-critical" facilities and unique procurement strategies for over twenty years. His dedication to innovative contracting methods, budget, risk, and schedule management drives the philosophy of the company, managing to simultaneously control the program, scope, quality, and value received on every project. Mr. Squadra is dedicated to involvement in the industry. He is a Certified Professional Estimator by the American Society of Professional Estimators, a member of the Construction Specification Institute, Urban Land Institute, Construction Management Association of America, and the Project Management Institute.

Kenneth Strong (Chapter 18) is a partner in the San Francisco office of Gordon & Rees, LLP, where he specializes in representing design professionals and contractors in litigation and transactional matters. He has over twenty-five years of experience in trials in California state and federal courts, and has participated in arbitrations and mediations in California and in other states, where his clients have included engineers, consultants, architects, general contractors, and subcontractors, as well as owners. Mr. Strong graduated from the University of California at Santa Barbara in 1975 and the Hastings College of Law in 1979. He is an active member of the San Francisco, California, and American Bar Associations, and has served in leadership positions with the Forum on Construction. He is also a member of the American Council of Engineering Companies Legal Counsel Forum.

Wendy Kennedy Venoit (Chapter 12) is a partner with the law firm, Pepe & Hazard, LLP in Hartford, Connecticut. Her practice is concentrated on construction litigation and dispute resolution. Ms. Venoit has litigated in state and federal forums throughout the country, and has mediated and arbitrated construction disputes both nationally and internationally. Ms. Venoit received a B.A. degree in economics and Spanish from the College of the Holy Cross, and a J.D. degree, *magna cum laude, Valedictorian*, from Pace University School of Law. While attending law school, Ms. Venoit was an editor on the Pace Law Review and authored multiple publications. Ms. Venoit is a member of the Construction Forum and Tort & Insurance Practice Section of the ABA, and is currently serving on the Steering Committee for Division 8 (International Contracting) of the Construction Forum. Ms. Venoit also serves as Secretary and General Counsel to the Professional Women in Construction Connecticut Chapter. Ms. Venoit is admitted to the bars of Connecticut, New York, and New Jersey.

CHAPTER 1

State Regulation of the Design Professional

G. WILLIAM MYERS, P.E., ESQ.

I. Introduction

This chapter discusses the regulation of the professions principally involved in the design of improvements to real estate—architecture and engineering. These same general legal concepts and regulatory requirements also apply to the closely allied professions of landscape architecture, interior design, geology, and land surveying, which may be similarly regulated by particular jurisdictions.

The practice of engineering or architecture as a professional requires a state-issued license or registration in all jurisdictions of the United States. The requirements for an individual to obtain and maintain professional licensure as an architect or engineer is found in the applicable statutes, regulations issued by licensing boards, and case law interpreting the statutes and regulations on a state-by-state basis. In addition to a requirement for individual licensure, any design professional who practices in the form of a corporation, limited liability company, professional corporation, professional association, or other similar limited liability fictitious business entity ("Fictitious Business Entity") must also obtain appropriate separate qualification, registration, or authorization for the Fictitious Business Entity. As a general rule, state jurisdictions do not license Fictitious Business Entities as design professionals—only individuals.

The same governmental department or agency that licenses corporations and limited partnerships typically handles the registration or licensing of business entities for Fictitious Business Entities. In contrast, the registration or licensing of individual design professionals is almost always handled by a separate professional board or registration agency that handles examination, disciplinary proceeding, and other professional matters for the affected professional.

Most states have legislation and supporting regulations that address the issue of firm practice for architecture as a separate and distinct matter from the firm practice of engineering. However, a few states, such as Illinois,[1] Virginia,[2] and Wisconsin,[3] take a more integrated approach to the licensure of design firms and issue one certificate or license to practice both architecture and engineering. These jurisdictions have a single board regulating both architecture and engineering; granting a single certificate of authorization or license to practice both architecture and engineering is a natural result of only one board regulating both professions.

Library References

C.J.S. *Architects* §§ 1–3, 7.
West's Key No. Digests, Licenses <KEY>11.

II. Individual Licensure

To practice architecture or engineering in the United States, an individual has to be licensed by the state in which the professional services are to be delivered. Typically, these services consist of drawings and/or specifications for the construction of a particular improvement to real estate. However, exactly what is included in the practice of architecture and engineering can be elusive.

Pennsylvania defines the practice of architecture as follows:

> The rendering or offering to render certain services, hereinafter described, in connection with the design and construction of a structure or group of structures which have as their principal purpose human habitation or use, and the utilization of space within and surrounding such structures. The services referred to in the previous sentence include planning, providing preliminary studies, designs, drawings, specifications, and other design documents, construction management and administration of construction contracts. The foregoing shall not be deemed to include the practice of engineering as such, for which separate registration is required . . . , excepting only engineering work incidental to the practice of architecture.[4]

Pennsylvania defines the practice of engineering as follows:

> (a)(1) "Practice of Engineering" shall mean the application of mathematical and physical sciences for the design of public and private buildings, structures, machines, equipment, processes, works or engineering systems, and the consultation, evaluation, engineering surveys, construction management,

---

1. 225 ILL. COMP. STAT. ANN. 305/21, 325/23 (West 2006).
2. VA. CODE ANN. § 54.1-411 (Michie 2006).
3. WIS. STAT. ANN. § 443.08 (West 2006).
4. 63 PA. CONS. STAT. ANN. § 34.3 (West 2006).

planning and inspection in connection therewith, the performance of the foregoing acts and services being prohibited to persons who are not licensed under this act as professional engineers unless exempt under other provisions of this act.

(2) The term "Practice of Engineering" shall also mean and include related acts and services that may be performed by other qualified persons, including but not limited to, municipal planning, incidental landscape architecture, teaching, construction, maintenance and research but licensure under this act to engage in or perform any such related services shall not be required.

(3) The foregoing shall not be deemed to include the practice of architecture as such, for which separate registration is required . . . , excepting only architectural work incidental to the "practice of engineering."

(4) The "Practice of Engineering" shall not preclude the practice of the sciences which shall include but not be limited to: soil science, geology, physics and chemistry.[5]

Because of the ambiguous and broad language of both definitions, which is similar to the statutory language in other states, issues arise as to what is included in the practice of architecture or engineering and what may be included in both. Where there are two separate boards regulating these professions, this problem seems to be exacerbated.

Rosen v. Bureau of Professional and Occupational Affairs, State Architects Licensure Board[6] is one example of this tension. In this case, the Architects Licensure Board imposed a civil penalty on an engineer, Rosen, for having signed and sealed plans to convert a four-story building in Philadelphia from a private club into law offices for the first three floors and an apartment for the fourth floor. The court noted that the most substantial aspects of the renovation project were the addition of an elevator shaft to the rear of the building and reinforcement of the first floor to accommodate a law library.

The Pennsylvania intermediate appellate court first noted that the matter appeared to be part of an ongoing turf war between the two learned professions over the design of buildings. The court then concluded that both statutes have to be read *in pari materia* and that there is an overlap between the two statutes, with neither statute establishing a clear, mutually exclusive delineation between the two professions.

The court examined the architecture and engineering statutes recounted above and reasoned that both addressed the design of buildings, there was no bright line between the two professions, and the engineering statute was broader than the architecture statute because the latter limited the practice of architecture to structures of human habitation and use.

---

5. 63 Pa. Cons. Stat. Ann. § 149(a) (West 2006).
6. 763 A.2d 962 (Pa. Commw. Ct. 2000), *appeal denied*, 781 A.2d 150 (Pa. Super. Ct. 2001).

The court also looked at three similar cases from Alabama,[7] Florida,[8] and Georgia[9] where architect licensure boards had cited engineers for the unlawful practice of architecture. Ultimately, the court concluded that the practice of engineering includes the design of buildings and structures; thus, Rosen was within the practice of engineering when he designed the renovations. Instead of interpreting the architects' law as limiting the practice of engineering, the court read it as limiting the practice of architecture to habitable structures.[10]

An individual's initial license is obtained by education, examination, and experience.[11] For engineers, this means: (1) graduating from an ABET[12] accredited engineering program at a college or university; (2) passing the fundamentals in engineering (FE) examination; (3) gaining the requisite years of experience in engineering under the supervision of professional engineers; and (4) passing the principles and practice of engineering (PE) examination. Those who pass the FE examination are usually referred to as an engineering intern (EI) or an engineer in training (EIT).[13]

Obtaining an architect's license requires: (1) graduating from an NAAB[14] accredited architecture program at a college or university; (2) gaining the required number of years of architectural experience under the supervision of registered architects; and (3) passing the architect registration examination.

An experienced architect or engineer who wants to become licensed in another state can often do so without taking an additional examination. This can be accomplished by comity, endorsement, or reciprocity. For example, one may obtain registration as an architect or an engineer in Michigan if the requirements of the other state for that person's registration are equivalent to the requirements of the Michigan board.[15]

In the past, many states had provisions called "grandfather" clauses or registration by eminence to address the issue of long-time practitioners who were licensed under older laws that did not require written examinations. Generally, these provisions no longer exist in most states. There may be some remnants of eminence registrations, but most of the states that have such provisions require the applicant to pass an

---

7. State v. Jones, 267 So. 2d 427 (Ala. 1972).

8. Verich v. Florida State Bd. of Architecture, 239 So. 2d 29 (Fla. Dist. Ct. App. 1970).

9. Georgia State Bd. for Examination, Qualification and Registration of Architects v. Arnold, 249 Ga. 593, 292 S.E.2d 830 (1982).

10. *Rosen*, 763 A.2d at 970.

11. Before modern licensure statutes, licensure could be obtained by merely graduating from an accredited college and gaining the requisite experience. These licenses typically date back to before the 1960s.

12. ABET stands for Accreditation Board of Engineering and Technology.

13. *See* the National Council of Examiners for Engineering and Surveying Web site, *at* www.ncees.org.

14. NAAB stands for National Architectural Accrediting Board.

15. Mich. Comp. Laws Ann. § 339.2013 (West 2006).

examination. For example, an engineer with more than twenty years of progressive engineering experience may obtain licensure in North Carolina after passing the principles and practice of engineering examination.[16]

Historically, professional licenses have been maintained by updating the applicable board on the registrant's current address, other contact information, and the payment of annual, biennial, or other periodic fees. However, requirements for continuing education have become more and more prevalent over the last decade. For example, in Florida, each biennial renewal of an engineering licensee requires the completion of four professional development hours in the licensee's area of practice and four professional development hours in professional responsibility.[17] Similarly, Florida architects have to show they have completed twenty hours of continuing education at each biennial renewal.[18]

In recent years, many jurisdictions have imposed continuing education requirements on architects and engineers. In those jurisdictions that currently do not impose such requirements, the trend is to impose such requirements as a condition of license renewal.

Library References
C.J.S. *Architects* §§ 1–3, 6–7; C.J.S. *Landlord and Tenant* §§ 337–339.
West's Key No. Digests, Licenses <KEY>11, 20, 25.

III. Firm Licensure

A. *Overview*

Once an individual architect or engineer is granted a license to practice in a jurisdiction, there are additional requirements that must be satisfied in order for that person to practice in the form of a Fictitious Business Entity. Just as individual licensing requirements and procedures vary from state to state, there are separate, but equally important requirements that apply to groups of professionals who wish to use a Fictitious Business Entity to practice the profession. The practice of engineering in some states does not require licensure at the firm level. The opposite is true with respect to the practice of architecture. The practice of architecture nearly always requires licensure at the firm level.

---

16. N.C. GEN. STAT. § 89C-13(a)(1)e. (2006).

17. FLA. STAT. ANN. § 471.017 (West 2006). Since 2000, Florida engineers participating in the design of structures covered by the Florida Building Code also must prove they have completed the core curriculum courses of the Building Code Training Program and any advanced or specialized courses on any portion of the Florida Building Code applicable to the licensee's area of practice. FLA. STAT. ANN. § 471.0195 (West 2006).

18. FLA. STAT. ANN. § 481.215(3) (West 2006). Florida architects also must prove completion of the core curriculum courses of the Building Code Training Program. FLA. STAT. ANN. § 481.215(5) (West 2006).

Pennsylvania demonstrates this dichotomy. Under the Engineer, Land Surveyor and Geologist Registration Law,[19] Pennsylvania does not require separate licensing of firms to practice engineering because it considers the practice of engineering, land surveying, and geology to be the function of individuals or individuals working in concert.

The only requirement for a firm to practice engineering is as follows:

[T]he directing heads and employees of such firm or corporation in responsible charge of its activities in the practice of such profession be licensed and registered in conformity with the requirements of this act, and whose name, seal, and signature, along with the date of signature, shall be stamped on all plans, specifications, plats, and reports issued by such firm or corporation.[20]

However, under the Pennsylvania Architects Licensure Law,[21] an individual architect or group of architects may only practice architecture as a sole proprietorship, partnership, professional association, business corporation, limited liability company, or limited liability partnership. Such entities are, in turn, subject to certain partnership, membership, directorship, and ownership restrictions.[22] For example, a business corporation may only practice architecture in Pennsylvania if it meets the following requirements:

(1) At least two-thirds of the directors are licensed under the laws of any state to practice architecture, engineering, or landscape architecture.
(2) At least one-third of its directors are licensed under the laws of any state to practice architecture.
(3) At least two-thirds of all classes of voting stock issued and outstanding at any one time shall be owned by an individual or individuals licensed under the laws of any state to practice architecture, engineering, or landscape architecture.
(4) At least one-third of all classes of voting stock issued and outstanding at any one time shall be owned by an individual or individuals licensed under the laws of any state to practice architecture.[23]

The first inquiry when considering practice in a new state is to determine whether licensure is required at the firm level. It may be necessary for certain individuals to be licensed in that state for the firm to qualify for a license.

Once it is determined that licensing is required at the firm level, the next inquiry is what requirements the firm has to meet to become licensed, which begins with an examination of the permissible legal forms for the business.

---

19. 63 PA. CONS. STAT. ANN. § 148 (West 2006).
20. 63 PA. CONS. STAT. ANN. § 153 (West 2006).
21. 63 PA. CONS. STAT. ANN. § 34.1 (West 2006).
22. 63 PA. CONS. STAT. ANN. § 34.13 (West 2006).
23. 63 PA. CONS. STAT. ANN. § 34.13(e) (West 2006).

In New York, a firm may not be a general business corporation unless the corporation was formed before April 15, 1935, and continuously thereafter lawfully practiced engineering in New York.[24] Thus, if the firm is a general business corporation, a new affiliate or subsidiary will have to be formed to gain firm licensure in New York. This affiliate or subsidiary will have to be a professional service corporation, a professional service limited liability company, a registered limited liability partnership, a partnership, or a joint enterprise.[25]

After determining the legal form the firm may take in a particular jurisdiction, the next step is to evaluate what other requirements, if any, are placed on the firm to gain licensure. As discussed above with respect to an architectural firm in Pennsylvania, there may be minimum ownership, directorship, or officership requirements for registrants in that state.

Also, there may be limitations on what can be used as the firm name, or the name presently used by the firm may not be available in that jurisdiction because it is already in use by another firm. The firm, in turn, will have to adopt a fictitious name for use in that state. For example, an architectural firm in Oregon may not include the surname of an individual who is not presently or was not previously associated in the practice of architecture or engineering in Oregon or in any jurisdiction recognized by the State Board of Architect Examiners or the State Board of Examiners for Engineering or Land Surveying.[26] In Alabama, a firm may not continue to use the name of an individual who by virtue of death, retirement, or resignation is no longer an active participant in the practice of architecture.[27]

In most jurisdictions, a design firm is required to register with the secretary of state in that jurisdiction to conduct intrastate business.[28] An argument can be made that the Commerce Clause[29] of the U.S. Constitution should shield firms from having to do this if no office has been established in that state;[30] others have suggested that registration with the secretary of state should be undertaken.[31]

However, this may not be in the design professional's best business or legal interests. If the firm wants to take the position that qualification with the secretary of

---

24. N.Y. EDUCATION LAW § 7209 subd. 6 (McKinney 2006).

25. N.Y. EDUCATION LAW § 7210 subd. 1 (McKinney 2006).

26. OR. REV. STAT. § 671.041(1)(b) (2006).

27. ALA. CODE § 34-2-37 (2006).

28. There is general confusion about the meaning of the term "doing business." It means doing intrastate business in that state versus providing services in interstate commerce. *See* CT CORPORATION, WHAT CONSTTITUTES DOING BUSINESS (2003), for an excellent discussion of this issue.

29. U.S. CONST. art. I, § 8, cl. 3. The Congress is to regulate commerce with foreign nations, and among the several states.

30. Cement Asbestos Prods. Co. v. Hartford Accident and Indem. Co., 592 F.2d 1144, 1146-48 (10th Cir. 1979).

31. ROBERT F. CUSHMAN & JAMES C. DOBBS, DESIGN PROFESSIONALS HANDBOOK OF BUSINESS AND LAW § 2.2 (John Wiley & Sons, Inc. 1991).

state is not required because its services are being provided in interstate commerce, a sample letter to the licensure board requesting or requiring this step is provided under the practice aides section below.

Although every state denies unqualified foreign corporations access to its courts,[32] there is an interstate commerce exception to this otherwise harsh result, as well as a method to cure this disability in most states. A state may not deny access to its courts if an unqualified corporation is only engaged in interstate commerce because of the impact of the Constitution's Commerce Clause.[33] Additionally, even if it is determined that the foreign corporation was involved in *intrastate* commerce instead of only *interstate* commerce, the corporation can simply register with the secretary of state and then bring a cause of action in that jurisdiction to enforce its contract rights (in all states, except Alabama).[34] In contrast, the failure to obtain a professional license may not be curable after the fact, as amply demonstrated in *Haith & Co., v. Ellers, Oakley, Chester & Rike, Inc.*,[35] discussed later in this section.

The application form and related instructions that are to be submitted for firm registration are a helpful guide to the statutory and regulatory framework for that state. These documents should be acquired early in the process of evaluation and are available from the licensure boards in each state. In many jurisdictions, they are available for printing from state-sponsored Web sites. In some cases, the practice of the state may be more liberal than suggested by the statutes under which it operates.

Section I of the North Carolina application for certificate of licensure as an engineering firm contains four check-off blocks: (1) business corporation; (2) partnership; (3) non-exempt partnership; and (4) other. Section II of the form states that if filing as a "business corporation," list the date and state of original incorporation. Thus, the application indicates that a business corporation may apply for a certificate of licensure to practice engineering in North Carolina. The North Carolina statute, Licensure of Corporations and Business Firms that Engage in the Practice of Engineering or Land Surveying,[36] provides "[a] corporation that is not exempt from Chapter 55B of the General Statutes by application of G.S. 55B-15 must be incorporated under that Chapter."[37] The term "that Chapter" refers to North Carolina General Statute 55B, the North Carolina Professional Corporation Act. The exemptions in Section 55B-15 of the North Carolina Professional Corporation Act are (1) for corporations formed prior to June 5, 1969, which were permitted by law to provide engineering or land surveying services, and (2) for design-build

---

32. *See* CT CORPORATION, WHAT CONSTITUTES DOING BUSINESS 7 (2003).
33. U.S. CONST. art. I, § 8, cl. 3.
34. See cases cited in CT CORPORATION, WHAT CONSTITUTES DOING BUSINESS 14-17 (2003).
35. Haith & Co. v. Ellers, Oakley, Chester & Rike, Inc., 778 S.W.2d 417 (Mo. Ct. App. 1989).
36. N.C. GEN. STAT. § 89C-24 (2006).
37. *Id.*

construction contractors. However, design-build construction contractors may not offer engineering as their primary service.[38] Based on the author's discussions with representatives of the North Carolina Board of Examiners for Engineering and Surveying, there is a third exception for corporations that were permitted by law to provide a special type of land surveying called photogrammetric mapping services prior to August 27, 1998—the effective date of the amendment to North Carolina law that expanded the definition of land surveying to include photogrammetry.[39]

B. *Noncompliance Has Its Price*

The consequences can be severe if a firm provides professional design services without properly registering the firm with the state's professional licensure board(s), which may also require registration of the individuals providing the professional services. This problem often arises in the context of a lawsuit to collect fees due to the firm. The Tennessee professional corporation of Ellers, Oakley, Chester & Rike, Inc. (Ellers Oakley) found this out the hard way when it tried to collect $275,000 in unpaid fees for engineering design- and construction-related services for air cargo facilities constructed at two airports in Missouri. Ellers Oakley first served a demand for arbitration pursuant to the arbitration clause in its contracts with its Missouri clients. Ellers Oakley's clients, however, filed suit in the Missouri court to stay the arbitration proceedings, contending that the contracts with Ellers Oakley and their arbitration clauses were invalid because engineering service contracts are unenforceable if the engineering firm was not authorized to provide engineering services in Missouri.[40] Although the engineers who had actually provided the design services were Missouri licensees, Ellers Oakley was not registered to offer engineering services in Missouri.[41]

The trial court issued the injunction and the Missouri Court of Appeals affirmed this decision.[42] The appellate court highlighted the statutory provision, which read as follows:

> [e]very contract for . . . engineering . . . services entered into by any person who is not a . . . registered or authorized professional engineer . . . shall be unenforceable by the unregistered or unauthorized professional engineer.[43]

---

38. N.C. GEN. STAT. § 55B-15(a)(2) (2006).
39. N.C. GEN. STAT. § 89C-3(7) (1999).
40. Haith & Co. v. Ellers, Oakley, Chester & Rike, Inc., 778 S.W.2d 417 (Mo. Ct. App. 1989).
41. MO. REV. STAT. §§ 327.011(6), 327.191, 327.401 (2006). These statutory provisions have remained virtually unchanged since the *Ellers Oakley* decision.
42. *Ellers Oakley*, 778 S.W.2d at 419.
43. *Id.* at 420 (citing MO. REV. STAT. § 327.461 (1986)). This statute remains substantially the same today, except the language has been broadened to state that such a contract "shall be unenforceable by the unlicensed person, whether in contract, quantum meruit, or other legal theory, regardless of whether a benefit has been conferred." MO. REV. STAT. § 327.461 (2006).

When Ellers Oakley entered into its first contract, it instructed its Tennessee counsel to "take the steps necessary to comply with Missouri law." Consequently, Ellers Oakley obtained a certificate of authority to do business in Missouri.[44] However, it did not obtain a certificate of authority to practice engineering from the Missouri Board of Architects, Professional Engineers and Land Surveyors until after Ellers Oakley's client filed the action to stay the arbitration proceedings.[45] That was too late.

Ellers Oakley argued that it had substantially complied with the statutory requirements; therefore its contracts should be enforced despite the Missouri statute cited above. The court rejected this argument based on the statutory mandate that contracts cannot be enforced when entered into by engineering firms that were not registered at the time the contract is formed.[46]

Of further interest, registering to do business with the Missouri Secretary of State, which Ellers Oakley did shortly after it entered into the first contract, was not actually necessary at that time in order to later enforce its contracts. This registration would have been required only before suit was filed. The Missouri appellate court compared the statute requiring registration with the secretary of state with the statute requiring registration with the Missouri Board of Architects, Professional Engineers and Land Surveyors and noted that registration with the secretary of state can be accomplished later, but registration with the board is required in advance of entering into a contract for engineering services. The court found that failure to register with the secretary of state can be cured, whereas failure to register with the Missouri Board of Architects, Professional Engineers and Land Surveyors cannot.[47]

Undaunted, Ellers Oakley filed a breach of contract and *quantum meruit* cause of action in the United States District Court for the District of Kansas to obtain its fees in the case of *Ellers, Oakley, Chester & Rike, Inc. v. St. Louis Air Cargo Services, Inc.*[48] The district court granted defendants' motion for summary judgment on the breach of contract claim based on the earlier state court decision discussed above, but allowed the *quantum meruit* claim to survive summary judgment.

The parties stipulated to the entry of final judgment in favor of Ellers Oakley and against the two defendants in the amount of $275,000, so the defendants could appeal the issue of Ellers Oakley's right to a *quantum meruit* recovery to the United State Court of Appeals for the Tenth Circuit. In making its decision, the federal appellate court first noted that the Missouri statute does not by its express terms preclude a *quantum meruit* recovery by an unregistered engineer. Nonetheless, the Tenth Circuit court rejected Ellers Oakley's *quantum meruit* cause of action and reversed the lower court, reasoning that to do otherwise would frustrate the Missouri statutory

---

44. *Ellers Oakley*, 778 S.W.2d at 418–419. This case serves as a good warning to attorneys counseling professionals who work in other jurisdictions.
45. *Id.* at 419.
46. *Id.* at 421.
47. *Id.*
48. 984 F.2d 1108 (10th Cir. 1993) (applying Missouri law).

scheme requiring engineering firms to obtain a certificate of authority before entering into contracts for engineering services.[49]

Library References
C.J.S. *Architects* §§ 1–3, 6–9, 12, 25; C.J.S. *Landlord and Tenant* §§ 337–339.
West's Key No. Digests, Licenses <KEY>11, 20, 22, 25, 39–40.

IV. Firm Practice

Once the firm's application has been approved and the firm is certified, licensed, or registered to practice architecture or engineering, the real work begins—ascertaining all the nuances of practice in that jurisdiction and meeting the challenge of complying with these requirements that are subtlety different from what is required in other jurisdictions. The following discussion will focus on some of these differences, but there are many more than are discussed here and they can only be found in the statutes and regulations of the fifty states.

At first blush, the act of sealing architectural and engineering work product would seem to be a reasonably straightforward undertaking. Nevertheless, this practice area can be problematic. The use of the professional seal is almost always addressed in the applicable licensing regulations. For this reason, one must always check regulations carefully.

A. *Who Is to Seal?*

The general answer is that the registered professional or professionals in responsible charge of the work reflected in the document should seal the architectural and engineering work product. Thus, it is important to consider the concept of who is "in responsible charge of the work." This term or similar terms are often defined in a state's statutes or regulations and may also be defined as or referred to as the architect or engineer of record.

The concept of the architect or engineer being in responsible charge or in supervisory control is one side of the coin and the architect or engineer of record is the other. One side of the coin focuses on the execution of professional services and the other focuses on the deliverable or work product resulting from the provision of professional services.

In New Jersey, "responsible charge" means the "rendering of regular and effective supervision by a competent professional engineer . . . to those individuals performing services which directly and materially affect the quality and competence of the professional services rendered by the licensee."[50] The statute goes on to specify what is not considered responsible charge as follows:

---

49. *Id.* at 1111–12.
50. N.J. STAT. ANN. § 45:8–28(g) (West 2006). *See* N.J. STAT. ANN. § 45:3–1.1.1 (West 2006) (nearly identical statutory language with respect to the practice of architecture in New Jersey).

A licensee engaged in any of the following acts or practices shall be deemed not to have rendered regular and effective supervision:

(1) The regular and continuous absence from principal office premises from which professional services are rendered, except for performance of field work or presence in the field office maintained exclusively for a specific project;

(2) The failure to personally inspect or review the work of subordinates where necessary and appropriate;

(3) The rendering of a limited, cursory or perfunctory review of plans or projects in lieu of appropriate detailed review;

(4) The failure to personally be available on a reasonable basis or with adequate advance notice for consultation and inspection where circumstances require personal availability.[51]

The term architect or engineer of record is not typically used in the licensing statutes and regulations, but is more often contained in statutes and regulations comprising building codes where the concept of architect or engineer of record is used to denote the architect or engineer who has in fact signed and sealed the construction documents under consideration. In Connecticut, prior to the issuance of a certificate of occupancy for certain structures, the local building official must require a statement signed by the architect or engineer of record responsible for the design of the structure affirming the professional opinion that the completed structure is in substantial compliance with the approved plans and specifications on file with the building official.[52]

Under the International Building Code (IBC), the registered design professional in responsible charge is the person designated by the building owner on the building permit application as such.[53] The IBC also states that the registered design professional in responsible charge is accountable for reviewing submittal documents prepared by others for compatibility with the design of the building,[54] as well as the preparation of the statement of required special inspections.[55]

In summary, architects and engineers may only seal documents prepared under their personal supervision. Once these documents are sealed, the sealing architect or engineer becomes the architect or engineer of record for that project, or at least for that portion of the project represented by the sealed documents.

However, there may be some exceptions to the direct supervision requirement. One such exception is sealing of documents produced by others, such as a subconsultant. Under Illinois law, an architect may seal documents prepared by or under

---

51. N.J. STAT. ANN. § 45:8–28(g) (West 2006).
52. CONN. GEN. STAT. ANN. § 29-276c(b) (West 2006).
53. International Building Code, § 106.3.4.1 (2003).
54. *Id.*
55. *Id.* at § 1704.1.1.

the responsible control of others who hold a license and have sealed the documents if the architect has reviewed in whole or in part such portions and has coordinated their preparation or integrated them into his or her own work.[56] Also in Illinois, a partner or a corporate officer of a professional design firm who has professional knowledge of the content of the technical submissions and intends to be responsible for the adequacy of the technical submission may sign and seal technical submissions that are prepared by or under the responsible control of architects who are licensed in Illinois and are in the regular employ of the professional design firm.[57]

The recommended practice is to have the architect or engineer in the most direct operational or supervisory role sign and seal the documents created, regardless of his or her corporate status in the firm and the exceptions discussed above. The identification of the licensed professional, who will apply a seal and signature to the design documents, at the appropriate time, must be addressed prior to the commencement of the creation of the design documents. This person supervises the development of the documents throughout the design process and normally would interface with the code compliance and building department officials when the design documents are submitted for plan check and permitting. Under the laws of most jurisdictions, there is potential for personal or individual liability to the professional licensing authorities for failure to properly discharge the duties associated with service as the responsible design professional. These duties are extremely important and must be taken seriously by properly qualified personnel.

Some states may require that multiple signatures and seals be placed on documents to comply with the applicable statutes and regulations. Pennsylvania law effectively requires multiple signatures and seals on architectural documents prepared by larger firms because, in addition to the signature and seal of the architect who prepared or personally supervised the creation of the documents,[58] the signature and seal of the corporate director under whose personal supervision the document was prepared are also required.[59]

In some jurisdictions, the firm also must obtain a seal and place it on documents it issues. South Carolina requires architectural firms to obtain a seal containing the firm's name, place of business, and the words "Registered Architect, State of South Carolina." Further, both the seal of the architect in responsible charge and the seal of the firm must be used.[60] A similar requirement exists for architectural firms in North Carolina.[61]

---

56. 225 ILL. COMP. STAT. ANN. 305/14 (West 2006).

57. *Id.* The statute further specifies that the architect exercising responsible control shall be identified on the documents or portions of documents by name and Illinois license number.

58. 49 PA. CODE § 9.141(b)(2) (West 2006).

59. 63 PA. CONS. STAT. ANN. § 34.13(i) (West 2006).

60. S.C. CODE ANN. § 40-3-280 (West 2006).

61. N.C. GEN. STAT. § 83A-10 (West 2006); N.C. ADMIN. CODE tit. 21, r. 2.206 (2006).

B. *What Is to Be Sealed and When?*

Florida law provides the generally accepted practice as to what and when a document is to be sealed:

> All final construction documents and instruments of service which include drawings, plans, specifications, and reports prepared or issued by the registered architect and being filed of public record shall bear the signature and seal of the registered architect who prepared or approved the document and date on which they were sealed.[62]

As a matter of general practice, plans and specifications, usually referred to as construction documents, are sealed when they are issued for permitting, public bid, or construction. South Carolina law provides that original seals shall appear on each print of the drawings, the index sheet(s) of each set of specifications offered to secure a building permit, and one record set for use on the construction site.[63]

Under Michigan law, architects are to seal plans, plats, drawings, maps, and the title sheet of specifications, addenda, bulletins, and reports. If a bound document is submitted, the index sheet of plans, specifications, and reports shall be stamped or sealed.[64]

Although it seems rather obvious that schematic, design development, partially complete construction documents, and other draft or incomplete documents should not be sealed, there appears to be little treatment of this issue in state statutes or regulations. Perhaps that is because the answer is obvious. Only documents intended for some significant purpose such as permitting, bidding, or construction should be sealed and then only when they are considered complete by the design professional. New Jersey, however, does address this issue in its regulation of engineers stating that incomplete or draft documents shall be conspicuously so identified and may be signed but shall not be sealed.[65]

C. *How Are Documents to Be Sealed?*

As noted above, the procedure for sealing and signing documents is generally addressed by the applicable regulations. Although there are variations, documents generally are sealed by the registrant applying a rubber stamp of the seal or impressing the document with an embossed metal seal followed by registrant's signature near or over the stamped or impressed seal. Also, the document is dated with the date of the stamping or sealing. In Pennsylvania, when a document is signed, the signature must be applied near or over the seal and without obliterating the registration number.[66] The

---

62. FLA. STAT. ANN. § 481.221(2) (West 2006).
63. S.C. CODE ANN. § 40-3-280(B) (West 2006).
64. MICH. COMP. LAWS ANN. § 339.2008(1) (West 2006).
65. N.J. ADMIN. CODE tit. 13, § 40-1.1(f) (2006).
66. 49 PA. CODE § 37.59 (2006).

purpose of the signature is to assure or validate that the seal has been personally and properly applied or used by the registrant.[67]

Some states may require additional certifications or statements. In addition to the date, seal, and signature, Iowa requires a certification that the work was done by the licensee or under the licensee's direct personal supervision.[68] New York law regulating the provision of engineering services requires a warning that it is illegal to alter the document in any way unless under the direction of a licensed engineer.[69]

The size and scale of today's larger and more sophisticated projects, as well as the advent of electronic documents, have raised the issue of facsimile seals and signatures. Not all states have addressed this issue. Engineers in New Jersey may only use impression seals—rubber stamps and digital seals are not permitted.[70] Illinois takes the middle ground in the regulation of its engineers: when computer or other electronic means create engineering documents, the registrant's seal may be computer-generated, but the engineer's signature may not.[71]

Oregon regulations pertaining to the practice of architecture state that the registrant's stamp and handwritten signature must appear on the original title page of specifications and on every sheet of drawings intended for permit or construction. However, the regulations then provide that originals may be reproduced for permit and construction purposes.[72] This statute appears to allow reproduction of additional copies for permitting and construction purposes without having to individually seal and sign each page of the drawings.

Pennsylvania allows the use of facsimile seals on all but the first page of drawing sets prepared by its architects and engineers.[73] However, Florida has more directly addressed the electronic documents issue in its architectural statutes and the Electronic Signature Act of 1996.[74] Final plans, specifications, or reports prepared by Florida architects may be dated, sealed, signed, and transmitted electronically.[75]

D. *Other Requirements Imposed by Some States*

Placing the firm's registration number on any documents prepared by the firm is another potential requirement. Illinois imposes such a responsibility on its architectural firms[76] and New Jersey imposes it on engineering firms.[77]

---

67. *See* MICH. ADMIN. CODE r. 339.16024 (2006).
68. IOWA CODE ANN. § 542B.16 (West 2006).
69. N.Y. EDUCATION LAW § 7209(2) (McKinney 2006).
70. N.J. ADMIN. CODE tit. 13, § 40-1.1(a) (2006).
71. 225 ILL. COMP. STAT. ANN. 325/14, 325/15 (West 2006).
72. OR. ADMIN. R. 806-010-0045 (2006).
73. 49 PA. CODE §§ 9.141, 37.59 (2006).
74. FLA. STAT. ANN. §§ 668.001–006 (West 2006).
75. FLA. STAT. ANN. § 481.221 (West 2006).
76. 225 ILL. COMP. STAT. ANN. 305/14 (West 2006).
77. N.J. ADMIN. CODE tit. 13, § 40-1.3(a)6. (2006).

Another aspect of firm practice, particularly for larger firms with several offices, is the requirement to have licensed staff in responsible charge at each office offering or providing professional services. There are many states, particularly in the southern United States, imposing this requirement for the practice of architecture, engineering, or both, such as Alabama,[78] Florida,[79] Kentucky,[80] Michigan,[81] North Carolina,[82] South Carolina,[83] and Virginia.[84]

Out-of-state firms offering or providing services in Georgia must indicate their name and license number as registered in Georgia on all firm letterheads, publicity, and instruments of service.[85] Contracts for architectural services to be provided in Massachusetts have to be signed by the officers who are Massachusetts licensees and who will exercise professional and supervisory control over the contracted services.[86]

Library References
C.J.S. *Seals* §§ 2–4; C.J.S. *Signatures* §§ 1–16.
West's Key No. Digests, Licenses <KEY>25; Seals <KEY>1, 3; Signatures <KEY>1, 2.

V. Conclusion

There is no quick and easy answer or solution to the many laws regulating the practice of architecture and engineering in the fifty states. It takes a concerted and dedicated effort to identify the licensure requirements for individuals and firms in each state, and to practice legally under the granted licenses.

---

78. ALA. CODE § 34-11-9(a)(5) (2006). Upon the death, retirement, or resignation of the individual included in the firm's name, the firm shall be renamed within two years. *Id.*
79. FLA. LA. STAT. ANN. § 481.225(1)(g) (West 2006).
80. KY. REV. STAT. ANN. § 322.060(1)(c) (Banks-Baldwin 2006).
81. MICH. COMP. LAWS ANN. § 339.2010(4) (West 2006).
82. N.C. ADMIN. CODE tit. 21, r. 56.901(a) (2006).
83. S.C. CODE ANN. § 40-22-260(C) (West 2006).
84. VA. REGS. §10-20-780 (2006).
85. GA. COMP. R. & REGS. r. 50-3-.03(f) (West 2006).
86. MASS. GEN. LAWS ch. 112, § 60L.8. (West 2006).

VI. Practice Aids

A. *Checklist for Practice in a New Jurisdiction*

Firm Professional Licensure and Practice Checklist and Analysis

State: _____	Area of Professional Practice: ❑ Architecture ❑ Engineering ❑ _____

Question	Answer	Comment
1. Does firm have owners, directors, officers, or employees currently licensed in state to provide this service?		
2. Who will have to become licensed in state to be in responsible charge of services to be provided?		
3. Must firm be registered to provide this service? If no, do not complete balance of form.		
4. Is firm's current legal form or that of one of its affiliates acceptable for firm registration?		
5. What ownership, directorship, officership, or employee requirements must be met to register firm?		
6. Which owners, directors, officers, or employees who are not already registered will have to become registered so firm can be become registered?		
7. Are there any restrictions on firm name?		
8. Does board require firm to register with secretary of state?		

Question	Answer	Comment
9. Decide whether firm will have branch office in state in near term. If yes, register with secretary of state. If no, consider sending interstate commerce letter.		
10. Does firm have to obtain its own seal and does it have to be applied along with licensee's seal on documents?		
11. Is there requirement for licensee to be resident in each office providing professional services to state?		
12. Who, what, and when are professionals to seal professional work products?		
13. Is there a requirement for certifications, warnings, or any other unusual insertions when sealing documents?		
14. Are there any unusual requirements for drawing title blocks?		
15. Are there any requirements for how professional contracts are to be entered?		
16. Are there any requirements imposed on how correspondence is issued to clients in state?		
17. Are there any requirements imposed to content of marketing materials issued to clients in state?		

B. NCARB Certificates and NCEES Records

The National Council of Architectural Registration Boards (NCARB) and the National Council of Examiners for Engineering and Surveying (NCEES) provide valuable assistance to firms involved in the multistate practice of architecture and engineering.

The NCARB Web site, at www.ncarb.com, provides a ready reference of registration requirements at both the individual and firm level for the fifty states based on information gathered by the NCARB from state architectural registration boards. The NCARB also offers a certification program that expedites reciprocal registration of individual architects among the 55 member boards in the United States and Canada. However, certification is not the same as registration and does not automatically qualify an individual to practice architecture in any particular jurisdiction. The present fee for the initial creation of a record is $300 and the cost to have the record certified to a registration board is $155 for each certification. For firms with multistate practices, the NCARB certification program is a must for those stockholders, directors, officers, and employees that have to be registered in various jurisdictions.

Similar registration information for engineers and engineering firms is available from the NCEES Web site, at www.ncees.org, based on a 2005 NCEES survey of engineering and land surveying licensing boards. The NCEES program to facilitate comity registrations is called its records program, but it works in the same way as the NCARB certification program. The cost to create the initial record is $150 and the cost of record transmittal is $60 per jurisdiction. The NCEES records program is also essential for stockholders, directors, officers, and employees of engineering firms with a large multistate practice, and for its key engineers who will have to become licensed in many states.

C. *Sample Interstate Commerce Letter*

_____ Board of Registration for
Professional Engineers and Land Surveyors
_____
_____, __ _____-____

                          Re: Application for Certificate of
                                      Authorization

Dear Board:

Our firm is interested in obtaining a certificate of authorization, but our firm will not open an office in _____ State and thus, our firm will not be doing business in _____. Instead, our firm will provide its services in interstate commerce.

In reviewing the application for a certificate of authorization, we believe the requirement that our corporation must first be registered with the _____ Secretary of State to obtain a certificate of authorization is contrary to the commerce clause of the U.S. Constitution, Article I, Section 8, Clause 3. This requirement is also contrary to _____'s own statute regarding the exceptions to the requirement that a certificate of authority has to be obtained by foreign corporations. In _____ statute _____, transacting business in interstate commerce is a specific exception to the requirement that a foreign corporation obtain a certificate of authority from _____'s Secretary of State. This specific exception is required of every state by the above-cited commerce clause in the U.S. Constitution.

Therefore, please find our application enclosed without our firm's registration with the _____ Secretary of State's Office because that is not properly required under the above discussed statutes in the situation where a firm is conducting business in interstate commerce.

                                                        Sincerely,

                                            _____
                                            General Counsel

References

Robert F. Cushman & James C. Dobbs, Design Professionals Handbook of Business and Law (John Wiley & Sons, Inc. 1991) (out of print). This book includes good discussions in chapter 2 of the topics of individual and corporate registrations to conduct business as a design professional.

Stephen G. Walker, et al., State-by-State Guide to Architect, Engineer and Contractor Licensing (Aspen Publishers, Inc. 1999 and updated annually). This book is focused squarely on the topics of individual and corporate registrations to conduct business as a design professional. This book should be part of any firm's library that is seriously going to pursue multijurisdictional practice.

Justin Sweet & Marc M. Schneier, Legal Aspects of Architecture, Engineering and the Construction Process (Thompson Canada Ltd., 2004). This book includes in chapter 10 a fine critical analysis and general discussion of the licensing process.

CHAPTER 2

Ethics and Rules of Conduct Governing Design Professionals

MOHAMMAD IQBAL, P.E., ESQ.

I. Introduction

Design professionals,[1] like lawyers, physicians, and accountants, are members of learned professions. The very essence of a profession is that its members cannot undertake their duties on behalf of a client without awareness of and responsibility for the public welfare.[2]

In the execution of professional services, it is well-accepted that a design professional's legal duty is to conform to the reasonable standard of care prevalent in the industry and regulations applicable to that profession.[3] Design professionals operate under the jurisdiction of state licensure laws and are subject to discipline for a violation thereof. Design professionals are also subject to discipline by their respective professional societies for a violation of the public trust.[4] The professional codes of ethics detail the aim, guidelines, duties, and obligations in serving the public good, and are deemed so fundamental that failure to fulfill them warrants reproach or formal sanctions.

---

1. In general, the phrase "design professional" refers to licensed engineers and architects; however, some authors also include surveyors, landscape architects, and interior designers as design professionals. *See* ROBERT F. CUSHMAN AND G. CHRISTIAN HEDERMANN, ARCHITECT AND ENGINEER LIABILITY: CLAIMS AGAINST DESIGN PROFESSIONALS (2d ed. 1995).

2. A. E. Investment Corp. v. Link Builders, Inc., 214 N.W.2d 764, 769 (Wis. 1974).

3. RESTATEMENT (SECOND) OF TORTS, § 288C (1963-1964); *see also* R. F. CUSHMAN AND G. C. HEDERMANN, *supra* note 1, at 136.

4. *See, e.g.*, AIA 2004 CODE OF ETHICS AND PROFESSIONAL CONDUCT, RULES OF APPLICATION, ENFORCEMENT, AND AMENDMENT, *at* http://www.aia.org/about_ethic.

A. *Ethical Issues in the Design Process*

The collapse of the Hyatt Regency sky walkways in 1981 (discussed in Section IX *infra*), which killed 114 people and injured 216, and the aftermath of destruction caused by Hurricane Andrew in 1991, brought to light that the practices of design professionals contributed to the loss of life, personal injuries, and property damage.[5] Although no court found a design professional guilty of any criminal act, these events serve as classic examples of the potential consequences of design professionals ignoring their ethical obligations—thus creating greater pressure on design professionals to strictly abide by their responsibilities to serve the public good.

Generally, "ethics" is synonymous with ordinary morality.[6] In broad terms, ethics is concerned with how to go about life, what it means to "live well," to accomplish "good" in the world, and be "just" or "fair" in one's personal and professional life.[7] Ethics are considered precatory and not mandatory; however, ethical commitments are of no less importance than legal obligations. As the late Chief Justice of the United States, Earl Warren, declared:

> Society would come to grief without ethics, which is unenforceable in the courts and cannot be made part of law.... Not only does law in a civilized society presuppose ethical commitment, it presupposes the existence of a broad area of human conduct controlled only by ethical norms and not subject to law at all....
>
> The individual citizen may engage in practices which, on the advice of counsel, he believes strictly within the letter of the law but which he also knows from his own conscience are outside the bounds of propriety and the right. Thus, when he engages in such practices, he does so not at his own peril—as when he violates the law—but at peril to the structure of civilization, involving greater stakes than any possible peril to himself.... This law beyond law, as distinct from law, is the creation of civilization and is indispensable to it....[8]

---

5. U.S. Department of Commerce, National Bureau of Standards Report on the Investigation of the Kansas City Hyatt Regency Walkways Collapse (Feb. 1982). In terms of loss of life and injuries, the collapse was the most devastating structural collapse in this country. For reference to Hurricane Andrew, *see* Steven G.M. Stein and Jeffery H. Winick, *The Straight and Narrow Path: Ethical Issues for Design Professionals*, RA&MCO Insurance Services Risk Management Library, *available at* http://www.Ramco-ins.com/resource-library/risk-management/12-stein-winick.htm.

6. Personal ethics is the set of one's own ethical commitments, which are usually acquired in early home or religious training and often modified by later reflection. Common morality or common sense is the set of moral ideals shared by most members of a culture or society. Barry Wasserman, Patrick Sullivan, & Gregory Palermo, Ethics and the Practice of Architecture (2000).

7. Wasserman, *supra* note 6.

8. Unidentified opinion cited in Milton F. Lynch, *Can the Professions Survive without Ethics?* Professional Eng'r 38, 39 (Oct. 1976).

Though ethical concerns pertain to all of us in our everyday lives, the design professional's ethics stem from the relationships and obligations with others and become paramount in the design and construction of the landscape in which we live.

The design professional's obligations arise in response to two sets of needs: individuals and other members of society require design professionals to design facilities to meet their needs; and, the public at large requires protection from potentially devastating effects of poor and insensitive building practices. In assuming these diverging obligations, society charges the design professional with dichotomous demands. This balancing process justifies the special status and distinguishes the design professional from an artist or a technician.[9]

Ethical issues arise as the design process moves from schematic, to design development, and on to the construction phase. The design professional encounters both moral and technical problems in dealing with clients; building codes; quality of work performed by coworkers, subconsultants and contractors; pressure imposed by time and superiors; and whims of the marketplace. The iterative nature of the design process presents ethical dilemmas. Changes made during one stage not only will affect subsequent stages but also may require assessment of prior decisions. The design professional must handle requests for changes during the construction phase with particular care or tragic consequences, such as the Hyatt Regency walkway failure, may result.[10]

Ethical responsibility means accountability or blameworthiness (or praiseworthiness) and is distinguishable from contractual or legal responsibility, although they overlap to some extent. Contractual or job responsibility consists of one's assigned tasks at his or her place of employment or under contract with the client. A particular law could be morally unjustified or inapplicable due to a contractually limited scope of the assignment undertaken.[11] Therefore, the study of ethics is integral to the design professional's ability to balance morality, law, and contractual obligations to the client.

The public requires professionals to meet certain ethical standards as a part of the social contract.[12] The media has published public opinion surveys on the image or perception of the ethical conduct of various professions.[13] In that respect, the code of ethics is an implicit contract between professionals and the public defining what professional conduct the public expects of design professionals. Design professionals agree among themselves to abide by uniform standards and with the public that these uniform standards will ensure competence and promote the well-being of the

---

9. TOM SPECTOR, THE ETHICAL ARCHITECT: THE DILEMMA OF CONTEMPORARY PRACTICE (2001).

10. M. LEVY & M. SALVADORI, WHY BUILDINGS FALL DOWN (1992). *See also* http://ethics.tamu.edu/hyatt/hyatt1.htm.

11. CHARLES B. FLEDDERMANN, ENGINEERING ETHICS (2004).

12. For a discussion on social contract theory and other ethical concepts and theories justifying a moral system, see HERMAN T. TAVANI, ETHICS AND TECHNOLOGY (2004).

13. *See Survey Says Engineers Image Nearly Divine*, ENGINEERING INC. (May–June 2002) (reporting Gallop honesty and ethics polls of various professions).

public as it relates to the professional's expertise.[14] It also enables a potential client to anticipate certain professional conduct.

Although codes of ethics may commend ideals that go beyond the legal standard of care, these commendations are abstract and vague, leaving it unclear how they apply to individual circumstances encountered by members of the profession.[15] Further, the day-to-day pressure associated with the practice of architecture and engineering in the marketplace is such that it is easy to lose sight of professional and ethical obligations.[16] "Architects live and work today in a functioning but weakened profession that lacks a dominant design ethics."[17] Unfortunately, a design professional who is a member of a professional society may not be aware of the existence of that society's code, or may not have read it. Quite often, design professionals learn ethical lessons only after something goes wrong. Responsible design professionals need to know what circumstances in practice call for ethical sensitivity and reflection.

B. *Rules of Professional Conduct*

Ethics and professional rules of conduct are the hallmark of all professions. While adherence to ethical standards is voluntary, abiding by the rules of professional conduct is mandatory for professionals and violation of a rule is grounds for disciplinary action by the professional society of which the professional is a member. In other words, ethics are goals to which the organization asks its members to aspire, but rules represent the minimum acceptable level of conduct below which the professional should not fall or face discipline from the professional society.

Additionally, pursuant to their police powers, states have created agencies, promulgated licensure laws, and appointed regulatory boards to oversee professionals such as architects, engineers, land surveyors, and landscape architects. State laws govern both individuals and entities offering design services. Every state has adopted its own rules and regulations, which may vary widely from one jurisdiction to the next for each profession or specialty. For lawyers representing design professionals, it is necessary to understand the interplay and hierarchy of ethics, rules of professional conduct, and the licensure laws.

This chapter reviews codes of ethics and rules of professional conduct for design professionals. It includes a list of available resources and Internet sites,[18] followed

---

14. U.S. DEPARTMENT OF COMMERCE REPORT, *supra* note 5, at 12.

15. For example, The American Institute of Architects (AIA) 2004 Code of Ethics and Professional Conduct, Ethical Standard No. 1.3, requires that "[m]embers should respect and help conserve their natural and cultural heritage while striving to improve the environment and the quality of life within it," *available at* http://www.aia.org.

16. Stein, *supra* note 5.

17. SPECTOR, *supra* note 9.

18. The list is an update of an earlier paper by this author, Mohammad Iqbal, *Researching the Internet: Ethics and Rules of Conduct Governing Professionals*, CONSTR. LAW., Winter 2001, at 37.

by a discussion on the state licensure laws that govern the practice of engineering and architecture and state-imposed disciplinary actions to which design professionals may be exposed.

Library References

C.J.S. *Architects* §§ 13–14.
West's Key No. Digests, Licenses <KEY>25, 38, 41

II. Code of Ethics

A. *Basic Concepts and Fundamental Principles*

A professional code of ethics is the standard adopted by professionals to govern their own professional conduct. It provides a framework for the professional's ethical judgment. The code simply reiterates principles and standards that constitute accepted and responsible practices, which define the roles and obligations of design professionals.[19] The code articulates the ways in which moral and ethical principles apply to unique situations encountered in professional practice. It indicates to others that the profession is seriously concerned about responsible and professional conduct.[20] For example, the American Council of Engineering Companies' (ACEC) guidelines on ethics and professional conduct state:

> Consulting engineering is an important and learned profession. The members of the profession recognize that their work has a direct and vital impact on the quality of life for all people. Accordingly, the services provided by consulting engineers require honesty, impartiality, fairness, and equity and must be dedicated to the protection of public health, safety, and welfare. In the practice of their profession, consulting engineers must perform under a standard of professional behavior which requires adherence to the highest principles of ethical conduct on behalf of the public, clients, employees and the profession.[21]

Although there is considerable overlap in the codes of various professional organizations, there is no unified code of ethics for design professionals.[22] There are many organizations devoted to the numerous specialties within the design profession. Virtually every occupational or specialty group claiming to have professional status has a code of ethics. For example, the practice of engineering comprises many disciplines and each has a professional society, such as the American Society of Civil Engineers (ASCE) for civil engineers, the Institute of Electrical and Electronics Engineers (IEEE) for electrical engineers, and the American Society of

---

19. CHARLES E. HARRIS JR., MICHAEL S. PITCHARD AND MICHAEL J. RABIN, ENGINEERING ETHICS—CONCEPTS AND CASES (2000).

20. *Id.*

21. *ACEC Professional and Ethical Conduct Guidelines*, ENGINEERING INC. (Sept.–Oct. 2003).

22. LEVY, *supra* note 10.

Mechanical Engineers (ASME) for mechanical engineers. There is no engineering society applicable to all engineers. The National Society of Professional Engineers (NSPE) attempts, like the American Bar Association (ABA) and the American Medical Association (AMA), to function as the umbrella professional society for engineers. Architects, though they may offer specialty services, identify with the American Institute of Architects (AIA) as their architectural society.

B. *Historical Development*

Design professionals began organizing their professional societies in the United States in the nineteenth century. In 1877, the board of the ASCE addressed the issue of whether it should have a code of ethics for its members. Initially, the ASCE resolved not to adopt a code of ethics, explaining "[t]hat it is inexpedient for the Society to instruct its members as to their duties in private professional matters."[23] However, in 1914 the ASCE membership approved the first version of its code of ethics.[24] The original code, like other professional codes in early part of the twentieth century, focused largely on issues related to conducting business. For example, it forbade advertising and price competition for design services among the ASCE members. The code gave less emphasis to issues of service and responsibilities to the public, including safety.[25]

One of the common provisions in the early codes was that members would not underbid other members of the same professional organization—a provision that triggered antitrust action against principal professional organizations. In particular, before 1971 the ASCE *Code of Ethics* contained a provision that made it unethical for an ASCE member "to invite or submit priced proposals under conditions that constitute price competition for professional services."[26] The ASCE enforced the anticompetition provision against its members. This triggered an action by the Department of Justice (DOJ) wherein the DOJ alleged that the ASCE *Code of Ethics* unreasonably restricted interstate trade and commerce by limiting price competition among its members and depriving customers of the benefits of free and open competition in the sale of services.[27] In order to resolve the controversy, ASCE voluntarily removed the anticompetitive provision from its code, annotated the code to state that the submission of fee quotations was permissible, and entered into a consent decree.[28]

---

23. Thomas W. Smith III & Mona B. Savino, *ASCE Ethics—Edict, Enforcement and Education* (2005), *available at* http://www.asce.org. *See also* NATIONAL INSTITUTE OF ENGINEERING ETHICS, ENGINEERING EHICS—CONCEPTS, VIEWPOINTS, CASES, AND CODES (Jimmy H. Smith & Patricia M. Harper eds., 2004).

24. Smith & Savino, *supra* note 23.

25. *Id.*; *See also* http://www.asce.org/inside/ethics_edict.cfm.

26. Smith & Savino, *supra* note 23.

27. United States v. American Soc'y of Civil Eng'rs, 1972 WL 551, No. 72-1776 (S.D.N.Y. May 1, 1972).

28. *Id.*

Like the ASCE *Code of Ethics*, the AIA code also contained an anticompetitive provision. The provision, known as Standard No. 9 of the AIA *Code of Ethics*, came into focus in *Mardirosian v. American Institute of Architects*.[29] In deciding whether the AIA violated antitrust laws, the court held that although the activities of professional societies have long been thought as "deserving of special treatment under the antitrust laws, it is now clear that an anti-competitive practice cannot be justified if it is unreasonable."[30]

The NSPE also had an anticompetitive provision in its rules of professional conduct. The DOJ challenged the provision as improper, and the Supreme Court agreed.[31] The Court held that the NSPE provision was unlawful. It reasoned that although professional services may differ significantly from other business services and the nature of the competition may vary, this difference does not create a broad exemption under the Rule of Reason for learned professions.[32]

Most contemporary codes emphasize commitments to safety, public health, and even environmental protection as the most important duties of a design professional. The rules of the free marketplace now govern design professionals' fees.

C. *Discussion of Professional Codes of Ethics*

Though the professional codes of ethics and rules of conduct have different origins, they overlap not only with themselves but also with moral ideals such as honesty and fairness. The Boston Society of Architects analyzed the 1997 AIA *Code of Ethics and Professional Conduct* and determined that the rules primarily require the members to follow the law and not lie. Specifically:

- Laws—29% of rules require members to obey the law;
- Lies—33% of rules require members not to lie; and
- Other—38% of the rules refer to some other admonition or standard of conduct.[33]

Although the ethical standards and rules profess conformity with laws and societal values, professional standards occasionally may differ from personal or common morality. For example, an engineer may refuse to design a military project believing that war is immoral based on personal conviction, and not on professional or common morality. An architect may refuse to undertake a project that is contrary to his or her individual professional ethics. An engineer, believing that certain design decisions might jeopardize public safety, may "whistle-blow" and violate the

---

29. 474 F.Supp. 628 (D.D.C. 1979).
30. *Id.* at 648.
31. National Soc'y of Prof'l Eng'rs v. United States, 435 U.S. 679 (1978).
32. *Id.* at 696.
33. Boston Society of Architects, *AIA Code of Ethics and Professional Conduct—Content Analysis*, ETHICS FORUM (2002).

employer's obligation to maintain confidentiality.[34] The design professional may not anticipate many questions regarding project purposes, client and personal values, kickbacks, corner-cutting, and diverse and multicultural perspectives, but the codes of ethics and rules of professional conduct of the professional societies generally address these issues.

A code of ethics can be brief, dealing with generalities, or detailed. For example, the IEEE code is short, but the codes of ethics of the NSPE, ASCE, and AIA are more explicit. Some argue that a short code lacking in specificity is more understandable and that members of a professional society are more likely to read it.

The Illinois Institute of Technology's Center for Study of Ethics in Professions (CSEP) has the largest collection of codes of ethics and related material on its Web site, including a search engine and an index organized by professional area.[35] The CSEP site also provides links to other ethics centers, organizations, and directories on the Web. Additional organizations that provide links to Web sites with ethics postings are the National Institute of Engineering Ethics (NIEE) (www.niee.org) and the Online Ethics Center for Engineering & Sciences (www.onlineethic.org). In addition to posting its code of ethics, the NSPE (www.nspe.org), makes available the cases and advance opinions of its Board of Ethical Review (BER). These cases provide an insight into NSPE's interpretation of its code of ethics. Similarly, several other professional organizations have their respective codes of ethics and related cases posted on their Web sites (see Appendix A).

The following is a brief discussion on selected codes of ethics.

Library References
West's Key No. Digests, Licenses <KEY>25.

III. AIA Ethical Standards and Rules of Professional Conduct

A. *Overview*

Founded in 1857, the AIA is the principal professional organization of architects. To persons engaged in the business and profession of architecture, membership in the AIA is a valuable asset that enhances their ability to compete for and obtain architectural business. AIA members normally use the insignia "AIA" following their names to denote membership.

Ethical issues arising in architecture develop during the design and construction processes as an architect deals with the motives and values of clients and society.[36]

---

34. FLEDDERMANN, *supra* note 11. (A good case study of political or institutional pressure affecting engineering performance is NASA's Challenger disaster.)

35. http://ethics.iit.edu.

36. WASSERMAN, *supra* note 6 (Architecture is an ethical discipline before it is an aesthetic one. When we build, we have not just the responsibility to our client, but to those who came before and those who will come after.).

The preamble to the AIA's *Code of Ethics and Professional Conduct* describes the principles on which the code is based. It states AIA members "are dedicated to the highest standards of professionalism, integrity, and competence," and that the code provides "guidelines for the conduct of Members in fulfilling those obligations."[37] The AIA has arranged its code of ethics in three tiers: Canons, Ethical Standards, and Rules of Conduct. Canons are broad principles of conduct. The Ethical Standards are specific goals to which AIA members aspire in their performance and behavior. The Rules of Conduct are mandatory and their violation is grounds for disciplinary action by the AIA.[38] To guide its members in using its code, the AIA has provided a commentary to its code and rules.

One of the most important issues facing the architectural profession is the attribution of credit to other architects in connection with work on a project. The AIA Web site provides guidelines on how to structure and provide that credit.[39]

B. *Enforcement*

The AIA administers its code of ethics through the National Ethics Council (NEC), which is appointed by the AIA Board of Directors.[40] An aggrieved party files formal charges with the NEC.[41] The NEC processes and disposes of complaints in accordance with "Rules of Procedure," which are intended to provide a fair and expeditious procedure for the disposition of cases. The Rules of Procedure include penalties ranging from admonition, suspension of membership, and expulsion from the AIA.[42] The NEC makes its decisions available with the names of the parties redacted. In addition, NEC publishes advisory opinions that illustrate its interpretations of the code. The AIA posts both types of decisions on its Web site.[43] The appeal procedures are confidential, as is the imposition of an admonishment; however, the AIA publicizes all other penalties.[44]

Library References

C.J.S. *Architects* §§ 13–14.
West's Key No. Digests, Licenses <KEY>25, 38, 41.

---

37. AIA 2004 CODE OF ETHICS AND PROFESSIONAL CONDUCT, RULES OF APPLICATION, ENFORCEMENT, AND AMENDMENT, *available at* http://www.aia.org/about_ethic.
38. *Id.*
39. *Id.*
40. *Id.*
41. *Id.*
42. *Id.*
43. *Id.*
44. *Id.*

IV. ASCE Code of Ethics and Rules of Professional Conduct

The American Society of Civil Engineers (ASCE) adopted its original code of ethics in 1914. Since then, other professional organizations have used the ASCE code as a primer in the development of their codes. The ASCE *Code of Ethics* has seven fundamental canons.[45] The code does not give equal weight to all canons—it considers the ethical duty to safety, health, and public welfare to be paramount.[46] Recognizing the ethical duty to the public as the highest level of responsibility, the ASCE's Board of Directors defined the civil engineering profession as a calling where an engineer in the service of humanity uses special knowledge and skill in a distinctly intellectual plane.[47]

In order to preserve the high ethical standards of the civil engineering profession and to govern ethics practices in the profession, the ASCE has developed its "Standards of Professional Conduct" which encompasses issues such as:

- conflicts of interest;
- ensuring legal compliance;
- employees and public safety;
- workplace quality;
- use and protection of employer assets;
- maintaining accurate and complete records;
- gifts, meals, services, and entertainment;
- confidential or proprietary information;
- outside employment/activities;
- purchases of goods and services;
- bribes and kickbacks;
- relationships with competitors;
- relationships with clients, outside contractors, and consultants;
- environmental protection; and
- whistle-blowing.[48]

An ASCE member who violates the ASCE's Standards of Professional Conduct is subject to a disciplinary action up to and including removal or suspension from the ASCE.[49] The ASCE maintains and enforces its code of ethics and allows anyone to bring charges of unethical conduct against a member.[50] It refers complaints to ASCE's Committee on Professional Conduct (CPC) for investigation. During the investigation, the CPC acts like a grand jury; if it finds sufficient evidence to

---

45. *ACEC Professional and Ethical Conduct Guidelines*, ENGINEERING INC. (Sept.–Oct. 2003).
46. *Id.*
47. *Id.*
48. http://www.asce.org.
49. http://www.asce.org/inside/ethics_enforce.cfm.
50. *Id.*

warrant disciplinary action, it refers the case for hearing before the Executive Committee.[51]

The Executive Committee holds proceedings to discipline an ASCE member upon the CPC's recommendation or by written request of ten or more ASCE members. The Executive Committee acts as an unbiased trier of fact and a CPC member acts as a prosecutor.[52] The Executive Committee conducts hearings in accordance with the procedure for professional conduct in cases. The organization affords additional due process to the accused member by serving a reasonable notice of the charges and the hearing, and by providing a fair opportunity to hear the evidence, question witnesses, and refute the evidence.[53] Upon finding a violation of the code, the Executive Committee may take disciplinary action such as a letter of admonition or suspension from membership. The Executive Committee cannot expel a member from the ASCE, but can make that recommendation to the Board of Direction.[54] If the Executive Committee votes to recommend expulsion, it refers the case to the Board of Direction, which maintains the same due process protection as afforded at the Executive Committee hearing. A decision to expel a member requires a 75 percent vote by the Board of Direction. Both the Executive Committee and the Board of Direction have authority to publish the action and to notify other professional organizations or registration boards of their decision.[55] The ASCE's legal staff publishes its opinions on ethical issues reviewed by the CPC in a monthly newsletter sent free to its members. The ASCE also posts opinions on its Web site.[56]

The most notable action taken by the ASCE for violation of its ethics was against Jack Gillum, the engineer of record for the Hyatt Regency Hotel in Kansas City where the sky walkways collapsed. The ASCE evaluated the evidence supplied by the Missouri licensing board and announced a policy of holding structural engineers responsible for all elements of structural safety in the buildings they design.[57] However, it decided that Gillum—the only ASCE member whose license was revoked by the Missouri licensing board—was not guilty of gross negligence or unprofessional conduct, and the ASCE merely suspended his membership for three years. Later, Gillum voluntarily relinquished his membership altogether.[58] Section IX of this chapter further discusses the Hyatt walkways collapse case.

---

51. *Id.*
52. *Id.*
53. *Id.*
54. *Id.*
55. *Id.*
56. http://www.asce.org.
57. http://studentweb.engr.utexs.edu/peterscj/Cover.htm (citing Sarah K.A. Pfatteicher, *The Hyatt Horror: Failure and Responsibility in American Engineering*, ASCE JOURNAL OF PERFORMANCE OF CONSTRUCTED FACILITIES, May 2000, 62–66).
58. Pfatteicher, *The Hyatt Horror: Failure and Responsibility in American Engineering*, ASCE JOURNAL OF PERFORMANCE OF CONSTRUCTED FACILITIES, May 2000, at 62–66.

Library References

C.J.S. *Architects* §§ 13–14.
West's Key No. Digests, Licenses <KEY>25, 38, 41.

V. NSPE Code of Ethics

A. *Overview*

In 1946, the NSPE adopted a *Code of Ethics for Engineers*, the canons of which provide that its members should:

1. hold paramount the safety, health, and welfare of the public;
2. perform services only in areas of their competence;
3. issue public statements only in an objective and truthful manner;
4. act for each employer or client as faithful agent or trustee;
5. avoid deceptive acts;
6. conduct themselves honorably, responsibly, ethically, and lawfully so as to enhance the honor, reputation, and usefulness of the profession.

The NSPE posts its code of ethics, rules of professional conduct, and enforcement procedures on its Web site.[59] The NSPE has formulated numerous hypothetical cases on ethics in order to stimulate the moral imagination and challenge design professionals to anticipate possible alternatives in resolving complex ethical issues. Although these case studies assist in understanding the code, they do not provide ready-made answers to all ethical questions that a design practice generates, and reasonable design professionals may disagree about the correct course of action.

B. *Indemnification, Limitation of Liability, and Ethics*

Design professionals often attempt to negotiate a limitation of liability (LOL) clause in their contracts with clients as a risk management technique. However, parties have challenged the LOL clause in numerous cases on public policy grounds.[60] One method of attacking a LOL clause is to allege that it violates the NSPE's *Code of Ethics for Engineers*, Section III.8, which requires professional engineers to "accept personal responsibility for their professional activities, provided, however, that engineers may seek indemnification for services arising out of their practice for other than gross negligence, where the engineer's interests cannot otherwise be protected."

---

59. http://www.nspe.org.

60. Cushman, *supra* note 1 at § 17.8; *see also* Howard W. Ashcroft, *Enforceability of Limitation of Liability*, 14 Constr. Law. 12 (1994).

The NSPE's Board of Ethical Review (BER) examined the issue of whether it is ethical to require clients to indemnify and hold an engineer harmless from and against any damages or legal expenses arising from negligence in the performance of pollution-related services.[61] The BER interpreted Section III.8 to mean that an engineer has an obligation to accept responsibility for professional activities and, where appropriate, obtain professional liability insurance or other protection available either in the market or in other spheres.[62] The BER opined that where such protection is commercially available at an affordable cost to the engineer and it protects the interests of both the client and the engineer, barring special or additional circumstances, the engineer has an ethical obligation to obtain such protection and not seek indemnification from the client for ordinary negligence. However, it reserved its right to modify its view as circumstances within the professional practice warrant due to the cyclical and fluid nature of the professional liability insurance environment.[63]

Three years later, the BER carved an exception to its earlier opinion on the indemnification clause.[64] The BER considered whether an engineer could include such a provision in an agreement with clients after "carefully surveying and analyzing the hazardous waste insurance market and determining that the hazardous waste liability insurance coverage currently available (was) inadequate to protect himself and his firm during the performance of hazardous waste remediation services for his clients."[65]

In distinguishing its earlier decision, the BER noted that "unlike traditional and customary engineering services, hazardous waste remediation services require engineers to perform actual physical work (sampling, testing) on the remediation site and such activities could easily expose professional engineers to significant third-party liability exposures."[66] Further, the BER noted that "the risk incurred in performing hazardous waste work may be quite significant, not necessarily based on fault and may significantly increase due to strict joint and several liability standards.... Because of these factors, engineers practicing in this challenging field should be allowed a degree of flexibility to explore a variety of risk management techniques which include insurance, contractual provisions (e.g., indemnification, limitation of liability) and other professional practice considerations."[67]

---

61. *Case No. 93-8*, Board of Review, National Soc'y of Prof'l Eng'rs, *available at* http://www.niee.org.
62. *Id.*
63. *Id.*
64. *Case No. 96-12*, Board of Review, National Soc'y of Prof'l Eng'rs, *available at* http://www.niee.org.
65. *Id.*
66. *Id.*
67. *Id.*

Library References
West's Key No. Digests, Licenses <KEY>25.

VI. Law, Ethics, and Professional Conduct

Legal requirements may establish minimum standards of performance; however, ethics provide a richer guide for conduct. Trying to determine the right thing to do is a value-driven quest. How does a design professional assist a community in evaluating alternate designs? How does the design professional account for a project's impact on the environment? Beyond technical and scientific knowledge and design talent, many of the design professional's activities require judgment, respect, and trust among the community of people participating in and affected by a project. When such issues arise, they are ethical ones defined in neither contract nor tort doctrines.[68]

A code of ethics is not a legal document. It applies only to the members of the professional society. For example, the AIA rules apply only to its members and violating its provisions may result in expulsion from that professional society. With the current state of professional societies, expulsion from a professional society generally will not necessarily result in an inability to practice engineering or architecture. Unless the state licensing board takes action, there may be no real consequence for violating professional ethical principles. However, the loss of credibility and lowered personal reputation could be significant for a design professional.

Very few courts have addressed the issue of how an ethical violation impacts a civil action for damages filed against a professional. Some courts have held that while the violation of a professional ethical rule may subject a design professional or a firm to disciplinary sanctions by the profession, a code of ethics does not create a civil cause of action.[69] Drawing on cases dealing with the legal profession, courts have also held that the violation of profession rule does not create a presumption that an attorney has breached a legal duty to the client.[70] However, the trier of fact may consider the construction of a relevant rule of professional conduct as evidence of the standard of care and a breach thereof.[71] Finally, some courts have held that an attorney's ethical violation creates a rebuttable presumption of actionable malpractice.[72] Thus, the design professional should anticipate that an ethics violation might have some effect in a suit brought by the injured client or third party.

In order for a professional society to subject a design professional to its code of ethics, he or she must be a member of that society. However, many states have

---

68. WASSERMAN, *supra* note 6, at 5.
69. Woodbridge Care LLC v. Englebrecht & Griffin Architects, 1997 WL 162808, 19 Conn. L. Rptr. 130 (Conn. Super. Ct. 1997).
70. Two Thirty-Nine Joint Venture v. Joe, 60 S.W.3d 896, 905 (Tx. Ct. App. 2001).
71. *Id.*; Lipton v. Boesky, 110 Mich. App. 589, 313 N.W.2d 163 (1981).
72. 110 Mich. App. at 598, *Lipton*, 313 N.W.2d at 167.

incorporated ethical codes into their licensing statutes, meaning that a violation of a provision of an ethical code would also constitute a violation of state licensing statute.[73] In addition, a client might incorporate the ethical code into its contract with a design professional. Two sample contractual provisions are:

§ 1.1 ETHICAL STANDARDS
The Owner has informed the Design Professional and the Design Professional hereby acknowledges that the Owner is an entity operated under the highest level of ethical standards. To ensure that the Project will be designed in a manner consistent with the Owner's high level of ethical standards, the Owner has requested that the Design Professional agree, and the Design Professional does hereby agree, to immediately disclose to the Owner in writing any financial connection, business relationship, contractual relationship, or other potential or actual conflict of interest between the Design Professional (including the owners, principals, directors, officers, and employees of the Design Professional) and any entity or individual (including the owners, principals, directors, officers, and employees of such party) which is under consideration for the award of a contract to perform work or provide services or materials which relate to the Project. The Design Professional further agrees that, at the Owner's option, the Design Professional and any entity affiliated or related to the Design Professional may be disqualified from providing materials, work, or services (other than the performance of the services by the Design Professional to be provided hereunder) for the Project during the term of this Agreement.

§ 1.2 NO BENEFIT
Neither the Design Professional nor any owner, principal, director, officer, shareholder, or employee of the Design Professional will accept for its or their own benefit any quantity discount, promotional discount, trade commission, rebate, discount or similar payment, or any personal or business benefit ("Benefit") in connection with activities pursuant to this Agreement or in the discharge of its obligations hereunder. Any such Benefit shall be provided to the Owner. In the event that the Design Professional has any agreement in existence prior to or after the effective date of this Agreement, with any entity or individual that provides for any Benefit of any kind to the Design Professional or to any owner, principal, director, officer, shareholder, or employee of the Design Professional if a client of the Design Professional, such as the Owner, purchases materials, equipment, or services from such entity or individual, the Design Professional shall identify such entity or individual, as well as the nature of such agreement to the Owner in

---

73. The Illinois Structural Engineering Practice Act has incorporated unethical or unprofessional conduct as cause for discipline. For example, the Structural Engineering Act states:

writing, within five (5) days after the execution of this Agreement, as well as within five (5) days after executing any such agreement after the effective date of this Agreement.

If the parties incorporate professional ethics into their contract, a court would undoubtedly hold that a violation thereof would constitute actionable breach of contract. In addition, to the extent that these contract provisions increase the applicable standard of care, the obligation may not be insurable under the design professional's errors and omissions coverage.

Library References

C.J.S. *Contracts* § 271.

West's Key No. Digests, Contracts <KEY>114; Indemnity <KEY>33(5).

VII. Ethical Dilemma

Ethical dilemmas arise when morals come into conflict or the application of moral values is unclear. As each project is unique, so are the ethical challenges that the design professional must face and resolve. The design professional's seal on a set of drawings is evidence that he or she has addressed and resolved all ethical considerations and dilemmas.[74]

There are two broad categories of ethical dilemmas: first, dilemmas that are traditionally resolved with a well-settled solution that codes of ethics clearly address, such as to tell the truth, obey the law, heed the standard of care, not offer or accept bribes, maintain confidentiality, etc.; and second, dilemmas that have several solutions or considerations. These include "gray areas" where it is often hard to differentiate right from wrong. The NIEE, AIA, NSPE, ASCE, and other professional organizations provide numerous case histories and hypothetical scenarios to illustrate the interplay of different moral values in the design process.[75] Case

---

Sec. 20. (a) The Department may, singularly or in combination, refuse to issue, renew, or restore, or may suspend or revoke any license or certificate of registration, or may place on probation, reprimand, or fine, with a civil penalty not to exceed $10,000 for each violation, any person, corporation, partnership, or professional design firm registered or licensed under this Act for any of the following reasons:

(2) Negligence, incompetence or misconduct in the practice of structural engineering;

...

(9) Engaging in dishonorable, unethical or unprofessional conduct of a character likely to deceive, defraud or harm the public ....

225 ILCS 340.

74. SPECTOR, *supra* note 9.

75. *See, e.g.*, NATIONAL INSTITUTE OF ENGINEERING ETHICS, ENGINEERING ETHICS—CONCEPTS, VIEWPOINT, CASE, AND CODES (Jimmy H. Smith & Patricia M. Harper eds., 2004); HARRIS ET AL, *supra* note 19.

studies illustrating ethical dilemmas serve a critical role in understanding professional ethics.[76]

Several case studies illustrate embedded ethical issues and a framework to solve ethical problems.[77] The first step in solving an ethical problem is to completely understand all the issues involved and divide them into three categories: factual, conceptual, and moral. After delineating the issues, frequently a solution or a range of solutions to the problem will become apparent.[78] Some may be better and others may be clearly wrong. This analytical approach aids the organization of solutions and is largely analogous to solving legal issues; however, the selected solution may rest with personal ethics.

The ASCE advises its members to follow the "PLUS" method to determine whether their acts in professional dealings conform to ethical standards. PLUS is an acronym that stands for Policies, Legal, Universal, and Self. The PLUS method requires the professional to answer the following questions about the act:

Policies	Is it consistent with your employer's policies, procedures, and guidelines and ASCE's code of ethics?
Legal	Is it acceptable under applicable laws and regulations?
Universal	Does it conform to the universal principles/values that your employer and the profession have adopted?
Self	Does it satisfy your own personal definition of right, good and fair?[79]

The ASCE also suggests that a member facing an ethical dilemma should discuss it with his or her supervisor or the CPC through ASCE's legal counsel.[80]

Library References

C.J.S. *Seals* §§ 2–3.
West's Key No. Digests, Licenses <KEY>25; Seals <KEY>1.

VIII. Licensing Boards Governing Design Professionals

A. *Licensure Laws*

A regulatory framework or state "licensure laws" govern professionals and restrict the practice of a profession to individuals who have demonstrated professional competency traditionally acquired through education, experience, and examination.

---

76. FLEDDERMANN, *supra* note 11.
77. *Id.*
78. *Id.*
79. AMERICAN SOCIETY OF CIVIL ENGINEERS, ETHICS—STANDARDS OF PROFESSIONAL CONDUCT FOR CIVIL ENGINEERS (2000), *available at* http://www.asce.org.
80. *Id.*

State licensure laws have been in effect for a very long time,[81] predicated upon the states' police power and justified as a mean to protect the public health, safety, and welfare.[82] The purpose of licensing statutes is to protect the public rather than to punish the licensed professional.[83]

State statutes use licensing and registration, often interchangeably, to establish licensing requirements. Individuals who satisfy specific criteria may receive a "license" to practice that profession. Registrations are distinguishable from licenses. A registration is a regulatory framework with substantially less or no criteria for competency. Registration may simply require an individual to enroll or register with the state in order to provide a regulated service. Many jurisdictions call the framework for regulating professional engineers "registration" instead of "licensure."[84]

Typically, a business entity or firm that desires to practice architecture, engineering, or surveying must obtain a license in its own name from the appropriate licensing board. In such jurisdictions, there are two levels of licensing requirements—one at the individual or professional level and the other at firm or business level.[85] Appendix B presents a list of the licensure boards and their Web sites. The boards' Web sites are an excellent source of licensure laws. Summaries of state licensure laws governing design professionals are also available.[86]

B. *Licensing Board Powers*

Licensing (or licensure) boards derive their authority from the various state statutes creating them and empowering them to act.[87] Fifty-five jurisdictions govern design professionals in the United States.[88] Every jurisdiction has a licensing board to administer laws and to regulate design professionals.[89] The licensure boards are executive branch entities and have investigative, disciplinary, and operational responsibilities as required by or authorized by the statute.[90] As a part of a regulatory

---

81. For example, the Illinois Architecture Act, enacted in 1897, was the first architectural licensing law in the nation.

82. *See, e.g.*, 225 Ill. Comp. Stat. Ann. 340 (2000).

83. State *ex rel.* Lentine v. State Board of Health, 334 Mo. 220, 65 S.W.2d 943 (1933).

84. National Soc'y of Prof'l Eng'rs, Engineering Licensing Laws (2004) (the terms "license" and "certificate of registration" are used interchangeably).

85. *Id.*

86. State-by-State Guide to Architects, Engineers, and Contractor Licensing (Stephan G. Walker, Richard A. Holderness & Stephen D. Butler eds., 1999).

87. 5 Bruner & O'Connor Construction Law § 16:25 (2002).

88. *Id.* The 55 jurisdictions include 50 states, District of Columbia, Guam, North Mariana Islands, Puerto Rico, and Virgin Islands.

89. The addresses and phone number of the licensing boards are published at many Web sites on the Internet. For example, visit NSPE and AIA Web sites.

90. 5 Bruner & O'Connor Construction Law § 16:25 (2002).

scheme, boards have enforcement and disciplinary power. The board uses the former against non-licensees and the latter against licensee violators.

The licensure laws, and thus the power of the boards, vary from one jurisdiction to the other. Some boards serve purely disciplinary and licensing functions, while others provide intervention and mechanisms for rehabilitation or correction of professionals rather than mere punishment. Licensure boards regulate professional engineers (PEs), PEs and land surveyors, or multiple professions; however, the joint boards are predominant.[91] The licensing system may divide the engineering profession into different categories for effective administration. For example, California licenses and regulates fifteen categories of engineers.[92]

Typically, a board makes decisions on policy, operational, investigative, and disciplinary issues at public meetings, subject to applicable state administrative and procedural laws. Additionally, a board may guide the building code officials on the building design requirements of the licensure act. Some boards have the authority to waive licensure requirements for case-specific purposes, while others play a role in reinforcing state laws governing the procurement of regulated services.[93]

In the interest of achieving uniformity in the licensure laws among all jurisdictions, the National Council of Examiners for Engineering and Surveying (NCEES) was formed. It is the umbrella organization of state and territorial engineering licensure authorities, and promulgates "Model Laws" and "Model Rules" as guides for engineering licensure boards and legislatures.[94] The NSPE endorses enactment of uniform licensure laws in all jurisdictions; however, it is unlikely that will ever occur. The National Council of Architectural Boards (NCARB), similar to NCEES, is the umbrella organization of architectural boards.[95]

To deal with the maze of state licensure laws, a compliance checklist can serve as a good starting point for evaluating licensure law issues.[96]

C. *Enforcement Powers*

State laws authorize their attorneys general to seek punishment of unlicensed individuals when they violate the law by engaging in the practice of engineering or architecture without a license. Normally, actionable conduct includes the unauthorized use of words such as "engineer," "engineering," "architect," or derivations thereof

---

91. *Id.* According to the NSPE analysis, there are 41 joint boards and 13 PE-only boards.

92. Board for Professional Engineers and Land Surveyors, 17-WTR CAL. REG. REP. 154 (Winter 2001).

93. NSPE, *supra* note 84.

94. http://www.ncees.org.

95. http://www.ncarb.org.

96. Paul W. Berning & John W. Ralls, *Complying with Licensing and Registration Laws: A Compliance Checklist*, *available at* http://www.constructionweblinks.com.

(except as authorized by the act); using the seal of another individual; providing false information to the licensing authority; falsely impersonating a licensee; using an expired, suspended, or revoked license; practicing engineering when not qualified; and falsely claiming a licensed status.[97] The engineering and architectural licensure laws generally consider these violations to be criminal offenses for which the state attorney general is responsible for prosecuting, and individuals found guilty are subject to criminal penalties, such as fines or prison terms.[98] Some states empower the licensure boards to issue letters of warning and to levy fines, including the costs of investigation, on unlicensed violators.[99]

Nearly all jurisdictions have empowered their respective licensing boards to subpoena witnesses and to require licensees and others to submit documents in connection with an investigation of an alleged violation.[100] The licensing boards petition the courts to enforce compliance with subpoenas and for injunctions against violators to compel them to cease the illegal practice.

Some boards are empowered to include the professional rules of conduct in their rules.[101] Licensure boards generally publish their decisions in their newsletters to licensees and on their Web sites. A published decision usually identifies the violator and briefly describes the complaint and the board's decision.[102]

D. *Investigative and Disciplinary Powers*

Because design professionals operate under the jurisdiction of state licensure laws, they are subject to discipline for violation of these laws. Discipline is not necessarily concerned with harm to the client, but with professional conduct and ethics in dealing with client.[103] Licensing boards open investigations for violations of licensure rules based on complaints filed with the board, personal knowledge of the board members or both. In many jurisdictions, a design professional has an ethical and/or legal obligation to report violations of rules to the board, including claims resolved through settlement or trial.[104]

A licensing board's statutory ability to investigate and prosecute violations of the rules varies considerably from one jurisdiction to another. For example, in some jurisdictions boards can directly regulate violations, while in others the boards must act through the state attorney general's office. The NSPE has published an

---

97. 5 BRUNER & O'CONNOR CONSTRUCTION LAW § 16:25.
98. *Id.*
99. *Id.*
100. *Id.*
101. *Id. See also* Stein, *supra* note 5.
102. *See, e.g.,* California Board's Web site, *at* http://www.dca.ca.gov/pels.
103. *Lipton,* 313 N.W.2d at 596.
104. *See, e.g.,* COLO. REV. STAT. § 12-25-108(1)(h) (2006).

analysis of the engineering licensure laws in the fifty-five jurisdictions comprising the United States.[105] That analysis reveals that the grounds for disciplinary action include engaging in behavior or conduct deemed dishonorable, unethical, or unprofessional such as negligence, gross negligence,[106] incompetence, misconduct, conviction, or pleas of no contest to any felony crimes, practice-related misdemeanor crime, action taken by another jurisdiction that led to disciplinary action there,[107] failure to provide information requested by the licensing authority, making false statements, attempting to mislead,[108] aiding others in violating the licensure laws, violating terms of probation or suspension, sealing or signing documents not prepared by the licensee or under the licensee's direct supervision, providing false testimony to the board, conviction of a drug offense, habitual intoxication or addiction to drugs or alcohol, failure to pay child support, delinquent student loans or taxes, and failure to comply with licensure laws.[109]

E. *Revocation and Suspension of Licenses*

If the state law authorizes a board to grant a professional license, it may also empower it to revoke or suspend a license for cause.[110] Boards have disciplined design professionals primarily under the rubric of gross negligence, incompetence, and misconduct for such activities as:

1. relying upon the judgment of untrained individuals;
2. certifying work as completed and knowing that it was not;
3. performing welding operations without being properly certified;

---

105. 5 BRUNER & O'CONNOR CONSTRUCTION LAW § 16:25 (2002).

106. The Missouri Supreme Court has stated that plaintiff gains nothing by branding negligence "gross" and has consistently refused to recognize differing categories of negligence. Boyer v. Tilzer, 831 S.W.2d 695 (Mo. Ct. App. 1992). However, the term "gross negligence" in a statute authorizing disciplinary action against architects and professional engineers did not render the statute unconstitutional because of vagueness. Duncan v. Missouri Bd. for Prof'l Eng'rs, Architects and Land Surveyors, 744 S.W.2d 524, 533 (Mo. Ct. App. 1988) (noting that the term "gross negligence" has been defined in various manners in different states, the court defined it as an act or course of conduct which demonstrates conscious indifference to a professional duty).

107. 5 BRUNER & O'CONNOR CONSTRUCTION LAW § 16:25 (2002). (37 jurisdictions consider a disciplinary action taken by one jurisdiction as a cause for disciplinary action in another).

108. *Sheils v. Florida Eng'rs Mgmt. Corp.*, 886 So. 2d 426 (Fla. Dist. Ct. App. 2004) (under Florida Administrative Code, a mere attempt to mislead the client, employer or the public constitutes misconduct in engineering and that success in misleading was not a required finding of professional misconduct against a professional engineer).

109. 5 BRUNER & O'CONNOR CONSTRUCTION LAW § 16:25 (2002).

110. 5 AM. JUR. 2D. *Architects* § 6 (2006).

4. preparing misleading plans;
5. permitting the licensee's seal to be affixed to plans neither prepared by the licensee nor under the licensee's revising control;
6. appropriating another's plans, removing that professional's seal, replacing the seal with one's own and then submitting the plans as one's own;
7. representing land surveys as one's own without performing an actual survey; and
8. offering a bribe to certain public officials in order to obtain a contract.[111]

Generally, the misconduct must relate to the profession to justify a disciplinary action, unless the conduct calls into question the professional's integrity.[112]

F. *Disciplinary Reciprocity*

Thirty-seven of fifty-five jurisdictions in the United States consider a disciplinary action taken by one jurisdiction as grounds for disciplinary action in another state.[113] For example, after the Hyatt walkway collapse, all twenty-eight states involved revoked Jack Gillum's PE license.[114] The California board revoked Gillum's civil and structural engineering licenses, but two years later granted Gillum's petition to reinstate the civil license. However, the board reinstated it with a four-year probationary period and specific terms and conditions requiring Gillum to provide 200 hours of pre-approved pro bono community service.[115]

Another issue is whether the law authorizes a licensure board to discipline a licensee for misconduct that occurred before the board granted a license to the accused. The California Supreme Court considered this issue in *Hughes v. Board of Architectural Examiners*.[116] There, the California Board of Architectural Examiners revoked Hughes's architectural license, accusing him of wrongful conduct that took place outside California and before he obtained a California license.[117] The Superior Court of Sacramento denied Hughes's petition for a writ of administrative mandamus seeking to overturn the board's decision, and the court of appeals reversed the lower court's decision.[118] The California Supreme Court reversed, holding that the California Architect's Practice Act did not limit discipline to post-licensure misconduct and no constitutional violation arises from the consideration of pre-licensure

---

111. 5 BRUNER & O'CONNOR CONSTRUCTION LAW § 16:25 (2002).

112. *Id.*

113. *Id.*

114. *See* http://www.euken.net/group/seaoc/mailarchive/2001a/msg00778.html; http://www.studentweb.engr.utexas.edy/peterscj/Cover.htm.

115. *Accusation 514-A*, at http://www.dca.ca.gov/pels.

116. 952 P.2d 641 (Cal. 1998).

117. *Id.*

118. *Id.*

conduct.[119] After remand, the court of appeals upheld the board's decision to revoke Hughes's license, and reasoned that as a preventive response the board could revoke his license based upon pre-licensure professional dishonesty and without a showing of actual harm to a California client.[120]

G. *Judicial Review*

It is well-accepted that the right of an individual to engage in any of the common occupations of life is among the fundamental liberties protected by the due process and equal protection clauses of the Fourteenth Amendment.[121] The courts have also recognized that an individual, having obtained a license to engage in a particular profession or vocation, has a "fundamental vested right" to continue in that activity.[122] A licensee is entitled to greater procedural protections than accorded to an applicant, requiring a licensure board to prove the allegations filed against a licensee by clear and convincing evidence rather than the preponderance of the evidence.[123]

Procedural due process demands a fair trial before a neutral or unbiased decision maker, but it does not require perfect error-free governmental decision making.[124] It also demands an appearance of fairness and absence of the probability of outside influence on the adjudication.[125] To avoid the appearance of unfairness in the licensing board's execution of its roles as police, prosecutor, jury, judge, and executioner, some state legislators have interposed a hearing commission into the process to hold hearings and issue finding of facts and conclusions.[126]

Design professionals have attacked the board's power to revoke or suspend a license on numerous grounds. However, most attempts to avoid disciplinary sanctions on the basis that the board lacked authority to take a particular action have met with little success. Another common challenge is to claim that statutory grant is unconstitutionally vague. However, the courts have upheld statutes containing undefined terms such as "unprofessional and dishonorable conduct," "bad moral character," "misconduct or dishonesty," "gross negligence," and other general terms.[127]

---

119. *Id.* at 763.

120. Hughes v. Board of Architectural Exam'rs, 68 Cal. Ct. App. 4th 685, 80 Cal. Rptr. 2d 317 (Cal. Ct. App. 1998).

121. *Hughes*, 952 P.2d 641, *citing* Schware v. Board of Bar Exam'rs, 353 U.S. 232, 238–239 (1957).

122. *Hughes*, 952 P.2d at 788-789.

123. *Id.* at 789.

124. Bracy v. Gamley, 520 U.S. 890, 904 (1997).

125. Utica Packing Co. v. Block, 781 F.2d 71, 77 (6th Cir. 1986).

126. Daniel R. E. Jordan, *Licensing Issues at the Administrative Hearing Commission*, 50 J. Mo. B. 23 (1994).

127. Holmes v. Missouri Dental Bd., 703 S.W.2d 11 (Mo. Ct. App. 1985); Duncan v. Missouri Bd. for Prof'l Eng'rs, Architects and Land Surveyors, 744 S.W.2d 524 (Mo. Ct. App. 1988).

Moreover, the courts have ruled that it is indeed impossible to define or categorize all the acts of impropriety.[128]

Based upon the specificity of charges in the board pleadings filed against that licensee, design professionals may also challenge an action taken by the board. One court held that "simply a statement that the accused has violated one or more of the statutory grounds for discipline without further elaboration, i.e., he has been grossly negligent" is insufficient to allow preparation of a viable defense and therefore does not comply with due process requirement.[129] Specificity is required in pleadings to set the course of conduct deemed to establish statutory ground for proposed disciplinary action.[130] Further, the board cannot find a violation if that charge does not appear in the board's complaint against the accused licensee.[131]

H. *Standard of Review*

A professional can challenge the licensing board decision exercising its disciplinary or enforcement power by appealing to a court.[132] After the trial court's independent review of the facts, an appellate court determines whether the trial court's findings are supported by substantial evidence.[133] The appellate court independently may exercise its ability to decide issues of law.[134] The function of a court in reviewing a challenge to a regulation is limited—the challenger of a regulation must establish that the regulation is arbitrary and capricious.[135] The courts presume that a board's decision is valid and the adversarial party has the burden of overcoming that presumption.[136] In *Moheet v. State Board of Registration for the Healing Arts*, a physician appealed the Missouri State Board of Registration for the Healing Arts' decision to publicly reprimand him. The Missouri Supreme Court held that it would affirm the board's decision unless it:

1. is in violation of constitutional provisions;
2. is in excess of the statutory authority or jurisdiction of the agency;
3. is unsupported by competent and substantial evidence;

---

128. Ray v. Dept. of Registration, 419 N.E.2d 413 (Ill. Ct. App. 1981); *Duncan*, 744 S.W.2d at 532.
129. *Duncan*, 744 S.W.2d 524.
130. *Id.* at 539.
131. *Id.*
132. *Id.*
133. Taylor, Thon, Thompson & Peterson v. Cannadays, 749 P.2d 63 (Mt. 1988).
134. Kelly v. Board of Registered Nurses, 2005 WL605459 (Cal. Ct. App. March 16, 2005).
135. General Bldg. Contractors of New York State, Inc. v. New York State Education Department, 670 N.Y.S.2d 697 (Sup. Ct. 1997); *see also* Adams v. North Carolina State Bd. of Registration for Prof'l Eng'rs and Land Surveyors, 501 S.E.2d 660, 663 (N.C. Ct. App. 1998).
136. Moheet v. State Bd. of Registration for the Healing Arts, 154 S.W.3d 393 (Mo. Ct. App. 2005).

4. is, for any other reason, unauthorized by law;
5. is made upon unlawful procedure or without a fair trial;
6. is arbitrary, capricious or unreasonable; or
7. involves an abuse of discretion.[137]

Library References

C.J.S. *Architects* §§ 2, 7, 10, 12–14; C.J.S. *Constitutional Law* §§ 1385–1387, 1391; C.J.S. *Trading Stamps and Coupons* § 6.

West's Key No. Digests, Constitutional Law <KEY>287.2(5); Licenses <KEY>1, 22, 24, 25, 38, 40.

IX. Hyatt Regency Sky Walkway Collapse

On July 17, 1981, the second and fourth floor walkways of the Hyatt Regency Hotel in Kansas City collapsed and fell to the floor of the main lobby, killing 114 people and injuring 216. In terms of loss of life and injuries, it was the most devastating structural collapse ever to take place in this country at that time.[138] The improper design of connections was the cause of collapse.[139] Due to a mammoth failure of communication, no one ever designed the rod hanger connections.[140]

In February 1984, the Missouri Board for Architects, Professional Engineers and Land Surveyors filed its complaint seeking a determination that the engineering certificates of registration of Daniel Duncan and Jack Gillum and the engineering certificate of authority of G.C.E. International were subject to discipline pursuant to Missouri licensing statute.[141] G.C.E. was a Missouri corporation holding a certificate of authority to perform professional engineering services in Missouri and had assumed contractual responsibility for performing all of such engineering services

---

137. *Id.* at 397.

138. U.S. Department of Commerce, National Bureau of Standards Report on the Investigation of the Kansas City Hyatt Regency Walkways Collapse (Feb. 1982).

139. Engineer Jack Gillum's first words when he saw the wreckage were, "[W]here are the stiffeners?" (G. William Quatman, *The Hyatt Skywalk Revisited: What Happened? Could It Happen Again?* Schinnerer's 42nd Annual Meeting of Invited Attorneys, *available at* http://www.schinnerer.com).

140. *Accusation 514-A, at* http://www.dca.ca.gov/pels (quoting Gillum); *see also* Administrative Law Judge Opinion of James B. Deutsch, Nov. 15, 1985, Case No. AR 84-0239; Duncan v. Missouri Bd. for Prof'l Eng'rs, Architects and Land Surveyors, 744 S.W.2d 524 (Mo. Ct. App. 1988). Numerous articles have been written on the causes of collapse. The ASCE devoted its entire May 2000 issue of the Journal of Performance of Constructed Facilities to the articles addressing the Hyatt collapse. Jack Gillum contributed *The Engineer of Record and Design Responsibility*.

141. Administrative Law Judge Opinion of James B. Deutsch, Nov. 15, 1985, Case No. AR 84-0239.

for the Hyatt. Gillum was a licensed PE in Missouri and president of G.C.E. Duncan was a licensed PE in Missouri and an employee of G.C.E. Gillum was the engineer of record for the Hyatt and supervisor of all professional engineering activities of G.C.E. in Missouri. Duncan was the project engineer for the Hyatt in direct charge of the actual structural engineering work on the project, under the direct supervision of Gillum.[142]

Judge Deutsch, Administrative Law Judge for the hearing commission, found Duncan guilty of gross negligence in the preparation and completion of a structural drawing and in failing to review shop drawings for the Hyatt project.[143] He also found Duncan guilty of misconduct in misrepresenting to the architects the safety of a connection (the double hanger rod-box beam connection), when in fact he had failed to perform engineering tests and calculations. The judge found that Gillum:

- was vicariously liable and responsible for the acts and omissions of Duncan because he had affixed his professional engineering seal on the structural drawings;
- was grossly negligent in failing to himself review or assure that someone had reviewed the connection drawing before affixing his seal thereto;
- had engaged in unprofessional conduct in failing and refusing to take responsibility for the entire engineering project; and
- was guilty of misconduct for his failure to perform a review of the Hyatt atrium design following a specific request by the architect for such review, and for continuing misrepresentations to the owner and architect concerning the atrium design review.

Judge Deutsch recommended that the Missouri Board for Architects, Engineers and Land Surveyors discipline Gillum, Duncan, and G.C.E. (which he found was vicariously responsible for the acts and omissions of Duncan and Gillum).[144] Upon this recommendation, the board revoked all three licenses. Gillum and Duncan petitioned the trial court challenging the board's decision, but that court affirmed the outcome. Duncan and Gillum appealed that decision to the Missouri appellate court.[145] The appellants challenged, in part, the validity of the engineering statute based on the definition of gross negligence and the fact that engineers customarily rely upon fabricators to design certain structural steel connections.[146] Affirming the revocation, the court held that the engineer was responsible for the design, reasoning that custom, practice, or "bottom line" necessity cannot alter that responsibility.[147]

---

142. *Duncan,* 744 S.W.2d 524.
143. *Id.*
144. *Id.*
145. *Id.*
146. *Id.* at 536.
147. *Id.*

The court held the statutory definition of gross negligence was not sufficiently vague to present a substantial constitutional challenge and affirmed the commission's definition of gross negligence as "an act or course of conduct which demonstrates a conscious indifference to a professional duty."[148] The court also noted that the structural engineer's duty was to determine whether the structural plans it designed and approved provided structural safety in order to prevent a strong probability of harm. The court said that indifference to the duty is indifference to the harm.[149] Reasoning that the purpose of disciplinary action against licensed professionals is the protection of the public, not the infliction of punishment, the court affirmed the revocation of all three licenses.[150] When the appellate court delivered its decision in 1988, the victims and their families had settled their lawsuits arising from the collapse, so the court's decision did not affect them.[151]

Although Gillum denied responsibility during the disciplinary hearing and appeal, he made three operational changes in G.C.E.:

1. A P.E. should prepare all steel shop drawings provided by fabricators; the fabricator's P.E. must seal the shop drawings before the engineer of record will approve them;
2. All project engineers within G.C.E. should personally seal and sign and take responsibility for all aspects of the project performed under their direction; and
3. All engineers, technicians, and draftsmen should follow the quality control/quality analysis manual.[152]

Gillum's first operational change aimed to shift design responsibility during the shop drawing phase from the design professional to the steel fabricator and erector did not meet full acceptance by the American Institute of Steel Construction (AISC). The AISC codified:

> 4.2 .... When the Owner's Designated Representative for Design provides the design, design drawings, and specifications, the Fabricator and Erector are not responsible for the suitability, adequacy, or building code conformance of the design....
> 1.5.1. When the Fabricator submits a request to change Connection details that are described in the Contract Documents, the Fabricator shall notify the Owner's Designated Representatives for Design and Construction *in writing in advance* of the submission of the Shop and Erection Drawings.

---

148. *Id.* at 541.
149. *Id.*
150. *Id.*
151. Mathew A. Victor, *The Hyatt Collapse—A Post Mortem*, CONSTR. LAW. (Nov. 1990).
152. Jack D. Gillum, *The Engineer of Record and Design Responsibility*, THE AMERICAN SOC'Y OF CIVIL ENG'RS J. OF PERFORMANCE OF CONSTRUCTED FACILITIES (May 2000), at 67–70.

The Owner's Designated Representative for Design shall review and approve or reject the request in a timely manner.[153]

Gillum's second change implied that a design professional sealing and signing the design had a nondelegable responsibility for the project.[154] In general, it presented a narrow view of the design process. Since then, it has been well-accepted that a design professional sealing and signing the design does not hold a monopoly on all design knowledge necessary to complete a successful, modern construction project, and it is often prudent for him or her to rely upon the expertise of experienced contractors or suppliers to design certain components of required work.[155] However, design delegation does not imply design abdication. The licensure statutes permitting design delegation regulate the process by which a licensee may delegate design tasks.[156]

Gillum's third operational change to have staff follow a quality control process has been an excellent risk management tool and a challenge for the design profession.

X. Conclusion

Professional ethics and rules of conduct is an expansive and complex discipline. It provides a basis for considering personal, professional, and communal values with respect to business practices and moral questions. Normally, codes of ethics do not have the force of law, but they can be binding on design professionals when incorporated into the licensure law. The violation of code may be admissible into evidence in a civil action brought by the client or an injured third party. To further the profession and protect the interests of the client and public safety, it is incumbent upon design professionals to be familiar with and abide by the rules of conduct adopted by applicable professional organizations and those embodied in the state's regulatory scheme.

---

153. AMERICAN INSTITUTE OF STEEL CONSTRUCTION, CODE OF STANDARD PRACTICE FOR STEEL BUILDINGS AND BRIDGES, at http://www.aisc.org (emphasis added).

154. Coincidently, *Duncan* reached the same conclusion. *Duncan*, 744 S.W.2d at 535.

155. Nancy A. Potter, *Design Delegation Provisions of AIA Document A201—1997 Edition*, 18-JUL CONSTRUCTION LAW 27.

156. *See, e.g.*, N.Y. COMP. CODES R. & REGS. tit. 8, § 29.3(b)(2)(vi), requiring that delegator determine in writing that the design prepared by delegatee conforms to the overall project design and can be integrated into it. *See* THE DESIGN/BUILD PROCESS (ABA 1997) for licensing and procurement requirements. *See also* 5 BRUNER & O'CONNOR CONSTRUCTION LAW § 16:25 (2002), for a summary of licensure laws on sealing and signing of design documents and definitions of "responsible charge" of design.

XI. Appendices

A. *Professional Organizations*

Profession	Organization	URL
Agricultural Engineer	American Society of Agricultural & Biological Engineers	http://www.asabe.org
Architect	American Institute of Architects	http://www.aia.org
Building Designer	American Institute of Building Design	http://www.aibd.org
Certified Public Accountant	American Institute of Certified Public Accountants	http://www.aicpa.org
Chemical Engineer	American Institute of Chemical Engineers	http://www.asche.org
Civil Engineer	American Society of Civil Engineers	http://www.asce.org
Computer Engineer	Association of Computing Machinery	http://acm.org
Consulting Engineer	American Consulting Engineers Council	http://www.acec.org
Cost Engineer	Association for the Advancement of Cost Engineers	http://aacei.org
Electrical Engineer	American Institute of Electrical Engineers	http://www.ieee.org
Engineering and technology	Accreditation Board of Engineering and Technology	http://www.abet.org
Environmental Engineer	American Academy of Environmental Engineers	www.aaee.org
Fire Protection Engineer	American Society of Fire Protection Engineers	http:/www.sfpe.org

B. *Licensure Boards Web Site Addresses*

Jurisdiction	WWW address	Jurisdiction	URL
Alabama	www.state.al.us	Iowa	www.state.ia.us/government.com/prof/pld.htm
Alaska	www.dced.state.ak.us	Kansas	www.ink.org
Arizona	www.btr.state.az.us	Kentucky	www.state.ky.us
Arkansas	www.state.ar.us/business_p2.html	Louisiana	www.state.la.us/state/dept.htm
California	www.dca.ca.gov	Maine	www.state.me.us/pfr/led/list.htm
Colorado	www.state.co.us/gov_dir/permits.html	Maryland	www.state.md.us/license/electron.html
Connecticut	www.state.ct.us/dcp	Massachusetts	www.state.ma.us/reg
Delaware	www.dape.org	Michigan	www.cis.state.mi.us
District of Columbia	www.dcra.org/opla.shtm	Minnesota	www.aelslag.state.mn.us
Florida	www.myfloridalicense.com/Default.asp www.fbpe.org	Mississippi	www.state.ms.us
Georgia	www.state.ga.us	Missouri	www.gov.state.mo.us/boards
Hawaii	www.hsba.org/Hawaii/Admin/DCCA/dcca.htm	Montana	www.com.state.mt.us/License/POL/licensing_boards.htm
Idaho	www.state.Id.us/working/licensing.htmlidaho.gov/business/licensing.html	Nebraska	www.nol.org/home/NBOP/law.html
Illinois	www.i.dfpr.state.il.us	Nevada	www.state.nv.us/index.htm
Indiana	www.state.in.us/state/agencies	New Hampshire	www.state.nh.us/jtboard/home.htm

Jurisdiction	WWW address	Jurisdiction	URL
New Jersey	www.state.nj.us/lps/ca/boards.htm	South Dakota	www.state.sd.us/state/executive/dcr/dcr.html
New Mexico	www.state.nm.us/state/agencies.html	Tennessee	www.state.tn.us/commerce/regbrdiv.html
New York	www.op.nysed.gov/proflist.htm	Texas	www.tded.state.tx.us/guide/REGULATORY-BODIES.html
North Carolina	www.state.nv.us	Utah	www.commerce.state.ut.us/
North Dakota	www.health.state.nd.us/gov/boards/boards.htm	Vermont	www.vtprofessionals.org/
Ohio	www.state.oh.us/ohio/agency.htm	Virgin Island	www.usvi.org/dlca
Oklahoma	www.state.ok.us	Virginia	www.state.va.us/dpor/indexie.html
Oregon	www.state.or.us/agencies.htm	Washington	www.wa.gov/dol/main/index.htm
Pennsylvania	www.dos.state.pa.us/bpoa/bpoa.html	West Virginia	http://www.state.wv.us/directory/default.htm
Rhode Island	www.state.ri.	Wisconsin	www.badger.state.wi.us/agencies/drl/
South Carolina	www.llr.state.sc.us/boards.htm	Wyoming	www.state.wy.us/governor/boards/boards.html

CHAPTER 3

Design Professional Contract Responsibilities—Pre-Design, Design, and Pre-Construction Phases

THEODORE D. LEVIN, P.E., ESQ.
MEHRDAD FARIVAR, ESQ., FAIA

I. Introduction

The contract is the lifeblood of the construction project. The parties in most commercial business transactions rely on a written agreement to memorialize and validate their intentions and rights. However, the contract seems to be more than this in a construction project. This probably stems from the fact that design and construction contracts are the vehicles for bringing to life the vision of the project that is often very personal to the owner, design professional, and others in the building process. Moreover, the fact that there are typically a number of parties involved in the construction process, its complexity, and that the owner is often a layperson, leads to the potential for disputes that may not be present in other commercial settings. In fact, it would be nearly impossible (and certainly inadvisable) to design and build a construction project of any complexity without written agreements memorializing the understanding between the parties. For all of these reasons, the details of the contracts are of paramount importance to a construction project and require specific attention.

This chapter is the first of several dealing with contractual and legal issues affecting design professionals. It focuses on the design professional's involvement with the process up to the time of actual construction, including contractual issues and risks that one must consider. Although it is concerned with the design professional, many of the lessons regarding contractual risk management discussed in this and other chapters could be applied universally to most of the participants in the construction process.

This chapter covers a broad range of topics, as the pre-design, design, and pre-construction phases of a project include an array of activities and obligations. Therefore, the authors provide a general overview of many of the issues involved, so that the reader may consider all of these issues and concerns. This is especially true with respect to the discussion of specific provisions of the design professional's contract, which are treated in varying depth and intended to provide the groundwork for understanding the import and impact of these contract clauses, and allow further investigation and consideration by the design professional and his counsel.

Library References

C.J.S. *Contracts* §§ 2–3, 9, 12.
West's Key No. Digests, Contracts <KEY>1.

II. Conceptual/Pre-Design Phase: Contractual and Extra-Contractual Roles of the Design Professional

The conceptual and "pre-design" phase of a design-bid-build project[1] typically starts when the owner forms the initial intent to build. Depending on the project, the owner may know "what" he or she intends to build, but not necessarily the scope or scale of the project or the complexities involved, including the process of design and construction, the project budget and schedule, and what may be involved in terms of owner's obligations. Even if the owner is experienced and has previously been through the process, the reality of modern construction projects (particularly in metropolitan areas) is that each is different and presents its own unique set of challenges. Within this framework, there are contractual responsibilities, as well as roles of the design professional that fall outside those that might be expressed in contract (i.e., extra-contractual obligations).

A. *The Process and the Product—Unique Aspects of the Design and Construction Process*

Although construction is conceptually similar to fabrication or manufacturing in the sense that a building or a construction project is ultimately a physical object, the

---

1. A design-bid-build project is considered the traditional form of "project delivery system" and is characterized by defined participants, roles, budget, and most importantly, design. Project delivery systems are methods by which construction projects are designed, financed, and built; organizing participants; marshalling resources; and giving direction for management. As the name suggests, under design-bid-build, once the design has gone through distinct stages of development, a complete set of bid documents (the "bid set" of plans and specifications) then are released for competitive bidding and a construction contract is awarded. Phillip G. Bernstein, FAIA, Chapter 10, *Delivery Methods and Compensation*, § 10.1, *Project Delivery Options*, in THE ARCHITECT'S HANDBOOK OF PROFESSIONAL PRACTICE, AMERICAN INSTITUTE OF ARCHITECTS (Joseph A. Demkin, AIA, ed., 13th ed., John Wiley & Sons, Inc. 2001). The publication may be considered the official handbook of the American Institute of Architects (AIA) for architects and other persons interested in the construction industry. Section 10.1, in part, quotes from Gordon H. Chong, FAIA, Editorial Task Group, HANDBOOK ON PROJECT DELIVERY, AMERICAN INSTITUTE OF ARCHITECTS (California Council: Sacramento, Cal. 1996).

process of design and construction is fundamentally different from the manufacturing process. After formulation of the initial concept, a manufacturer first develops a prototype, extensively tests it in an experimental mode, and then modifies and refines it. Thereafter, the manufacturer streamlines the sequence and mass produces the item in bulk, in a controlled factory setting. The finished product is then transported, and can be sold anywhere. In the construction industry, each project is unique. It is a prototype *and* the end product. The opportunity for experimentation is limited or nonexistent. Each project is "manufactured" one at a time, on location, and cannot be transported. It is always sold on location. The distinction is more than factual. It also has legal significance, which has been recognized and given particular treatment by the courts in determining the underlying liabilities of participants in the construction industry.

The design process is where "conceptual" prototype(s) must be developed, tested, debugged, and streamlined (on paper) by design professionals for "manufacturing" on location. Once construction (manufacturing) begins, the result cannot be discarded, as it can be in a factory. It must be finished and delivered. Also, the physical construction does not occur in a controlled setting. Hence, there is a much greater reliance on the building plans as a means of controlling in detail what is built. Therefore, the consequences of a faulty design process can be drastic for the owner, the contractor, and of course, the design professional.

The design professional is typically the first in the line of many professionals and tradespeople hired by the owners. The design professional participates in the realization of owner's intent through the design process and customarily in the "delivery" of the completed project to the owner and other users. Therefore, the design professional has the opportunity to win the owner's trust and confidence at the inception of the project, and become his advisor for the duration of the project. In traditional design-bid-build projects, the design professional's unique position as owner's advisor at times is transferred from the architect to the contractor at the start of construction because once construction begins, the contractor has more opportunities for interacting with the owner as compared to the architect or other design professional—out of sight, out of mind. Construction managers, who are a relatively new breed, may also assume this initial role, particularly with certain public and institutional clients, or with projects involving multiple bid packages.

1. *The Design Professional as Advisor: To Be or Not to Be*

When an owner approaches a design professional at the conceptual stage of a project, the design professional must decide whether to accept the position of "advisor" to the owner rather than the more limited (or more precisely defined) technical and professional role. This consideration must be carefully weighed by the design professional, based on the circumstances surrounding the project and the design professional's ability to render necessary advice based on the owner's needs. The design professional and the owner should enter into a written contract providing the scope of this initial assignment, so that there is no confusion as to the nature and extent of the investigation and services the design professional will provide.

Typically, at this stage of the project, the design professional is either operating under an oral "hourly" contract or is providing these services gratuitously without a contract in anticipation of a formal written contract, pending greater clarity in the scope of the project.

Practice Tip

It is advisable that the design professional operate under a written contract with the owner even when acting in an advisory capacity. The hourly contract is suitable for this purpose due to its inherent flexibility. The advantage of operating under a written contract at this stage is to apply specific terms and disclaimers to the relationship that would otherwise not be applicable. The purpose of these terms would be to reduce and control risks assumed by the design professional, and to prevent unfulfilled expectations on both sides. This chapter will later discuss some of these terms.

The design professional who accepts the role of advisor to the owner must act in a prudent manner. He should have the ability to educate the owner as to the complexities of the design and construction process, without making this complex process seem too overwhelming, and must also have the knowledge, time, and resources to research particular issues presented by the project or owner and recommend appropriate courses of action. For example, if the project is contingent on a particular event such as a governmental permit or approval, or a particular use (e.g., a particular type of machinery or process around which a building must be designed), the design professional must devise and recommend a sequence of actions (such as securing entitlements, discussed *infra*) to meet those needs. The danger is that when the design professional serves as an advisor, he may inadvertently assume responsibilities beyond the terms of his existing contract with the owner.[2]

---

2. *See* Sweet v. Gribaldo, Jones & Assoc., 40 Cal. Ct. App. 3d 573, 567, 115 Cal. Rptr. 99 (1974) (where a soils engineering firm, employed solely in an advisory capacity and paid by the hour with no interest in the property being improved, nor its sale, nevertheless faced allegations of express warranty and strict liability for damages from cracks in the constructed home, resulting from an instability of the improved lot. It was undisputed that the firm was at no time consulted as to final grading or excavation of the finished lot, never had the opportunity to approve the foundations for the house thereon, and never was even shown the plans for the final grading or foundation. The court thus rejected the allegations and held that the engineering firm would be responsible only for its advisory services that were not shown to have breached applicable professional standards or provided with guaranteed results).

2. Understanding the Owner's Initial Concept

The design professional involved in a project during the conceptual stage should be aware that the owner might have difficulty clearly communicating goals and objectives for the project, because he or she has not yet formulated many of these details. The design professional can assist the owner in formulating and refining key concepts by asking the right questions and focusing the owner's attention in the right direction. Through this process, the missing details should emerge.

a. Project Use

Perhaps the most fundamental aspect of the owner's concept is the intended use of the facility or project. If the design professional has prior experience in designing for the particular use, he would be able to ask the right questions of the owner. On the other hand, if the design professional is unfamiliar with the particular use, he might study the use to become familiar with it, or associate with another design professional with the requisite experience. The professional will design the project either to meet the specific needs of a known user group or to address a general need in the marketplace. In either case, the "use" drives the project and sets the initial parameter. If the use is complex, and the design professional is inexperienced in the particular use, the design professional may want to consider associating with another design professional experienced in that use.

b. Size of the Project

The next parameter is the size of the project, which is as significant as the use of the project. Even if the design professional is qualified with respect to a use, he should also be certain he has the experience and qualifications with respect to the size of the project. Designing a single house is vastly different than designing a subdivision of hundreds of houses, although the use is identical. Similarly, designing a single retail store is very different from designing a large retail center, although the use is the same. It would not be prudent for a design professional to assume the role of an advisor to the owner with respect to a project that is too large or small for the design professional. Lack of experience with the size or scale of the project might cause the design professional to misquote a fee for design services, incorrectly estimate the schedule for the performance of the design services, or commit other errors and omissions.

3. Initial Formation of the Project Team

The project team usually consists of multiple design professionals and other nondesign consultants. For ease of reference, we will refer to these as "design consultants" and "project consultants," respectively.

a. Design Consultants

Design consultants closely integrate their work product with that of the lead design professional. The most frequent design professionals are the structural, mechanical, or electrical engineer; code and elevator consultants; as well as others who work closely with the design professional to prepare design documents. Design consultants traditionally operate under a contract with the design professional, providing the design professional an opportunity to select its consultants. In addition, a direct contractual relationship between the design professional and design consultants facilitates design coordination between them. The design professional can also recommend that the owner select and contract with the design consultants.[3]

Practice Tip

If the design professional recommends that the design consultants be hired by the owner, he or she may want to consider pre-qualifying the consultants, providing the owner with more than one name in each discipline, and informing the owner that, because of the lack of contractual privity between the design professional and the design consultants, the design professional will not have control or oversight responsibilities for such consultants. However, the design professional would endeavor to coordinate his own work toward the overall design of the project with the work of such consultants.

---

b. *Project Consultants*

The owner usually hires other consultants who provide services concerning the project but do not necessarily serve a design role. These include the land surveyor, the geotechnical consultant, environmental consultant, and the land use consultant or attorney. Such consultants do not have a direct role in the design of the structures for the project, and thus are not required to closely coordinate and integrate their work product with that of the design professional.

Ordinarily, the design professional may rely on the accuracy and completeness of the information provided by or through owner's consultants. For example, a design professional would not generally be responsible for losses arising from an erroneous

---

3. It should be noted that the architect is typically at the center of the design team, but other design professionals such as civil engineers have assumed that central role, providing similar services. *See* Lehman v. Dalis, 119 Cal. Ct. App. 2d 152, 154, 259 P.2d 727 (1953) (to the extent that architectural services and civil engineering services overlap, they may be rendered either by a licensed architect or by a registered civil engineer) (the case involved plans drawn by a civil engineer for the erection of a bowling alley building). *See also* Wynner v. Buxton, 97 Cal. Ct. App. 3d 166, 158 Cal. Rptr. 587 (1979) equating the functional role of engineers and architects on a construction project in terms of their contractual undertakings.

survey prepared by the land surveyor under contract with the owner. When the design professional is in contractual privity with a consultant, he may assume vicarious liability for the errors and omissions of these consultants. (See section III, A, *infra*.)

Practice Tip

The provision of initial information about the project site is ordinarily the owner's responsibility. This is because the owner may have superior knowledge as to the history of the project site and its prior uses (such as possible contamination). Therefore, as a matter of risk management, it is not prudent for the design professional to take on the responsibility of hiring and coordinating consultants whose responsibility it is to provide information about the site, such as the land surveyor, geotechnical engineer, and/or environmental consultant. The design professional's written contract should confirm this division of responsibility.

---

4. *Issues and Obligations in Dealing with Governmental Agencies/Approvals*

Prior to obtaining the building permit, projects are subject to governmental permits and approvals at various stages of their planning and design ("Land Use Entitlements" or "Entitlements"). On the other hand, "Building Permit and Inspection Approvals" are those required in connection with the issuance of the building permit and inspection of the construction at various stages. Therefore, entitlements regulate "what" is built, and health and safety and building codes regulate "how" the project is built.

a. *Entitlements*

Entitlements secure the owner's right to a real property improvement consisting of a certain size, specific permitted use(s), and sometimes a specific design on a particular legal parcel of land. In urban areas, entitlements involve obtaining multiple approvals from governmental agencies and political bodies, many of which are discretionary and subject to public comment and public hearings. These range from the ever-proliferating federal environmental statutes, such as the National Environmental Policy Act of 1969 (NEPA)[4] and the Clean Air Act of 1970,[5] and various state regulatory controls and procedures, such as Environmental Impact Reports under the California Environmental Quality Act of 1970[6] to local

---

4. 42 U.S.C. § 4321 (1970).
5. 42 U.S.C. § 7401 (1990).
6. Cal. Pub. Res. Code § 21000 (1979).
7. For an analytical review of state and federal land use laws, *see* Peter W. Salsich, Jr. and Timothy J. Tryniecki, *Land Use Regulation: A Legal Analysis & Practical Application* (American Bar Association, Section of Real Property, Probate and Trust Law: Chicago 1998).

regulatory reviews and land use controls.[7] The latter includes zoning and land use permits, conditional use permits, planning approvals and permits, subdivision map approvals, traffic approvals, and planning commission and architectural review approvals.

Due to the political environment in most urban jurisdictions, particularly in coastal areas, metropolitan areas, and historic cities and communities, the entitlement process is fraught with risk and uncertainty. Complex urban projects can take years. In fact, in some projects the entitlement phase is far longer and perhaps riskier than the detailed design and construction phases that follow.

b. *Building Permits and Approvals/Inspections*

Governmental agencies exercise comparatively less discretion in reviewing plans for issuance of a building permit. Here, the law (often the model building code that has been adopted in whole or in part by the local governing body for application in its jurisdiction) is more prescriptive and quantified, and is not dependent on the political judgment of elected officials.

5. *Implied Obligations of the Design Professional in the Conceptual/ Pre-Design Phase: Dealing with Governmental Agencies and Approvals*

The design professional should be familiar with the model code and laws governing the object or system that he will design. Typically, issues relating to compliance with locally adopted model codes and the issuance of building permits and approvals do not arise during the conceptual phase of the project. In some states, failure to comply with a building code or standard is deemed "negligence per se" and, depending on the jurisdiction, represents either conclusive proof of negligence or enough evidence of negligence to justify submitting the question to the jury.[8] In either application, the question is whether the legislature intended to protect the person injured and prevent the type of harm that occurred.[9]

During the entitlement phase, the design professional should work closely with the expediter or land use lawyer in securing discretionary permits. When a

---

8. JUSTIN SWEET & MARC M. SCHNEIER, LEGAL ASPECTS OF ARCHITECTURE, ENGINEERING AND THE CONSTRUCTION PROCESS 59-60 (7th ed. 2004), generally applying the tort concept of *negligence per se* when based on violation of code to the construction industry. Section 14.03 of this treatise provides multiple illustrations taken from judicial opinions of types of claims that have been made against design professionals during the design phase, including *inter alia*, specifying material that did not comply with building codes, positioning a building in violation of setback requirements, and failing to know local building codes and safety laws; and, during the construction phase, allowing material not approved by code to be installed, and failing to make changes needed to comply with codes (citations omitted). *Id.* at 232-234.

9. *Id.* at 59-60; *see also* Alber v. Owens, 66 Cal.2d 790, 59 Cal. Rptr. 117 (1967) (safety regulations imposed on a prime contractor were for the benefit of a subcontractor).

subdivision of land is involved, the law in some states, such as California, requires that other design professionals such as civil engineers review and verify the adequacy of the infrastructure for a proposed project.[10]

During the pre-design phase, the architect's role is to guide the owner to assemble the proper team to test the overall feasibility of the project. If a design professional practices extensively in a certain jurisdiction, he may become a "local" expert in the entitlement process. In that situation, the design professional takes on the role of the entitlement consultant or land use advisor. The design professional must properly allocate and control the risks and rewards associated with the assumption of this role. Also, he must be careful not to give legal advice, and should recommend that a lawyer represent the client during public hearings as procedural missteps could result in the loss to the owner of valuable entitlement rights.

Practice Tip

It is not prudent for a design professional to advise an owner during the entitlement phase of the project without a written contract delineating the role of the design professional in relationship to other project participants, including limitations and disclaimers applicable to the design professional's services. Counsel for the design professional should carefully negotiate these services and draft the professional services agreement so that a court can enforce these limitations and disclaimers.

B. *Potential Liabilities of the Design Professional in the Pre-Design Phase*

There are competing needs in the pre-design relationship between the design professional and the owner. The owner needs to gather information and evaluate the project, the design professional, the design process, costs, time involved, and the overall feasibility of the project. From the design professional's point of view, the pre-design phase involves the need to evaluate the project and the client and, if the result of this evaluation is favorable, secure a professional services contract that may include services beyond design, such as contract administration which is performed during construction of the project (discussed in Chapter 4).

1. *Liability Arising from Initial Proposals*

During the initial contact with the design professional, the owner will likely ask for a multitude of information and rough cost estimates. The design professional can only answer such questions based on prior experience with similar projects. The opinions of the design professional in the early stages of the project may later prove

---

10. *See generally* The Subdivision Map Act, CAL. GOVT CODE § 66410 (1975).

erroneous due to an aspect of the project that he may not have known at the time. Equally, certain project parameters may change in the course of the design and construction of the project, which may render early opinions of the design professional erroneous.

Usually, the parties at this early stage are rather optimistic in their assumptions concerning the project but for different reasons. The design professional may be trying to impress the owner with his capabilities, while being careful not to intimidate or overwhelm the owner with his initial conceptualizations for the project.

The owner, on the other hand, may be overly optimistic in hearing the design concepts from the design professional. Ultimately, the owner may be disappointed with the project because the design professional failed to achieve certain results due to a change in the key parameters of the project. The owner will typically focus on the unattained result, and will ignore the change in the project parameters.

On some projects, if the owner becomes disappointed with the project or design professional, he or she may assert a claim against the design professional based on what was an avoidable miscommunication. A common practice among design professionals is to attach to their written contract the initial proposal that was presented to the client. Sometimes the owner or the owner's attorney wants to attach the proposal. However, this could actually produce internal inconsistencies in the contract, particularly if there is a large time lag between the initial proposal and the contract. Certain project parameters may have changed since the preparation of the proposal, and that may create inconsistencies between the contract and the attached proposal.

Practice Tip

It is generally inadvisable to attach to the final written design contract the initial proposal or similar document, prepared at the inception of the parties' relationship, when the project parameters may have changed.

---

2. *Setting the Project Parameters*
   a. *Generally*

Before a contract can be negotiated and finalized between an owner and a design professional, the parties must agree on certain project parameters that form the basis of the contract, including the design professional's compensation. These parameters include, but are not limited to, project scope, the owner's budget, and schedule for the project.

Therefore, project parameters must be ascertained as carefully as possible by the design professional in the exercise of all due diligence, and confirmed by the owner. Although the owner has an obligation to accurately represent the project parameters to the design professional, this is not necessarily a linear process. For the sake of the owner, the design professional, and the integrity of the project, the design professional may have to take the lead to ascertain the project parameters by conducting some preliminary investigation and asking the right questions of the owner. Although the owner may not necessarily know what information the design professional needs in order to arrive at the essential contract terms, the design professional should be adequately familiar with possible areas for foundational information.

Ultimately, it is not the finality of the project parameters that is critical to the negotiation and finalization of a successful contract between the owner and the design professional. Instead, the clarity and ability of the contract to deal with the uncertain, dynamic nature of such parameters, and their effect on the nature and progress of the design, determine the quality of the contract.[11] One should expect, and provide in the contract, that the project parameters will continue to change after the parties sign the contract.[12] The important issues are: (1) reaching a consensus of the parties as to what potential project parameters exist at the time they enter into the contract; and (2) identifying which parameters they will address and incorporate into the contract. Equally important is the recognition by the owner that changes in project parameters can lead to changes in the essential terms of the contract, such as the design professional's compensation and the project schedule.

b. *The Initial Program*

The initial program is the "script" for the project and the first indication of its scope. Design professionals approach the program in different ways based upon the type of project. For example, with single-family dwellings, the design professional is more likely to incorporate the initial program into a conceptual plan or layout, rather than reducing it to writing. However, with hospitals, courts, and criminal justice facilities, some design professionals primarily devote their careers to preparation of the program.

The challenge for the design professional is rather unique. It is to document changes in the program and the project clearly and over time, and have a mechanism in his contract with the owner to adjust the design professional's compensation and schedule of performance because of these program changes. Not every change in the program warrants a change in the compensation and schedule of performance of the design professional. Only those changes that impact the design professional's scope of work are significant. However, recognizing which changes are significant is easier said than done because changes in the program and the project often occur incrementally. These incremental changes may not appear to be significant to the owner or to the design professional. However, taken as a whole and over a significant period of time, these incremental changes can be significant.

---

11. American Institute of Architects, Document B141-1997, Article 1.2, Responsibilities of the Parties, § 1.2.3 at ¶ 1.2.3.2 (recognizing the need for and providing for flexibility in the project schedule).

12. *Id.* at Article 1.3, Terms and Conditions, § 1.3.3 at ¶¶ 1.3.3.1 and 1.3.3.2 (recognizing the possible impact of and making provision for changes in project parameters, and their effect on the services of the architect).

Working with the owner, the design professional must document the project program and attempt to identify all major contingencies that can affect the scope of the project. Ultimately, the parties to a design agreement must agree to a "snapshot" of the scope of the project at the time they sign the contract. In addition, the parties must understand that changes in project parameters, such as the program, may impact the project scope and possibly alter the design professional's fee and schedule of performance. The parties must clearly express this understanding in the contract.

3. *Initial Budget Estimates*

Many lawsuits and claims against a design professional grow from the owner's unmet expectation that the project would be completed within a specific budget or unit cost established at the onset of the project. There are a couple of likely scenarios.

Scenario 1: The owner asks the design professional for a preliminary unit cost or budget at the inception of the project. The design professional (usually against his better judgment) yields to the pressure and supplies a number based on unverified assumptions, and usually a past project. Subsequently, the owner claims that he proceeded with the project relying on the erroneous budget. The owner makes a claim against the design professional for bad advice, which allegedly caused a financial loss to the owner.

Scenario 2: At the inception of the project, the owner provides the budget to the design professional. The design professional either makes no representation as to the adequacy of the budget or expresses a guarded opinion as to the adequacy of the budget given the project scope. The budget proves to be incorrect, either because it was inadequate from the beginning or because of intervening changes in the project parameters. The owner makes a claim against the design professional for not informing the owner at the onset that his budget was insufficient, or failing to notify the owner that his budget became inadequate due to specific changes in the project. The owner claims financial loss because the project exceeded the budget.

In Scenario 1, the claim against the design professional is for an erroneous estimate. In Scenario 2, the claim against the design professional is for failure to notify the owner that his budget was inadequate or became so during the course of the design of the project. Both scenarios can occur during the pre-contract phase of the project.

Practice Tip 1

The design professional must resist pressure from the owner in the conceptual phase of the project to provide an unqualified estimate of the likely project cost. Such estimates are usually based on similar past projects of the design professional. However, as the design professional knows, each project is unique. Inflation and changing market conditions alone can significantly affect the cost of the project, even if all other variables remained unchanged.

Practice Tip 2

If the owner unilaterally announces his budget for the project during the pre-contract phase, the design professional should not remain silent for fear of offending or losing a potential client. The owner may interpret silence as acquiescence. The design professional must inform the owner that he is in no position to endorse or reject the owner's proposed budget; by agreeing to further involvement on the project, he is not endorsing the owner's budget. The design professional must educate the owner that he will not guarantee the adequacy of the owner's budget. A contractor can objectively test the owner's budget after the parties identify and adequately document the specifics of the project scope.

---

4. *Initial Schedule Estimates*

Similar to cost estimates, a forecast of the construction schedule is problematic for the design professional. Again, the design professional must not remain silent in the face of a major schedule contingency unilaterally set or announced by the owner. Additionally, the design professional must resist pressures to provide a preliminary schedule to the owner in the conceptual phase of the project. Any proposed schedule agreed to by the design professional must be carefully qualified. If possible, the design professional should not agree to or warrant any fixed deadline in the initial stages of the project.

Library References

C.J.S. *Agency* §§ 12–13, 15–16, 18–19, 21–22; C.J.S. *Architects* §§ 19, 21–23; C.J.S. *Contracts* §§ 2–3, 9, 12, 271, 333, 346–347, 407–420, 434–435, 450–455, 507–508, 531–532, 534–544, 557–576, 584, 592–593, 596; C.J.S. *Health and Environment* §§ 106, 116, 131, 163–165, 173; C.J.S. *Negligence* §§ 162–164, 399, 595–596, 600; C.J.S. *Zoning and Land Planning* §§ 70, 191, 193–203.

West's Key No. Digests, Alternative Dispute Resolution <KEY>111, 441; Contracts <KEY>1, 114, 168, 236–246, 283, 312–321; Damages <KEY>118; Environmental Law <KEY>129, 265, 531, 580; Indemnity <KEY>26–31, 33(5); Negligence <KEY>321, 322, 1205(4); Principal and Agent <KEY>3(1); Zoning and Planning <KEY>371–394.

III. Contract Negotiation and Formation

As discussed above, the design professional provides initial design services during the conceptual/pre-design phase on an hourly basis and based on an initial proposal to the owner.[13] Once the client and design professional have sufficiently defined the project in this phase and established a working relationship, they can consider a contract for the design of the project.[14]

A series of judicial decisions have explained the nature of the contractual undertaking of design professionals and the duty of care expected when rendering professional services. The cases are instructive for understanding the nuances of the various roles of the design professional in the construction project.

For example, *Huber, Hunt and Nichols v. Moore*[15] explained that, under AIA Document B141, Standard Owner/Architect Agreement, an architect might have three roles to perform in the overall construction project. An architect serves as *independent contractor* to the owner in preparation of plans and specifications for the project. Then, during the course of construction, the architect serves as the *owner's representative* performing a number of administrative duties under the contract (the Contract Administration or "CA" phase under AIA Document B141). Lastly, during the course of construction, the architect also serves as a *quasi-judicial officer* resolving disputes between owner and contractor. One or more of these roles could change depending on the project and the participants therein. Accordingly, the liability exposure of the architect can vary greatly when performing services in these various and distinct roles.

Courts evaluate the services of an architect in the role of independent contractor in terms of compliance with the professional standard of care for architects in that locale. The Maine Supreme Court first applied the standard of care to design professionals over a hundred years ago in *Coombs v. Beede*.[16] The holding in *Coombs* echoed across the country with *Paxton v. County of Alameda*.[17] It is interesting to note that while the design professional's compliance with the standard of care may vary by local practices within any given jurisdiction, the courts across the nation have consistently applied the standard since *Coombs*.

Wynner v. Buxton[18] examined the variations in the scope of work undertaken by a design professional on a construction project. The court explained that the contract controls the design professional's scope of work on a project, not the general notions of what a design professional does or ought to do. It stated that an engineer who

---

13. JAMES R. FRANKLIN, ARCHITECT'S PROFESSIONAL PRACTICE MANUAL, § 1.72 (McGraw-Hill 2000).
14. *Id.*
15. 67 Cal. Ct. App. 3d 278, 136 Cal. Rptr. 603 (1977).
16. 89 Me. 187, 36 A. 104 (1896).
17. 119 Cal. Ct. App. 2d 393, 259 P.2d 934 (1953).
18. 97 Cal. Ct. App. 3d 166, 158 Cal. Rptr. 587 (1979).

prepared civil engineering plans required for three steel structures was not absolutely responsible for the plans covering all aspects of the project.[19] The fact that the civil engineering plans did not include all details governing every aspect of the three steel structures did not render them defective, based on a California statute allowing nonengineers to design certain building components.[20]

The *Wynner* rationale may also apply to projects involving a sharing of design responsibility by construction professionals (such as trade contractors providing design services for their portion of the project). For this type of project, the drawings and specifications prepared by the design professional are not considered to be a complete "bid set" since negotiated agreements fill in various design details. In this type of project, design responsibility is *shared* rather than delegated, since the owner does not retain the architect to furnish the entire design.[21]

A. *The Roles, Responsibilities, and Obligations of the Parties in Entering into the Contract*

The negotiation and preparation of the contract may be the most important aspects of the design professional/client relationship. Some states have enacted legislation requiring written contracts between the design professional and the client.[22] The owner, design professional, and the rest of the design team should commit to paper their exact intentions, responsibilities, and expectations for a project.

It is at this time that the design professional should be most aware of liabilities, real and potential. Normally, the type of project will govern, at least in part, the nature and form of the contract the parties will use for a specific project. If the project is residential, then most certainly the architect will propose the contract, since homeowners usually lack the experience to suggest the form of agreement. On the other hand, on commercial projects the owner proposes the form of contract as often as the design professional. Finally, in the public works arena, the contact form is most often, if not exclusively, that of the public entity, and often not subject to significant negotiation.

In the traditional "design-bid-build" project delivery system, the owner will contract separately with the project contractor and the design professional. The design professional's contract will typically provide that he is to include in his services some range of consultants' design services, which may include those such as structural engineering, mechanical engineering, and other similar services. Although, as discussed in *Wynner*,[23] the design professional does not always retain each consultant,

---

19. *Id.* at 174.
20. *Id.*
21. *See* Arthur F. O'Leary, *Third Party Legal Claims against Architects: Pitfalls for the Architect Who Prepares Plans and Specifications for Owner-Builder-Developers*, DESIGN & COST DATA (Jan.-Feb. 1980) (discussion on this type of project by noted forensic architect).
22. CAL. BUS. & PROF. CODE § 5536.22 (1995).
23. *See supra* text accompanying note 3.

but usually does this in a typical design-bid-build project. The reason is that the owner wants the design professional to provide a complete design for the project and be the head of the "design team." The owner does not want to be responsible for and may not even be aware of all of the consultants that may be involved in the design process.

However, the responsibility of the design professional to assemble the entire design team does not come without risk. Several legal concepts affect the design professional's potential liability with respect to the actions (or inactions) of his consultants. The first is the doctrine of vicarious liability. Under this principle, one who delegates contractual responsibility to another is generally liable for the actions of their hiree. Therefore, if the consultant makes an error, irrespective of whether the design professional incorporates that error into his design or communicates it directly to the owner or other user, the design professional will generally be liable for the damage that results.[24]

Practice Tip

If the owner will agree, it may be preferable for the design professional to have the owner contract directly with the project consultants. This may relieve the design professional of certain risks and obligations, including the obligation of payments directly to the consultants and the potential vicarious liability for the consultants' acts or omissions. At the same time, the design professional who contracts directly with consultants should always make certain that the consultant(s) have adequate professional liability insurance.

In addition, the design professional might be liable for the negligent hiring of a consultant if the owner could show that he hired, recommended, or referred a consultant who was not qualified to do the work.[25] Although this may be difficult to

---

24. *See generally* Johnson v. Salem Title Co., 246 Or. 409, 425 P.2d 519 (1967) (architect was held liable for the negligence of his consulting engineer; this concept applies equally to a consultant that might hire a subconsultant).

25. *See, e.g.*, Gonzalzles v. American Exp. Credit Corp., 315 Ill. Ct. App. 3d 199, 733 N.E.2d 345 (2000) (claim was brought against an Illinois attorney for the "tort of negligent referral" of a client to another attorney); Burlington Ins. Co. v. Oceanic Design & Const., Inc., 383 F.3d 940 (9th Cir., 2004) (claim was brought against a contractor for negligently recommending an unqualified mason to do excavation work); Capel v. Allstate Ins. Co., 78 Ark. Ct. App. 27, 77 S.W.3d 533 (2002) (property insurer was charged with making a negligent recommendation of a roof repair contractor); Sapp v. State Farm Fire & Cas. Co., 226 Ga. Ct. App. 200, 486 S.E.2d 71 (1997) (claim based on contractor's negligent recommendation and installation of hardwood flooring).

prove in many situations, some courts have allowed this type of claim against design professionals.[26]

B. *The Contract*

Considerations for the scope and terms of the contract for professional services, particularly the terms that will most affect the design professional, are against the backdrop of his relationship to the project (discussed above). Although a detailed analysis of all possible contract clauses and their ultimate effect is beyond the scope of this chapter, some provisions that the design professional must consider are discussed below.

1. *Standard Form Contracts—the Test of Time?*

Several organizations promulgate standard form contracts for the construction industry. The better known are the American Institute of Architects (AIA), the Associated General Contractors of America (AGC), the Engineering Joint Contracts Design Committee (EJCDC), and the Design-Build Institute of America (DBIA). Certainly, each of these organizations has its own view and perceived bias in the contracts it creates. However, these groups have generally attempted to create their "family" of documents to be as fair as possible, so the widest audience within the construction industry would use them. This chapter will primarily focus on the

---

26. *See, e.g.*, Hubert v. Aiken, 15 Daly 237, 2 N.Y.S. 711 (1888) (Architect was found liable for negligently recommending use of a product or material, rather than a consultant or subcontractor. The architect had specified a particular chimney design based exclusively upon the representations of the installing contractor without his own independent investigation as to its performance characteristics. The court explained that "[r]esponsibility could not be shifted in that way." Interestingly, with implications for the specification of solar power and other "green technologies," the 1888 court also said that "when, in the progress of civilization, new conveniences are introduced into our homes, and become, not curious novelties, but the customary means of securing the comfort of the unpretentious citizen, why should not the architect be expected to possess the technical learning respecting them that is exacted of him with respect to other and older branches of his professional studies?" [meaning, an exercise of due diligence and compliance with professional standards]). *Id.* at 712-13; *see also* Scott v. Potomac Ins. Co., 217 Or. 323, 341 P.2d 1083 (1959) (citing *Hubert* and explaining in dicta that an architect would have been found liable for violation of professional standards when he negligently recommended use of an alternative imbedded piping system for heating a hospital without appropriate changes in the flooring to allow for expansion of the piping. The argument of the architect had been that he had the right to rely without independent investigation on the representation of the manufacturer and of the approval of its use by the heating engineer employed by the architect to design the heating system. Although not specifically at issue in the appeal, the *Scott* court let it be known that the architect would have been found liable for professional negligence for such blind reliance, citing the *Hubert* decision.).

provisions and risks contained in the AIA family of documents,[27] with a secondary reference to the EJCDC and AGC contract forms.

What are the current benefits and risks of the use of industry form agreements, including AIA form contracts? From their inception, the use of the AIA form contracts, as well as form documents by any number of other publishers, has been the subject of debate. Some have criticized the documents, not because of the appropriateness or sufficiency of a form document for a given project, but on the perceived bias or favoritism to one or more project participants, especially the architect. A number of publications with suggested revisions to these forms have been offered to the construction industry, often with alternative revisions (additional and/or

---

27. As a brief background, the AIA was founded in 1857 when a group of American architects met together and formed what would become the American Institute of Architects. The stated purpose for the professional organization was to "promote the scientific and practical perfection of its members" and to "elevate the standing of the profession." Up to this point, no American schools of architecture, or architectural licensing laws, standards, regulations, or formal organization had existed in any state of the union. Anyone up to this point could call himself or herself an architect, including masons, bricklayers, and various other members of the building trades. *See History of the American Institute of Architects*, on the AIA Web site, *at* http://www.aia.org; and DANA CUFF, ARCHITECTURE: THE STORY OF PRACTICE (1991).

As part of this effort, it was believed that standard construction documents could assist by providing guidance and regularity to the system of building. In fact, one of three standing committees formed by the AIA in its founding year was the Committee on Papers, dealing with construction documents. The first ever construction document promulgated by the AIA focused on the contractual needs of its client community in the building process, indicating an early sensitivity to its clientele and a professional response to the realities of the construction industry. The history of the AIA Contract Documents dates to 1888, when the AIA first published its *Uniform Contract* for use between an owner and a contractor. *See History of Contract Documents at* http://www.aia.org. In keeping with this broad, industry-wide vision, the organization in 1911 published its first standardized general conditions for construction. *Id.* However, it was not until 1917 that the AIA first published its B141 owner/architect agreement. AIA Documents B141, Standard Form of Agreement Between Owner and Architect (now in its 1997 edition), along with A201, General Conditions of the Contract for Construction, have since become documentary workhorses of the construction industry, providing common documentary references for design and construction practice if not (off-the-shelf) useable, model agreements. *See* Howard G. Goldberg, Esq., Hon. AIA, *Commentary on AIA Document B141-1997*, at 12, on the AIA Web site, *at* http://www.aia.org. Since the inception of standard forms, the AIA has adhered to a document *Revision Policy*, revising its standard documents every ten years, generally. The current 1997 edition of AIA Document A201 is the fifteenth edition of the general conditions. The stated purpose of its document revision policy is to maintain a high level of accuracy and timeliness in AIA contract documents, given ever-changing social and construction industry conditions, and to ensure that AIA contract documents are current and fair to all participating parties.

substitute language) tailored to the interests of each party.[28] Even form documents related to design-build have been the subject of debate and suggested revisions.[29]

2. *The Use of Custom Contracts*

In addition to form contracts, design professionals and other project participants use custom contracts specifically prepared by or for the users. Business or public agencies often draft custom contracts, either with or without legal counsel.

Custom contracts are advisable or essential when the requirements of a project, or the nature or the number of the project participants (such as multiple primes for project construction or project design), cannot fit into the standard terms of the form contract. The drafter can modify a form contract from any publisher to fit a given project only to a point; beyond that, he will need a custom contract.

Justin Sweet, law professor and noted author, states this caution with regard to the modification of form contracts, "[w]e cannot emphasize sufficiently the need for *careful* modification of any AIA document. [emphasis added.] If this cannot be done, it is better to use the document unchanged or to use a custom-made contract."[30] Overall, the use of a form contract properly modified to fit the project's particular characteristics is the best alternative.

Another difficulty in modifying form contracts is that they are often *interrelated* to other form documents. Thus, modifying one form without changing other related forms can create conflicts. Similarly, it may be problematic to use both form and custom agreements on a project without appropriate modification. For example, AIA Document B141-1997 (Owner-Architect Agreement) references AIA Document A201 (General Conditions) for construction of the project. The blind use of such a form contract for design services "off the shelf" may create legal inconsistencies and operational problems with custom-made construction contracts not utilizing A201 for the same project.[31]

---

28. Most notably, *see* Justin Sweet & Jonathan J. Sweet, Sweet on Construction Industry Documents: Major AIA Documents (4th ed. 1999); Gower W. Jones, Alternative Clauses to Standard Construction Contracts (1998); Werner Sabo, Legal Guide to AIA Documents (4th ed. 1998); and James Acret, Construction Industry Formbook (2d ed. 1979).

29. G. William Quatman, Design-Build for the Design Professional (2001), including Appendixes with Supplementary Conditions to current design-build documents by various publishers.

30. Sweet & Sweet, *supra* note 28, at 17 (emphasis added); *Relevance: Source of Pertinent Information*, AIA California Council Newsletter (April 2004) (the AIACC reports that the most common complaint it receives from consumers of architectural services [the client] regarding owner/architect agreements is not that there was no agreement, but rather, that the agreement was silent on important issues).

31. Andrew M. Civitello, Construction Operations Manual of Policies and Procedures, at § 3.2 (3d ed. 2000).

It is also dangerous to mix different editions of AIA documents on the same project. For example, if the architect utilizes AIA Document B141/CMa-1992 (for construction management projects) for his agreement with the owner and AIA Document C141-1997 for the agreement with his consultant, the architect could be liable for consequential damages sustained by the owner because of the consultant's design error. However, he would not be able to pass those damages through to the consultant due to the waiver of consequential damages found in the 1997 family of documents.[32]

Sweet has observed, "AIA documents [as well as form documents by other publishers] are most useful for small and middle-priced projects as well as large-scale residential and commercial projects in which design and construction are separated, and in which the architect [rather than a construction manager or multiple primes] plays a central administrative role."[33]

3. *Specific Contract Provisions—the Risks Defined*

Contracts of any length and complexity may contain hundreds of provisions, terms, and conditions. It would exceed the scope of this chapter to delve into all of these provisions, let alone provide counsel with sufficient information as to the interpretation of and liabilities involved with all of this contract language. However, there are several key provisions in design professional contracts that frequently arise and which the parties often heavily negotiate. It is no surprise that these contract provisions are almost exclusively "business terms" (as opposed to project "scope terms" that delineate what work the design professional is or is not going to perform). In addition, "business terms" typically involve the most risk considerations for the design professional; e.g., indemnity provisions, discussed *infra*.

The authors do not intend this discussion to be exhaustive, but to provide a general background discussion on these topics, including practice tips to handle these issues and increase the reader's awareness of liability traps. The professional and legal practitioner who requires a more detailed discussion and analysis should consult the many treatises and sources cited herein for additional consideration. However, one excellent source for discussion of many of these topics is *The Construction Contracts Book: How to Find Common Ground in Negotiating Design and Construction Contract Clauses*, published by the ABA's Forum Committee on the Construction Industry.[34]

Practice Tip

Whether a form or custom contract is used, the architect should make certain that there is consistency in the contract between the owner and architect, and between the architect and consultant(s) (if the owner contracts directly

---

32. AIA Document C141-1997, § 9.3.
33. SWEET & SWEET, *supra* note 28, at 18.
34. THE CONSTRUCTION CONTRACTS BOOK: HOW TO FIND COMMON GROUND IN NEGOTIATING DESIGN AND CONSTRUCTION CONTRACT CLAUSES (Daniel S. Brennan et al. eds. 2004).

with the consultants; see discussion, *supra*). This is necessary to avoid any inconsistent procedures or results, for example, those involving the use of dispute resolution services, copyright issues, and termination.

4. *Indemnity*

The indemnity provision could be argued to be the most important "business term" of any design professional's contract (or for that matter, any commercial contract). Negotiated indemnity rights can ultimately have the greatest impact on the handling and resolution of claims arising after completion of the construction project.

First, what is "indemnity"? One court defines it as "a contract by which one engages to save another party from a legal consequence of the conduct of one of the parties, or of some other person."[35] The court in *Myers Building Industry, Ltd. v. Interface Tech., Inc.* further explained that "an indemnity agreement is to be interpreted according to the language and contents of the contract as well as the intention of the parties as indicated by the contract."[36] However, this is overly complicated wording. Simply put, indemnity is the allocation of loss.[37] It can arise in many ways, including by statute, operation of law, or contract.

It is noteworthy that the "B" series of the AIA form contracts (owner-architect agreements) do not include an indemnity provision, as contrasted with the "A" series (owner-contractor agreements). Of course, in the absence of contractual indemnification the design professional will nevertheless be liable for *negligent* errors and omissions, which violate the standard of care.[38]

The owner and architect are always free to enter into an agreement to allocate loss between them without resort to standardized language. However, indemnity provisions contain many dangers for the design professional. For example, professional liability insurance policies may not provide coverage for a contractual undertaking beyond the liability imposed by law, such as indemnity for losses caused by others.[39] Thus, by indemnifying an owner against "any and all losses," the design professional might become liable for injuries and losses that were not his fault

---

35. Myers Bldg. Indus., Ltd v. Interface Tech., Inc., 13 Cal. Ct. App. 4th 949, 968, 17 Cal. Rptr. 2d 242 (1993), *quoting* CAL. CIV. CODE § 2772.

36. *Id.*

37. For a more complete discussion of the different types of indemnity that might be considered or applied (such as noncontractual, equitable indemnity) in the construction context, *see generally* SWEET & SCHNEIER, *supra* note 8, Chapter 31.

38. John R. Clark, *Indemnification by Design Professionals*, THE CONSTRUCTION SPECIFIER, 23-24 (Sept. 1989), offers an excellent discussion of these issues. Mr. Clark was serving at the time of this article as special counsel to the Engineers Joint Contract Documents Committee.

39. *Id.*; AVA J. ABRAMOWITZ, ARCHITECT'S ESSENTIALS OF CONTRACT NEGOTIATION 215-221 (2002).

(usually including the cost of defense incurred by the owner) and caused by the mistakes of others, such as the contractor and subcontractors, over whose activities he had no authority or control.[40] This language may also be sufficient to subject the architect to liability for economic loss resulting from design or construction defects, contrary to the economic loss rule applied in many jurisdictions (providing there is no cause of action in tort to recover solely economic loss).[41]

The owner often asks the design professional to indemnify it from "errors, omissions, or negligent acts" (in that particular sequence). However, this proposed language could extend the liability of the design professional beyond negligent acts, and include errors and omissions where there was no violation of the standard of care. Thus, this provision might not be insurable since it would increase the liability of the design professional beyond the duty imposed by law.[42]

The purpose of indemnity is to create a duty by contract where none may have existed. However, a design professional has always had a common law duty to meet the professional standard of care. Therefore, as seen in the foregoing paragraphs, *depending on the language used*, an indemnity clause may simply be a redundant recitation of that preexisting duty or an uninsurable provision creating through contract an additional *basis* of liability on the part of the design professional.[43]

5. *Basic vs. Additional Services*

A dichotomy between "Basic" and "Additional" services has been prevalent in design practice from the earliest editions of the AIA documents. This concept of Basic and Additional services is derived from the nucleus of traditional services provided by the design professional versus those additional services he might provide to the owner. Although the present edition of the AIA Document B141-1997 does not draw a distinction between Basic and Additional services, other versions of the AIA

---

40. John R. Clark, *Indemnification by Design Professionals*, THE CONSTRUCTION SPECIFIER (Sept. 1989).

41. *See, e.g.*, Moorman Mfg. Co. v. National Tank Co., 91 Ill.2d 69, 435 N.E.2d 443 (1982); Seely v. White Motor Co., 63 Cal.2d 9, 403 P.2d 145 (1965); Aas v. Superior Court, 24 Cal. 4th 627, 101 Cal. Rptr. 2d 718 (2000).

42. Abramowitz, *supra* note 39, at 217-218.

43. *See* Paxton v. County of Alameda, 119 Cal. Ct. App. 2d 393, 399, 259 P.2d 934 (1953) (the professional standard of care and whether it has been violated can only be established through professionals from that discipline); *see also* SWEET & SCHNEIER, *supra* note 8, § 14.05 B. and § 14.06, citing, *inter alia*, Eugene J. Farrug, *The Necessity of Expert Testimony in Establishing the Standard of Care for Design Professionals*, 38 DEPAUL L. REV. 873 (1989) (attacking Illinois law which does not require expert testimony in cases of professional malpractice). Contractual indemnity would be more immediate and readily available, especially when containing indemnification from any "claims" since only mere allegations could trigger it.

form contracts do, such as the AIA B151-1997 Edition.[44] Article 2 of the B151-1997 contains five Basic Services: the Schematic Design Phase, the Design Development Phase, the Construction Documents Phase, the Bidding or Negotiating Phase, and the Construction Phase (also typically known as the Contract Administration Phase).[45]

At the same time, Article 3 of the B151-1997 addresses Additional Services. It is separated into three different categories: Project Representation Beyond Basic Services, Contingent Additional Services, and Optional Additional Services. Also, § 3.1 entitled "General" discusses the concepts behind and usages of the three delineated Additional Services.

What are the risks involved with these distinctions? They generally arise from the fact that the Basic Services are those that are included in the general compensation that the design professional is to receive for their work (regardless of the manner in which that is determined, such as on a flat fee, a percentage of construction, or other method).[46] However, the Additional Services are those that are within the control of the owner, in that the owner must specifically order or request them for them to be paid. (*See* B151, § 3.1.1, providing that the owner must make a written request for the Additional Services.) Still, disputes may arise as to what services the owner requested, and whether the services are truly Additional Services or part of the Basic Services agreement. Moreover, even if the owner did not specifically request any Additional Services, many claims arise when there are changes to the scope of the work that the design professional performs and for which he wants to be paid. At the same time, the owner may dispute that the design work is an Additional Service, believing that it is part of what the owner should receive as part of Basic Services. Complications may also arise when there is a contractual requirement of a *written* authorization before the design professional may perform such Additional Services. The owner can explicitly or implicitly waive this requirement, especially under the pressure of completing a project (e.g., the owner's verbal promise to the design professional to "go ahead, we will take care of payment for that when the project is done").

What can the design professional do to reduce these risks and the potential for claims? The key is better business practices through such techniques as appropriate contract terms, confirming letters throughout the life of the project, and

---

44. The B151-1997 is used for this discussion, rather than the more common AIA B141-1997, because it continues the use of Basic versus Additional Services dichotomy (B151-1997, Articles 2 and 3), contained until very recently in earlier editions of AIA B141 (*see* B141-1987, Articles 2 and 3) and still adhered to by various participants in the construction process. Compare the revised description of services contained in Part 2 of the reformatted B141-1997, Articles 2.1 through 2.8.

45. The details of the architect's obligations under the contract administration phase (during construction of the project) are discussed elsewhere in this text, *see* Chapter 4.

46. *See* AIA B151-1997, at §§ 11.2.2 and 11.3.3.

structuring interactions of project participants through methods such as periodic "partnering" meetings, as poor communication precipitates most claims. In this connection, one commentator and author has welcomed the B141-1997 form agreement, since "nowhere in the new B141 does the *red flag phrase* 'Additional Services' appear. You and the client just agree on project specific services and [related] fees!" (emphasis added).[47]

6. *Payments to the Design Professional*

Architects are usually paid for their services in various ways (at least with respect to Basic Services), including on a fixed fee, an hourly fee, or a percentage of the construction cost (the project consultants are usually paid on an hourly basis).[48] In this sense, the concept of Basic versus Additional Services, as reflected in the B151-1997 owner architect agreement discussed above, is really a subpart of the larger issue of the appropriate compensation the owner should pay to the design professional.

7. *Termination Issues*

The importance of a termination provision in the design professional contract cannot be underestimated. At the beginning of the project, the prospect of disputes that might require contract termination seems distant. Disputes certainly arise, however, and it is important that there be some mechanism for the contractual relationship to end if necessary.[49]

Termination provisions usually fall into two categories: termination for convenience and for cause. Termination for convenience is often thought of as an owner's remedy, and in fact, the AIA B141-1997 allows the owner to terminate without cause.[50] This is due to the nature of the owner/architect relationship, in that, if the owner were to terminate for convenience, the architect would still be entitled to payment for the work performed.[51] On the other hand, the AIA has reasoned

---

47. FRANKLIN, *supra* note 13, at 1.15 (favors avoiding the distinction and misunderstandings between basic and additional services altogether).

48. Complaints by design professionals based on fee disputes and misunderstandings with owners have a way of leading to retaliatory cross-complaints for professional negligence. THE ARCHITECT'S HANDBOOK OF PROFESSIONAL PRACTICE, *supra* note 1, at § 14.3.

49. In some states, there is a statutory requirement that a design professional's contract include a termination provision. *See, e.g.*, CAL. BUS. & PROF. CODE § 5536.22(a)(5) (1995).

50. *See* AIA B141-1997, § 1.3.8.5.

51. The architect who is terminated for convenience and without cause is entitled to "Termination Expenses" that include the architect's anticipated contract profits. (*See* AIA B141-1997, § 1.3.8.7.)

on behalf of the owner that if the architect had a *unilateral* right to terminate, this could lead to abuses in that an architect who has been paid for his work could then "walk" from the project even if the owner has not technically breached the agreement.[52]

Termination for cause clauses, under any of the form contracts, permit termination if the other side has materially breached the agreement.[53]

There are certain issues the design professional should consider when drafting or reviewing termination clauses. First, if the owner's agreement with the design professional contains a termination for convenience clause, then the design professional should include a similar provision in agreements with his consultants. This will avoid the unpleasant situation where the owner has terminated the design professional, but the design professional must still go forward with and pay the consultants because they have not breached their contract(s). Second, despite the concerns of abuse by the design professional discussed above, if the owner will accept it, the design professional should consider including a clause permitting him to terminate for convenience. Most design professionals would not abuse the provision (and some have suggested that the design professional's right to terminate for convenience is not absolute and requires some degree of cause[54]). The termination for convenience provision gives the design professional an "out" if there were valid reasons to terminate even if they do not rise to the level of a substantial nonperformance by the owner. Finally, the design professional should pay close attention to the interplay between ownership of the Instruments of Service used for the project and termination, since such ownership can be an important consideration in the decision (especially by the owner) to terminate. If the design professional transfers ownership of the Instruments of Service to the owner upon termination, the owner might have a greater incentive to terminate; the opposite would be true if ownership of the Instruments of Service stays with or reverts to the design professional.[55]

---

52. SABO, *supra* note 28, at 83 n.146, discussing Gunter Hotel v. Buck, 775 S.W.2d 689 (Tex. Ct. App. 1989) (termination for convenience by owner not absolute).

53. Both the AIA and EJCDC schemes refer to a "substantial" failure of the design professional to perform.

54. *See generally*, SABO, *supra* note 28, at 81–84, citing Herbert Shaffer Assocs. v. First Bank, 30 Ill. Ct. App. 3d 647, 332 N.E.2d 703 (1975), (construing § 1.3.8.4 of AIA B141-1997 as allowing either owner or architect to send a written termination notice if the other breaches the agreement in a "substantial" manner (an oral disagreement between owner and architect was not a material breach by either party)).

55. The AIA scheme contained in B141-1997, § 1.3.2, provides that the owner's nonexclusive license to use the Instruments of Service is terminated upon termination of the contract. *See* Chapter 7 of this book for a much more extensive analysis of ownership of project Instruments of Service, including upon contract termination.

> **Practice Tip**
>
> The parties should not take lightly the decision to terminate, since that can frequently lead to disputes and lawsuits. Most contracts (including the AIA documents) include a notice provision to terminate, which one might construe as a "cooling-off" period so the parties may evaluate and negotiate.[56]

8. *Budget Estimates*

Section II above discussed budget and schedule estimates prior to the design professional entering into a contract, and the resulting liability issues. However, what can the design professional do to protect from claims arising when the project runs over budget and schedule?

The answer lies in the use of proper contract language and provisions to define the services provided by the design professional when such estimates are given, and presumably placing obligations on the owner to establish a budget so that there are no misunderstandings. For example, the AIA B141-1997 (as well as the form consultant agreements under the "C" series) devotes an entire section to "Evaluation of Budget and Cost of the Work."[57] The AIA documents also provide requirements for the project owner to maintain and update a budget.

Specifically, B141-1997, § 1.2.2.2, provides that the owner shall maintain an updated budget for the project, and shall not significantly change the budget, including contingencies, without the agreement of the architect. This provision works in conjunction with B141-1997, § 2.1.7, which has six subsections relating to the project budget and cost of the work. This scheme requires the architect to prepare a preliminary project cost estimate (the "preliminary estimate of the Cost of the Work"). Thereafter, the design professional compares the estimate to the owner's budget, and makes refinements at that time and through the bidding process. The document includes provisions that address increases in the "general level of prices in the construction industry" and if the design professional's estimate exceeds the cost of the work. The AIA provisions also make it clear that the design professional has no control over the contractor or construction costs, and does not warrant the estimate.

The AIA "C" series documents include provisions parroting those of the "B" series.[58] Other form contracts, such as the AGC and EJCDC documents, include provisions related to budgeting, but they are not as detailed as the AIA documents.

There are reported cases involving disputes arising from budget estimates, some where form contracts were used and others with custom contracts. However, the

---

56. *See* SWEET & SCHNEIER, *supra* note 8, at 213.
57. *See* § 2.1.7 of AIA B141-1997; § 3.2 of AIA C141-1997 (regarding consultants).
58. *Id.*

common theme (not surprisingly) appears to be that there is confusion between the design professional and the client as to what the project budget may be (a "hard" budget or "soft" budget),[59] and a misunderstanding as to the design professional's expertise in and responsibility for accurately predicting construction costs.

One way to minimize these problems is to carefully draft the contract to define explicitly the design professional's responsibility and qualifications for estimating services, and limits thereon. Another option is to avoid the problem by simply eliminating the design professional's contractual involvement with construction budgeting and/or estimating and leaving those services to others better equipped to provide the services—the contractor, construction manager, and/or a separately retained project estimator.

Finally, the owner may attempt to bind the design professional to cost estimates provided before signing the contract. The best method to eliminate this exposure is to include an "integration" clause in the contract, stating that the written contract represents the full and complete contract between the parties and supersedes all prior oral or written agreements.[60]

9. Schedule Estimates

Design professionals should generally avoid providing schedule estimates for the design phase. The uncertainties and vagaries of the plan check and entitlement processes—and *the numerous personnel involved*, including code enforcement officers who have discretion to interpret and apply provisions of the building code—caution against assignment of dates for completion of the design.[61] At most, milestones and an estimated time for completion of the various stages of design (i.e., an estimated time for each phase, such as six weeks for schematic design, four weeks for plan check, etc.), can be set out in advance.

Wisely, the AIA Document B141-1997 acknowledges this reality and confirms the understanding of the parties that the process is multifaceted.[62]

10. Copyright Issues

Chapter 7 of this book contains an in-depth discussion of copyright issues affecting design professionals.

---

59. *See generally* SWEET & SCHNEIER, *supra* note 8, at § 12.03.

60. *See* AIA B141-1997, § 1.4.1 ("This Agreement represents the entire and integrated agreement between the Owner and Architect and supersedes all prior negotiations, representations or agreements, either written or oral...."). However, there have been cases where an integration clause has been found to be inapplicable or limited in this instance.

61. RICHARD T. CONRAD & STEVEN R. WINKEL, DESIGN GUIDE TO THE 1997 UNIFORM BUILDING CODE, Article 2, § 2A, RELATIONSHIP OF THE DESIGNER TO THE ENFORCEMENT OFFICIAL (1998).

62. AIA Document B141-1997, Article 1.3, Terms and Conditions, § 1.3.3, Change in Services, ¶ 1.3.3.1 and, especially, ¶ 1.3.3.2.

The ownership of the project documents (what the AIA calls the "Instruments of Service"), as well as the design itself (which is also protected by copyright law), is important to the design professional because the work represents his work and livelihood above all else. However, the design professional should also be aware that ownership of the project documents can have enormous implications when a dispute arises between the design professional and client, including termination of the contract. The terms of the contract between the owner and design professional (and between the design professional and its consultants) will affect how the project proceeds upon termination, including how it can be completed, and greatly affect potential resolution of the dispute.

11. *Dispute Resolution*

Like many professional relationships, some have likened the building construction process to a "honeymoon."[63] The parties start out on their best behavior, not anticipating the difficulties that may lie ahead. However, as with many relationships, problems may develop in construction project, and certainly, design professionals are not immune to these disputes. Unlike the typical marriage, the parties to most construction agreements often include in their contract a "pre-nuptial provision" for resolving disputes, typically known as a "Dispute Resolution" provision.[64] Dispute Resolution provisions can take many forms. They usually include a few key elements:

- the "forum" for the resolution of the dispute—(typically either litigation or arbitration);
- the use of nonbinding mediation, which usually occurs prior to arbitration or litigation; and
- the allocation of attorneys' fees and other expenses (such as the cost of the arbitration expense) to the prevailing party in the dispute.

Arbitration and mediation are the typical methods of alternative dispute resolution (ADR). Private ADR providers have been in existence for decades. For instance, the American Arbitration Association (AAA) has been conducting construction arbitrations and other disputes since 1926. However, it has been within approximately the past ten years that the use of other ADR forms has come to the forefront, most notably mediation. Because of this transformation, many national and local dispute resolution providers have been formed.[65] The AAA now offers mediation services,

---

63. *See* JUSTIN SWEET, SWEET ON CONSTRUCTION LAW § 2.21 (1997).

64. This is the term that most form contracts use, including those of the AIA and the Associated General Contractors of America (AGC).

65. Examples of such providers, in addition to the American Arbitration Association (AAA), include the Judicial Arbitration and Mediation Service (JAMS), the National Arbitration Forum, and the Alternative Resolution Centers (ARC).

and resources for advance dispute avoidance techniques, such as its publication titled *A Guide to Partnering in the Construction Industry*.[66]

The construction industry appears to have embraced ADR, especially mediation. The AIA first included mediation as a condition precedent to arbitration of a dispute in the 1997 edition of its form documents. On the other hand, the AGC and EJCDC form contracts are not as stringent on requiring mediation prior to arbitration/litigation, generally making it an elective between the parties.

Practice Tip

The design professional contract should include a clause requiring mediation prior to arbitration or litigation, even if this requires modification of an existing (including a form) contract. The authors have found that many disputes, no matter how far apart that parties may seem to be, can be resolved through good, strong mediation efforts at the onset of a dispute, particularly with a mediator experienced in construction matters.

A detailed discussion of all of the considerations surrounding the allocation of litigation expenses is beyond the scope of this chapter. However, one should carefully consider these clauses, since the attorney's and expert's fees frequently exceed the amount in dispute.

12. *Insurance*

The availability of insurance is an important aspect of the construction industry and how it conducts business. At a bare minimum, every design professional should understand that regardless of working in an incorporated business, he can still be held *personally* liable for the consequences of his negligent actions or failure to act. This fact alone should compel the design professional to obtain professional liability insurance.[67] Many construction contracts, including the form contracts discussed in this book, contain clauses providing certain insurance requirements for the design professional.

For the design professional, having the proper professional liability insurance (as well as general liability and other necessary coverage) is a must, even if not required by the terms of a contract. The construction process is inherently risky. Claims and lawsuits against the design professional can come from many

---

66. *See* the AAA's Web site, *at* http://www.adr.org for this and other resources.

67. *See* the "Professional Liability" section of the *Practice Management Guide* at the CNA/Victor O. Schinnerer Web site, *at* http://www.schinnerer.com.

directions, including the owner, contractors, and third parties to the construction project.[68]

The AIA form contracts between the owner and architect do not include provisions requiring insurance of the design professional.[69] At the same time, interestingly, § 12.8.1 of the AIA C141-1997 Standard Form of Agreement between Architect and Consultant requires the consultant to maintain professional liability insurance "if required by the Architect." The AGC Form Contract No. 240 between the Owner and Architect/Engineer also contains professional liability insurance requirements. This difference may indicate a bias by the AIA in favor of the design professional on this subject, to avoid requiring that the design professional meet financially onerous insurance requirements. However, the design professional should not resist obtaining a reasonable professional liability policy, and meeting other insurance requirements that the owner may require, while consulting with his insurance professional and/or legal counsel if the owner makes what appears to be an unreasonable request to procure insurance.

13. *Consequential Damages—Waiver and Related Issues*

The 1997 edition of the AIA form documents added a provision where the contracting parties (the owner, architect, or contractor) waived the right against the other

---

68. While B141-1997 at ¶ 1.3.7.5 provides that "[n]othing contained in this Agreement shall create a contractual relationship with or a cause of action in favor of a third party against either the Owner or Architect[,]" it is the inherent nature of design which can give rise to a number of such third party claims, depending on the type of project constructed and the underlying basis of the claim. More than 100 separate lawsuits were pending in the year following the July 17, 1981, collapse of two suspended walkways in the Kansas City Hyatt Regency Hotel where 113 persons eventually died from injuries sustained as they fell with the walkways or were crushed underneath; 180 more were injured, many of them maimed for life. A detailed explanation of the design, the construction (containing fatal changes to the initial design), and the ultimate tragedy can be found at Steven S. Ross, *Construction Disasters: Design Failures, Causes and Prevention*, ENGINEERING NEWS RECORD, at 388-406 (McGraw-Hill Book Company 1984). The cause of the collapse was traced to a single change in the design of a support detail apparently made to facilitate the erection of the walkways. All parties denied having approved the change in the walkways suspension, apparently from various court filings, blaming any and all other project principals for the fatal change. *Id.* at 406. However, the *Kansas City Star* published a steel fabricator's drawing in mid-August, indicating that all of the project principals had known of the design change. *Id.* at 397. The Hyatt Regency case illustrates the continuing liability exposure to third party claims long after a commercial project is designed and constructed. It also involves another, potential area of liability exposure when pre-engineered components are integrated into the architectural and/or structural design of a building, involving various participants in the fabrication of the component, over whom the design professional may have little or no control.

69. However, B151-1997 at § 10.2.1.6 does state that reimbursable expenses may include professional liability insurance dedicated exclusively to the project (project insurance), as well as insurance coverage requested by the owner "in excess of that normally carried by the

contracting party to recover "consequential damages." Most practitioners and commentators consider this a very significant development in the evolution of these form documents, including the secondary impact on construction contracting generally. We will examine the nature of consequential damages, how and why this new development occurred, and why it has had such an impact.

The AIA's consequential damages waiver arose primarily from a New Jersey case where the court found a construction manager liable to a casino owner for an arbitration award of $14.5 million, when its fee was in the range of $600,000.[70] This decision, which in part dealt with whether the arbitrator had made a mistake of law in the award, created such a sufficient concern and uproar that the AIA inserted consequential damages waivers into its 1997 AIA form contracts.

The impact of this addition to the AIA documents has been widespread for all concerned, with the design professional[71] probably gaining a greater benefit. With respect to the AIA owner/contractor agreements, the contractor obtains more of a benefit. Although a design professional could potentially assert consequential damages in a claim (such as lost profits, lost business opportunities, etc.), it is more likely that it will be the owner (or contractor, for that matter) who will be claiming consequential damages. The owner's claims will probably be of a much greater dollar figure (such as the casino owner in *Perini Corp.*[72]). Most construction professionals can envision a plethora of situations where the most substantial damage claim in a construction-related dispute is the consequential damage claim by the owner, which could easily overshadow any claims of repairs for defective construction, or even delays in construction (casinos, malls, stadiums are but a few of many examples).[73]

Therefore, as a risk management tool the design professional should include a waiver of consequential damages provision in all of his contracts, including form agreements and personally drafted contracts.

14. *Limitations of Liability*

A Limitation of Liability (LOL) provision is an important risk control mechanism for design professionals. LOL clauses typically arise where a disparity exists

---

Architect and the Architect's consultants." The revised B141-1997 does not contain such an insurance provision.

70. Perini Corp. v. Greate Bay Hotel & Casino, Inc., 129 N.J. 479, 610 A.2d 364 (1992).

71. It is generally considered that contractors gain a greater benefit over owners from waiver of consequential damages provisions, even though contractors give a broad waiver of such claims under the specific definition of consequential damages found in AIA A201-1997, ¶ 4.3.10.1. This is because of the seemingly broad scope and extent of the waiver of the owner's potential claims and damages, as opposed to amounts that might be more limited for the contractor (and the design professional, as discussed above).

72. *Supra* note 70.

73. A good discussion of the impact of consequential damages appears in THE CONSTRUCTION CONTRACTS BOOK: HOW TO FIND COMMON GROUND IN NEGOTIATING DESIGN AND CONSTRUCTION CONTRACT CLAUSES, *supra* note 34, at Chapter 6.

between the fees charged by design professional in comparison with the potential risk assumed on a project (based on the project's size, complexity, and other factors). Many design professional firms are small and may not be able to absorb the risks. Other factors that justify an LOL are the pressures of the professional liability insurance market, and the fact that in many jurisdictions the architect's or engineer's signing and stamping of plans can render them personally liable for claims that may be asserted.

The design professional's LOL clauses come in many different forms, but always have the objective of placing a financial limit on the design professional's responsibility to the owner or some other party with whom the design professional has contracted. The limits that are typically used are a limitation up to a specified amount, to the design professional's fee, or to the amount of the insurance proceeds available to the design professional.[74]

Jurisdictions in the country have treated the issue of enforceability of LOL clauses in various ways. Courts typically struggle to balance the rights of parties to enter into contracts on their own terms (i.e., the right to "freedom of contract") versus the possibility that the limitation of liability clause could be considered a "contract of adhesion," where the party providing the limitation is not able to properly assess or negotiate an allocation of the risks involved.[75]

---

74. Most all designer's professional liability insurance policies are written on a "burning limits" basis; i.e., the policy limits are reduced by each application of coverage under the policy to defend a claim, including the payment of attorney's fees. Although the subject is beyond the scope of this discussion, if the LOL clause is limited to coverage remaining under the design professional's insurance policy limits, the careful design professional and his counsel will make certain that the LOL clause reflects this reality. *See generally* Victor O. Schinnerer & Company, Inc., *Practice Management Guide, at* http://www.schinnerer.com.

75. A good discussion of limitation of liability clauses with citations to differing state court rulings is contained in SWEET & SCHNEIER, *supra* note 8, at § 15.03D. The difference in approach by the various states to the enforceability of such clauses is illustrated by two cases involving design professionals, cited in this treatise. The facts in the cases were similar but the outcome differed. The New York case of Ricciardi v. Frank, 655 N.Y.S.2d 242, 170 Misc.2d 777 (1996), involved a claim by a consumer against an inspecting engineer who had agreed to perform a house inspection for $375. *Id.* While the New York trial court refused to enforce the LOL clause in the contract at issue that had limited liability to $375 (compared to actual damages in the case that were close to $3,000), the New York Appellate Term enforced the clause on appeal. In Estey v. MacKenzie Eng'g, Inc., 927 P.2d 86 (Ore. 1996), a structural engineer was retained to make a limited, visual review of a house that his client was considering purchasing. The contract limited liability to $200 for any negligence by the engineer, the amount of his fee for the review. The client prevailed in a negligence action against the engineer, proving losses at trial of $190,000 for needed repairs with an estimated cost of $150,000 to be incurred. While the intermediate Oregon court enforced the clause, the Oregon Supreme Court refused to enforce the limitation of liability.

15. *Modifications and Changes to the Contract: Changes in Services*

A contract modification is one that the parties agree to by mutual written assent.[76] In contrast is a change to the contract terms, which typically one party to the agreement makes unilaterally.[77] The AIA B141-1997 form contract deals explicitly with changes in services in § 1.3.3. This section provides that the parties can change the services in separate ways: by mutual agreement in writing; because of circumstances beyond the control of the design professional; or because of any of the numerous circumstances discussed in § 1.3.3.2. Many of these situations are ones that might arise during the project construction, but just as easily could involve changes during the design phase. Examples of these project changes include changes in codes, laws, or regulations, or changes in the project size, quality, or budget.

It would appear that the conditions of § 1.3.3.2 would be easier to apply than a determination of "circumstances beyond the Architect's control." On its face, this would seem to include calamities such as fire or other disaster. More tenuous would be

---

Several factors can be gleaned from Professor Sweet's work which the courts may be considering in deciding whether to enforce LOL's such as relative bargaining strengths, the opportunity for informed consent, type of client (public, private, business, lay), freedom of contract, market realities for desired services, dampening effect on willingness to undertake risky projects, insurance rates, risk compared to compensation, opportunity for counsel, and whether the LOL clause is sought to be applied in a consumer or a commercial context. JUSTIN SWEET & MARC M. SCHNEIER, LEGAL ASPECTS OF ARCHITECTURE, ENGINEERING AND THE CONSTRUCTION PROCESS (7th ed. 2004). It could be said, especially in view of the Oregon and New York cases, that the kind of services sought as well as the structure of the process when entering into a contract can be a factor. In these two cases, the work did not involve design services open to continual input by clients but a predetermined bundle of services to be rendered by the engineers in performing the inspections. Thus, in the context of Pre-Design, Design and Pre-Construction Phase Services—where a client would have greater opportunity to participate in, control, or influence the eventual outcome of the professional services bargained for—limitation of liability clauses may be more enforceable. Indeed, as pointed out by Professor Sweet, a number of jurisdictions have enforced LOL clauses. The California appellate court decision of Markborough v. Superior Court, 227 Cal. Ct. App. 3d 705 (1991), held that a LOL clause in a preprinted contract was valid, even though it had not been specifically negotiated or initialed. An important consideration for the court was the existence of an opportunity to negotiate, taking the form of a cover letter from the design professional, which gave the client the option to accept, reject, or modify any element of the contract. Another consideration was the relative bargaining power of the parties. *Markborough* also involved design services in which client interaction is inherent, and not inspection or other type of "off the shelf" consumer service bundles.

76. AIA Document B141-1997, Article 1.4, Scope of Services, ¶ 1.4.1.

77. *See* SWEET, *supra* note 8, § 6.11(a). Sweet explains that the owner may typically make changes to the contract for construction by way of a "change order."

78. § 6.14. This procedure is unique to the contractor's contract, although it may have impact on the design professional's contract, as discussed in Chapter 4, *infra*.

project changes to scope or schedule, which might appear to be contractor issues but could also allegedly include design professional involvement. Disputes may arise in this manner. However, at the same time, it would be impossible for an agreement to list each and every possible unforeseen condition or calamity that might require a contract change. Therefore, it is necessary and fair to leave the broad language of the right to change the contract if there are conditions beyond the design professional's control.

The AIA C141-1997 form contract between the architect and consultant includes an even more detailed listing of the potential methods for the consultant to receive compensation for changes in the work (thirteen versus the seven contained in AIA B141-1997, § 1.3.3.2). The C141-1997 also provides that the consultant is entitled to compensation for circumstances "beyond the Consultant's control."

It is also notable that the AGC form contracts, including the AGC Document No. 240, do not provide a *change in services* type clause, leaving to operation of law the right of the design professional to obtain payment for changes in the project.

IV. The Design Services Phase: Roles and Responsibilities

A. *Contractual Roles and Liabilities*

Under the AIA model, there are three design phases: Schematic Design, Design Development, and Construction Documents. Although there are no precise definitions for these phases, there is an understanding in the design professions as to the general meaning of these terms.

1. *Schematic Design*

Schematic Design is the stage of design where the architect finalizes and confirms the project size and scope with the owner. If a clearly defined program does not exist before the commencement of this phase, the Schematic Design phase may be long and tedious. Otherwise, this phase involves only the lead design professional, usually the architect, and the owner. The end product is the outline of the project, with confirmation of certain fundamentals.

In the case of architects, the Schematic Design may contain the seeds of potential design liability that may surface much later in the other phases of the design or during construction. These liabilities can be divided broadly into two categories: (1) liability arising from the failure of the design to conform to owner's requirements; and (2) liability arising from code violations or nonconformance with expected and applicable standards.

If the Schematic Design fails to conform to the owner's requirements, and the owner does not discover the failure until a later phase of the project (through no fault of his own), the design professional incurs liability to the owner. The owner can assert liability under the legal theories of breach of contract, negligence, negligent misrepresentation, and perhaps others.

The Schematic Design is, of course, a work in progress. Therefore, one cannot necessarily scrutinize it for code violations. However, there may be aspects of

Schematic Design that may be fixed and final and subject to an evaluation for conformance with applicable law. For example, a schematic architectural design must conform to applicable land use and zoning laws, such as height limits and setbacks. Failure of the Schematic Design to conform with such laws is potentially actionable under the theories of negligence, breach of contract, and perhaps negligence per se (discussed *infra*).

2. *Design Development*

Design Development is a significant phase in the development of the project, in that it involves the integration of multiple design disciplines. In addition to areas of potential liability highlighted under Schematic Design above, design professionals face potential liability during this phase for failure to coordinate the various design disciplines into a coherent and functional whole. AIA B141-1997, § 2.1.1 provides that the architect shall coordinate his and his consultants' services with the services provided by the owner and his respective consultants.

Various commentators and publications have discussed coordination of consultants, and the accompanying issues that arise.[78] In short, responsibility for the coordination of consultants is one that the design professional should not take lightly. Therefore, the prime design professional who undertakes this responsibility should make sure that he has the proper contractual agreements and protections in place with the owner and consultants as to responsibilities. Coordination is, of course, an ongoing responsibility of all design professionals throughout the project. It is particularly important during Design Development because in this phase, the project can still undergo significant changes in scope and parameters. It is because of such changes that failure to coordinate occurs, usually through communication failure or untimely communication among design professionals.

It is in the best interest of both the owner and the prime design professional to have a coordination protocol in place, including a proper clause in the contract to outline the obligations and limitations.[79] In addition, it is recommended that if the owner requests the architect, who is typically the consultant coordinator, not to perform the coordination function, then the architect should obtain appropriate indemnity.

3. *Construction Documents*

The theories of liability against the design professional arising from the Construction Documents Phase are the same as those highlighted in the preceding

---

78. *See* THE ARCHITECT'S HANDBOOK OF PROFESSIONAL PRACTICE, *supra* note 1, at § 11.2.

79. *See* Sheila Dixon & Richard D. Crowell, *Multiple Prime Contracts*, *in* THE CONTRACT GUIDE: DPIC'S RISK MANAGEMENT HANDBOOK FOR ARCHITECTS AND ENGINEERS 213-216 (DPIC Companies, Inc. 1993). An updated version of DPIC's CONTRACT GUIDE, THE XL INSURANCE CONTRACT GUIDE FOR DESIGN PROFESSIONALS: A RISK MANAGEMENT GUIDE FOR ARCHITECTS, is currently being written.

two sections. However, many questions arise in this phase as to whether the conduct of the design professional met the applicable professional standard of care. During this phase, issues arise as to clarity and organization of the documents, adequacy of detailing, the quality and thoroughness of the specifications, completeness of the documents, and constructability.

Although technically not a part of the Construction Documents Phase, the administration of bidding and answering questions of bidders is another area fraught with potential liability, primarily arising from miscommunication with bidders. This phase, if called for by the design professional's contract, would usually be termed the "Bidding & Negotiations Phase." Issues arising from design professional's services in this phase will usually be subject to expert opinion as to whether the conduct of the design professional met the prevailing standard of care in the profession at the time and in the locale of where the professional performed the services.

B. *Implied Obligations of the Design Professional during the Design Services Phases*

Beyond the liabilities and obligations that may be expressly stated in the design professional's contract, there are also potential implied obligations of which the design professional should be aware. Such implied obligations may begin to materialize during the design services phase, but may also be present during the pre-design phase discussed above and other phases of the project discussed elsewhere in this handbook.

The first and foremost implied obligation in any design professional contract is that the design professional will act with professional care, typically known as the "standard of care." The courts normally define the standard of care as that care and skill ordinarily exercised by design professionals practicing under the same circumstances in the same locale and at the same time as the design professional. Thus, in *Paxton v. County of Alameda*,[80] the court stated that an architect "impliedly represents that he possesses, and it is his duty to possess, that degree of learning and skill ordinarily possessed by architects of good standing, practicing in the same locality."[81]

A further implied obligation could be that the design professional will comply with all codes and laws (certainly including building codes) applicable to the design. In fact, AIA B141-1997, § 1.2.3.6 provides that the architect will "review laws, codes and regulations applicable to the Architect's services." However, it is notable that more than one commentator has suggested that a court might construe these

---

80. *Paxton*, 259 P.2d at 938.
81. Of course, many design professional contracts include an express provision that the design professional will act according to the standard of care. *See*, *e.g.*, AIA B141-1997, ¶ 1.2.3.2.

provisions as imposing a standard of care much higher than that otherwise created by the contract documents.[82]

The courts also imply in every contract a requirement of "good faith and fair dealing." This covenant provides that the parties to the contract will act and work with each other in good faith and not act to frustrate or harm the others' rights and expectations. Design professional agreements are no different, and the design professional and his contracting party must work with one another to accomplish the intended result of the contract.

Finally, it is worth noting that a warranty is one obligation that the courts should not imply in the design professional's contract. Although counsel for claimants often attempt to clothe the design professional with such a warranty obligation, most jurisdictions have refused to imply such a warranty obligation on the design professional, most often based on the fact that the design professional is providing a service rather than a product.[83]

---

82. *See* Douglas M. Folk, *Handling an Unfamiliar Role: Ethical Considerations for the Design Professional and Its Attorney*, in AMERICAN BAR ASSOCIATION FORUM ON THE CONSTRUCTION INDUSTRY (2000), *available at* http://www.folklaw.com/ethics.htm.

83. *See* Gagne v. Bertran, 43 Cal. 2d 481, 275 P.2d 15 (1954); *cf.* Tamarac Dev. Co. v. Delamater, Freund & Assoc., 234 Kan. 618, 675 P.2d 361 (1984).

CHAPTER 4

Design Professional Legal Responsibilities—Construction and Completion Phases

DAVID HATEM, ESQ.

I. Introduction

This chapter examines the legal responsibilities of design professionals during the construction and completion phases of a construction project. Although the contract primarily defines these responsibilities, principles of tort and statutory law, as well as the actual conduct of the design professional, influence the determination of whether the design professional may be legally responsible for claims asserted against it. Tort and statutory law determinations are also impacted by contract terms.[1] Further, the delivery method context in which the design professional performs its services plays an important role in defining the nature and extent of its legal responsibilities. This chapter primarily reviews the general principles of design professional responsibility based upon relatively standard contract terms and conventional roles of the design professional in the design-bid-build delivery method. Of course, there are always variations on these standard contract terms and the design professional's roles which can and do affect the legal duties, responsibilities, and liabilities of the design professional.

II. Construction Phase: Roles and Responsibilities

As previously noted, the contractual terms of engagement typically define the design professional's duties during the construction phase. These contractual terms

---

1. *See, e.g.,* Parent v. Stone & Webster Eng'g Corp., 556 N.E. 2d 1009 (Mass. 1990); Note, *Architectural Malpractice: A Contract-Based Approach*, 92 HARV. L. REV. 1075 (1979); P. Jackson, *The Role of Contract in Architectural and Engineering Malpractice*, 51 INS. COUNSEL J. 517 (1984).

include scope of service definition, as well as terms pertaining to standard of care, indemnification, and the allocation of risk and responsibility to other project participants (e.g., construction means, methods, and safety responsibility typically are allocated to the contractor).

A. *Roles and Responsibilities of Project Participants in Design-Bid-Build*

1. *Traditional Roles and Responsibilities*

Traditionally, the owner is responsible for establishing the project program and budget, securing project funding, and procuring and contracting with the major project participants, notably the design professional and the contractor. In the design-bid-build delivery method, the project owner typically engages the design professional to design the permanent components of the project and to serve as the owner's representative during construction. The owner approves the design as the designer is developing it. During the construction phase, the design professional's services primarily focus on providing the owner with confidence that the work, when fully completed, will conform to the requirements of the contract documents.

The owner engages the contractor to construct the project in accordance with the contract documents (which embody the design developed by the design professional and approved by the owner). In addition, the contractor (directly or through others) is responsible for the design and implementation of the construction means, methods, procedures, sequences, and techniques, as well as safety precautions and programs (construction means/methods and safety) employed in the performance of the work.

Most standard agreements—e.g., published by the American Institute of Architects (AIA) and the Engineers Joint Contract Documents Committee (EJCDC)—describe the roles and responsibilities of the owner, the design professional, and the contractor in a manner generally consistent with the traditional roles and responsibilities previously summarized.

2. *Relationship of Roles and Responsibilities to Risk Allocation*

Logically and appropriately, there should be a correlation between (1) the contractually defined roles and responsibilities of the respective project participants; (2) their ability to control and manage risk variables; and (3) risk allocation among the project participants.[2]

Application of these basic principles supports the widely accepted obligation of the owner to impliedly warrant the project design adequacy—the so-called *Spearin Doctrine*[3]—predicated on the rationale that the owner selects the design professional,

---

2. *See, e.g.,* R. J. Smith, *Allocation of Risk—The Case for Manageability*, 13 INTL. CONSTRUCTION L. REV. 549 (1996).

3. United States v. Spearin, 248 U.S. 132, 39 S. Ct. 59, 63 L. Ed. 166 (1918). A contractor was entitled to damages after the government annulled his contract to build a dry dock. The government breached its implied warranty that a defective sewer, which he was required to move, would be adequate if he followed the plans.

generally approves the design, and has the responsibility to fund project costs. The design professional typically is required to perform its design and other services, including professional services during construction, in accordance with the degree of care and skill exercised by other similarly qualified and experienced design professionals performing under similar circumstances and terms of engagement (the professional standard of care). The professional standard of care does not require perfection. The Supreme Judicial Court of Massachusetts has stated:

> As a general rule, '[a]n architect's efficiency in preparing plans and specifications is tested by the rule of ordinary and reasonable skill usually exercised by one of that profession .... [I]n the absence of a special agreement he does not imply or guaranty a perfect plan or satisfactory result ....
> 'Architects, doctors, engineers, attorneys and others deal in somewhat inexact sciences and are continually called upon to exercise their skilled judgment in order to anticipate and provide for random factors which are incapable of precise measurement. The indeterminable nature of these factors makes it impossible for professional service people to gauge them with complete accuracy in every instance.... Because of the inescapable possibility of error which inheres in these services, the law has traditionally required, not perfect results, but rather the exercise of that skill and judgment which can be reasonably expected from similarly situated professionals.[4]

The contractor typically assumes the risk and responsibility for liability, claims, damages, or losses due to construction means/methods and safety, since those areas generally fall in the province of the contractor's contractual role, qualification, and actual ability to control.

Based on these standard roles and responsibilities and their underlying risk allocation rationale, most standard contracts define the design professional's performance standard in a manner consistent with the professional standard of care. Those same contracts and standard General Conditions of the Construction Contract state that the contractor is exclusively responsible for construction means/methods and safety, and is obligated to indemnify the owner and design professional for claims and damages due to deficiencies in construction means/methods and safety.[5]

3. *Variations on Traditional Roles and Responsibilities in Design-Bid-Build*

On design-bid-build projects, the respective roles and responsibilities of the design professional and contractor may vary from their traditionally defined roles and responsibilities. For example, while a design professional retained by the owner

---

4. Klein v. Catalano, 437 N.E. 2d 514, 525 (Mass. 1982).
5. AIA Document A201-1997, ¶ 3.18.1.

is normally not responsible for the design of construction means/methods, or administrative roles during the construction phase pertaining to that area, owners may prefer that design professionals become involved in providing design and other professional services related to construction means/methods and other aspects of construction execution and methodology. This role transformation stands in contrast to the traditional role of the noninvolvement of the design professional in construction means/methods. Design professionals may become involved in construction means/methods issues in different ways—such as, preparation of design criteria or performance specifications, or preparation of detailed and mandatory construction means/methods design. Each of these different types and degrees of role transformation carries different legal responsibilities and potential liability exposures for the design professional. When this type of role transformation occurs, the design professional is likely to assume some degree (more or less depending upon the scope and contractually defined level of involvement) of risk and legal responsibility for deficiencies in affected construction means/methods.

Conversely, while a contractor is typically not responsible (directly or vicariously) for the design of *permanent* project work (as distinct from design of construction means/methods), the contractor may assume such responsibility by contractually accepting design delegation responsibilities or proposing and becoming responsible for value engineering modifications to that work.[6] In such circumstances, the contractor typically may assume some degree of risk and responsibility for deficiencies in the design.[7]

These variations in roles and responsibilities are becoming more common in the design-bid-build delivery approach since they may afford reasonable options (if appropriately undertaken) to shift to other delivery method approaches, such as design-build. From the perspective of the design professional, there is nothing especially or absolutely problematic about contractual assumption of these nontraditional roles and responsibilities. The key is to understand the potential risk allocation and resulting liability.

B. *Roles and Responsibilities of the Design Professional during the Construction Phase*

1. *Roles and Responsibilities Defined by Contract and Conduct*

Undoubtedly, contracts are the principal source for definition of the nature and extent of the design professional's obligations during the construction phase. Contracts serve a central function in defining the roles and responsibilities for both contract-based and tort claims against the design professional arising out of construction

---

6. *See, e.g.*, David. J. Hatem, *Design Delegation: Risk Management/Allocation Considerations for Design Professionals*, 4 THE CA/T PROFESSIONAL LIABILITY REPORTER No. 1, 7 (Nov. 1998).

7. *See, e.g.*, AIA Document A201-1997, ¶ 3.2.10, General Conditions of the Contract for Construction.

phase services and occurrences. Succeeding sections of this chapter will discuss some of the more important of these contract provisions.

In addition, the design professional's conduct during the construction phase may have a significant impact upon the design professional's responsibility. For example, if the design professional directs the contractor in the field or limits the contractor's discretion in the areas of work performance, construction means/methods, or safety, the design professional may face potential professional liability exposure despite the clear terms of the contract negating any role or responsibility in that area.[8] Thus, the design professional's prospects of achieving summary judgment based on such contract language may be significantly diminished, if not eliminated entirely.

Courts will generally respect clear contract terms defining roles and limiting responsibilities of the design professional. However, if credible evidence exists that the design professional by its conduct contradicted and did not adhere to its own role as contractually defined, a court may disregard or de-emphasize the contract terms as dispositive in determining responsibility, and resort to more generally applicable tort principles to determine liability. Typical factors that a court would consider to determine whether the design professional's actions evinced an intent to expand his responsibilities beyond the contours of the contract with the owner include: "(1) actual supervision and control of the work; (2) retention of the *right* to supervise and control; (3) constant participation in ongoing activities at the construction site; (4) supervision and coordination of the subcontractors; (5) responsibility for taking safety precautions at the job site; (6) authority to issue change orders; (7) the right to stop the work."[9]

2. Modification of Roles and Responsibilities during the Construction Phase

There are circumstances where otherwise clearly defined contractual roles and obligations of the design professional and other project participants are contractually modified due to some evolving project requirement or circumstance (often, owner driven) that develops during the construction phase; for example, an owner's request that its design professional become involved in a construction means/methods issues due to concern about claims involving the contractor. Alternately, public safety or stability concerns related to abutting property could cause owners to closely monitor or even direct the contractor's operations. Similarly, an obstinate contractor taking advantage of a required change to its contract may engage in brinksmanship negotiations causing the owner to turn to its design professional to rescue it by assuming a duty that the contractor refuses. The design professional all too often is willing to step outside its contractual obligations to assist the client. When previously defined roles and responsibilities change during construction, it is important for the design professional to (1) understand the potential risk and responsibility implications of

---

8. *See, e.g.*, Balagna v. Shawnee County, 233 Kan. 1068, 1074 (1983); Peck v. Horrocks Engineers, Inc., 106 F.3d 949, 954 (10th Cir. 1997).

9. Westerfield v. Arjack Co., 78 Ill. Ct. App. 3d 137 (Ill. Ct. App. Ct. 1979).

the proposed change in scope of services to make an informed decision whether to assume it; (2) evidence any change by written modification or amendment to the owner-design professional agreement; and (3) make certain that the new or evolving role of the design professional is clearly communicated and understood by all affected project participants (e.g., the contractor).[10]

3. Standard Contract Terms Regarding Design Professional Roles and Responsibilities during the Construction Phase

The AIA form contracts define the traditional duties of design professionals during the construction phase.

a. *Site Visits*

AIA Document B141-1997, ¶ 2.6.2.1 (*see also* AIA Document A201-1997, ¶ 4.2.2) defines the architect's role regarding site visits during construction. Paragraph 2.6.2.1, "Evaluations of the Work," states:

> The Architect, as a representative of the owner, shall visit the site at intervals appropriate to the stage of the Contractor's operations, or as otherwise agreed by the Owner and the Architect in Article 2.8, (1) to become generally familiar with and to keep the Owner informed about the progress and quality of the portion of the Work completed, (2) to endeavor to guard the Owner against defects and deficiencies in the Work and (3) to determine in general if the Work is being performed in a manner indicating that the Work, when fully completed, will be in accordance with the Contract Documents. However, the Architect shall not be required to make exhaustive or continuous on-site inspections to check the quality or quantity of the Work. The Architect shall neither have control over or charge of, nor be responsible for, the construction means, methods, techniques, sequences or procedures, or for safety precautions and programs in connection with the Work, since these are solely the Contractor's rights and responsibilities under the Contract Documents.

Regarding the architect's evaluative role during construction, ¶ 2.6.2.1 makes clear the following:

- It is performed "as a representative of the Owner."
- The frequency of site visits is defined by intervals appropriate to the stage of Contractor's operations.
- The purpose of the site visits is to evaluate whether the work, when fully completed, will conform to the requirements of the Contract Documents.
- The architect is not responsible for construction means/methods and safety.

---

10. *See* SUBSURFACE CONDITIONS: RISK MANAGEMENT FOR DESIGN AND CONSTRUCTION MANAGEMENT PROFESSIONALS ¶ 10.3.4, at 364-67 (David Hatem ed., 1998).

Paragraph 2.6.2.1 represents a typical contractual definition and accepted industry understanding of the design professional's function in making site visits during construction. It emphasizes the intent that the design professional's role is focused on *permanent* project work that he or she has designed, and generally does not extend to responsibility for observations or evaluations of construction means/methods or safety. In addition, ¶ 2.6.2.1 makes it clear that the architect's evaluative role is not "exhaustive" or "continuous." In this regard, ¶ 2.6.2.1 reinforces other provisions of the contract documents that assign to the contractor primary responsibility for supervision, inspection, and quality of the work.

The responsibility for construction defects must always be viewed in the light of the contractor's plenary obligation to perform the work in accordance with contract documents. Statutes often reiterate this requirement for public construction projects.[11] The courts will look to the terms and conditions of the design professional's and the construction contractor's contracts to define and distinguish the respective duties of these two parties on a construction project. For example, in an appeal of an OSHA finding related to a fatal accident, the Seventh Circuit Court of Appeals overturned OSHA's determination that a design professional was "engaged in construction."[12] The appeals court ruled that the design professional had not exercised "substantial supervision over the construction" and expressed its concern that in reaching its conclusion OSHA had ignored the contract language defining the design professional's duties.[13]

Notwithstanding this traditional distinction between the responsibilities of the designer and contractor, owners often argue that their design professional is in a position to detect and prevent defects in the constructed project. A seminal case discussing the design professional's potential liability for its site observations as compared to the more extensive clerk-of-the-works type of continuous presence and inspection services is *Watson, Watson, Rutlan/Architects, Inc. v. Montgomery County Board of Education.*[14] The owner declined its contract option for the design professional to provide clerk-of-the-works inspection services, but still attempted to hold it liable for not detecting contractor defects that led to a leaking roof. Noting the importance of the owner/design professional agreement's language, the *Watson* court reviewed the decisions from other jurisdictions. Ultimately, it concluded that the owner had not carried its burden of proof and ruled in favor of the architect, but also held that a design professional has a duty to report known deficiencies to the owner and could not turn a blind eye to contractor failings that might affect the finished constructed project.

---

11. *See, e.g.*, MASS. GEN. LAWS ch. 30, § 39I (1960), prohibiting any willful, substantial deviation from contract documents without specific written authorization from the public owner.

12. CH2M Hill, Inc. v. Herman, 192 F.3d 711 (7th Cir. 1999).

13. *Id.* at 722-23.

14. Watson, Watson, Rutland/Architects, Inc., v. Montgomery County Bd. of Educ., 559 So. 2d 168 (1990).

The AIA documents also provide that the design professional performs its site visits as a representative of the owner and, thus, for the benefit of the owner. Put another way, the design professional's site visit role must be understood in the context of the contractor's fundamental role to (a) perform the work in accordance with the contract documents and (b) be primarily responsible for construction means/methods and safety. A split of authority exists in the jurisdictions that have considered and decided whether to hold a design professional responsible for worker injuries resulting from unsafe working conditions.[15] The design professional need also be cognizant of any statutory provisions, such as the Illinois Structural Work Act (repealed by the General Assembly in 1995 but lately the object of a groundswell of support for reinstatement), that could heighten its liability potential for worker injuries.

b. *Submittal Review*

Generally, contractor submittals, such as shop drawings, are contractually required. The submittals serve an important function in validating the contractor's understanding of the design and Contract Document requirements and provide an opportunity for the contractor to communicate in a more detailed executory way its proposed plan for fulfilling the requirements of the contract documents.

AIA Document B141-1997, ¶ 2.6.4.1 defines the architect's role in the submittal process as follows:

> The Architect shall review and approve or take other appropriate action upon the Contractor's submittals such as Shop Drawings, Product Data and Samples, but only for the limited purpose of checking for conformance with information given and the design concept expressed in the Contract Documents. The Architect's actions shall be taken with such reasonable promptness as to cause no delay in the Work or in the activities of the Owner, Contractor or separate Contractors, while allowing sufficient time in the Architect's professional judgment to permit adequate review. Review of such submittals is not conducted for the purpose of determining the accuracy and completeness of other details such as dimensions and quantities, or for substantiated instructions for installation or performance of equipment or systems, all of which remain the responsibility of the Contractor as required by the Contract Documents. The Architect's review shall not constitute approval of safety precautions or, unless otherwise specifically stated by the Architect, of any construction means, methods, techniques, sequences or procedures. The Architect's approval of a specific item shall not indicate approval of an assembly of which the item is a component.

The intent of ¶ 2.6.4.1, like that of virtually all other standard contract terms addressing the design professional's submittal review role, is to define in a concentrated

---

15. *See., e.g.,* Hanna v. Huer, Johns, Neel, Rivers & Webb, 233 Kan. 206, 218 (1983).

manner that the architect's review focuses on submittal content relevant to the design requirements of permanent project work in a manner distinct from the contractor's contemplated construction means/methods. The distinction between design of permanent project work and construction means and methods may be blurred in submittal content, and the design professional should be clear to limit the scope of his or her review to the former. The contract should always require that contractors stamp submittals prior to forwarding them to the design professional for review. The contract should similarly prohibit the contractor from commencing any item of work addressed in that submittal until the contractor has satisfactorily addressed the design professional concerns as indicated by the design professional's acceptance or approval of the submittal.

Design professionals face exposure to professional liability claims from project owners if they fail to review in a timely manner a contractor's submittal, resulting in contractor delay claims against the owner. Indeed, contractors may deliberately attempt to overwhelm the design professional's capacity to review submittals by simultaneously unleashing a barrage of submittals, essentially orchestrating the delay. By including construction contract requirements that the contractor submit a schedule of submittals, the design professional can muster the required resources to perform the expected review and combat any delay allegations caused by deviations from that schedule.[16] Generally, it is good practice for the design professional to maintain a log documenting dates of receipt and review return of contractor submittals, and then monitor that log during the project.

c. *Construction Means/Methods*

As previously noted, design professionals generally do not design or otherwise become involved in construction means/methods, the latter of which typically falls within the expertise and control of the contractor. However, there are special circumstances in which a design professional may, in varying degrees, become involved in the design of construction means/methods; for example, by preparing a very detailed method specification or design that can only be constructed in a sequence that later proves impossible. The extent of design professional responsibility and potential professional liability exposure that attaches to these varying degrees of construction means/methods involvement of the design professional typically depends upon a complex and fact-intensive analysis. For example, in *CH2M Hill, Inc., v. Herman*,[17] the Seventh Circuit Court of Appeals stated, "whether or not the construction standards apply has previously been a fact-specific inquiry that appears to turn on the responsibilities assumed by the firms in question."[18]

AIA Document A201-1997, ¶ 3.3.1, acknowledges this type of special circumstance in stating:

---

16. *See, e.g.,* AIA Document A201-1997, ¶ 3.10.2.
17. CH2M Hill, Inc., v. Herman, 192 F.3d 711 (7th Cir. 1999).
18. *Id.* at 719.

The Contractor shall supervise and direct the Work, using the Contractor's best skill and attention. The Contractor shall be solely responsible for and have control over construction means, methods, techniques, sequences and procedures and for coordinating all portions of the Work under the Contract, unless the Contract Documents give other specific instructions concerning these matters. If the Contract Documents give specific instructions concerning construction means, methods, techniques, sequences or procedures, the Contractor shall evaluate the jobsite safety thereof and, except as stated below, shall be fully and solely responsible for the jobsite safety of such means, methods, techniques, sequences or procedures. If the Contractor determines that such means, methods, techniques, sequences or procedures may not be safe, the Contractor shall give timely written notice to the Owner and Architect and shall not proceed with that portion of the Work without further written instructions from the Architect. If the Contractor is then instructed to proceed with the required means, methods, techniques, sequences or procedures without acceptance of changes proposed by the Contractor, the Owner shall be solely responsible for any resulting loss or damage.

The author recommends that when a design professional becomes involved in the design of construction means/methods, the extent of that involvement and the allocation of risk should be clearly and consistently stated in both the owner-design professional agreement (or any appropriate amendment thereto) and the general conditions of the construction contract, or in the applicable technical provisions of the specifications or notes on the drawings.

d. *Safety*

It is widely recognized and accepted in the design and construction industry that since the contractor controls the site and has discretion as to the means/methods by which he executes the work, the contractor should be exclusively responsible for construction means/methods and safety. In circumstances where the contractor's control of those means/methods is limited or otherwise curtailed by provisions of the contract documents or directives issued by others during construction, this general rule may be modified.

This general rule supports the logic of contractual indemnification provisions in favor of the owner and design professional for personal injury and/or property damage claims and liabilities arising out of the contractor's construction means/methods and/or failure to implement or monitor adequate safety precautions or programs.

e. *Quality Assurance*

Quality Assurance refers to the design professional's role in generally monitoring or auditing the contractor's performance, including the latter's quality control, in the performance of permanent project work. As distinct from the design professional's quality assurance role, the contractor is responsible for quality control including supervision and inspection in the performance of its work and that of the trade subcontractors. Quality control generally includes a plan for detailed supervision and inspection of work and materials furnished by the prime contractor, all of its subcontractors, and their respective materials providers. In performing its quality assurance role, the design professional serves as a representative of the owner; performance of that role does not relieve the contractor of its responsibility for performing the work in accordance with the requirements of the contract documents and for adherence to its own contractual obligations. However, the design professional may face liability from the owner for failing to detect an ineffective contractor quality control program. For example, if a contractor's quality control procedure contains hold points for self-inspection during a particular construction sequence and the design professional witnesses but fails to recognize or report that the contractor is ignoring or only superficially performing the self-inspection, then the design professional could share liability for defects in that constructed item that might have been prevented had the hold point inspection been implemented.

f. Rejection of Work

Typically, the design professional is empowered under the contract documents to reject work performed by the contractor that does not conform to the requirements of the contract documents. AIA Document B141-1997, ¶ 2.6.2.5, provides that the architect's authority to reject work or any judgmental decision made in good faith by the architect to exercise or not to exercise that authority shall not give rise to a duty or responsibility of the architect to the contractor.

Of course, as in the performance of all construction phase services, the design professional exercises its authority to reject work as a representative of the owner. If the design professional fails to exercise sound judgment in that regard, especially judgment that fails to meet the professional standard of care, the design professional may face liability to the owner. For example, if the design professional accepts the incorporation of certain mechanical equipment into the work that later proves to be defective or incapable of fulfilling stated performance requirements, the design professional will likely be a target of cost recovery by the owner. Although the construction contract undoubtedly assigns responsibility and therefore primary liability to the contractor for this deficiency, if the damages are substantial, the design professional may become embroiled in the dispute.

Wrongfully rejecting contractor work that turns out to be adequate can also lead to potential liability. Contractors sometimes attempt to recover their additional costs from design professionals for rejected substitutions under unfair business practice statutes or under a theory of negligent misrepresentation. In addition, most construction contracts contain a provision for "uncovering the work" that requires the owner to pay the costs of that uncovering if the item of work proves to be acceptable. If these owner-incurred costs are the result of the design professional's malfeasance, then that design professional can expect to face additional liability to the owner.

Similarly, the design professional needs to avoid an overly aggressive or hypertechnical inspection that is incompatible with the intent of that aspect of the work. For example, if the industry standard for determining the acceptability of repairs to minor damage of concrete is to "sound" the patch with a mason's hammer, then requiring the contractor to adopt a more elaborate and expensive approach such as ultrasonic testing must be resisted. Such an "over-inspection" could give rise to successful contractor claims of a "constructive change" for which the owner could seek recovery from the design professional.[19] Moreover, strict compliance with the contract requirements is not always the measure of acceptability of the installed work. Sometimes the doctrine of economic waste will mandate the acceptance of technically noncompliant work.

The limitations on an owner's right to insist on strict compliance with contract provisions were discussed in *Granite Construction Co. v. United States*,[20] which involved the construction of a lock and dam on the Mississippi River for the Corps of Engineers. The contractor in *Granite* was required to comply strictly with the government specifications for materials. When the government learned that Granite had embedded a noncompliant waterstop in portions of the cast-in-place concrete, it ordered Granite to remove and replace the noncompliant material with the approved waterstop, despite the fact that all evidence suggested that the material would function as well as the one specified. The appeals court recognized the government's right to insist on strict compliance, but found that where the cost of replacement is economically wasteful and the work is otherwise adequate, the government should allow the noncompliant material to remain in place. However, the court held that the contractor owed the government a reduction in contract price.

g. *Contractor Claims, Payment Issues, and Interpretation of the Contract Documents*

In addition to performing services during construction as the owner's representative, the design professional is usually required to perform certain administrative services with respect to investigating, evaluating, and making recommendations concerning contractor claims and disputes. A routine part of the design professional's services typically includes the review, evaluation, and recommendations regarding contractor payment requests. Similarly, the construction contract nearly always assigns the design professional the responsibility for interpretation of the contract documents. In performing these services, the design professional is required to act impartially, in good faith, and diligently.

The design professional should undertake any investigation and evaluation of contractor claims in accordance with the contractually stipulated procedures, and recommendations should be governed by and consistent with the intent and requirements of the contract documents. The design professional's evaluations and recommendations of a contractor's payment request should also be based upon reasonably objective data and observations of the contractor's work performance and the diligent and good faith exercise of sound judgment. The design professional

---

19. *See e.g.,* Southwest Welding & Mfg. v. United States, 413 F.2d 1167 (Ct. Cl. 1969).

20. Granite Construction Co. v. United States, 962 F.2d 998 (1992), *cert. denied,* 506 U.S. 1048 (1993).

should always make clear in any payment certification that its representations are based on the limited degree of contractually defined site visits of the design professional. The standard AIA general conditions provisions clearly describe the limitations of the Architect's Certificate for Payment.[21] The AIA agreement between owner and architect also reflects those limitations.[22] Claimants may attempt to imply an expanded role of the design professional in such certifications, resulting in liability problems for design professionals.

When the owner is a public entity, the potential liability for inappropriate payment certifications takes on a new dimension in the form of potential liability under a false claims statute as discussed later in this chapter.

The contractual and industrial context in which the design professional performs its role and review of contractor pay applications is a significant consideration in this regard. The contractor is primarily responsible for the adequacy of its work, warranty, and indemnity obligations. In addition, the system of payment withholding and retention evidences an intention to provide protection to the owner and opportunities for recourse against the contractor for defective work or overpayment. However, the design professional faces potential liability from all sides. Architects may also be subject to suit by sureties complaining that their principals were overpaid by virtue of negligently issued certificates.[23]

Obviously, the original intent and specific requirements of the contract documents should substantively guide the design professional's interpretation of those documents. Rendering these interpretations consistent with contract protocol and the prescribed time limitations will minimize contractor claims of ineffective contract administration.

Professional liability problems arise when the design professional does not reasonably adhere to the substantive and procedural requirements of the contract documents or exercise impartial and reasonable diligence in the interpretive process. Given the quasi-judicial nature of these construction administration services—especially interpretation of the contract documents and the making of recommendations and/or decisions regarding construction contractor claims—the design professional may be entitled to a qualified immunity defense in connection with claims arising out of the performance of such services.

For example, in fulfilling its assigned arbitral role, the design professional is exercising an authority that has been assigned to him by the construction contract and agreed to by both the owner and the contractor; he is immune from subsequent action by either party as long as he exercises that authority in good faith.[24] If the

---

21. AIA Document A201-1997, ¶¶ 9.4.2, 9.5.1.
22. AIA Document B141-1997, ¶¶ 2.6.3.1, 2.6.3.2.
23. *See* Peerless Ins. Co. v. Cerny & Assoc., Inc., 199 F.Supp. 951 (1961).
24. Lundgren v. Freeman, 307 F.2d 104, 118 (1962) ("We think he [the architect] should be protected when he acts in good faith, however erroneously, and that such protection is enough. If he acts fraudulently, or with willful and malicious intent to injure the contractor, he should be liable.").

design professional acts beyond the scope of its contractually assigned role, fails to act in good faith, or uses its arbitral powers to cover its own defective design, that immunity is no longer available.

In *Craviolini v. Scholer & Fuller Associate Architects*,[25] a contractor attempted to overcome the architect's presumed immunity for its decisions on disputes between the owner and contractor. The contractor claimed that the architect attempted to destroy the contractor's business after it complained about defects in the plans and specifications, and that the architect wrongfully disallowed material substitutions. The appeals court ruled that the "cloak of immunity" was available to the architect only when performing in its role as arbitrator, deciding a dispute between the owner and contractor.[26] However, the court considered those decisions made by the architect in its role as construction administrator to be fair game for the contractor's claims of misfeasance. The availability of this quasi-judicial immunity is dependent upon the jurisdiction of the dispute.[27] However, to be entitled to such a defense, the design professional must establish that it acted in good faith, impartially, and with reasonable diligence.[28]

h. *Termination of the Contractor*

The terms of the construction contract and general common law principles define the circumstances, duties, rights, and remedies pertaining to contract termination. The law favors the formation of contracts and disfavors their dissolution. Accordingly, when a design professional is involved in an owner's effort to terminate a construction contractor, the design professional must rigidly adhere to the principles of good faith and fair dealing. Under the AIA agreement, the owner has a contractual obligation to provide an impartial unbiased architect to perform the construction phase administrative duties. The design professional's role, if any, in the process of an owner's termination of a construction contract likely will be defined in the contract documents.[29] It is critical that the design professional carry out those duties in an impartial and unbiased manner, resisting any pressure by the owner to bolster its case for termination by overstating the record of alleged contractor transgressions or failings.

First and foremost, the authority and decision to terminate rests with the owner. The terms of the owner/contractor construction contract may require that the design

---

25. Craviolini v. Scholer & Fuller Assoc. Architects, 89 Ariz. 24, 357 P.2d 611 (1960).
26. *Id.* at 614.
27. City of Durham v. Reidsville Eng'g Co., 255 N.C. 98, 120 S.E.2d 564 (1961).
28. *See*, AIA Document B141-1997, ¶ 2.6.18; JUSTIN SWEET & MARC M. SCHNEIER, LEGAL ASPECTS OF ARCHITECTURE, ENGINEERING AND THE CONSTRUCTION PROCESS 283-85 (7th ed. 2004).
29. *See* Appeals of E.L. David Const. Co., Inc., 89-3 B.C.A. (CCH) ¶ 22140, WL 90328 (Armed Serv. B.C.A. 1989).

professional certify that sufficient cause exists to support the owner's termination of the contractor. In such circumstances, the design professional should base its certification on a reasonable investigation that is factually based and well documented. While the design professional should not hesitate to state its opinion and recommendations as discussed above, the ultimate decision to terminate must remain with the owner. The ultimate determination or judgment of whether sufficient cause exists to terminate (whether there has been a material breach of contract) is legal in nature, and therefore best left to the owner and its attorneys. The design professional may present subsidiary facts or objective evidence that provides the basis upon which a legal determination or judgment may be exercised by others (typically, the owner's legal counsel).[30] Finally, the owner or its counsel—not the design professional—should communicate with the contractor or its surety regarding termination.

In the case of an owner terminating its contractor based in part upon the design professional's certification, that design professional may find itself subjected to contractor claims of tortious interference with an advantageous business relationship or violation of an unfair business practices statute. The contractor could pursue this type of claim independent of the owner-contractor agreement, possibly eviscerating the design professional's attempt to invoke quasi-judicial immunity.

4. Commonly Recognized Implied Obligations of the Design Professional in the Performance of Construction Phase Services

This chapter has appropriately focused on the central role of the contract in defining the obligations of the design professional during the construction phase. However, the case law amply demonstrates that there are certain implied obligations to which the design professional must adhere in the performance of those services.

A review of case decisions reveals the importance of the design professional's exercise of good faith in its evaluations and recommendations regarding contractor performance and payment issues, review of and action upon contractor submittals, resolution of claims and disputes, and interpretation of the contract documents.[31] In this context, proof of good faith generally means that the design professional can objectively demonstrate that it acted in accordance with the procedural and substantive provisions of the contract documents and in accordance with sound professional judgment and industry custom. In addition to good faith, many case decisions have demonstrated the importance of the design professional's diligent and timely performance of its construction administration services.[32]

---

30. See SWEET & SCHNEIER, *supra* note 28, at 721-22.
31. *See, e.g.,* General Trading Corp. v. Burnup & Sims, Inc., 523 F.2d 98 (1975) (in addition to wrongfully issuing certificates when the work is either defective or incomplete, architects have been held responsible for not issuing these submittals in a timely manner.)
32. *See, e.g.,* Nelson v. Commonwealth, 235 Va. 228, 368 S.E.2d 239 (1988); French v. Jinright & Ryan, P.C. Architects, LEXIS 19446 (U.S. Dist. 1983).

5. *Areas of Potential Professional Liability Exposure*

The range of professional liability exposures and claims against design professionals arising out of construction phase services may be classified in a number of ways—by the identity of the claimant, the nature of the legal theory supporting the claim (whether predicated on common law or statutory grounds), and the type of damages or injuries alleged.

a. *Owner Claims*

Professional liability claims asserted by the owner (typically, the design professional's client) against the design professional arising out of construction phase services represent, by far, the highest percentage of such claims. As a general matter, an owner may predicate its claims on the basis that the design professional failed to meet certain specific contractual obligations and/or failed to perform in accordance with the professional standard of care. In most claims, there is substantial overlap between these two theories.

In professional liability claims arising out of construction phase services, an owner may assert that the design professional failed to satisfy specific contractual terms, e.g., the frequency of site visits, timeliness of shop drawing reviews, responses to Requests for Information (RFIs), quality assurance, and inadequate or inappropriate certification of contractor applications for payment. To prevail, the owner must establish a breach of the specific contract term or requirement *and* that such breach proximately caused the owner to sustain damage. In a substantial number of such claims, the owner's primary recourse for any damage sustained is and should be against the contractor. Specifically, under the standard contract documents the contractor is responsible for defective work (whether or not the design professional observed or otherwise detected the defect).[33] On public projects in some jurisdictions, statutory requirements reinforce this contract provision.[34] When the owner overpays the contractor based upon the design professional's recommendations, the owner's primary (if not exclusive) remedy should be against the contractor. However, in certain circumstances, when the owner's recourse against the contractor is foreclosed due to the passage of time or warranty periods, and/or the financial insolvency of the contractor, the design professional may have more exposure to a professional liability claim from the owner. Overpayment to a contractor based on a design professional's negligently prepared payment certification could also lead to liability to the contractor's surety that is forced to complete a project for an insolvent contractor that had collected more contract funds than it was entitled to.

The owner must present adequate proof that the design professional failed to meet the professional standard of care. Typically, this would require an expert

---

33. AIA Document A201-1997, ¶ 12.2.1; EJCDC Standard General Conditions of the Construction Contract, 1996, ¶ 6.19, B.
34. Mass. Gen. Laws ch. 30, § 39I (1960).

opinion that the design professional failed to meet the professional standard of care.[35] The scope and terms of the design professional's contract with the owner, as well as the relevant facts and circumstances, will determine whether the design professional met the appropriate standard of care. Clearly, the most significant factor in determining the design professional's liability is whether its contract with the owner required a heightened standard of care; in other words, one requiring a better performance than required by the common law. Similarly, any contractual provision that guarantees a certain result or owner satisfaction will expose the design professional to additional liability.

Project cost overruns and schedule impacts represent an important area of professional liability exposure common to both owner and contractor claims often asserted against the design professional. The extent to which such claims pose a substantial risk for the design professional depends on a number of factors and considerations; such as the provisions in the owner-design professional agreement regarding responsibility for indirect or consequential damage, and the status of the economic loss doctrine in the particular jurisdiction. In many instances, the underlying causes of a professional liability claim arising out of cost overruns or schedule delays that manifest during construction originate or relate back to the design phase services of the design professional and, as such, may not be strictly viewed within the scope of this chapter. However, the expression of these claims invariably involves some allegation or complaint regarding the design professional's performance of construction phase services (e.g., failure to review shop drawings or respond to RFIs in a timely manner).[36]

The assertion of cost overrun and schedule delay claims by third parties such as general contractors and subcontractors against the design professional are on an upward trend.[37] In many situations, these claims are asserted against the design professional because the third party's opportunity for cost or time equitable adjustment against its contracting partner is foreclosed by virtue of a "no damage for delay" or related provision.[38]

b. *Third Party Claims*

Depending upon the governing law, there are a variety of third parties (or non-clients) who may assert claims against a design professional arising out of the latter's

---

35. *See, e.g.,* Atlas Tack v. Donabed, 47 Mass Ct. App. Ct. 221, 712 N.E.2d 617 (1999).

36. *See, e.g.,* Nelson v. Commonwealth, 235 Va. 228, 368 S.E.2d 239 (1988).

37. *See, e.g.,* A. Meagher & M. O'Day, *Who is Going to Pay for My Impact? A Contractor's Ability to Sue Third Parties for Purely Economic Loss*, THE CONSTRUCTION LAWYER, AMERICAN BAR ASSOCIATION, at 27 (Fall 2005); Justin Sweet, *Contractor Postcompletion Claims: Advice to Engineer*, JOURNAL OF PROFESSIONAL ISSUES IN ENGINEERING EDUCATION AND PRACTICE, ASCE, 298 (Oct. 2004).

38. *See* David. J. Hatem, *The Relevance and Potential Impact of Risk Allocation Provisions in Owner-Contractor Agreements on Professional Liability Exposure of Design Professionals*, DESIGN AND CONSTRUCTION MANAGEMENT PROFESSIONAL REPORTER 1 (Oct. 2003).

construction phase services. These claimants include general contractors, subcontractors, construction lenders, surety companies, injured construction workers, state and federal governmental agencies, and others, who may predicate claims based upon common law theories of negligence, third party beneficiary, or violation of state or federal laws. These claimants, asserting bodily injury, property damage, or purely economic loss, may also seek statutory fines and penalties. Regardless of the legal theory, type of injury or damage, or relief sought, the terms of the design professional's agreement with its client and the applicable provisions of the contract documents often play an important role in conjunction with the facts surrounding the design professional's performance in determining legal responsibility of the design professional for these claims.

c. *Contractor/Subcontractor Claims*

As noted previously, contractors and subcontractors may assert a variety of claims against the design professional arising out of the latter's performance during the construction phase. A variety of legal theories, including negligence and third party beneficiary, are available as predicate for these claims. In many respects, the assertion of such claims both contradicts and undermines contractual risk allocation embodied in contract documents, as well as the structure of project participant relationships established in those documents.

More specifically, in the traditional design-bid-build method, the design professional serves during the construction phase under an agreement with and as a representative and an agent of the owner. Any performance deficiencies of the design professional should be addressed and remedied in the context of the direct (privity) relationship between the owner and design professional, and in accordance with risk allocation (i.e., standard of care, indemnification) and other relevant provisions of the owner-design professional agreement. Similar principles of risk allocation dictate that cost or time impacts incurred by the contractor or its subcontractors should be addressed and equitably adjusted, as appropriate, between the contractor and owner consistent with their agreements embodied in the contract documents. The allowance of direct claims by contractors and subcontractors against the design professional serves to circumvent and subvert this structure of project contractual relationships and risk allocation.

The relevance of the design professional's status as the owner's representative and agent during construction is an important factor in the context of such third party claims. For example, in observing and evaluating work performance by the contractor for conformance with contract documents requirements, the design professional represents and serves the interests of the owner in providing the latter with greater confidence that the work, when completed, will meet those requirements. As such, there is often a tension and conflict between the owner's interests to obtain the requisite work quality and performance, and the contractor who may be advocating for the acceptance of a lesser standard. To allow a contractor's claim against the design professional in such circumstances would, in most cases, subject the design professional to an unreasonable burden of defending its performance as the owner's agent in attempting to secure contractor performance in accordance with the requirements of the contract documents.[39]

On the other hand, all parties must ultimately acknowledge that design professionals have an obligation to perform their services in accordance with the professional standard of care, which requires reasonable evaluations, judgments, and recommendations by the design professional. If those evaluations are arbitrary or not supported by the requirements of the contract documents, this could lead to a professional liability claim by third parties who sustain damages as a result thereof (assuming the economic loss doctrine does not preclude such claims).

Perhaps the most commonly pled contractor claim against a design professional that is not precluded by the economic loss doctrine is negligent misrepresentation.[40] There the contractor/plaintiff will allege that it relied to its detriment on the design professionals negligently prepared work product. In the case of *Craig v. Everett M. Brooks Co.*,[41] the Massachusetts Supreme Court discussed the requirements for a negligent misrepresentation claim to qualify as an exception to the economic loss doctrine. *Craig* dealt with a claim related to a contractor's reliance on survey stakes erroneously placed by the defendant surveyor—who placed those stakes for the sole benefit of the contractor. *Craig* recognized the right of a plaintiff to sue the surveyor for misrepresentation based upon the *staking* of catch basins and a road. The court did not permit recovery for the negligent design services such as errors in drawings. However, it did permit recovery for improper surveying services in the nature of staking the land.[42] The court took great care to distinguish the claims based upon "misrepresentations" in the "plans" (which formed *no* basis for recovery) as opposed to the "placing of stakes."[43]

d. Personal Injury Claims

Injured construction workers often assert personal injury claims against the design professional. Standard contract documents provide substantial legal defenses to such claims on the basis that the design professional, as discussed earlier in this chapter, is generally not responsible for construction means/methods or safety. In addition, in many instances the design professional is entitled to contractual indemnification

---

39. *See* Bernard Johnson, Inc. v. Continental Constructors, Inc., 630 S.W.2d 365 (Tex. Ct. App. 1982).

40. The majority of jurisdictions follow the Second Restatement of Torts, § 552, for the elements of a negligent misrepresentation claim. With various caveats, § 552 provides: "One who, in the course of his business, profession or employment, or in any other transaction in which he has a pecuniary interest, supplies false information for the guidance of others in their business transactions, is subject to liability for pecuniary loss caused to them by their justifiable reliance upon the information, if he fails to exercise reasonable care or competence in obtaining or communicating the information."

41. Craig v. Everett M. Brooks Co., 351 Mass 497 (1967).

42. *Id.* at 501.

43. *Id.* at 499-501.

from the contractor and its subcontractors for such claims and is specifically named as an additional insured under general liability insurance policies procured and maintained by the latter.[44] Note, however, that this contractor-supplied general liability insurance will likely exclude coverage for claims of professional malpractice.

Notwithstanding these formidable contract-based defenses which, in many situations, provide the basis for a legally dispositive summary judgment motion in favor of the design professional, the design professional may assume increased liability exposure for construction worker claims by supervising or directing the work, construction means/methods, or safety precautions and programs. Moreover, if a design professional has actual knowledge of a contractor employing an unsafe practice on a worksite, it is incumbent upon the design professional to take some action to correct or mitigate the danger. The most obvious and direct step that a design professional can take is to call the unsafe situation to the attention of the contractor who is in a position to rectify the situation. For the most part, however, injured construction worker claims against the design professional are consistent with basic risk allocation principles embodied in standard contract documents, and should be highly defensible. Additionally, many states provide statutory protection to design professionals from worker injury claims. For example, the Immunity for Third-Party Design Professional statute in Alaska provides:

> (a) A person entitled to compensation under this chapter as a result of injury occurring at the job site of a construction project may not bring a civil action to recover damages for that injury against a design professional or an employee of a design professional who provides professional services for the construction project.
> (b) This section does not apply to a person receiving compensation under this chapter who is injured at a job site at which the design professional or employee of the design professional
>> (1) specifically assumed responsibility for job site safety practices under a contract;
>> (2) actually exercises control over the premises where the injury occurred; or
>> (3) prepared design plans or specifications, the plans or specifications contributed to the injury, and the plans or specifications were prepared negligently, recklessly, or with intentional misconduct.
> (c) In this section,
>> (1) "design professional" means a person registered under AS 08.48 as an architect, engineer, or land surveyor;
>> (2) "professional services" means services provided by a design professional that are within the scope of services for which the design professional is registered.[45]

---

44. *See, e.g,* AIA Document A201-1997, ¶ 3.18.1.
45. AS 23.30.017 ("Immunity For Third-Party Design Professional").

e. *OSHA*

The design professional's potential exposure to fines and penalties under the Occupational Safety and Health Administration (OSHA) regulations is dependent upon the degree and level of the design professional's involvement during the construction phase. A design professional intimately involved with a construction contractor's means/methods, and safety precautions or programs should expect heightened scrutiny and potential fines and penalties from OSHA. OSHA assigns regulatory enforcement to its regional offices, each of which has somewhat different interpretations and expectations of design professionals involved in construction. The degree of the design professional's apparent or actual engagement in construction work and its power to control the process and/or substantial supervision of construction work will determine whether OSHA has authority over the design professional's activities.[46] The case law supports the proposition that if the design professional's contractual scope of services and actual conduct do not involve physical construction work, or substantial control or supervision over the contractor's construction means/methods, safety precautions or programs, or work performance, the design professional should be able to successfully defend claims predicated upon alleged violations of OSHA and regulations promulgated thereunder.[47]

f. *False Claims Act*

In the performance of services on publicly funded projects, the design professional may provide express or implied certifications or representations regarding a contractor's entitlement to payment. The third party prosecuting the false claim act claim may be the sovereign or it may be an individual in a *qui tam* action wherein a whistleblower, informer, or otherwise interested person brings the claim on behalf of himself and the sovereign. In general, federal and state[48] false claims legislation provides the government with a remedy for actual and punitive damages against parties who submit or cause to be submitted a false claim for payment.[49] Under a typical false claims act statute, a person is liable if he/she (1) knowingly presents, or causes to be presented, a false or fraudulent claim for payment or approval, or (2) knowingly makes, uses, or causes to be made or used, a false record or statement to obtain

---

46. *See, e.g.,* CH2M Hill, Inc., v. Herman, 192 F.3d 711 (1999).

47. *See* SUBSURFACE CONDITIONS: RISK MANAGEMENT FOR DESIGN AND CONSTRUCTION MANAGEMENT PROFESSIONALS, at 305-12 (David Hatem ed., 1998); SWEET & SCHNEIER, *supra* note 28, at 274-79; Reich v. Simpson, Gumpertz & Heger, Inc., 3 F.3d 1 (U.S. Ct. App. 1993).

48. California, Delaware, Florida, Hawaii, Illinois, Massachusetts, Nevada, New Mexico, Virginia, and Washington, D.C. have enacted false claims statutes similar to the federal statute.

49. The False Claims Act, 31 U.S.C. §§ 3729-3310 (1994); MASS. GEN. LAWS ch. 12, § 5B (2000); CAL. GOV'T CODE § 12650 (1997); 740 ILL. COMP. STAT. ANN. 175/3(a)(2) (2006).

payment or approval of a claim by the public entity.[50] Cases interpreting the federal False Claims Act have generally refused to find liability where the certification, "reflects an engineering judgment and recommendation."[51] Nevertheless, aggressive owners have recently attempted to mischaracterize payment certifications in order to invoke the statutes and threaten design professionals with the associated penalties which, in some cases, include treble damages.[52] Thus, design professionals face potential exposure in connection with certifications, representations, or recommendations that they make concerning contractor payment applications or requests.[53] The federal False Claims Act has been in existence since the Civil War and the law is well developed. Many of the state statutes, however, are more recent, and the courts have not had the opportunity to rule on a large number of construction cases. Based upon claim trends on public construction projects, this will likely be an area of increasing professional liability exposure.[54]

Library References

C.J.S. *Agency* §§ 344, 346–349, 352, 369, 384–385, 389, 392, 410; C.J.S. *Architects* §§ 4, 16–17, 19, 21–24, 32; C.J.S. *Contracts* §§ 11, 341, 343, 347, 355, 359, 407–420, 434–435, 557, 563–576, 579–581; C.J.S. *Negligence* §§ 163–164, 304, 311, 399, 532–535, 561; C.J.S. *Public Contracts* §§ 25–41; C.J.S. *United States* §§ 205, 212–215, 242, 270.

West's Key No. Digests, Contracts <KEY>196–199, 205, 213(2), 220, 236–246, 283; Indemnity <KEY>33(5); Labor and Employment <KEY>2586; Negligence <KEY>322, 1204(4); Principal and Agent <KEY>136; Public Contracts <KEY>19–33; States <KEY>188; Torts <KEY>242; United States <KEY>122.

III. Completion and Commissioning Phase: Roles and Responsibilities[55]

As a typical construction project nears completion, a heightened sense of urgency tends to envelope all parties. For the contractor, schedule and budget concerns are approaching the day of reckoning. The owner begins to realize that it will soon take possession of the facility and assume responsibility for operation and maintenance. The design professional is increasingly called upon to exercise its role as unbiased arbiter, making decisions of great consequence to the owner and the contractor.

---

50. *See* The Massachusetts False Claims Act, Mass. Gen. Laws ch. 12, § 5B (2000).
51. Boisjoly v. Morton Thiokol, Inc., 706 F.Supp. 795, 809 (D. Utah 1988).
52. *See, e.g.,* The Massachusetts False Claims Act, Mass. Gen. Laws ch. 12, § 5B (2000).
53. *See, e.g.,* David J. Hatem, *Federal and Massachusetts False Claims Act Exposure for Design and Construction Management Professionals*, The CA/T Professional Liability Reporter 1 (July 2001).
54. *See, e.g.,* Erick M. Kerness, *False Claims Can Bring Real Pain*, Design-Build (Oct. 2004).
55. I am indebted to my partner, David H. Corkum, for his significant contribution to this portion of Chapter 4.

Moreover, new players often emerge. For example, equipment vendors planning to start up, test, and obtain sign off on their particular contributions to the facility; the start-up group assuring that the various pieces of equipment are properly integrated; and regulatory agencies determining compliance with permits and regulations.

Clearly, a project's completion phase requires the close cooperation of all the parties involved as well as a mutual understanding of each party's responsibility. The design professional is often in a position of orchestrating the actions of the various parties and requiring close adherence to contractual terms, with the flexibility to adapt to unforeseen circumstances. In the midst of this flurry of activity, meticulous documentation of progress, accomplishments, and commitments will help to insulate the design professional from subsequent liability.

The completion phase begins with the contractor's request for a determination of substantial completion and runs through final completion, the warranty period, and the resolution of any contractor/owner claims. The completion phase exposes the design professional to certain unique liabilities that must be recognized and managed. The theories of liability that a project owner or contractor might assert against the design professional are generally the same as discussed earlier in this chapter, namely negligence, negligent misrepresentation, and breach of contract. Other more creative theories include tortious interference with a contractual relationship, defamation, and unfair business practices. Proper planning and preparation with experienced personnel and sufficient resources will allow the design professional to successfully navigate the conclusion of the project.

For a design-bid-build construction project where an owner has opted to employ standard form contract agreements from a family of documents such as AIA or EJCDC, it is less likely that a misalignment or misunderstanding of roles and responsibilities exists. The owner-design professional agreement seldom addresses the specific responsibilities of the design professional during the completion phase of a construction contract. However, the owner-contractor agreement likely contains a detailed listing of the owner's and contractor's expectation of the design professional's role. Accordingly, the design professional needs to understand and plan its participation with that specific role in mind. Problems often arise when the past experiences of one party on prior projects, or based on the course of conduct during the present project, lead to erroneous expectations of the design professional's role and scope of authority. Not only must the design professional closely adhere to the procedure and time constraints provided, but successful project completion also requires that it be in a position to encourage and in some cases enforce both the owner's and the contractor's compliance with their respective obligations.

A. *Substantial Completion*

In addition to being an important symbolic achievement, "Substantial Completion" is a critical contractual milestone. Often found as a defined term in the construction contract's general conditions, many state statutes and the Federal

Acquisition Regulations also choose to define substantial completion for publicly procured construction projects.[56] The design professional needs to understand the interplay between the statutory and contractual provisions on those projects. Generally, substantial completion means that the owner has taken beneficial use of the project.

Substantial completion is closely linked with the common law doctrine of substantial performance, which recognizes that contracts are rarely if ever performed perfectly and attempts to protect the promisor (contractor) when it does not perform exactly as promised. The seminal case describing the doctrine of substantial completion is *Jacobs & Young v. Kent*.[57] A general contractor's plumbing subcontractor installed an adequate piping material that was not the brand called for in the owner's material specifications. When the owner learned of this noncompliance, it directed the contractor to replace the piping with the specified "Reading" pipe, notwithstanding the fact that the house was completed, the pipe was encased in the finished walls, and the owner was occupying the house. Because of the noncompliant pipe, the owner refused to make its final payment to the contractor, giving rise to a lawsuit in which the owner prevailed at trial. On appeal, the trial court's decision was overturned, and Judge Cardozo explained the reversal as follows:

> We must weigh the purpose to be served, the desire to be gratified, the excuse for deviation from the letter, the cruelty of enforced adherence. Then only can we tell whether literal fulfillment is to be implied by law as a condition. This is not to say that the parties are not free by apt and certain words to effectuate a purpose that performance of every term shall be a condition of recovery. That question is not here.[58]

The court found that while the owner was entitled to the difference in value, if any, between the noncompliant pipe and the specified pipe, to require replacement of the pipe from attic to cellar would be economic waste. Thus, the doctrine of substantial performance developed by the courts has been adopted "as an instrument of justice"[59] in order to protect a contractor from the harsh reality that contracts are usually not performed exactly as originally envisioned. The construction industry generally understands that after a contractor has achieved substantial completion on a project, it cannot, under the common law, be found to have materially breached a contract for some subsequent act or failure to act.[60] While achieving substantial completion in no way relieves a contractor from the obligation of full performance,

---

56. 48 C.F.R. § 552.270-4(k)(2005).
57. Jacobs & Young v. Kent, 230 N.Y. 239, 129 N.E. 889 (1921).
58. *Id.* at 243.
59. *Id.* at 245.
60. Huguet v. Musso P'ship, 509 So. 2d 91, 92 (La. Ct. App. 1987) ("the law is clear that a building contract may not be dissolved after substantial performance has been performed").

it does remove the severe material breach of contract action and its harsh remedies for the failure to fully perform. Most public contracts, however, and many form contracts such as the AIA documents, contain provisions that allow the owner to terminate the contractor either for cause or for convenience. These provisions tend to supplement and enlarge the owner's common law rights to end a contractual relationship whether or not a default has occurred.[61]

Achieving substantial completion on a project often affects several important processes; such as, the achievement of a bonus or the assessment of liquidated damages, commencement of the warranty and callback period, partial release of retention, countdown to final completion, and commencement of the running of statutes of limitations and/or repose.[62] Because of its significance in redefining the rights and responsibilities between the owner and contractor, the owner-contractor agreement generally provides a step-by-step procedure with specific time constraints for ushering a project into substantial completion and final completion.[63]

The design professional plays a key role in the determination and declaration that the contractor has achieved substantial completion. The determination of substantial completion on a particular project can range from a very objective to a very subjective exercise by the design professional. For example, the construction contract may define substantial completion as the issuance of a Certificate of Occupancy. Alternately and more frequently, the determination involves the design professional's substantive and subjective determination of the progress of the work. In either case, the design professional is responsible for making that determination and certifying that substantial completion has occurred.[64] Absent fraud or gross mistake, the owner-contractor construction contract generally states that the design professional's certification is final and binding. The design professional must exercise its duty with respect to this determination in a fair and impartial manner. The design professional must resist any pressure from the owner to delay or defer the declaration of substantial completion based upon the owner's desire to influence the outcome of bonus or liquidated damage obligations, or other improper influence. Likewise, the design professional must resist pressure to certify substantial completion motivated by the owner's desire to obtain a certificate of occupancy as soon as possible even though the certification may not be warranted. The former could expose the design professional to contractor claims of fraud, unfair business practices, or tortious interference, while the latter could expose the design professional to claims by third parties based on deficiencies in the work.[65]

---

61. *See, e.g.*, AIA Document A201-1997, Article 14.
62. *See, e.g.*, AIA Document A201-1997, ¶¶ 9.8.5, 12.2.2.
63. *See, e.g.*, AIA Document A201-1997, ¶ 9.10.
64. *See, e.g.*, AIA Document A201-1997, ¶ 9.8.4.
65. *See generally* Richard C. Lowe & Elise H. Walthall, *When Architects Withhold Certificates of Substantial Completion and Other Problems*, 19 CONSTRUCTION LAWYER 5 (Oct. 1999).

B. *Punchlists*

In the vernacular of the construction industry, a punchlist is a list of all incomplete or unsatisfactory elements of the constructed project. The punchlist is developed concurrently with the application for and declaration of substantial completion. The process of working off the punchlist items often becomes a ritualized event on the project, leading to a modification of the attitudes and dispositions of key players as the project approaches completion. Either the contractor or the design professional may prepare the punchlist. Its purpose is the documentation of the number, character, and value of incomplete work. A typical scenario begins with the contractor submitting a punchlist to the design professional; the design professional then augments and agrees to that list, and assigns estimated values for correcting or completing the listed items. The design professional then utilizes that list as the basis of subsequent inspections.

From a best practices standpoint, the seemingly simple task of making and keeping a list requires close management and control. A complex project will likely require multiple disciplines of engineers to participate in the development and inspection of numerous punchlist items, and the creation and revision of these punchlists can lead to confusion. Contractors tend to view the punchlist as a goal line and the addition of items to subsequent versions of a punchlist as the design professional's attempt to move that goal line. On public construction projects in at least one state, Mississippi, multiple lists are not allowed, and the owner is limited to two punchlists, one at substantial completion and one just prior to final completion.[66] Communication, coordination, planning, and scheduling help to avoid contractor hardships, yet allow the owner the assurance it needs that its contract is being completed in accordance with the contract. Very few punchlist disputes revolve around objectively measurable elements of the project such as missing doors, nonfunctioning lights, or torn carpets. However, disputes can easily arise out of differences in subjectively determined and aesthetically judged elements of the work.[67] The design professional's contractually designated role as sole and indisputable judge and final decision maker on all aesthetic elements during the course of a project carries over to its decisions on punchlist items. Moreover, terms such as "to the satisfaction of the engineer" transfer a broad, but not unfettered, license to the design professional to judge the quality of the work. As mentioned previously in this chapter, the design professional generally enjoys a degree of immunity from owner or contractor claims related to these types of decisions on punchlist items, as long as the design professional renders those decisions in good faith.

As with inspecting and rejecting the work discussed earlier in this chapter, the design professional must be mindful of the appropriateness of its acceptance/rejection criteria for punchlist work. The design professional should avoid over-inspection

---

66. MISS. CODE ANN. §31-7-13(t) (2006).
67. *See, e.g.,* Varner Constr. Co. v. Marrs, LEXIS 275 (Tenn. Ct. App. 2002).

that could lead to delays and or contractor claims of a constructive change or charges of bad faith.

Toward the end of a project when all parties focus on completion, the design professional needs to maintain its vigilance and resist pressure by the owner or contractor to exceed its authority in accepting or rejecting noncompliant work. On private construction projects, the design professional's agency relationship, supported by its apparent and/or implied authority, will impute all of the design professional acts, approvals, or acceptance to the owner unless explicitly prohibited by contract. Thus, a design professional's oral acceptance of work that the owner subsequently rejects will often result in a dispute between the owner and contractor. If the contractor prevails in that dispute, then the design professional will be exposed to owner claims of negligence for knowingly accepting noncompliant work. Misunderstandings related to apparent authority are less of a concern for the design professional on public construction projects. Federal Acquisition Regulations, reflected in contracts with the United States, provide specific limitations on the authority of the design professional to accept deviations from or make modifications to the contract documents.[68] Similarly, the terms and conditions of state construction contracts typically contain explicit limitations on design professional's authority in keeping with that state's procurement regulations.[69] Thus, a design professional's commitment to grant a change order or time extension or to accept noncompliant work is not binding on the public owner where the contract or applicable law provides that only the owner can approve changes in contract time and price. Accordingly, any contractor reliance on the design professional's commitment would be patently unreasonable, and a negligent misrepresentation claim by that contractor should fail.

C. *Final Completion*

Like the application for and achievement of substantial completion, the owner-contractor agreement generally spells out the steps and procedures to obtain final

---

68. 70 F.A.R 43.102 (2005) provides that: "(a) Only contracting officers acting within the scope of their authority are empowered to execute contract modifications on behalf of the Government. Other Government personnel shall not ... (1) Execute contract modifications; (2) Act in such a manner as to cause the contractor to believe that they have authority to bind the Government; or (3) Direct or encourage the contractor to perform work that should be the subject of a contract modification ...."

69. MASS. GEN. LAWS ch. 30, § 39I (1960) provides: "Every contractor having a contract for the construction, alteration, maintenance, repair or demolition of, or addition to, any public building or public works for the commonwealth, or of any political subdivision thereof, shall perform all the work required by such contract in conformity with the plans and specifications contained therein. **No willful and substantial deviation from said plans and specifications shall be made unless authorized in writing by the awarding authority or by the engineer or architect in charge of the work who is duly authorized by the awarding authority to approve such deviations**.... " (Emphasis added).

completion. Most agreements call for the design professional to issue a Certificate of Final Completion or a final Certificate for Payment. For example, AIA Document A201, § 9.10.1 states:

> [U]pon receipt of an Application for Final Payment the Architect will promptly make such inspection and when the Architect finds the Work acceptable under the Contract Documents and the Contract fully performed, the Architect will promptly issue a final Certificate for Payment stating that to the best of the Architect's knowledge, information and belief, and on the basis of the Architect's on-site visits and inspections, the work has been completed in accordance with the terms and conditions of the Contract Documents and that the entire balance found to be due to the Contractor and noted in the final Certificate is due and payable.

Under this very typical certification, the design professional represents to the owner that the contract is complete. This includes not just the physical aspects of the project, but all of the paperwork, record drawings, photo documentation, maintenance manuals, lien waivers, etc., as required by the contract. This certificate is a document that the owner, lending institutions, the contractor's surety, and regulatory agencies will rely upon and take certain actions; such as releasing bonds. That reliance, particularly by the owner but also by third parties such as the surety in certain cases, exposes the design professional to liability for a negligently issued certification.[70] In one example of improper certification, the court found an architect liable to a contractor's surety because the architect failed to obtain and review evidence that the contractor had paid suppliers and material men as required before releasing retainage. Once the contractor received the retainage, it defaulted and the surety was required to satisfy the outstanding claims.[71]

D. *Warranties*

In addition to the array of implied warranties, a contractor is usually required to provide certain express warranties. For example, AIA Document A201 requires the contractor to provide three basic warranties: (1) that the material and equipment incorporated into the project are new (unless otherwise allowed) and of good quality; (2) that the work will be free from defects; and (3) that the work will conform to the contract documents.[72] The technical specifications normally contain other special warranties for equipment and other incorporated elements of the project.

Warranty issues between the owner and contractor give rise to design professional liability in several different ways. For example, under the AIA family of documents, the architect is required to obtain special warranties for certain enumerated systems

---

70. *See, e.g.,* the British case of Sutcliffe v. Thacker, A.C. 727 (1974).
71. National Surety Corp. v. Malvaney, 72 So. 2d 424 (Miss. 1954).
72. *See, e.g.,* AIA Document A201-1997, ¶ 3.5.

and equipment from the contractor.[73] As discussed above, the architect's approval of a Certificate of Final Completion means that the contractor has performed all of the work, including supplying warranties in accordance with the contract documents. The erroneous certification that the contractor supplied the required warranties could result in allegations of negligence or negligent misrepresentation against the design professional if the owner is unable to obtain satisfactory resolution of a construction defect, for example, a leaking roof or underperforming piece of equipment.

E. *Project Close-Out Documentation*

Close-out documentation is considered the "dotting of i's and crossing of t's" of the construction project. In addition to special equipment warrantees and lien waivers, the contractor is required to submit maintenance manuals, operating instructions, and as-built or record drawings at project completion. Collecting and reviewing these documents for compliance is a tedious and thankless task. If all goes well with the project, these documents will never be removed from their file folder. However, if a subcontractor or material supplier files a lien or an insurance issue arises, the design professional may find itself exposed to liability to the owner for failing to secure these documents or for accepting defective documents.

F. *Support to Owner in Evaluation/Defense of Contractor Claims*

Throughout the course of a construction project, the design professional is required to make decisions that affect the economic balance between the owner and contractor. Additionally the owner-contractor agreement often assigns the design professional the role of arbiter of construction claims and disputes between the owner and contractor. In this role, the design professional must exercise its decision-making power in contrast to its responsibility of interpreting the contract in a fair, unbiased manner, based on the facts and circumstances. Clearly, however, the resolution of disputes often results in one embittered party who may seek redress from the design professional for its loss. This problem can be exacerbated when owner or contractor suspects that the design professional's decision is influenced by a desire to cover up for defects in the design documents. Moreover, if the contractor files litigation or arbitration challenging the design professional's decisions, that will likely occur after project completion. At that point the owner will often seek the design professional's assistance as an advocate in its defense of contractor claims at which point the cloak of arbitral immunity is, of course, removed.

---

73. *See, e.g.*, AIA Document A201-1997, ¶ 4.2.9.

CHAPTER 5

Design Professionals and the Design-Build Project

PETER HALLS, ESQ.
BERNARD (B.J.) NODZON, ESQ.
NATHAN A. BRENNAMAN, ESQ.

I. Introduction

The design-build delivery system is becoming increasingly popular.[1] Some commentators predict that design-build projects will be the predominant form of delivery system within the next few years.[2] Many states throughout the country have passed design-build legislation, allowing state agencies to enter into design-build contracts with construction professionals.[3]

As the popularity of design-build increases, so will the need for design professionals to be familiar with their role and potential liabilities in the design-build process. This chapter will examine the different roles of design professionals in design-build delivery systems.

Library References

C.J.S. *States* § 261.
West's Key No. Digests, States <KEY>86.

---

1. 2 BRUNER & O'CONNOR ON CONSTRUCTION LAW § 6:22 (2005 update). *See also* A.H. Gaede, Jr., *Risk Management in Design-Build*, 467 PLI/REAL ESTATE 319 (Apr. 2001) (noting that the design-build delivery system continues to be one of the fastest developing construction project delivery systems and is expected to continue its growth throughout the twenty-first century).

2. *See* G. WILLIAM QUATMAN, DESIGN-BUILD FOR THE DESIGN PROFESSIONAL, § 1.07 (2001) (reporting that design-build will soon be the dominant delivery method).

3. According to the Design-Build Institute of America (DBIA) as many as thirty states introduced new design-build legislation in 2005 with half of those states actually enacting some

II. Differing Roles for Design Professionals in Design-Build Projects

This section introduces the four primary roles design professionals can play in design-build projects: 1) acting as a design-builder; 2) acting as a subcontractor to a design-builder; 3) participating with another company in a joint venture or other business entity as a design-builder; and 4) serving as a bridging consultant.

A. *Design-Build Led by the Design Professional*

When the design professional is the design-builder, the design professional contracts directly with the owner and is responsible for all aspects of the project, including the scheduling, design, estimating, construction, and project management.[4] Design professionals that act as design-builders often do so by forming separate but related construction companies that in turn subcontract the design services to the design professional. Alternatively, the design professional contracts as the design-builder

---

form of design-build law. *See* www.dbia.org. Minnesota enacted new design-build legislation for use on state and university projects. *See* MINN. STAT. CHS. 16B and 16C (2005). Arizona, a state which has been labeled by some as the leader in design-build legislation, added statutes allowing the Arizona Department of Transportation to enter into design-build contracts. *See* ARIZ. REV. STAT. § 28-6924–28-7367 (1997). In Arkansas, the legislature enacted legislation on the use of design-build in K-12 school projects. *See* ARK. CODE § 19-11-807 (2005). Similarly, in Indiana, public education and state agencies may now use the design-build delivery method. *See* IND. CODE § 5-30 (2005). Other states to enact some form of design-build legislation in 2005 include Georgia, Illinois, Louisiana, Mississippi, Montana, Nevada, New Hampshire, New Mexico, North Dakota, Oklahoma, Virginia, Washington, and West Virginia. *See* www.dbia.org. Additionally, design-build legislation for use on federal highway projects was passed by the United States Congress in 2005. *See* P.L. 109-59, 119 STAT. 1764 (Aug. 10, 2005) § 1503.

4. It should be noted that designer-led design-build teams are somewhat rare. *See* Friedlander, *Designer-Led Design-Build Teams*, CONSTRUCTION BRIEFINGS No. 97-9 (Aug. 1997), which notes:

> In design-build contracting, in which one organization is retained for both design and construction, it is rare for a design professional—the architect or engineer (A/E)—to contract directly with the owner both to design and to construct a project. Except for heavy industrial projects using the "engineer, procure, and construct" design-build method, an area dominated by the largest engineering firms, the lead member of the design-build team usually is the contractor or the developer. The situation may be simply a historical accident, or it may have occurred because contractors have greater financial capitalization, are more entrepreneurial, or are perceived as more disciplined about issues of cost and constructability than design professionals. Whatever the reason, the construction community seems to have adopted the assumption that a contractor should lead the design-build team with the A/E as a subcontractor or joint venture partner.

and performs the design with its own employees, possibly subcontracting certain engineering services to separate engineering firms.

B. *Design Professional as a Subcontractor to Design-Builder*

Most commonly, the design professional participates in design-build projects as a subcontractor to the design-builder. The design professional is designing the project as it would on a traditional design-bid-build project, but now the designer answers to the design-builder instead of the owner. Because a design-builder's interests are not identical to those of an owner, some designers report pressure, either real or imagined, to compromise their professional judgment in favor of cheaper construction.

C. *Design Professional as a Member of a Joint Venture or Other Business Entity*

Another means for design professionals to participate in design-build projects is to participate with another company, often a contractor, in the formation of a joint venture, new corporation, limited liability company, or other legal entity that acts as the design-builder. Under this model, the architect and the contractor are either joint venturers or partners, members, or shareholders of an entirely new contractual or legal entity.[5] The joint venture or newly formed entity holds the design-build contract with the owner and is responsible for both the design and construction of the project. Participating in design-build projects in this manner allows design professionals to share in anticipated profits, but also requires the design professional to share the risks. In addition, the design professional might be unable to act as the design-builder itself because bonds may be required and the design professional may not have sufficient bonding capacity. By joint venturing or forming a new legal entity with a contractor possessing the requisite bonding capacity, the design professional can meet the bonding requirements and participate in the design-builder's prospective profits and control of the project. Further, if the new legal entity is a corporation, limited liability company, or a limited partnership and is properly founded and maintained, the design professional can shield itself from the liability of the design-builder. In some cases, owners may find it difficult to recover from the contractor and design professional when they form a new entity.[6] They may not have sufficiently capitalized these entities to pay significant damages to the owner, which may shield the contractor and design professional from direct liability.[7] As a result, owners and their lenders often insist that the design firm and the contractor guaranty the liabilities of the new entity.

---

5. *See generally* 2 BRUNER & O'CONNOR ON CONSTRUCTION LAW § 6.37 (2005 update).

6. *See id.* (discussing the difficulties that may arise in attempting to recover from the design-builder that operates as a separate entity).

7. *Id.*

D. *Bridging*

Under the bridging approach to design-build projects, the owner hires an architect or engineer as a bridging consultant to develop a program and partially design the project. The level of the bridging design varies by project, but generally ranges from a detailed program to schematic design level documents. The owner then uses the bridging consultant's partial design in a competitive bidding or request for proposal process to select the design-builder for the project. After the owner has selected a design-builder, the owner's architect or engineer usually remains involved with the project and might provide services, such as reviewing the design-builder's design, construction work, pay applications, and performing other administrative duties on behalf of the owner.

Library References

C.J.S. *Contracts* §§ 2–3, 9, 12.
West's Key No. Digests, Contracts <KEY>1.

III. General Considerations

No matter what role the design professional plays, certain aspects of the design-build process impose increased risks for design professionals. This section will discuss some aspects of design-build projects that generally heighten risks for design professionals. Subsequent sections will examine the liability of design professionals in the varying roles they play in the design-build process.

A. *Fast-Track Construction*

Design-build projects are typically constructed on a fast-track basis. Under the fast-track construction process, construction starts before the entire design is completed.[8] Essentially, the construction phase and the design phase overlap. The theory is that this overlap will shorten the overall schedule and save time in the end. As one commentator notes, "[t]his phased approach to construction has been used in private construction for years as a means of accelerating construction."[9] To accomplish the fast-track procedure, construction work is often divided into "packages." Once the design of a particular construction package is completed, construction begins while the design of the remaining packages continues. Parties can use fast-track construction on any type of project, but it is particularly suited to design-build projects.

There are some disadvantages to the fast-track process. Completing the overall design for a project before construction begins allows the design professional to discover errors that were not evident earlier. Desirable refinements in the later

---

8. Loulakis *et al*, *Design-Build*, CONSTRUCTION BRIEFINGS No. 2003-12 (Dec. 2003); *see also* Nash, Jr. & Love, Jr., *Innovations in Federal Government Construction Contracting*, 45 GEO. WASH. L. REV. 309, 384 (1977).

9. Nash, *supra* note 8, at 384.

phases of design can require adjustments to design decisions made earlier in the project. When construction begins before the overall design is completed, there is less opportunity for the discovery and correction of design errors before damage results, and it may be difficult or impossible to make adjustments in constructed work to accommodate later refinements. Owners or even design-builders sometimes view the early design's failure to accommodate these later desired refinements as negligence on the part of the design professional.

Finally, scheduling and coordinating the project is often very difficult because the construction progress schedule must be modified to account for design changes.[10] As one court explained:

> [The Owner's] fast track approach resulted in frequent and significant modifications to both the architectural plans and the progress schedule. These modifications prevented [the contractor] from proceeding in the orderly sequence that was anticipated by the original CPM [Critical Path Method Schedule], thereby causing [the contractor] to incur significant additional costs due to inefficiency and under-utilized labor.[11]

In addition, it is more expensive to make design changes after completion of some phases of construction, increasing the amount of damages for which a design professional is liable if negligent design caused the problem. As a result, the fast-track method, often used on design-build projects, can potentially lead to increased design liability for the designer.[12]

B. *Less Complete Design*

Often the design provided by a design-builder is less complete than one on other types of projects. In the traditional design-bid-build project delivery system, because the entity providing the design does not contract directly with the entity that uses the design to construct the project, there can be a greater need to communicate the design with complete and detailed design documents than on a design-build project. One of the advantages for design-build delivery is that there are cost savings because the design documents do not need to be as complete. On the other hand, the less

---

10. *See* Marriott Corp. v. Dasta Const. Co., 26 F.3d 1057, 1059 (11th Cir. 1994) (discussing the fast-track construction method and listing some advantages and disadvantages).

11. *Id.*

12. *See* O'Connor, *The Rights and Responsibilities of Design Professionals, Part II*, CONSTRUCTION BRIEFINGS 2002-5 (May 2002) (stating that the fast-track system along with the increasing popularity of the design-build delivery method, has made "the practice of architecture and engineering more dangerous than ever"); *see also* Block, *As the Walls Came Tumbling Down: Architects Expanded Liability under Design-Build Construction Contracting*, 17 J. MARSHALL L. REV. 1 (1984); Nichwitz, *The Crumbling Tower of Architectural Immunity: Evolution and Expansion of Liability to Third Parties*, 45 OHIO ST. L.J. 217 (1984).

complete the documents, the less documentation there is of the designer's design intent and the greater the risk there is of a dispute in the event there is a problem.

Library References

C.J.S. *Contracts* §§ 407–420, 434–435; C.J.S. *Public Contracts* § 13; C.J.S. *States* § 272.

West's Key No. Digests, Contracts <KEY>236–246; Public Contracts <KEY>18; States <KEY>106.

IV. Design-Build Led by Design Professional

A. *Reduction of Claims by Practice*

Although some aspects of design-build may have greater risks than traditional work, there are some hard data to suggest the opposite, especially in designer-led design-build delivery.[13]

While designers acting as design-builders necessarily take on more responsibility and therefore more risk than when they perform only design services, this does not necessarily mean a consequent increase in claims. There are fundamental advantages to the design-build structure that can reduce claims. First, design professionals who are design-builders may be able to catch and resolve problems early in the process before they cause serious damage. Second, the design professional acting as design-builder has more control over the resolution of the problem and the cost thereof. Third, due to the absence of the traditionally adversarial relationship between the architect and the contractor, design errors are not as frequently exploited by contractors and turned into claims with large mark-ups, although that may still happen at the subcontractor level.

Therefore, due to the inherent structure of the design-build project delivery method, there is less finger pointing, and many traditional architect-contractor claims and grievances will be resolved without involving the owner.

B. *Warranty of Design*

Generally, courts do not hold architects and engineers to a standard of perfection, but are only required to exercise reasonable care and skill.[14] Professional liability insurance for design professionals typically excludes coverage for any liability beyond the reasonable care standard. Consequently, design professionals generally avoid

---

13. *See* QUATMAN, *supra* note 2, at § 21.01. *See generally* Konchar & Sanvido, *Comparison of U.S. Project Delivery Systems*, 124 J. CONSTR. ENG. & MGT. 435 (Nov./Dec. 1998).

14. *See generally* James Acret, *Architect and Engineer Malpractice*, 3 CONSTRUCTION LAW DIGESTS § 25:17 (updated Nov. 2005); *see also* Corcoran v. Sanner, 854 P.2d 1376, 1378 (Colo. Ct. App. 1993); Fox v. Stanley J. Howard & Assocs., Inc., 309 N.W.2d 520, 525 (Iowa Ct. App. 1981) (holding architect will be held liable for negligence if it fails to exercise ordinary skill of its profession); Klein v. Catalano, 437 N.E.2d 514, 516 (Mass. 1982) (stating architect

warranting that their design will be free from defects. Design-builders, on the other hand, often provide explicit contractual guaranties to the owner that the design will be free from defects. The most recent American Institute of Architects (AIA) form of design-build contract, AIA Document A141-2004, contains an express contractual guaranty of the design by the design-builder.[15]

Design-builders are willing to provide a warranty that the design is free from defects in part because owners often insist upon it, and in part because many courts would imply that liability unless the design-builder clearly disclaims it. Although it may be possible to avoid this implied liability by disclaiming it with a carefully drafted contract clause, those clauses are not always easy to negotiate.

There are two types of implied warranties of design that can apply to design-builders. The first is an implied warranty of fitness for a particular purpose. Under this doctrine, courts read into every design-build contract an obligation by the design-builder that the building will be fit for the owner's intended use.[16] Several courts have found design-builders liable under this theory, despite the lack of any express warranty.[17] For example, in *Robertson Lumber Co. v. Stephens Farmers Cooperative Elevator Co.*,[18] a property owner entered into a design-build contract for the design and construction of a grain storage facility. The storage facility, however, collapsed after the owner began storing grain in it. The court held the contractor liable, stating that the design-build contractor impliedly warranted that the building would be fit for its intended purpose.[19] The court noted that the owner furnished no plans, design, specifications, details, or blueprints and relied upon the experience and skill of the design-build contractor.[20] Under these facts, the court found that the design-builder impliedly warranted that the building would be fit for its intended purpose.

Similarly, in *O'Dell v. Custom Builders Corp.*,[21] the owner sued a design-build contractor for damages after the foundation of its home cracked. A jury determined

---

impliedly promises to exercise that standard of reasonable care required of members of the profession); Prichard Bros., Inc. v. Grady Co., 436 N.W.2d 460, 463 (Minn. Ct. App. 1989) (standard of reasonable care that applies to conduct of architects is same as that applied to lawyers, doctors, engineers, and other professionals engaged in furnishing skilled services for compensation).

15. *See* AIA Document A141-2004, Exhibit A, Terms and Conditions, § A.1.1.6.
16. *See, e.g.,* Robertson Lumber Co. v. Stephens Farmers Coop. Elevator Co., 143 N.W.2d 622, 625-27 (Minn. 1966).
17. *See id. See also* O'Dell v. Custom Builders Corp., 560 S.W.2d 862, 867 (Mo. 1978) (holding design-builder impliedly warrants the building will be fit for its intended purpose); Prier v. Refrigeration Eng'g, Co., 442 P.2d 621 (Wash. 1968) (same).
18. 143 N.W.2d 622 (Minn. 1966).
19. *Id.* at 625.
20. 143 N.W.2d 622 (Minn. 1966).
21. 560 S.W.2d 862 (Mo. 1978).

that the contractor breached its duty to design the project so that it would be fit for the owner's intended use. The Supreme Court of Missouri affirmed, stating that the parties' contract impliedly required the design-build contractor to construct the building fit for its intended use.[22]

Prier v. Refrigeration Engineering Co.[23] reached a similar conclusion. There, the owner hired a contractor to design and install a portion of an ice rink. The contractor's design melted the ice, forcing the rink to close. The owner sued the contractor and the district court awarded the owner a portion of its damages. The owner appealed and the Washington Supreme Court reversed, holding that the owner was entitled to greater damages.[24] While the court's decision focused on damages, it stated that a design-builder who provides drawings and specifications for the construction of a building or system impliedly warrants that the system or building will be fit for its intended purpose.

The second type of implied warranty is the implied warranty of the quality of the drawings and specifications. In traditional construction where the design professional prepares drawings and specifications for the owner and the owner, in turn, provides drawings and specifications to the contractor, the courts have held that the owner warrants the accuracy and adequacy of those drawings and specifications. This concept is widely known as the *Spearin* doctrine, stemming from a famous U.S. Supreme Court decision of that name from the early twentieth century.[25] The *Spearin* doctrine holds that when an owner furnishes detailed drawings and specifications to a contractor, the owner impliedly warrants that those documents are accurate and suitable for their intended use. The principle is that a contractor is entitled to rely on the accuracy of the drawings and specifications provided by the owner, and later, if there is a problem caused by an error in the drawings and specifications, the owner, and not the contractor, will be liable for the consequences.

Design-build turns this famous doctrine on its head.[26] The owner is not furnishing any design—the design-builder furnishes the drawings and specifications to the owner. This not only eliminates the owner's implied warranty, but courts have also held that the design-builder impliedly warrants to the owner that the drawings and specifications are free from defects. Some have aptly dubbed this shift in liability from the owner to the design-builder the "reverse *Spearin* doctrine."

Designer-led design-builders must appreciate the increased risks they face in design-build. Compounding these risks is the fact that the designer's typical professional liability insurance program will not provide coverage for such a warranty of

---

22. *Id.* at 867.
23. 442 P.2d 621 (Wash. 1968).
24. *Id.* at 624.
25. *See* United States v. Spearin, 248 U.S. 132 (1918).
26. *See generally* Robert L. Meyers, III, CONSTRUCTION BRIEFINGS NO. 2004-5, *Owner's Design Liability Revisited: Spearin Today* (May 2004).

defect-free design, and there is usually no general liability insurance coverage available for the correction of one's own defective workmanship.[27]

However, there are some strategies to help limit this risk. First, design-builders can attempt to disclaim the implied warranties of defect-free design and fitness for intended use and negotiate express warranties that apply only to the construction work and not to the design. Second, one can attempt to buy insurance covering the risk of a claim under a warranty of the design beyond the standard of reasonable care. Finally, for any portion of the design done by others, the designer-led design-builders can try to pass whatever warranties that exist in the prime contract through to the design contracts and design-build subcontracts with a suitable indemnification clause.[28]

C. *Strict Liability for the Design-Builder*

Manufacturers of defective products are strictly liable to consumers if their defective products pose an unreasonable risk to the user, consumer, or property. This liability applies regardless of the level of care used by the manufacturer. This law has developed over time to allow consumers to recover from manufacturers with whom they do not have a contract, since the product may have proceeded down a chain of wholesalers, distributors, and retailers.

With the increased use of design-build, there has been speculation in recent years as to whether the courts would apply this strict liability standard to building defects causing injuries on design-build projects. Unlike traditional delivery methods, where the roles and responsibilities are independent and compartmentalized, design-builders arguably provide services more like those of a manufacturer, which creates a product.

Fortunately for design-builders, thus far, the courts have not held design-builders strictly liable in tort for construction defects.

D. *Contractor-Type Liability*

Designers acting as design-builders will have the same liability as a general contractor—liability far beyond what designers are accustomed to or for which they are insured. Contractors typically guaranty to provide work free of defects and have responsibility for on-site safety, disposal of materials, scheduling multiple subcontractors and suppliers, permitting and licensing typically provided by contractors, and testing and inspections of construction work.[29]

---

27. *See generally* Michael G. Welbel, *Insurance Coverage for Design Errors: The Owner's Perspective*, ABA, INSURANCE COVERAGE FOR DEFECTIVE CONSTRUCTION II (June 3-5, 1998) (outlining the types of exclusions commonly found in professional liability policies); Robert M. Chemmers & James A. Knox, *When Designs Don't Work: Professional Liability Coverage for Architects and Engineers,* 15 CONSTRUCTION LAWYER 17, 22 (Aug. 1995);

28. *See* QUATMAN, *supra* note 2, at § 21.04.

29. *See generally* Thomas H. Asselin & L. Bruce Stout, *Legal Exposure of the Design/Build Participants: The View of the General Contractor*, 15 CONSTRUCTION LAWYER 8, 8-10 (Aug. 1995).

E. *Insurance Coverage and Bonding*

A design firm's insurance needs will depend on whether it is working as a prime contractor, subcontractor, or joint venturer. While insurance is relatively straightforward for design firms providing design services as a subcontractor to a design-builder, insurance issues arise when the design firm is responsible for both design and construction. In order to adequately mitigate the additional risks of design-build construction, it is important for a design professional to understand the differences between contractor's and architect/engineer's insurance. For example, professional liability insurance might exclude coverage for services beyond normal design services. Acting as a design-builder could jeopardize coverage, and an appropriate endorsement may be necessary. For example, a design firm's insurance may exclude coverage for joint ventures. The designer should implement a careful review of its insurance to adapt to being a design-builder.

Design professionals acting as design-builders must become familiar with bonds. Nearly all public projects require bonding. Additionally, private owners often require bonding. Design professionals new to design-build often have never provided bonds and do not have bonding capacity or an established relationship with a surety. Design professionals should also be aware that typically sureties require personal guaranties and indemnities.

F. *Increased ADA Liability*

Design professionals that act as design-builders risk greater exposure to claims under the Americans with Disabilities Act (ADA). Section 302(a) provides that:

> No individual shall be discriminated against on the basis of disability in the full and equal enjoyment of the goods, services, facilities, privileges, advantages, or accommodations of any place of public accommodation by any person who owns, leases (or leases to), or operates a place of public accommodation.[30]

Since the legislature enacted the ADA in 1990, there has been substantial litigation over whether the courts can hold architects and contractors directly liable for violations under this federal statute. Today, the courts are split on whether architects, contractors, and design-builders are liable under the ADA. This exposure turns on § 303(a), which provides:

> Except as provided in subsection (b), as applied to public accommodations and commercial facilities, discrimination for purposes of section 302(a) includes—(1) a failure to ***design and construct facilities*** for first occupancy ... that are readily accessible to and usable by individuals with

---

30. 42 U.S.C. § 12182(a) (1990).

disabilities, except where an entity can demonstrate that it is structurally impracticable to meet the requirements of such subsection in accordance with standards set forth or incorporated by reference in regulations issued under this title.[31]

Design professionals and contractors operating in their traditional roles have argued against liability under the ADA because they do not perform both design *and* construction and therefore do not fall under this statute. Some courts have embraced this argument, agreeing that this phrase of the statute is clearly "conjunctive."[32]

Design-builders, however, are unable to make this argument. Therefore, courts have found that this "design and construct facilities" language of the statute allows design-builders to be sued under the statute.[33] For instance, in *United States of America v. Days Inn*,[34] the court held that it mattered not whether the phrase "design and construct" was conjunctive or disjunctive, because the defendants were involved in both the design and the construction of the facility.[35]

Design-builders can argue that they are not within the scope of the persons responsible for meeting the ADA requirements as defined in the statute, but not all courts accept this argument. Section 303(a) specifically references, and is limited by, § 302(a) (quoted above), which obligates only the "person who owns, leases (or leases to), or operates a place of public accommodation" to not discriminate. Since the enactment of the ADA, many builders and architects have argued that they are not liable under the ADA because design builders do not own, lease, or operate a place of public accommodation.[36] The hallmark case supporting this proposition is *Paralyzed Veterans of America v. Ellerbe Becket*.[37] In that case, the court held that the design professional could not be liable because "the limitation in § 302 to owners, operators, and lessors also applies to § 303 and thereby excludes architects from liability under the section."[38]

Other courts, however, have disagreed. For instance, in *Johanson v. Ellerbe Becket*,[39] the court denied the design professional's motion to dismiss because "limiting § 303 to the parties covered by § 302 would effectively eliminate § 302's coverage of commercial facilities.... If architects are not liable under the ADA, then

---

31. 42 U.S.C. § 12183(a) (emphasis added) (1990).
32. Paralyzed Veterans of Am. v. Ellerbe Becket, 945 F. Supp. 1, 2 (D.D.C. 1996); *but see* Johanson v. Huizenga Holdings, Inc., 963 F. Supp. 1175, 1177-78 (S.D. Fla. 1997).
33. *See, e.g.,* United States v. Days Inn of Am., Inc., 997 F. Supp. 1080, 1084 (C.D. Ill. 1998).
34. *Id.*
35. *Id.* at 1084.
36. *See Paralyzed Veterans*, 945 F. Supp. at 2; United States v. Days Inn of Am., *Inc.*, 22 F. Supp. 2d 612, 615 (E.D. Ky. 1998).
37. 945 F. Supp. 1 (D.D.C. 1996).
38. *Paralyzed Veterans*, 945 F. Supp. at 2.
39. *Johanson*, 963 F. Supp. 1175 (S.D. Fla. 1997).

it is conceivable that no entity would be liable for construction of a new commercial facility which violates the ADA."[40]

Thus, because of the volatility of the law in this area, design professionals should be mindful of the added exposure to liability for violations of the ADA that may arise when operating as a design-builder.

G. *Economic Loss Doctrine Eroded*

Under a traditional delivery method, an architect contracts with an owner but not with contractors or subcontractors. Consequently, claims against architects by contractors or subcontractors are limited in many states by the economic loss doctrine. Under this doctrine, a contractor or subcontractor can recover only for damages to persons or property, and cannot recover purely economic loss without privity of contract. Thus, if the floor of a building collapses due to an architect's negligent design and causes injury to a contractor's workers, the contractor would clearly have a claim against the architect even though the contractor did not have a contract with the architect. However, in jurisdictions that enforce the economic loss doctrine, if a project is delayed due to a design error, a contractor or subcontractor cannot file suit against the architect for the architect's negligence because the contractor and subcontractors do not have a contract with the architect and suffered only economic loss—not bodily injury or property damage.

Design professionals in design-build construction are therefore open to a wider range of claims for purely economic damages in many jurisdictions. Indeed, designer-led design builders may have contracts with numerous contractors and suppliers. Consequently, because there is privity of contract with these contractors and subcontractors, they can now sue the design professional for economic losses, such as delay damages and lost profits, even in jurisdictions recognizing the economic loss doctrine.

H. *Contractor Licensing*

Many jurisdictions require contractors to be licensed. A design-builder, because he or she undertakes responsibility for construction, must generally be licensed or a contractor in those jurisdictions. This requirement can easily surprise a design professional and the failure to be in complete compliance can be devastating.

Generally, jurisdictions that require a contractor's license will find that merely bidding or submitting an offer to work as a contractor or design-builder is a violation of their licensing laws. The consequences can be severe. Many jurisdictions hold that a contractor without a license has no right to payment for work performed, even under theories of *quantum meruit* or unjust enrichment.[41] California law even

---

40. *Johanson*, 963 F. Supp. at 1178; *see also* United States v. Ellerbe Becket, 976 F. Supp. 1262, 1267-68 (D. Minn. 1997).

41. 5 BRUNER & O'CONNOR ON CONSTRUCTION LAW § 16:13 (2005 update).

provides that an owner can sue and recover payments previously made to the contractor.[42]

Library References

C.J.S. *Architects* §§ 1–4, 6–9, 16–17, 19, 21–25, 32; C.J.S. *Contracts* §§ 341, 343, 347, 359; C.J.S. *Damages* §§ 53, 58–61; C.J.S. *Insurance* §§ 54, 956–957; C.J.S. *Negligence* §§ 162–164; C.J.S. *Public Contracts* §§ 10, 25–28, 42, 45, 48; C.J.S. *States* §§ 289–294, 305–310; C.J.S. *Torts* § 26; C.J.S. *United States* §§ 117, 120–122, 127, 130, 139–148.

West's Key No. Digests, Civil Rights <KEY>1021, 1338; Contracts <KEY>196, 205; Damages <KEY>40; Insurance <KEY>2391(7); Licenses <KEY>11(1), 39.40; Negligence <KEY>321, 322; Public Contracts <KEY>16, 19, 41; States <KEY>101, 104, 107; United States <KEY>67, 70(29), 70(30), 73(17).

V. Contractor-Led Design-Builder

The contractor-led design-build approach is by far the most popular form of design-build delivery method. This section will explore the risks, roles, and responsibilities of design professionals in the contractor-led design-build process.

A. *Different Dynamics*

Although the contractor may have an opportunity to comment on the design, when the designer works for an owner, the owner makes the design decisions. The contractor does not control payment to the designer and the designer communicates directly with the owner. When the designer works for a contractor-led design builder, the dynamics are different. Some designers find that a contractor-led design-builder is more willing than an owner to challenge the design as being excessive. Now the party that has committed to a price to perform the construction work is the one writing the checks to the designer and is between the designer and the owner in the chain of communication. However, as long as the designer and contractor-led design-builder maintain their professionalism, having the design challenged should result in a better design without compromising safety.

Another result of the different dynamics is that with a direct contractual relationship between the designer and the contractor-led design builder, issues between the two can often be resolved without involving the owner. This may help explain why designers appear to experience fewer claims on design-build projects.[43]

B. *Murder Clauses*

Murder clauses are onerous contract clauses that can create devastating results. A designer who is under contract to a design-builder might see a new set of murder

---

42. CAL. BUS. & PROF. CODE § 7031 (2003); *see also* MW Erectors, Inc. v. Niederhauser Ornamental & Metal Works Co., Inc., 36 Cal. 4th 412, 115 P.3d 41 (Cal. 2005).

43. *See supra* text accompanying note 13.

clauses compared to those an owner might attempt to impose. This section examines some of the onerous clauses that often appear in contractor-led design-builder's subcontracts.

1. *Pay-If-Paid or Pay-When-Paid Clauses in Design-Build Contracts*

Contractors will often include "pay-when-paid" or "pay-if-paid" clauses in their subcontracts. These clauses state that the contractor will not pay its subcontracts until the contractor receives payment from the owner. Typically, courts disfavor these clauses and strictly construe them against the contractor.[44] Generally, courts will enforce these clauses only if the provision makes payment contingent upon receipt of payment from the owner.[45] Nevertheless, design professionals should be aware of these clauses and avoid including them in their contracts.

2. *Warranty Clauses*

The law does not require that design professionals perform error-free services. Instead, design professionals are only required to comply with the professional negligence standard, meaning the design professional must perform its work to the minimally acceptable standard of care in the profession under the circumstances. The design-builder, however, will often make express warranties in its design-build contract with the owner, guaranteeing that certain aspects of the project—or even the entire project, including the design—will be error free. As referenced above, the AIA design-build forms, AIA Document A141-2004, include such a warranty. Under these forms, the design-builder expressly warrants that its work will be free of defects. The document defines the term "Work" to include both design and construction.[46] This contract provision thus expressly warrants an error free design—a warranty that is higher than the ordinary standard of negligence the law imposes upon design professionals. When design-builders warrant the design to be error free, they will have a strong incentive to impose the same standard on the designer in the subcontract with the designer.

---

44. *See, e.g.,* Bonavist v. Inner City Carpentry, Inc., 244 F. Supp. 2d 154 (E.D. N.Y. 2003) (stating pay-if-paid or pay-when-paid clauses are against public policy); Mrozik Const., Inc. v. Lovering Assocs., Inc., 461 N.W.2d 49, 52 (Minn. Ct. App. 1990) (strictly construing a pay-if-paid clause). *See also* L. Harvey Concrete, Inc. v. Argo Const. & Supply Co., 939 P.2d 811 (Ariz. Ct. App. 1997) (pay-if-paid clauses will be strictly construed); Blakeslee Arpaia Chapman, Inc. v. EI Constructors, Inc., 687 A.2d 506 (Conn. 1997) (same).

45. *See* Mrozik, 461 N.W.2d at 52.

46. *See* AIA Document A141-2004, Exhibit A, Terms and Conditions, § A.1.1.6 (defining work as "the design, construction and services required by the Design-Build Documents, whether completed or partially completed, and includes all other labor, materials, equipment, and services provided or to be provided by the Design-Builder to fulfill the Design-Builder's obligations").

Design professionals must be aware of any clauses in their contracts with the design-builder that give rise to a higher standard of care because the design professional's insurance policy will typically exclude liability coverage for breaching a higher standard of care.

3. *Indemnification Clauses*

Many contractors will include onerous indemnification clauses in design-build contracts with design professionals. These clauses sometimes require the designer to indemnify the design-builder for 100 percent of the damages even though the designer's percentage of fault was small as compared to that of the design-builder. The indemnification clause could also require the designer to indemnify the design-builder for any design error. Such an indemnification creates a guaranty of design and goes beyond the negligence standard.

Some clauses require that the design professional indemnify the owner and contractor for claims of payment or liens by subconsultants without first requiring that the contractor make payment to the design professional.[47] Further, many of these indemnity clauses do not make any exceptions for when the owner or contractor provides the initial design. In those cases, the owner and the contractor should indemnify the design professional—not the other way around.[48]

4. *Flow-Down Clauses*

Some design-build subcontracts contain flow-down clauses, which provide that the subcontractor has the same obligations to the design-builder that the design-builder has to the owner. Thus, if the design-builder warrants to the owner that the design will be free from defects, a flow-down clause would require the design professional to make the same warranty to the design-builder. As discussed above, this is beyond the reasonable care standard normally applicable to designers and typically beyond the coverage of professional liability insurance.

Another flow-down obligation relates to scheduling. Typically, designers promise no more than to perform within a reasonable time. The design-builder's contract with the owner, however, will normally provide that time is of the essence and impose delay liability on the design-builder for missing time deadlines. If this obligation is flowed down to the designer, the designer will have more responsibility and liability for delay than the designer might otherwise expect, and delay damages can be devastating. In addition, although the design-builder's contract with the owner will usually adjust deadlines for delays beyond the control of the design-builder, often the design-builder is required to provide written notice within a set period of time in order to receive the extension. Design-builders can also flow this notice

---

47. *See* QUATMAN, *supra* note 2, § 5.05[M], at 68 (reviewing various common indemnity provisions in design-build contracts).
48. *Id.*

requirement down to designers, and designers often will not be accustomed to meeting these requirements.

C. *Licensing Issues—Illegality of Design-Builder Not Having a Design Professional License*

As noted in Chapter 2, it is important to be aware of the licensing laws of the state in which the project is located. Some states make it illegal for a design-builder to enter into a design-build contract unless it is properly licensed as a design professional in that state.[49] Design professionals acting as a subcontractor must confirm that the design-builder is properly licensed because it is often a violation of state licensing laws to assist an unlicensed party in the practice of architecture or engineering. Accordingly, a state could sanction a licensed design professional who subcontracts with an unlicensed contractor in a design-build project for assisting an unlicensed design professional. Further, some states hold that design-build contracts with unlicensed contractors are unenforceable, which means the owner does not need to pay the contractor for its work on the project and may even disgorge amounts already paid to the contractor.[50] A design professional acting as a subcontractor in this situation will find it difficult to collect from a design-builder who cannot receive payment from the owner because of an unenforceable prime contract.

D. *Bond Claim Limitations*

In some jurisdictions, the contractor's payment bond on a contractor-led design-build project might not protect the design professional. Some states have held that since payment bonds usually cover only "labor, materials, or construction equipment," there is no coverage under the bond for the design professional's architectural services.[51] If the designer is relying upon a payment bond to protect it in the event of nonpayment, it is important to make certain that the design-builder's payment bond covers the designer's services or that there is equivalent protection under the state's mechanic's lien laws and adequate equity in the property.

Library References

C.J.S. *Architects* §§ 1–3, 6–9, 16, 19, 21–23, 25; C.J.S. *Bonds* § 52; C.J.S. *Contracts* §§ 341, 347, 356, 359, 442–445, 450, 583; C.J.S. *Damages* §§ 62, 64–65, 113.

West's Key No. Digests, Bonds <KEY>62; Contracts <KEY>205, 211, 221(3); Damages <KEY>45, 122; Indemnity <KEY>25–39; Licenses <KEY>11(1), 25, 39–40.

---

49. *See, e.g.*, CAL. BUS. & PROF. CODE §§ 5536 (2001), 7065 (1989); COLO. REV. STAT. § 12-4-102(1); FLA. STAT. chs. 471.031 (2004), 481.223 (2006); KY. REV. STAT. § 323.020 (1960).

50. *See, e.g.*, CAL. BUS. & PROF. CODE § 7031 (2003); FLA. STAT. ch. 489.128 (2006); NEV. REV. STAT. 624.320 (1953).

51. *See, e.g.*, Fields Hartwick Architects v. Capitol Indemn. Corp., 884 P.2d 198, 199 (Ariz. Ct. App. 1994).

VI. Member of a Joint Venture or Other Business Entity

A. *Choosing a Business Form*

Design professionals have several options in choosing what type of business entity it would like to create when teaming with a construction contractor to perform design-build construction. As discussed below, each business form has its advantages and disadvantages.

1. *Joint Ventures*

A joint venture is the most common business form chosen by design professionals and contractors who wish to join forces to be the design-builder for a project. A joint venture is a contractual relationship created between two persons or companies (or possibly more, but not usually) to share profits from an enterprise similar in many respects to a partnership. One important difference is that a joint venture is limited in scope and duration. Joint ventures in the design-build context are often between a design firm and a contractor for the design and construction of a particular project. The most important similarity of joint ventures to partnerships is that in the event the joint venture faces liability, the claimant can sue the design professional and contractor jointly or severally, just like partners in a partnership. In other words, in a joint venture, the design firm will be liable to the owner and others for the mistakes and negligence of its joint venture partner. Generally, the designer would be entitled to recover contribution from its joint venture partner for the joint venture partner's share, but this is of no benefit if it has insufficient assets to pay its share.

While the term joint venture means a contractual relationship with a sharing of profits, sometimes the term is used loosely to mean creating some type of joint endeavor that the parties have not clearly defined, such as a joint venture, new corporation, or other legal entity owned jointly by the two or more firms. This section deals with the technical meaning of joint venture. In any case, joint ventures can put an architect at risk for construction defects by the contractor partner. As professional liability insurance policies typically exclude coverage for work done as a joint venture or construction work, design professionals might well be uninsured for losses relating to these construction defects. Design-builders can attempt to limit their risk for these construction defect losses by including indemnity agreements or limitation of liability clauses in the joint venture agreements. Design firms should think carefully about managing the risk for construction defect liabilities, either by obtaining insurance to cover that exposure to the limited extent that insurance is available, or by making sure they draft their joint venture agreements and construction subcontracts in such a way to protect their interests.

2. *Limited Liability Companies*

Limited Liability Companies, or LLCs, are a somewhat new business form that allows the owner of the LLC to realize both the protection from personal liability offered by a corporation and the tax benefits of a partnership. Not every state has permitted design professionals to practice as LLCs.

LLCs are a good business form for design-build joint endeavors. In addition to the tax benefits and limited personal liability, LLCs also have flexible membership requirements. Corporations and general partnerships can be members of an LLC, making it a desirable alternative to a joint venture arrangement for single project associations.

3. *Partnerships*

States normally define a general partnership as "an association of two or more persons to carry on as co-owners of business for profit."[52] Although it is advisable, partners do not need to enter into a written agreement to form a legal partnership and there are no formal filing requirements or licensing requirements in most states. If there is no written agreement, the applicable partnership laws of the state in which they are doing business often govern the rights of the parties.

Partnerships can be general or limited partnerships. In a general partnership, all the partners are general partners. In a limited partnership, there is at least one general partner, and the other partners are limited partners who do not have control over the limited partnership.

As discussed above in the joint venture section, each general partner (whether in a general partnership or a limited partnership) is jointly and severally liable for the liability of the partnership. Limited partners are more like shareholders in a corporation; their investment is at risk, but they do not have personal liability for the debts and liabilities of the limited partnership.

Partnerships have one significant advantage—they are not required to pay taxes. The law imparts profits and losses to the partners. Individual partners, of course, must declare their income and pay taxes on their personal returns; however, the partnership itself does not pay taxes on the income of the partnership, and unlike general corporations, partnerships can avoid double taxation.

4. *Limited Liability Partnerships*

A limited liability partnership or LLP is different from a limited partnership despite the similar name. A limited liability partnership is at its root a general partnership, and all the partners are general partners and permitted to exercise control. It is a business entity that retains all the benefits of a partnership for tax purposes, but not the joint and several liability typical of a general partnership. Thus, while partners of the limited liability partnership are still liable for their own negligence, they are not liable for the negligence or other wrongful acts of their partners, and they are not individually liable to creditors, owners, subcontractors, suppliers, or others for general business debts.

5. *Corporations*

While a corporation is by far the most popular business form for designers and contractors individually, designers and contractors do not often use it when they team

---

52. 59A AM. JUR. 2d, Partnership § 2 (2006).

to perform design-build work. The legal formalities for the formation and maintenance of a corporation are more burdensome than for the other available options.

The obvious advantage of forming a corporation is that its officers, directors, and shareholders are not personally liable for the acts or debts of the business. In other words, this form shields the personal assets of the owners from the creditors and only the corporate assets are at risk. However, design professionals should note that they cannot shed personal liability for their professional design negligence merely by incorporating as a design-builder. Most state licensing statutes provide that architects and engineers remain individually responsible and liable for professional negligence that they personally commit despite their role as an officer or employee of a corporation.

B. *Insurance and Bonding*

The different roles of a design firm and the different legal arrangements for participating in design-build projects have different insurance requirements and restrictions. Design firms need to review their insurance programs carefully to make sure they afford proper coverage.

In particular, there are problems with insurance coverage when a design professional participates in a joint venture or other legal entity as design-builder. A careful review of the insurance program for the design professional, the design-builder, and the design firm's joint venturer or partner is important, since there are several potential shortfalls in insurance coverage. For example, most professional liability policies for designers contain an exclusion barring coverage when a design firm participates in a joint venture or other legal entity. Naming the joint venture or separate legal entity as an additional insured under the professional liability policy is not a complete solution because most professional liability policies exclude coverage for claims made by one insured against another. As a result, the design firm would not have coverage for claims made against it by the joint venture or other legal entity that is serving as the design-builder. The joint venture or joint ownership exclusion typically bars claims made against the design firm by any entity in which the design firm has an ownership interest in excess of a set percentage, often ranging from 10 to 50 percent. As a result, the design firm must ensure that there is coverage for the joint venture or other legal entity that is acting as the design-builder *and* that it covers claims made against it by the design-builder.

C. *Non-Compete and Exclusivity*

When a design professional and a contractor join together to pursue a design-build project, one issue to address is to what extent their cooperation on that project will prevent them from competing against each other on other projects, with or without other joint venturers, or even whether they will work with others on the targeted project. Careful parties will address these issues in the written agreements. Contract provisions might prohibit the two parties that are cooperating from hedging their bets by working with a competing offeror, explicitly allow the parties to compete against each other on other projects, and prevent the parties from hiring away or poaching each other's employees.

D. *Teaming Agreements*

Teaming refers to parties forming strategic alliances to pursue work on a project or projects. Design professionals and contractors have often found it advantageous to team together to pursue design-build projects. Teaming allows firms with complementary strengths to win contracts that they would lose acting alone. Teaming relationships are often (and should be) memorialized in written contracts between designers and contractors that establish basic rules for the working relationship.

Teaming agreements can range from nonbinding letters of intent to detailed binding contracts and can include other key participants in the prospective work; such as key trade contractors or suppliers.

The teaming agreement should define the project or type of projects it covers and the term for which the agreement will be effective and how the parties can terminate it. It should define the roles of the team members in marketing and preparing proposals and how the costs of doing so will be shared and then should address what their roles will be if they are successful in winning the contract—who will sign the contract, what business entity they will form, how it will be controlled, how profits and losses will be shared, what scope of work each will perform, and how liabilities for failure to perform will be handled. The agreement should specify whether it is binding. For design professionals, it is should address ownership of and copyrights in the design. As discussed previously, it may also be advisable to address exclusivity, noncompetition, and poaching of the other's employees. Often, the parties will share confidential information, and confidentiality agreements are useful to facilitate the process.

Library References

C.J.S. *Contracts* §§ 249–268; C.J.S. *Corporations* §§ 414, 417, 425; C.J.S. *Insurance* §§ 54, 956–957; C.J.S. *Joint Ventures* §§ 1–2, 4, 9, 14, 18, 56–67.

West's Key No. Digests, Contracts <KEY>115; Corporations <KEY>215, 225; Indemnity <KEY>33(5); Insurance <KEY>2391(7); Joint Adventures <KEY>1.1, 1.12, 6; Limited Liability Companies <KEY>1, 17, 24; Partnership <KEY>1, 22, 349, 353, 371.

VII. Bridging

Because most design-builders are contractors who subcontract the design component to design professionals, owners do not have a design professional protecting their interests as they would in the traditional delivery method.[53] Moreover, the time and resources required for design-builders to prepare designs sufficient to submit cost proposals eliminate many capable design-builders from the proposal process. To remedy these problems, some owners engage an architect or engineer to prepare preliminary designs that can be used as the basis for issuing requests for proposals from competing design-builders and to assist with a request for proposal and

---

53. QUATMAN, *supra* note 2, at § 6.01.

selection process. "Bridging" refers to the engagement and retention of an architect by the owner in this design-build context.

The AIA and AGC[54] define bridging as:

> Bridging is a hybrid of the traditional design-bid-build process and design-build. An owner selects an A-E to develop a project design through design development (approximately 30% to 50% of the design work), and prepares scope of work documents which form the basis for competitive selection of the project delivery team. The A-E specifies the project's functional and aesthetic requirements but leaves the details of construction technology up to the contractor. Construction technology is specified with the performance specifications. The project delivery team then has single-point responsibility for final design in constructing the project.[55]

The owner commonly continues to retain the bridging consultant to monitor the completion of the design by the design-builder, review construction, and help with the administration of the design-build contract. Public owners most often use bridging.

Bridging allows owners to have an architect or engineer consultant loyal to the owner while retaining the benefits of the design-build delivery method. Bridging also gives an owner more control over the end product. The criteria and plan created by a bridging consultant are more detailed and provide greater definition of what the design-builder is to deliver than the typical design-build project. On the other hand, having the greater level of definition can also act to limit the creativity and flexibility of the design-builder, lengthen the design process and increase costs, diminishing some of the benefits of design-build. For public owners, the greater definition of the project allows for meaningful competitive bidding and a greater ability to fairly compare one design-builder's proposal with others.[56]

While work as a bridging consultant can be a marketing opportunity, design professionals who participate as a bridging consultant typically cannot bid or work on the projects for which they have served as the bridging consultant. Many owners would prevent the bridging consultant from being or working for the design-builder. If the bridging consultant could also propose to be the design-builder, others might feel they are at a disadvantage and not submit a proposal. That would also deny the owner the services of the bridging consultant as an independent and impartial overseer of the design-builder. While the law is not fully settled in every state, several

---

54. The Associated General Contractors of America.

55. AIA/AGC, RECOMMENDED GUIDELINES FOR PROCUREMENT OF DESIGN-BUILD PROJECTS IN THE PUBLIC SECTOR 5 (Jan. 1995).

56. Some states have statutorily required the use of bridging on public design-build procurement. *See, e.g.*, FLA. STAT. ch. 287.055 (2005).

states have laws that prohibit architects and engineers from bidding on the plans they prepare for public construction projects.[57]

Finally, bridging can cause confusion and uncertainty as to the responsibility and liability for the plans and specifications on any given project. For instance, bridging creates a conflict with regard to who is the architect of record or the engineer of record for the project. Similarly, bridging creates a dilemma, both for the bridging consultant and the design-build designer, neither of whom can say that all the plans were prepared under their direct supervision and control. Of course, the consequence of this uncertainty is that both the bridging consultant and the design-build designer potentially face liability for errors that were not their own.

Library References

C.J.S. *Counties* §§ 151, 165–168; C.J.S. *Municipal Corporations* §§ 911, 917–933; C.J.S. *Public Contracts* §§ 6, 14–24; C.J.S. *States* §§ 270, 274, 278–288; C.J.S. *United States* § 113.

West's Key No. Digests, Counties <KEY>115–121; Municipal Corporations <KEY>234–243; Public Contracts <KEY>5; States <KEY>97–99; United States <KEY>64.

---

57. For example, Florida, Oklahoma, and South Carolina have laws that prevent the architect or engineer who performed bridging to perform any work on the same project as the contractor. FLA. STAT. ch. 287.055 (2005); OKLA. STAT. tit. 59, § 46.27 (1986); S.C. CODE § 11-35-3245 (1997).

CHAPTER 6

The Design Professional and Tort Liability

TERENCE J. SCANLAN, ESQ.
LINDSEY E. MALONE, ESQ.

I. Claims against Design Professionals

A. *Strict Liability*

Tort liability for design professionals typically is defined in terms of negligence, rather than strict liability.[1] The latter usually is imposed in cases involving the mass production of goods or a large body of distant consumers for whom it would be unfair to require tracing the article to the original manufacturer.[2] For this reason, professional services cases are distinguished from consumer products cases.[3] Even in those jurisdictions that impose a rule of strict liability for products, a majority of court decisions decline to apply strict liability to professional services.[4] The rationale for the distinction is stated succinctly in *Gagne v. Bertran*:[5]

> The general rule is applicable that those who sell their services for the guidance of others in their economic, financial, and personal affairs are not liable in the absence of negligence or intentional misconduct. Those who hire [experts] are not justified in expecting infallibility, but can expect only reasonable care and competence. They purchase service, not insurance.[6]

---

1. La Rossa v. Scientific Design Co., 402 F.2d 937, 942-43 (3d Cir. 1968) (wrongful death suit against engineering firm alleging that pellets used in manufacturing phthalic anhydride were defective).
2. *Id.*
3. *Id.*
4. *Id.*
5. 275 P.2d 15, 20–21 (Cal. 1954).
6. *Id.*

In those jurisdictions where a design professional may be subject to strict liability, two theories apply: a products liability claim arising from a defective condition in the structure; or a breach of warranty claim based on affirmative representations made by the design professional.[7]

A number of jurisdictions have held that strict liability in tort is inapplicable to design professionals.[8] Some jurisdictions have recognized, or expressed a willingness to recognize, a narrow exception and impose strict liability on companies that design, manufacture, and construct homes *en masse*.[9] One can differentiate easily between mass-produced homes (which are akin to other manufactured goods) and structures built from a unique set of plans and specifications.

B. *Negligence*

A negligence claim against a design professional is composed of four elements: duty to exercise an accepted standard of care, breach of the standard of care, causation, and damages.[10] "The test [for negligence] is, would the ordinary man in the

---

7. Constance Frisby Frain, *Architect and Engineer Liability*, 35 WASHBURN L.J. 32, 36 (1995). *See also* K-Mart Corp. v. Midcon Realty Group, Ltd., 489 F. Supp. 813 (Conn. 1980) (theory of strict liability does not extend so far as to include the claim of a building's user that the architect of the building is liable, as a consequence of his sale of allegedly "defective" designs to the person who constructed the building); Abdul-Warith v. Arthur G. McKee & Co., 488 F. Supp. 306 (Pa. 1980), *aff'd*, 642 F.2d 440 (3d Cir. 1981) (design professionals who provide services and participate in assembly or construction of a product may be held strictly liable, while architects or engineers who provide design or supervisory services but do not participate in the construction of the challenged product will not be subject to strict liability); Laukkanen v. Jewel Tea Co., 222 N.E.2d 584, 589-90 (Ill. 1966) (court denied imposition of strict liability on architects and engineers who designed a concrete pylon that fell during a windstorm and caused harm to retail customer plaintiff); Reber v. Chandler High Sch. Dist. #20, 474 P.2d 852, 854 (Ariz. 1970) (court denied imposition of strict liability on an architect where the contractor's employees sustained injuries due to alleged defects in plans and specifications for a school gymnasium).

8. City of Mounds View v. Walijarvi, 263 N.W.2d 420, 423-25 (Minn. 1978); Sears, Roebuck & Co. v. Enco Assocs., Inc., 372 N.E.2d 555, 559 (N.Y. 1977); Queensbury Union Free Sch. Dist. v. Jim Walter Corp., 398 N.Y.S.2d 832 (Sup. Ct. 1977); Stuart v. Crestview Mut. Water Co., 34 Cal. Ct. App. 3d, 802, 811-12, 110 Cal. Rptr. 543, 549-50 (1973); Board of Trustees v. Kennerly, Slomanson & Smith, 400 A.2d 850 (N.J. Super. 1979).

9. *See* Schipper v. Levitt & Sons, Inc., 207 A.2d 314, 321-328 (N.J. 1965); Wingfield v. Page, 278 Ark. 276, 281, 644 S.W.2d 940 (1983) (plaintiffs did not seek recovery under strict liability theory, but court expressed support for a broad strict liability policy reaching "the party best able to shoulder it").

10. Wells v. Vancouver, 77 Wash. 2d 800, 803, 467 P.2d 292 (1970); Nicholson v. Turner/Cargile, 107 Ohio Ct. App. 3d 797, 801, 669 N.E.2d 529 (1995); *see also* Brown v. Koulizakis, 331 S.E.2d 440 (Va. 1985); Eastern Steel Constructors, Inc. v. City of Salem, 209 W. Va. 392, 396–97, 549 S.E.2d 266, 270–71 (2001).

defendant's position, knowing what he knew or should have known, anticipate that harm of the general nature of that suffered was likely to result?"[11]

1. *Standard of Care for Design Professionals*

A design professional is required to apply the skill and learning, as it exists at the time, required of a similarly situated design professional in his or her community.[12] A design professional has the duty "to exercise reasonable care, or the ordinary skill of the profession, for the protection of anyone lawfully upon the premises whose injury is reasonably foreseeable as the result of negligent design, plans, orders, or directions."[13]

2. *Scope of Design Professional's Duty*

Claimants often allege that a design professional failed to take some action required by the standard of care. Thus, an allegation of negligent "omission" necessitates review of the contract between the design professional and client. The scope of services that the design professional agreed to perform within the contract defines the scope of that professional's duties.[14]

---

11. Insurance Co. of N. Am. v. Town of Manchester, 17 F. Supp. 2d 81, 84-85 (D. Conn. 1998); *Eastern Steel,* 209 W. Va. at 397, 549 S.E.2d at 271.

12. Weston v. New Bethel Missionary Baptist Church, 23 Wn. Ct. App. 747, 752, 598 P.2d 411, 414 (1979); RESTATEMENT (SECOND) OF TORTS § 299A (1965). *See also* Mears Park Holding Corp. v. Morse/Diesel, Inc., 427 N.W.2d 281, 285 (Minn. 1988); Winrick v. Halco, Inc., 827 P.2d 1238 (Kan. 1992); Colbert v. B.F. Carvin Constr. Co., 600 So. 2d 719, 729 (La. 1992); Cain v. Springfield Eng'g Corp., No. LAW 135358, 1996 Va. Cir. LEXIS 616 (Mar. 18, 1996); Adobe Masters v. Downey, 118 N.M. 547, 548, 883 P.2d 133, 134 (1994).

13. *See* Chaney Bldg. Co. v. Tucson, 148 Ariz. 571, 574, 716 P.2d 28, 31 (1986); Lavy v. W&M Constr. Corp., No. X08 CV 010187185S, 2003 Conn. Super. LEXIS 1719 (June 10, 2003); Council of Unit Owners v. Carl M. Freeman Assoc., Inc., Nos. 86C-A8-49, 86C-AU-50, 86C-AU-51, 1990 Del. Super. LEXIS 469 (Dec. 20, 1990); Klein v. Catalano, 386 Mass. 701, 718, 437 N.E.2d 514 (1982); Waldor Pump & Equip. Co. v. Orr-Schelen-Mayeron & Assocs., Inc., 386 N.W.2d 375, 377 (Minn. 1986).

14. *See* Seattle W. Indus., Inc. v. David Mowat Co., 110 Wash. 2d 1, 10, 750 P.2d 245, 249–50 (1988) (the scope of an engineer's common law duty of care extends at least as far as the duties assumed by him in the contract with the owner); Esposito v. Shell Oil Co., No. X01 CV970159182S, 2001 Conn. Super. LEXIS 1476 (May 29, 2001) (court dismissed wrongful death suit against firm that designed car wash, because defective features were outside the engineer's scope of work); Wiekhorst Bros. Excavating & Equip. Co. v. Ludewig, 529 N.W.2d 33 (Neb. 1995) (citing Dehnert v. Arrow Sprinklers, 705 P.2d 846 (Wyo. 1985)) (design professionals have a duty to give honest advice to homeowner if it is within the scope of their contract with the homeowner).

A design professional may assume duties beyond the scope of any contract through affirmative conduct:

> It is not true ... that the scope of the [engineer's] duty is always limited [by its contract with the owner]. Additional duties might be assumed by affirmative conduct.[15]

3. *Necessity of Expert Witness Testimony*

Generally, the plaintiff must establish the standard of care applicable to a design professional through expert testimony from a design professional practicing within the defendant's area of expertise. "The duty of physicians must be set forth by a physician, the duty of structural engineers by a structural engineer and that of any expert must be proven by one practicing in the same field—by one's peers."[16]

In some instances, however, if the issues are comprehensible to the average juror, then expert testimony may not be required to establish the standard of care required of the design professional.[17] Nevertheless, it is important to note that even where

---

15. *Seattle Western*, 110 Wash. 2d at 10, 750 P.2d 245, 249–50 (rejecting jury instruction stating that engineer's duty was only defined by contract); Jones v. Logansport, 436 N.E.2d 1138 (Ind. 1982) (summary judgment proper because defendant construction administrator had neither a contractual duty to ensure plaintiff's safety nor had it assumed a duty through its affirmative conduct); Magnolia Constr. Co. v. Mississippi Gulf S. Eng'rs, Inc., 518 So. 2d 1194 (Miss. 1988) (summary judgment in favor of engineer reversed on appeal, because it was not clear what the contract contemplated as far as the scope of the engineer's duties).

16. McKee v. Am. Home Prods., 113 Wn. 2d 701, 706-707, 782 P.2d 1045 (1989) (standard of care for pharmacist must be established by expert testimony from pharmacist familiar with the standard of care in Washington); *see also* AA Mech. v. Superior Court of Arizona, 190 Ariz. 364, 367, 948 P.2d 492 (1997); Doe v. Yale Univ., 252 Conn. 641, 686-687, 748 A.2d 834 (2000); City Express, Inc. v. Express Partners, 87 Haw. 466, 468, 959 P.2d 836, 838 (1998) (architect's or engineer's standard of care must be established by expert testimony; failing such expert testimony, a prima facie case of negligence has not been established); *Winrick*, 827 P.2d 1238; Martin v. Sizemore, 78 S.W.3d 249 (Tenn. 2001) (architectural disciplinary board's finding was not supported by substantial evidence, i.e., expert testimony); Palladian Bldg. Co. v. Nortex Found. Designs, Inc., 165 S.W.3d 430, 432 (Tex. 2005); Robert M. Shoenhaus, *Necessity of Expert Testimony to Show Malpractice of Architect*, 3 A.L.R.4th 1023 (1981).

17. *See* Hull v. Enger Constr. Co., 550 P.2d 692 (Wash. Ct. App. 1976), *rev. denied*, No. 44381, 1976 Wash. LEXIS 822 (Nov. 23, 1976); *Colbert*, 600 So. 2d at 729 (if lay persons can infer negligence by applying a common sense standard in the case of physicians, then an architect's negligence may on occasion be established without reference to expert testimony); D & O Contractors, Inc. v. Terrebonne Parish Sch. Bd., 545 So. 2d 588, 591 (La. 1989); *Adobe Masters*, 118 N.M. at 549, 883 P.2d 133, 135; Mid-Western Elec. v. DeWild Grant Reckert & Assocs. Co., 500 N.W.2d 250, 255 (S.D. 1993); John's Heating Serv. v. Lamb, 46 P.3d 1024, 1038 (Alaska 2002).

expert testimony is not required to prove negligence against a design professional, it is often the evidence the court relies upon most heavily in deciding whether the design professional breached the standard of care.[18]

4. Causation

To establish negligence, a plaintiff is required to prove that the alleged error on the part of the design professional actually caused the plaintiff's damage. "Proximate cause" is an act or omission that in a natural and continuous sequence, unbroken by any new, independent causes, produces the injury and without which the injury would not have occurred.[19] Thus, a court held that an engineering firm was entitled to summary judgment against homeowners because the developers did not implement the drainage system it had designed, and there was no proximate cause under the circumstances.[20]

To recover under a theory of breach of express warranty, the plaintiff must prove that the existence of the warranty and its breach were the "direct and proximate cause of the damage."[21]

C. Gross Negligence

1. Definition

"[M]ost courts consider that 'gross negligence' falls short of a reckless disregard of the consequences, and differs from ordinary negligence only in degree, not in kind."[22] Put another way, "if the course of conduct is such that the likelihood of

---

18. Stults & Assocs. v. Neidhart, Nos. 99 CA 11 and 99 CA 17, 1999 Ohio Ct. App. LEXIS 5400 (Nov. 15, 1999) (defendants refused to pay design professional for services; court held that defendants failed to raise issue of material fact to avoid summary judgment and provided no expert testimony to support conclusory statements in affidavits); Martin v. Garner & Law, No. 03A01-9102-CV-00080, 1991 Tenn. Ct. App. LEXIS 768 (Sept. 13, 1991), *rev. denied*, 1992 Tenn. LEXIS 225 (Mar. 9, 1992) (plaintiff's inclusion of expert affidavit creates issue of material fact to survive summary judgment).

19. Brickman v. Walter Schoel Eng'g Co., 630 So. 2d 424, 425 (Ala. 1993); *see also* Union Pac. R.R. v. Sharp, 330 Ark. 174, 181, 952 S.W.2d 658, 662 (1997); *Weston*, 23 Wn. Ct. App. at 752, 598 P.2d at 414 (architect not liable for failure of rockery, when it was undisputed that the wall, as built, substantially deviated from the architect's design).

20. *Brickman*, 630 So. 2d at 425.

21. *Martin*, 1991 Tenn. Ct. App. LEXIS 768 (citing Mitchell v. Pearson Enters., 697 P.2d 240, 247 (Utah 1985)); Interwest Constr. v. Palmer, 923 P.2d 1350, 1357 (Utah 1996); *Cain*, 1996 Va. Cir. LEXIS 616 (plaintiff required to show proximate cause where she alleged design professional was negligent in failing to provide an initial site plan that met city requirements for site approval).

22. Ross v. Rossi, No. CV 89-010491S, 1990 Conn. Super. LEXIS 40 (June 29, 1990) (claim for gross negligence relating to failure of a septic system); Johns v. Council of the

injury to other persons or property is known by the actor to be imminent or 'clear and present,' then the negligence is gross, whereas other negligence would be simple negligence."[23]

"Gross negligence" is the failure to exercise slight care.[24] Failure to exercise slight care does not mean a total absence of care, but care substantially less than ordinary care.[25] A Virginia court described slight care as "such care as careless or inattentive persons usually exercise, while gross negligence is carelessness manifestly and materially greater than want of common prudence."[26] Gross negligence is "that degree of negligence which shows an utter disregard of prudence amounting to complete neglect of the safety of another."[27]

The case law involving design professionals and gross negligence is sparse. However, gross negligence claims generally arise in cases relating to host-guest statutes, state statutory immunity under involuntary commitment laws, personal injuries incurred in the course of adventure sporting activities, railroad injuries, and damage to personal property.

To establish gross negligence, one must offer evidence that the defendant knew, or should have known, facts that would lead a reasonable person to conclude that his or her conduct would lead to an unreasonable risk of harm to others.[28] This requirement

---

Delaware Assn. of Prof'l Eng'rs, No. 03A-07-001 WLW, 2004 Del. Super. LEXIS 257 (July 27, 2004) (engineer grossly negligent in designing on-site wastewater disposal systems, because design was "an extreme deviation" from the standard of care); Courtney v. Florida Transformer, Inc., 549 So. 2d 1061, 1064 (Fla. 1989).

23. White v. Godwin, 124 So. 2d 525, 526 (Fla. 1960).

24. Driggers v. Southern Ry. Co., 169 S.C. 157, 160, 168 S.E. 185, 185 (1933) (widow could not recover for decedent's death if the decedent had been grossly negligent, which it defined as willfulness or wantonness and a lack of slight care for his own protection when he was hit by rail car); Jones v. Atchison Topeka & Santa Fe R.R., 98 Kan. 133, 137, 157 P. 399 (1916) (passenger who fell while trying to jump on a freight train could not hold carrier liable, because those who venture on freight trains have no right to ask more than slight care and can hold the carriers for nothing less than gross negligence).

25. RESTATEMENT (SECOND) OF TORTS § 500 (1965).

26. Wallower v. Martin, 206 Va. 493, 497-98, 144 S.E.2d 289 (1965) (passenger action against driver one and driver two of automobile accident).

27. *Id. But see* City of Middlesboro v. Brown, 63 S.W.3d 179, 181 (Ky. 2001) (gross negligence defined as "something more than the failure to exercise slight care . . . there must be an element either of malice or willfulness or such an utter and wanton disregard of the rights of others as from which it may be assumed the act was malicious or willful").

28. Johnson v. Schafer, 110 Wash. 2d 546, 549-50, 756 P.2d 134 (1988) (quoting RESTATEMENT (SECOND) OF TORTS § 500 (1965)); Wal-Mart Stores v. Berry, 833 S.W.2d 587, 593 (Tex. Ct. App. 1992) (defendant's conscious indifference to the rights, welfare, and safety of others may be inferred when the evidence demonstrates that a reasonable person would have realized that his conduct created an extreme degree of risk to the safety of others); Eister v. Hahn, 420 N.W.2d 443, 446 (Iowa 1988).

of foreseeable risk to persons or the property of others appears to exclude the possibility of a gross negligence claim in cases where the plaintiff has incurred only economic losses.[29]

2. *Effect on Contractual Limitations of Liability*

Most states hold that a person cannot contract against liability for his own gross negligence (or intentional conduct). Even in jurisdictions that allow parties to contractually limit or completely exculpate themselves from liability due to their own negligence, public policy may prohibit parties from shielding themselves from liability for their own gross negligence.[30]

Following that approach, courts in some jurisdictions have held that contract clauses capping a design professional's liability will not apply to damages caused by that design professional's gross negligence.[31] Logically, this would also apply to any contractual provision limiting the liability of a design professional, i.e., a waiver of consequential damages, indemnity for one's own conduct, etc. In contrast, Missouri allows sophisticated businesses to limit their liability for gross negligence.[32]

D. *Breach of Fiduciary Duty*

Plaintiffs have also alleged tort liability against design professionals by claiming that a design professional, by virtue of his or her knowledge, expertise, and skill, owed a fiduciary duty to the plaintiff. Specifically, they allege that this special duty is breached when the design professional fails to apply this knowledge or expertise in a manner consistent with the plaintiff's best interests.

The elements in a cause of action for breach of fiduciary duty are: the existence of a fiduciary duty owed by the design professional to the plaintiff; a breach of that duty; and damages proximately caused by the breach.[33] A fiduciary or confidential

---

29. See discussion below regarding the economic loss doctrine.

30. *See* Vodopest v. MacGregor, 128 Wn. 2d 840, 848, 913 P.2d 779, 782 (1996); Royal Ins. Co. of Am. v. Southwest Marine, 194 F.3d 1009, 1016 (9th Cir. 1999); Donlon Bros. v. Southern Pac. Co., 151 Cal. 763, 767, 91 P. 603, 604 (1907) (a common carrier may limit its liability for ordinary negligence, but is prohibited from doing so as to its gross negligence); Liberty Furniture v. Sonitrol, 770 P.2d 1086 (1989), *rev. denied*, 777 P.2d 1050 (1989) (contractual limitation of liability clause invalidated by security company's gross negligence in failing to notify building owner that fire alarm had been set off); RESTATEMENT (SECOND) OF TORTS § 195 (1981).

31. *See* Sear-Brown Group v. Jay Builders, Inc., 665 N.Y.S.2d 162 (1997); Factory Ins. Ass'n v. American Dist. Tel. Co., 277 So. 2d 569, 570 (Fla. 1973).

32. In re NHB, LLC, 287 B.R. 475, 477 (E.D. Mo. 2002) (sophisticated businesses may legally enter into contracts that limit liability for gross negligence or willful injury arising out of the performance of a contract).

33. Senn v. Northwest Underwriters, 74 Wash. Ct. App. 408, 414, 875 P.2d 637, 639 (1994); United States ex rel. Hill v. Teledyne, Inc., No. 95-55866, 1996 U.S. Ct. App. LEXIS 30178 (9th Cir. Nov. 18, 1996); Griffin v. Fowler, 260 Ga. Ct. App. 443, 445, 579 S.E.2d 845 (2003) (client

relationship is characterized by a unique degree of trust and confidence between parties, where one party possesses superior knowledge, skill, or expertise and has a duty to represent the interests of the other.[34] The superior position of the fiduciary or dominant party affords great opportunity for abuse of the confidence reposed in him or her.[35] Breaches of fiduciary duties have been found where fraud, self-dealing, or conflicts of interest were present, although the absence of these elements does not preclude a finding of that breach.[36]

Some jurisdictions may require an agency relationship between the design professional and plaintiff before a fiduciary duty will be imposed.[37] To establish that a defendant is an agent, the plaintiff must prove "that the principal controlled the manner of performance."[38] Simply placing trust or confidence in another is not sufficient to establish a fiduciary relationship.[39] "Fiduciary duties, which often produce the

---

contended attorney breached his fiduciary duty, but court dismissed the claim as duplicative of the malpractice claim).

34. Winsted Land Dev. v. Design Collaborative, No. CV 960071571, 1999 Conn. Super. LEXIS 2180, 47-48 (Aug. 12, 1999) (design professional firm found liable for breach of fiduciary duty where it filled in wetlands pursuant to contract with developer without obtaining the proper permits).

35. *Id., citing* Murphy v. Wakelee, 247 Conn. 396, 400, 721 A.2d 1181, 1183 (1998).

36. *Id., citing* Murphy v. Wakelee, 247 Conn. 396, 400, 721 A.2d 1181, 1183 (1998).

37. *See* Mullaney, Wells & Co. v. Savage, 78 Ill. 2d 534, 547, 402 N.E.2d 574, 580 (1980); Moss v. Vadman, 77 Wn. 2d 396, 404, 463 P.2d 159, 164 (1969) (accountant who acted as plaintiff's real estate advisor was not agent and, therefore, owed no fiduciary duty to plaintiff); BancOklahoma Mortgage Corp. v. Capital Title Co., 194 F.3d 1089, 1108 (10th Cir. 1999) (in order to establish a breach of fiduciary duty to the homeowners, mortgage company would have to show that agency relationship existed between the title companies and the homeowners).

38. Uni-Com N.W., Ltd. v. Argus Publ'g Co., 47 Wash. Ct. App. 787, 796–97, 737 P.2d 304, 309–310 (1987), *rev. denied*, No. 54108-6, 1987 Wash. LEXIS 1387 (May 26, 1987); State v. Superior Court, 120 Ariz. 501, 504, 586 P.2d 1313, 1316 (1978) (insurance adjuster's promises of confidentiality and of providing defense counsel with the fruits of the investigation did not make the insurance adjuster defense counsel's agent); Flores v. Brown, 39 Cal. 2d 622, 628, 248 P.2d 922, 925 (1952) (inference of agency drawn between owner of automobile and driver of automobile, because owner retains power over how his automobile should be operated).

39. Moon v. Phipps, 67 Wash. 2d 948, 954, 411 P.2d 157, 160–61 (1966) (property owner sued defendant husband and wife to rescind an option to purchase her property, which she alleged was obtained from her by fraud and deceit; court found husband was the property owner's agent, that he breached a fiduciary relationship, and that property owner was entitled to rescind the option agreement); State v. Superior Court, 120 Ariz. 501, 504, 586 P.2d 1313, 1316 (1978).

duty to disclose, 'are not inherent in normal arm's length business relationship, and arise only when one undertakes to act primarily for another's benefit.'"[40]

In the absence of a fiduciary relationship, allegations of a breach of fiduciary duty through fraudulent concealment will not stand simply because a cause of action remains undiscovered, but will stand if the defendant affirmatively prevents discovery.[41]

Whether a fiduciary duty exists and the scope of that duty are normally questions of law for the court.[42]

E. Fraud—Elements

The plaintiff must establish each element of fraud by clear, cogent, and convincing evidence. Most jurisdictions require proof of nine elements: representation of an existing fact, materiality, falsity, the speaker's knowledge of its falsity, the speaker's intent that the plaintiff should act on it, the plaintiff's ignorance of its falsity, plaintiff's reliance on the truth of the representation, plaintiff's right to rely upon it; and damages suffered by the plaintiff.[43] Failure to prove any component typically defeats a claim.[44]

---

40. Cleveland v. City of Lead, 2003 S.D. 54, P18, 663 N.W.2d 212, 217–218 (2003); George Cohen Agency, Inc. v. Donald S. Perlman Agency, Inc., 114 A.D.2d 930, 931, 495 N.Y.S.2d 408, 409–410 (1985) (in the absence of a confidential relationship or fiduciary relationship between two parties to a contract, no duty to disclose exists).

41. Cleveland v. City of Lead, 2003 S.D. 54, 218, 663 N.W.2d 212, 217 (2003), *citing* Purdy v. Fleming, 2002 S.D. 156, P20, 655 N.W.2d 424, 431 (2002) (homeowners sued developer and engineers of shopping mall for fraudulent concealment for removing slope, which allegedly caused damage to the homeowners' houses); *see also* Koenig v. Lambert, 527 N.W.2d 903, 905–906 (S.D. 1995); Conway v. Conway, 487 N.W.2d 21, 23 (S.D. 1992); DeLuna v. Burciaga, 359 Ill. Ct. App. 3d 544, 550, 834 N.E.2d 478, 483 (2005).

42. Fuqua v. Taylor, 683 S.W.2d 735, 737 (Tex. Ct. App. 1984) (geologist owed fellow investors a fiduciary duty as a matter of law); Schneider v. Plymouth State College, 144 N.H. 458, 462, 744 A.2d 101, 105 (1999) (professor who sexually harassed student and college sued for breach of fiduciary duty; court said whether a duty can be imposed upon an entity for the care and protection of a person is a question of law). *But see* Brickner v. Federal Deposit Ins. Corp., 747 F.2d 1198, 1201 (8th Cir. 1984) (courts have variously characterized the determination that a party breached his fiduciary duty as a question of fact and a question of law; in the present case, the issue might also be described as a mixed question of fact and law).

43. Stiley v. Block, 925 P.2d 194 (Wash. 1996); Hewlett v. Hewlett, 845 S.W.2d 717 (Mo. Ct. App. 1993) (trial court set aside decree of dissolution based upon fraud); Anderson Dunham, Inc. v. Aiken, 133 So. 2d 527 (Miss. 1961) (dump truck buyers filed actions against seller for fraud and deceit based on the ground that the seller falsely represented to them the maximum capacity of the trailers' axles).

44. Shayne v. Julien, Schlesinger & Finz, P. C., 131 A.D.2d 655, 656-57 (N.Y. Ct. App. Div. 1987).

An owner might claim fraud when it believes that the design professional intentionally deceived it by affirmatively misrepresenting or concealing material facts related to the contract or the project.

F. *Negligent Misrepresentation*

If a person in the course of his or her profession or business supplies false information for the guidance of others in their business transactions, he or she can be held liable for any pecuniary loss caused by others' justifiable reliance on the information if he or she failed to exercise reasonable care in obtaining or communicating that information.[45] Claims of negligent misrepresentation against design professionals have included allegations that an engineer supplied specifications containing false statements of fact or opinion,[46] that an engineer provided false information in an inspection or geotechnical report,[47] or that an engineer's or architect's representations contained errors as to the cost of a project.[48]

Jurisdictions are divided on the issue of whether clients may sue a design professional for negligent misrepresentation.[49]

Library References

C.J.S. *Architects* §§ 19, 21–23; C.J.S. *Contracts* § 271; C.J.S. *Negligence* §§ 91–97, 162–164, 188, 191–192, 197, 803, 805, 808, 810–811; C.J.S. *Products Liability* §§ 1–2.

West's Key No. Digests, Contracts <KEY>114; Negligence <KEY>273, 321, 322, 370, 1657, 1662; Products Liability <KEY>1.

---

45. RESTATEMENT (SECOND) OF TORTS § 552 (1977). *See also* Ritter v. Custom Chemicides, 912 S.W.2d 128, 130 (Tenn. 1995) (tort of negligent misrepresentation not limited to professionals and privity of contract not essential to a tort action of negligent misrepresentation based on information negligently supplied for the guidance of others); Savings Bank of Manchester v. Ralion Fin. Services, 881 A.2d 1035 (Conn. Ct. App. Ct. 2005) (company not liable for misrepresenting status of lessor to bank, because bank failed to prove company made false statements on which bank relied).

46. Burns Philp, Inc. v. Cox, Kliewer & Co., P.C., No. 4-99-CV-90033, 2000 U.S. Dist. LEXIS 21653 (S.D. Iowa Nov. 2, 2000) (plaintiffs sued architecture firm, architects and employee, claiming they made false misrepresentations regarding a mezzanine they designed for plaintiffs).

47. Green Constr. Co. v. Kansas Power & Light Co., 717 F. Supp. 738, 743 (Kan. 1989) (plaintiff contractor sued engineer for misrepresentation when contractor expended more money excavating due to unexpected water levels in the soil not reported in the geotechnical report); Howell v. Fisher, 272 S.E.2d 19 (N.C. Ct. App. 1980) (geological engineer liable to corporate shareholders for negligent misrepresentation in preparation of soil testing for the purpose of inducing their investment in the corporation).

48. Charles L. Hazelton & Son, Inc. v. Teel, 211 N.E.2d 253 (Mass. 1965) (wharf owners sued contractor for alleged misrepresentation regarding cost of repairing wharf; verdict properly directed for contractor where there was no evidence that he knew his estimate to be false).

49. *See* Air Prods. & Chemicals, Inc. v. Eaton Metal Prods. Co., 2003 WL 21697376 (Pa. 2003) (court refuses to recognize causes of action in negligence and strict liability for purely

II. Issues Affecting the Tort Liability of Design Professionals

A. *Assumption of a Higher Degree of Care*

The design professional may be held to a higher standard of care based upon contract clauses in which the design professional agreed to satisfy the subjective standard of the client, meet the highest standard of the profession, or achieve specific, objective criteria. Thus, a design professional can assume a contractual duty to perform services under a heightened standard of care, greater than what is required by common law.[50] In addition to increased liability, the design professional may not have coverage for this greater exposure under the typical professional liability policy.

---

economic injury); Smiley v. S&J Invs., Inc., 580 S.E.2d 283 (Ga. 2003) (homeowner can pursue negligent misrepresentation claim against inspection company based on errors in inspection report); Marcellus Constr. Co., Inc. v. Village of Broadalbin, 755 N.Y.S.2d 474 (2003) (contractor cannot sue engineer for negligent misrepresentation arising from errors in subsurface site condition report); David Pflumm Paving & Excavation, Inc. v. Foundation Servs., Co., 816 A.2d 1164 (Pa. 2003) (excavation subcontractor cannot sue engineering firm or township engineer for negligent misrepresentation for errors in report on subsurface conditions). *But see* Koontz v. Thomas, 511 S.E.2d 407 (S.C. 1999) (client could sue architectural firm for negligent misrepresentation, but alleged misrepresentations relating to anticipated construction costs and project duration insufficient to state a claim); Plantation House & Garden Prods., Inc. v. R-Three Investors, 248 A.D.2d 606, 670 N.Y.S.2d 505 (1998) (warehouse owner could pursue negligent misrepresentation claim against engineers based on representation that concrete floor had been completed and installed in accordance with plans and specifications); Empire W. Cos., Inc. v. Albuquerque Testing Labs., Inc., 800 P.2d 725 (N.M. 1990) (negligent misrepresentation claim against geotechnical firm based on statements in report as to suitability of soil allowed to go to trial; verdict for defense); Buchanan v. Scottsdale Envtl. Constr. & Dev. Co., Inc., 787 P.2d 1081 (Ariz. 1990) (homeowner could proceed to trial on claim of negligent misrepresentation against soils engineer based on alleged misrepresentation as to the suitability of the site for residential construction). *See also* Valhal Corp. v. Sullivan Assocs., 44 F.3d 195 (3rd Cir. 1995) (client alleged negligent misrepresentation based on engineer's failure to disclose height restriction on parcel being purchased by client; court did not discuss issue of whether such a claim could be pursued).

50. Town of Breckenridge v. Golforce, Inc., 851 P.2d 214, 216 (Colo. 1992) (court held expert testimony regarding standard of care required of design professional not necessary where contract defines the standard of care); Ferentchak v. Village of Frankfort, 105 Ill. 2d 474, 483, 475 N.E.2d 822, 826 (1985) (engineer not negligent in failing to set the foundation grade elevations, because the professional standard did not obligate him to do so and the contract did not obligate him to a heightened standard of care); Strauss Veal Feeds, Inc. v. Mead & Hunt, Inc., 538 N.E.2d 299, 302 (Ind. 1989) (engineering firm not liable for negligence when building complied with all building and safety codes even though it created a biohazard, because engineer complied with applicable industry standards and did not assume independent contractual duty to deal with industrial waste treatment and disposal); R.J. Reagan Co. v. Kent, 654 S.W.2d 532, 533

B. *Warranty Disclaimers*

Warranty disclaimers are routine in agreements for the sale of goods. However, design professional contracts, along with other construction-related agreements, are typically considered contracts for services, not goods. Therefore, Uniform Commercial Code (U.C.C.) Article 2 warranty provisions do not apply.[51]

There is no clear consensus between jurisdictions as to the enforceability of disclaimers of express or implied warranties. As a common form of the mechanism to shift risk, warranty disclaimers are more likely to be enforced when they appear in construction contracts between commercially sophisticated parties. For example, a court has held that the jury should be permitted to consider a clause in an engineer's contract disclaiming any warranty as to the accuracy of cost estimates.[52] Thus, in transactions between sophisticated entities, warranty disclaimers are presumed to be conscionable and enforceable, unless the plaintiff can establish some type of "unfair surprise in the negotiations."[53] In *American Nursery Products, Inc. v. Indian Wells Orchards*,[54] the court stated that the U.C.C. requirement mandating such disclaimers be "specifically

---

(Tex. 1983) (where no evidence was produced that architect undertook a contractual obligation regarding proper roof construction methods, architect cannot be held responsible for defective construction); Todd County v. Barlow Projects, Inc., Nos. 04-4218 ADM/RLE and 04-4220 ADM/RLE, 2005 U.S. Dist. LEXIS 8648 (Minn. May 11, 2005) (engineering consultant liable only if he did not exercise the care contemplated by the contract with the client); CH2M Hill Southeast v. Pinellas County, 698 So. 2d 1238, 1240 (Fla. 1997) (if the professional contracts to perform duties beyond those required by ordinary standards of care, the quality of that performance must comport with the contractual terms).

51. Kirkpatrick v. Introspect Healthcare Corp., 845 P.2d 800 (N.M. 1992) (interior designer's predominantly service-oriented contract not subject to U.C.C. Article 2); Pass v. Shelby Aviation Inc., No. W1999-00018-COA-R9-CV, 2000 Tenn. Ct. App. LEXIS 247 (Apr. 13, 2000) (if a contract is predominantly for the sale of goods, it falls under the U.C.C. and the warranty provisions of Article 2 apply; if a contract is predominantly for services, it falls outside the U.C.C. and the warranty provisions of Article 2 are inapplicable).

52. Post, Buckley, Schuh & Jernigan, Inc. v. Monroe County, 851 So. 2d 908, 909-10 (Fla. 2003); *see also* Berschauer/Phillips Constr. Co. v. Seattle Sch. Dist. No. 1, 881 P.2d 986 (Wash. 1994).

53. M.A. Mortenson Co. v. Timberline Software Corp., 998 P.2d 305 (Wash. 2000) (discussing substantive and procedural unconscionability in commercial sales contract); Hewlett-Packard Co. v. Intergraph Corp., No. C 03-2517 MJJ, 2004 U.S. Dist. LEXIS 17601 (N.D. Cal. Sept. 6, 2003) (where parties are commercial entities, warranty disclaimers are presumptively conscionable and the burden is on the party seeking to invalidate the provision to prove it is unconscionable); Alpha One v. Nynex Info. Res. Co., 2 Mass. L. Rep. 568 (Mass. Super. Ct. 1994) ("unconscionability must be determined on a case-by-case basis . . . giving particular attention to whether, at the time of the execution of the agreement, the contract provision could result in unfair surprise and was oppressive to the allegedly disadvantaged party").

54. 115 Wn. 2d 217, 797 P.2d 477 (Wash. 1990).

negotiated" would not be imposed in the commercial context, because to do so "would unnecessarily interfere with the freedom to contract in the commercial context."[55]

However, when contracting with an unsophisticated client (i.e., retail consumer), disclaimers may be subject to challenge. Under the U.C.C., disclaimers in consumer contracts are presumed unconscionable, and it is up to the defendant to prove otherwise.[56] Although the U.C.C. does not apply to service contracts, courts may look to it for guidance in determining the conscionability of such disclaimers in contracts with design professionals.[57]

C. *Contract Modifications*

1. *Use of a Master Agreement*

Complex, long-term projects require that particular attention be paid to the original owner-design professional agreement. Ideally, where the parties realize that the project will be performed in parts over time, the seminal contract document will identify itself as a *master* or *controlling* agreement. If it is the parties' intent, the contract should explicitly state that the parties intend for the master contract to provide controlling authority for all subsequent agreements, including an explicit statement that the general terms and conditions will control it and all subsequent contracts.

2. *Amendments and Modifications*

Subsequent phases of work should be identified explicitly as an *amendment* or *modification* to the original master agreement and, likewise, should explicitly refer back to and incorporate the general terms and conditions contained within the master contract.

3. *Scope Creep*

Design professionals frequently encounter projects that grow in scope, either by plan, necessity, or circumstance. In some instances, the project is intended to be done in phases for which separate written agreements are executed. In other instances, an initial phase of work is all that was originally contemplated, but subsequent events cause the design professional to perform additional services not anticipated in the contract. Both situations require attention to the contract documents to ensure that all significant terms are enforceable.

---

55. *American Nursery Products,* 115 Wash. 2d at 224, 797 P.2d at 481–82.
56. U.C.C. § 2-302 (2005).
57. *See* Sprague v. Household Int'l, No. 04-0106-CV-W-NKL, 2005 U.S. Dist. LEXIS 11694 (W.D. Mo. June 15, 2005) (principles underlying the common law concept of unconscionability do not differ from those codified in the U.C.C.); Puget Sound Fin. v. Unisearch, Inc., 146 Wash. 2d 428, 440, 47 P.3d 940, 945–46 (2002) (Washington courts may, by analogy, use U.C.C. sections to interpret service contracts).

It is common for a design firm to be drawn into work not contemplated by the scope of services in the written contract. This may happen when the client asks for additional work, often prompted by situations encountered in the field or when an unexpected problem arises. Frequently, design professionals find themselves performing these add-on services without the benefit of a contract amendment. This may occur because the additional scope seems relatively innocuous and minor in nature, or because the individuals actually performing the extracontractual work are often field personnel not usually involved in the contract process. In any event, providing this type of service without the safety net of a written contract puts the design professional at risk. Although bothersome and frequently at odds with the immediate nature of many requests for add-on services, it remains imperative that all work be subject to a written agreement. Ideally this takes the form of an amendment that explicitly refers back to the original contract, including the general terms and conditions.

D. *Preparation of Plans and Specifications*

In adjudicating a claim for defective design, a court must consider the plans and specifications.[58] Plans and specifications do not have to be perfect to protect the design professional from liability;[59] however, they must conform to the standard of care exercised by other design professionals in the community.[60]

Contractors or subcontractors attributing cost increases or delays due to defective design documents often make claims against design professionals. Design professionals

---

58. Cady v. E. I. DuPont de Nemours & Co., 437 F. Supp. 1030, 1032 (D.C. Tex. 1977).

59. Ramey Constr. Co. v. Apache Tribe of Mescalero Reservation, 673 F.2d 315, 322 (10th Cir. 1982) (architect's plans and specifications adequate for purposes of bidding and construction, even though they contained more than 100 design errors and omissions, given that there were more than 330 drawings and 500 pages of calculations); *Chaney Building*, 148 Ariz. at 574, 716 P.2d 280 at 31 (architect's work can be inaccurate or imperfect without being an actionable deviation from the standard of care observed by design professionals).

60. *Id.; Mounds View*, 263 N.W.2d at 424 (undertaking by architect implied he possessed skill and ability sufficient to enable him to perform the required services at least ordinarily and reasonably well, and that he would exercise and apply his skills without neglect; it did not imply or warrant a satisfactory result); St. Joseph Hosp. v. Corbetta Constr. Co., 21 Ill. Ct. App. 3d 925, 942–943, 316 N.E.2d 51 (1974) (architect always held to an implied contract that the work shall be suitable and capable of being used for the purpose for which it is prepared); Seiler v. Levitz Furniture Co. of E. Region, Inc., 367 A.2d 999, 1007-08 (Del. 1976) (while an architect does not guarantee a perfect plan or a satisfactory result, he does by his contract imply that he enjoys ordinary skill and ability in his profession, and that he will exercise these attributes without neglect and with a certain exactness of performance to effectuate work properly done).

may defend against these claims by proving that the contractor failed to follow the plans and specifications.[61]

Owners also may claim that the project exceeded the quoted cost due to defective plans and specifications.[62]

E. *Environmental Liability*

Engineers have been held liable as "operators" under the Comprehensive Environmental Response, Compensation and Liability Act (CERCLA).[63] In *Ganton Technologies v. Quadion Corp.*,[64] a third party plaintiff (owner) successfully brought a complaint against a supervisor and a contractor who had been hired to clean up PCB contamination, alleging that they further contaminated the site.[65] The court held that liability for violations of the Act as an operator attaches "if the defendant had authority to control the cause of the contamination at the time the hazardous substances were released into the environment."[66]

The Resource Conservation Recovery Act (RCRA) concerns the handling, transportation, treatment, and disposal of hazardous waste.[67] It is another regulatory statute that may impose liability on design professionals. The language of the RCRA provides for injunctive relief, as well as civil fines, penalties, and criminal sanctions for violations of the Act.[68]

---

61. Goette v. Press Bar & Cafe, Inc., 413 N.W.2d 854, 856 (Minn. 1987); *see also* Friederick v. County of Redwood, 153 Minn. 450, 451, 190 N.W. 801, 802 (1922) (contractor is not permitted to vary from plans even if he deems them improper and insufficient and, therefore, cannot be held to guarantee that work performed according to the plans will be free from defects).

62. Department of Natural Res. v. Transamerica Premier Ins. Co., 856 P.2d 766, 772 (Alaska 1993) (project owner may sue design professional in tort for economic losses arising from the professional's malpractice, despite existence of a contractual relationship between the parties); Cluett, Peabody & Co. v. Campbell, Rea, Hayes & Large, 492 F. Supp. 67, 74 (Pa. 1980) (plaintiff lessor of building charged architect with breach of its contractual obligations to design and supervise construction of a warehouse, to exercise the skill in its possession in designing and supervising construction of the warehouse, and negligence in the design, selection of material, supervision of construction, and arranging for adequate repairs to the roof resulting in cost incursions by plaintiff to replace roof).

63. CERCLA § 107(a)(2), 42 U.S.C.S. § 9607 (2002).

64. 834 F. Supp. 1018, 1022 (N.D. Ill. 1993).

65. *Id.* at 1018.

66. *Id.* at 1022 (quoting Kaiser Aluminum v. Catellus Dev., 976 F.2d 1338, 1342 (9th Cir. 1992) (excavator grading a section of land for a proposed housing development who spread some displaced contaminated soil from the property over other parts of the property was held liable under CERCLA)).

67. 42 U.S.C. § 6972 (1984).

68. *Id.*

Other environmental regulations, such as the Clean Air or Clean Water Acts, also may pose a risk to design professionals.[69] However, there is little case law to indicate how these regulations might affect them.

Library References

C.J.S. *Architects* §§ 16, 19, 21–23; C.J.S. *Contracts* §§ 359, 407–420, 434–435; C.J.S. *Damages* §§ 53–61; C.J.S. *Fraud* §§ 4–8, 10–17, 23–25, 28–31, 34–36, 59–60, 62–63, 67–75, 77–78, 80; C.J.S. *Health and Environment* § 170; C.J.S. *Negligence* §§ 162–164; C.J.S. *Torts* § 26.

West's Key No. Digests, Contracts <KEY>205.25, 205.35, 236–246; Damages <KEY>35; Environmental Law <KEY>445; Fraud <KEY>2–13, 64; Indemnity <KEY>33(5); Labor and Employment <KEY>2607, 2832, 2856; Negligence <KEY>321, 322.

III. Who Can Sue Design Professionals: Economic Loss Rule

The economic loss doctrine may bar tort claims against the design professional when the damages are purely monetary. This doctrine may apply to actions brought by one in privity of contract with the designer or by a third party who is a stranger to the contract.

A. *Application to Parties in Privity of Contract*

Parties in privity with design professionals may bring contract actions to recover economic losses.[70] Generally, it is assumed that a design professional owes a common law duty to its clients to exercise the skill and expertise ordinarily used by members of that profession and that a breach of this duty results in liability for any resulting damages. However, some jurisdictions also bar tort claims for purely economic losses between parties in contractual privity.[71]

The practical effect of this doctrine is that the law applicable to tort actions will not apply to the owner's claim against the design professional; for example, comparison of negligence, joint and several liability, contribution, and the tort statute of limitations.

For an excellent discussion of the economic loss doctrine, see *Calloway v. City of Reno*.[72]

---

69. *See* Clean Water Act, 33 U.S.C.A. §§ 1251–1387 (1948), and Clean Air Act, 42 U.S.C.A. §§ 7401–7671 (1955).

70. *Berschauer/Phillip*, 881 P.2d 986.

71. *See* Aas v. Superior Court, 24 Cal. 4th 627, 12 P.3d 1125, 101 Cal. Rptr. 2d 718 (2000); *City Express*, 959 P.2d 836.

72. 993 P.2d 1259 (Nev. 2000).

B. *Application to Third Parties Not in Privity*

Under the economic loss doctrine, a third party not in privity of contract cannot sue a design professional in tort if the only damages sustained are monetary losses.[73] This principle is a major bar to lawsuits brought against design professionals.

"Economic loss" has been defined as a "diminution in the value of a particular product caused by the product itself or by the failure of the product to function as represented; in other words, to fail economic expectations."[74] Damages for delay, remediation, repair, and diminution in value are economic losses.[75]

The economic loss doctrine is rooted in products liability law. It provides that certain types of "economic loss" damages are recoverable only in contract, generally between parties in privity. The rule marks the fundamental boundary between contract and tort; the former addresses expectancy interests of the parties, while the latter addresses the duty of all persons to exercise reasonable care. It bars recovery in tort for purely economic losses.

An early, seminal case discussing the doctrine is *Seely v. White Motor Co.*[76] (still widely cited, especially in design and construction cases). There, the court noted:

> The distinction that the law has drawn between tort recovery for physical injuries and warranty recovery for economic loss is not arbitrary and does not rest on the "luck" of one plaintiff in having an accident causing physical injury. The distinction rests, rather, on an understanding of the nature of the responsibility a manufacturer must undertake in distributing his products ... A consumer should not be charged at the will of the manufacturer

---

73. *Berschauer/Phillip*, 881 P.2d 986 (economic loss rule prevented a general contractor from recovering purely economic damages from a structural engineer under any tort theory); Hayden Bus. Ctr. Condos. Assn. v. Pegasus Dev. Corp., 209 Ariz. 511, 516, 105 P.3d 157, 162 (2005) (homeowner's breach of warranty suit against developer sounded in contract and was barred, because of lack of privity between the parties); Spancrete, Inc. v. Ronald E. Frazier Assocs., 630 So. 2d 1197 (Fla. 1994) (subcontractor suing for economic losses not in contractual privity with architect could not state claim under either tort or contract); Floor Craft Floor Covering, Inc. v. Parma Cmty. Gen. Hosp. Ass'n, 560 N.E.2d 206 (Ohio 1990) (in absence of contractual privity, no cause of action exists to recover economic damages against design professional who drafted plans and specifications); Atherton Condo. Apartment-Owners Ass'n Bd. of Dirs. v. Blume Dev. Co., 115 Wn. 2d 506, 534, 799 P.2d 250, 266 (1990) (no cause of action in tort by a third party seeking economic loss damages for negligent design against an architect); *but see* Beachwalk Villas v. Martin, 406 S.E.2d 372 (S.C. 1991) (architect may be liable to homebuyer for negligence in connection with home construction even when no contractual privity exists between homebuyer and architect).

74. Wausau Paper Mills Co. v. Charles T. Main, Inc., 789 F. Supp. 968, 971 (W.D. Wis. 1992).

75. *See* East River S.S. Corp. v. Transamerica Delaval, Inc., 476 U.S. 858, 870, 106 S. Ct. 2295, 2301–2302, 90 L. Ed. 2d 865, 876–877 (1986) (economic losses include lost profits, repair costs, and decreased value).

76. 63 Cal. 2d 9, 45 Cal. Rptr. 17, 403 P.2d 145 (1965).

with bearing the risk of physical injury when he buys a product on the market. He can, however, be fairly charged with the risk that the product will not match his economic expectations unless the manufacturer agrees it will.[77]

Put simply, the question becomes whether the aggrieved plaintiff is among "the disappointed users ... [or] the endangered ones."[78] Generally, recovery may be had in contract for the former, and in tort for the latter. In *East River Steamship Corp. v. Transamerica Delaval, Inc.*,[79] the United States Supreme Court approved the distinction drawn between commercial expectation interests reflected in contractual relationships and traditional concepts of duty and foreseeable risk contemplated under tort theory. The Court noted that commercial controversies were particularly well-suited to application of the doctrine, and observed that commercial parties have the ability to negotiate contract terms reflecting the economic expectations of the parties.

Numerous jurisdictions have considered the application of the economic loss rule to the construction/design professional industry. The results are mixed and largely dependent on a given state's choice of standards for evaluating economic loss claims and how that state's respective products liability laws apply.[80]

1. *Two Tests*

States typically follow two schools of thought as to how the economic loss doctrine is perceived and applied.

a. *Damages-Driven Test*

This perspective approaches the economic loss doctrine by asking what type of damage is alleged to have occurred, and determining whether the claim properly lies

---

77. 63 Cal. 2d at 18, 45 Cal. Rptr. at 23, 403 P.2d at 151.

78. Russell v. Ford Motor Co., 281 Or. 587, 595, 575 P.2d 1383, 1386 (1978).

79. *Supra* note 75.

80. *See Aas*, 12 P.3d 1125; A.C. Excavating v. Yacht Club II Homeowners Ass'n, 114 P.3d 862, 865 (Colo. 2005) (court reaffirms adoption of the economic loss rule in Colorado, but holds that builders have an independent duty to construct a home without negligence, and tort claims may be brought based on breach of a duty independent of the contract); Richards v. Midland Brick Sales Co., 551 N.W.2d 649 (Iowa 1996) (homeowner claim against contractor for negligent brickwork barred by economic loss doctrine); Koss Constr. v. Caterpillar, Inc., 960 P.2d 255 (Kan. Ct. App. 1998), *rev. denied*, No. 96-78054-AS, 1998 Kan. LEXIS 541 (Sept. 10, 1998) (purchaser of hydraulic roller sued manufacturer for product liability; economic loss rule barred negligence or strict liability claims); Young v. City of Plaquemine, 818 So. 2d 892 (La. 2002) (economic loss doctrine bars claim against architect for defective work in the absence of personal injury); Maine Rubber Int'l. v. Environmental Mgmt. Group, 298 F. Supp. 2d 133 (Me. 2004) (land buyer's claim against engineering firm for negligence and negligent misrepresentation in connection with Phase I ESA barred by economic loss rule). *But see* Forte Bros. v. Nat'l Amusements, Inc., 525 A.2d 1301

in tort or contract. This view more clearly defines the nature of the theory of liability by working backward to determine the source of duty owed—either by contractual assumption or pursuant to a standard of care imposed for the activity. A majority of jurisdictions follow this approach.[81]

b. *Foreseeable Duty Test*

The second perspective begins its analysis by asking what duty of care is owed by the design professional and/or if the damage was "foreseeable" even as it relates to a third party not in contractual privity with it. Using this approach virtually guarantees that the court's conclusion will allow a tort theory to proceed against the design professional. Unfortunately for architects and engineers, this perspective has gained currency in some jurisdictions.

Florida is one state that has adopted the forseeability test. In *Moransais v. Heathman*,[82] the Florida Supreme Court reviewed a negligence claim against an individual engineer, who performed a pre-purchase home inspection. The court held that by Florida statute, engineers are to be grouped with other professionals, such as attorneys, physicians, and accountants, and may not avoid personal liability for "professional malpractice." The court further ruled that the economic loss doctrine in Florida essentially was limited to consumer products liability cases, and not applicable to claims against professionals who knew or should have known that the claimant would be injured by their negligence.

This view was reaffirmed in *Stone's Throw Condominium Assoc., Inc. v. Sand Cove Apartments, Inc.*,[83] which involved claims for negligence and negligent misrepresentation

---

(R.I. 1987) (contractor allowed to sue architect/site engineer for negligent performance of oversight duties); Goose Creek Consol. Indep. Sch. Dist. v. Jarrar's Plumbing, Inc., 74 S.W.3d 486 (Tex. 2002), *rev. denied*, Nov. 21, 2002) (school district's tort claim against architect and contractor for plumbing defects barred by economic loss rule); *Eastern Steel*, 209 W. Va. at 392, 549 S.E.2d at 266 (contractor claim against design professional not barred by economic loss rule; design firm may owe duty of care to contractor despite absence of privity of contract); Rowe v. Akin Flanders, Inc., 525 S.E.2d 123 (Ga. Ct. App. 1999) (landowner not barred from suing contractor for negligent construction on the basis of economic loss rule); *Waldor Pump*, 386 N.W.2d 375 (engineer liable to subcontractor for negligence in drafting and interpreting specifications; rejected application of economic loss doctrine); Glacier Tennis Club at the Summit, LLC v. Treweek Constr. Co., 87 P.3d 431 (Mont. 2004) (contractor may recover for economic loss against project engineer or architect when design professional knew or should have known that plaintiff was at risk in relying on information supplied); Tommy L. Griffin Plumbing & Heating v. Jordan, Jones & Goulding, Inc., 463 S.E.2d 85 (S.C. 1995) (contractor allowed to sue design professional for economic losses, because design professional has duty not to negligently design or supervise project).

81. *See Aas*, 12 P.3d at 1125; *Tommy L. Griffin*, 463 S.E.2d at 85 (economic loss doctrine analysis rests first on what kind of damages were caused, and second on what kind of duty was owed).

82. 744 So. 2d 973 (Fla. 1999).

83. 749 So. 2d 520 (Fla. 1999).

against an architect. A condominium association that had not been in privity with the design professional hired by the developer brought the claim against the architect. Citing *Moransais*, the court held that the economic loss doctrine does not bar a claim against a design professional, even where the only damages are economic. Florida recognizes a common law cause of action against design professionals for negligence, even in the absence of contractual privity.

Thus, when the court's analysis begins by inquiring, "what duty was owed by an engineer" or "was it foreseeable that some third party would necessarily rely upon the architect's work," the outcome may be a foregone conclusion. In *Jim's Excavating Service, Inc. v. HKM Associates*,[84] the Montana Supreme Court held that "a third party contractor may recover for purely economic loss against a project engineer or architect when the design professional knew or should have foreseen that the particular plaintiff or an identifiable class of plaintiffs were at risk in relying on the information supplied."[85] Other jurisdictions reaching similar conclusions include Arizona,[86] Connecticut,[87] Michigan,[88] and Alabama.[89]

C. *Claims against Individual Design Professionals*

Very few courts have analyzed whether the economic loss doctrine bars tort claims against an individual design professional in the absence of contractual privity between the claimant and that person. Contrast *Moransais v. Heathman*[90] and *Gerald M. Moore & Sons, Inc. v. Drewery*.[91]

In *Moransais*, the homeowner signed a contract with an engineering firm, but not its individual employees, to perform a pre-purchase inspection of a private home. The engineers argued that absent contractual privity, the economic loss doctrine barred recovery against them individually for a tort claim. The court found that

---

84. 878 P.2d 248, 252 (1994).

85. *Id*. at 255.

86. Donnelly Constr. Co. v. Oberg/Hunt/Gilleland, 137 Ariz. 184, 187–188, 677 P.2d 1292, 1296 (Ariz. 1984) (the potential for imposing a duty owed to third parties not in privity will be evaluated on a case-by-case basis, using a reasonable "forseeability of risk and plaintiff" analysis).

87. *Insurance Co. of N. Am.,* 17 F. Supp. 2d 81; Carolina Cas. v. 60 Gregory Boulevard, No. CV 980169383S, 2000 Conn. Super. LEXIS 739 (Mar. 20, 2000).

88. Bacco Const. Co. v American Colloid Co., 384 N.W.2d 427 (Mich. Ct. App. 1986) (contractor allowed to bring negligence action to recover purely economic losses from project engineer; court applied foreseeable risk of harm analysis).

89. Keck v. Dryvit Sys., Inc., 830 So. 2d 1 (Ala. 2002) (where one party assumes a duty to another party in contract and it is foreseeable that injury to a third party could occur as result of breach, the promissor owes a duty to all within foreseeable area of risk).

90. *Moransais*, 744 So. 2d at 973.

91. 467 S.E. 811 (Va. 1996).

Florida statutes explicitly provide for the liability of individual professionals and that they are not shielded by being employees of a corporation.[92]

The contrary view was expressed in *Drewery*. An owner brought a negligence claim against both the engineering firm *and* its employee. The Virginia Supreme Court held that even if negligence by the individual engineer could be proven, in the absence of contractual privity, the economic loss doctrine barred recovery from that person for purely economic losses.

D. *Exceptions to Application/Enforcement of the Doctrine*

1. *Negligent Misrepresentation*

While generally recognizing the economic loss doctrine as a bar to purely economic damages claims, certain jurisdictions have carved out an exception to this rule when the tort claimed is negligent misrepresentation. Illinois probably has the most extensively developed line of cases dealing with this issue.

Illinois originally adopted the economic loss rule in *Moorman Manufacturing Co. v. National Tank Co.*,[93] but it also recognized two exceptions—intentional fraud and where a defendant "*who is in the business of supplying information* for the guidance of others" makes a negligent misrepresentation.[94]

By subsequent decisions, the *Moorman* rule and its exceptions have been refined and advanced. In *2314 Lincoln Park West Condominium Assn. v. Mann, Gin, Ebel & Frazier*[95] and *Fireman's Fund Insurance Co. v. SEC Donohue, Inc.*,[96] the courts found that the typical object of engineering and architectural efforts is to produce something tangible (a building or structure) and, in those instances, design professionals are not engaged in the business of "supplying information." Thus, in those instances, the negligent misrepresentation exception does not apply to design professionals.

The Illinois Supreme Court has also determined that the rulings in *2314 Lincoln Park West* and *Fireman's Fund Insurance Co.* are not absolute. In *Tolan & Son, Inc. v. KILL Architects, Inc.*,[97] design professionals prepared plans and specifications and provided consulting services during construction. The plaintiff urged the court to bifurcate the two roles filled by the design professional defendants and find that the consulting services performed during construction fell within the negligent misrepresentation exception recognized in *Moorman*. Although the court declined to do so (noting that the consulting services were ancillary to design services and, therefore, a continuation of design), it observed that in a purely evaluative or advisory role, a design professional could well "engage in the business of providing information"

---

92. *Moransais*, 744 So. 2d at 979.
93. 435 N.E. 2d 443 (Ill. 1982).
94. *Id.* at 452 (emphasis added).
95. 555 N.E. 2d 346 (Ill. 1990).
96. 679 N.E. 2d 1197 (Ill. 1997).
97. 719 N.E.2d 288 (Ill. Ct. App. 1999).

and, therefore, fall within the exception. This decision clearly raises implications for design firms that engage in purely construction management services on a particular project. The *Tolan* court explicitly allowed for the possibility that design professionals could be sued for negligent misrepresentation when only economic damages are sought.

The negligent misrepresentation "exception" has been adopted in other jurisdictions.[98] An odd result was reached in *Council of the Dorset Condominium Apartments v. Dorset Apartments*.[99] In *Dorset*, engineers hired to perform a pre-purchase inspection report were sued when the estimated $35,000 repairs ballooned to over $700,000. The court inexplicably noted that engineers (but not architects!) fall within the exception to the economic loss rule: "Unlike an architect who prepares plans which are converted into a tangible product—a building—engineers prepare information to be used as information."[100] Therefore, in Illinois an engineer, but not an architect, is exposed to a claim for negligent misrepresentation where the damages are economic in nature.

2. Sudden/Calamitous Event

Where the economic loss doctrine otherwise might apply, some jurisdictions nonetheless permit tort claims when there is an allegation of threat to physical safety. In Washington, an exception arises when the plaintiff has suffered economic damage caused by a "sudden and dangerous" event as described in *Touchet Valley Grain Growers, Inc. v. Opp & Seibold Construction, Inc.*[101] In *Touchet*, the owner of a grain storage building sued the general contractor and metal fabrication subcontractor after its building collapsed. The Washington Supreme Court allowed the plaintiff to recover economic losses from the defendants under a tort theory of liability based

---

98. *See* Resell Constr. Managers v. EH Constr., LLC, 134 S.W.3d 575 (Ky. 2004) (contractor claim of negligent misrepresentation against construction manager not barred by economic loss doctrine); Note Constr. Corp. v. Keyes Assocs., 694 N.E.2d 401 (Mass. 1998) (economic losses resulting from negligent misrepresentation are exception to the doctrine, when claimant reasonably relies upon design professional's work product); Belt-Rite Contractors v. Architectural Studio, No. 74 MAP 2002, 2005 Pa. LEXIS 99 (Jan. 19, 2005) (economic loss doctrine does not bar contractor's negligent misrepresentation claim against architect [not in privity], when contractor reasonably relied upon defective plans and specifications); AmSouth Erectors, LLC v. Skaggs Iron Works, Inc., No. W 2002-01944-COA-R3-CV, 2003 Tenn. Ct. App. LEXIS 551 (Aug. 5, 2003) (subcontractor's claim against construction management firm dismissed, because of failure to prove negligent misrepresentation, an exception to economic loss doctrine).

99. CIV.A.90C-10-269, 1992 WL 240444 (1992) (unpublished but cited in subsequent decisions in that jurisdiction).

100. *But see* Blake Constr. Co., Inc. v. Alley, 353 S.E.2d 724 (Va. 1987) (refusing to recognize negligent misrepresentation claim for purely economic loss in construction context).

101. 119 Wash. 2d 334, 831 P.2d 724 (1992).

on a "risk of harm" analysis previously adopted by the court in a products liability case.[102] The *Touchet* court held that in cases where the otherwise purely economic harm arises out of a "sudden and dangerous" or "calamitous" event, a plaintiff would be allowed to seek damages in tort.[103]

This narrow exception is justified on the theory that "sudden and dangerous" events are more likely to result in bodily injury and calamitous property damage. Accordingly, the availability of a tort remedy, even when the plaintiff has suffered only economic losses, provides an important incentive for design professionals to exercise reasonable care whenever there may be this kind of damage.

Maryland has adopted a similar approach. In *Milton Co. v. Council of Unit Owners of Bentley Place Condominium*,[104] the court found that some of the claimed construction defects "clearly ... pose a risk of serious physical injury" and, therefore, "a plaintiff may recover in tort for purely economic loss where the defect creates a substantial risk of death or personal injury."[105]

IV. Site Safety

A. Background

Generally, workers' compensation statutes grant immunity to the contractor from suit for injuries sustained by the contractor's employees in the course of working on a construction project. These statutes represent a tradeoff of interests: the injured worker or the deceased worker's family receive access to a pool of funds available for a defined amount of compensation following an injury; and the contractor/employer, having paid into the workers' compensation fund, has purchased certainty by being made immune from suit by his injured worker or the deceased worker's family. These statutes provide compensation for injured workers without requiring proof of fault or negligence of any kind, and that compensation is generally the exclusive remedy of the employee against his employer.

However, in many instances, the recovery from state workers' compensation funds is insufficient compared to the damages that can be obtained in a lawsuit. As the employer, the contractor is immune from such tort suits, and in some states, the owner may be immune as a statutory employer. In addition, owners are less likely targets of such suits since the contract between owner and contractor normally

---

102. Washington Water Power Co. v. Graybar Elec. Co., 774 P.2d 1199 (Wash. 1989).
103. *Touchet*, 119 Wn. 2d at 352, 831 P.2d at 733.
104. 121 Md. Ct. App. 100, 708 A.2d 1047 (1998).
105. *Id.* at 115, 708 A.2d at 1054; *see also* Danforth v. Acorn Structures, Inc., No. 90C-JN-30, 1991 Del. Super. LEXIS 390 (Aug. 27, 1991) (court ruled against homeowner who sued contractor for negligence, because damages were purely economic; sudden, calamitous event is an exception to the economic loss doctrine, but no such event occurred in this case).

assigns control over the site and job site safety to the contractor. If an employee of a subcontractor is injured, the law may shield the subcontractor, general contractor, and owner from liability. However, third parties, such as design professionals and suppliers, may still be exposed to these claims.[106] While an injured worker may not sue his own employer for a job site injury, he or she may be authorized statutorily to sue any person not hired by the same employer who is at fault for an injury.[107]

The unintended consequence of workers' compensation laws has been to encourage injured workers to seek recovery from parties not shielded by these statutes. Some courts have recognized that the "proliferation of such suits may conceivably be based upon the desire of the injured workman to find a 'deep pocket' to more adequately compensate for serious injury than may be recovered from the employer or general contractor under the limitations of workers' compensation statutes."[108] In response, a minority of states have extended the immunity of their workers' compensation statutes to design professionals (see below).

Typically, claims against design professionals either implicate the engineer's design as having created the hazard leading to injury or, more commonly, implicate the construction site services provided by the engineer. Most state statutes impose principal responsibility for site safety on the contractor, but in some jurisdictions, burdens are placed on owners, who in turn may seek to shift the task to the design professional. Design professionals should know the law concerning site safety in the jurisdiction where the project will be constructed, especially as to the duties imposed on design professionals.

B. *Responsibility of the Contractor and Owner*

Although the contractor and/or owner may be immune from suit under workers' compensation statutes, it is important to understand their legal and contractual responsibilities for safety. These issues also may affect their liability in suits by injured third parties or bystanders who are not employed by any project participant.

---

106. Sonners, Inc. v. Department of Labor & Indus., 3 P.3d 756 (Wash. Ct. App. 2000); Jansen v. Aaron Process Equip. Co., 207 F.3d 1001, 1003 (7th Cir. 2000) (employee recovered workers' compensation benefits from employer and subsequently recovered a judgment from manufacturer of the equipment involved in employee's injury).

107. Frazier v. St. Paul Ins. Co., 880 So. 2d 406 (Ala. 2003) (Under Ala. Code § 25-5-14 (1984), a workers' compensation exclusivity-of-remedy provision provided complete immunity to employers and limited immunity to others closely affiliated with the employment relationship from liability, but court did not extend immunity beyond this.); Williams v. Int'l Paper Co., 129 Cal. Ct. App. 3d 810, 816 (1982) (third party tortfeasor not immune from common law tort action; self-insured employer does not fall within definition of "person other than employer").

108. Hanna v. Huer, Johns, Neel, Rivers & Webb, 662 P.2d 243, 249 (Kan. 1983).

Under the workplace safety statutes, many jurisdictions assign a nondelegable duty to the general contractor to ensure the safety of all workers on a job site.[109] A number of jurisdictions also recognize that a general contractor owes a duty to all employees at a work site to comply with, or ensure compliance with, safety regulations.[110] This duty to keep the premises in a safe condition may subject the general contractor to direct liability for negligence in two situations: those arising from a premises defect; and those arising from an activity or instrumentality. This duty is based upon the general contractor's "innate supervisory authority" that constitutes "per se control over the workplace" and statutory authority.[111]

Absent some form of control over the subcontractor's method of operation, the general contractor and owner of the construction project are not liable for injuries to the subcontractor's employees.[112] The subcontractor, despite the general contractor's duty for workplace safety, retains concurrent responsibility to meet workplace safety standards in the areas under its control.[113]

---

109. Brennan v. RCP Assocs., 257 A.D.2d 389 (N.Y. Ct. App. Div. 1999) (Under McKinney's N.Y. Lab. Law § 240(1) (1981), a nondelegable duty is imposed upon owners and general contractors to furnish proper safety devices and protection in order to ensure the safety of workers exposed to elevation-related risks during the construction, repair, demolition, painting, and alteration of a building or structure.); Nave v. Harlan Jones Drilling, 827 P.2d 1239 (Mont. 1992) (general contractor had contractual duty to provide for the safety of all employees at the work site, which it could not delegate to the subcontractor); Pfenninger v. Hunterdon Cent. Reg'l High Sch., 338 N.J. Super. Ct. App. Div. 572, 581, 770 A.2d 1173, 1179 (1999) (decedent died when the trench he was working in caved in; court held a landowner or general contractor has a nondelegable duty to ensure the safety of the premises upon which work is to be performed).

110. Stute v. P.B.M.C., Inc., 114 Wash. 2d 454, 464, 788 P.2d 545, 550–51 (1990) (it was the general contractor's responsibility to furnish safety equipment or to contractually require subcontractors to furnish adequate safety equipment relevant to their responsibilities at the job site); Arias v. MHI P'ship, 978 S.W.2d 660, 662 (Tex. Ct. App. 1998) (a general contractor on a construction site, who is in control of the premises, is charged with the same duty as an owner or occupier, who has a duty to use reasonable care to keep the premises under its control in a safe condition).

111. *Stute*, 114 Wash. 2d at 457–60, 788 P.2d at 546–49.

112. Stepanek v. Kober Constr., 191 Mont. 430, 433, 625 P.2d 51, 53 (1981) (contractor found liable to subcontractor's injured employee, because contractor exercised control over safety conditions at the work site); *but see* Jones v. Logansport, 436 N.E.2d at 1147 (Indiana recognizes exception to this rule if work being performed is intrinsically dangerous).

113. Gilbert H. Moen Co. v. Island Steel Erectors, Inc., 128 Wash. 2d 745, 757, 912 P.2d 472, 478–79 (1996); Lewis v. N.J. Riebe Enters., 170 Ariz. 384, 388, 825 P.2d 5 (1992) (although general contractor has a general duty to provide a reasonably safe workplace for the employees of subcontractors, the scope of this duty extends only as far as the amount of control the general contractor retains over the work of the subcontractor).

It may be difficult to determine whether an owner maintained control over the site. If the owner retains "only the right to inspect the construction work to see that the contract specifications are met, while the independent contractor controls how and when the work is to be done, there is probably not sufficient retained control to subject [the owner] to liability."[114]

Merely retaining the power to forbid or stop work from being done in an unsafe manner is not a sufficient degree of retained control.[115] "Similarly, if the employer retains only standard 'boilerplate' provisions with respect to safety inspections and requirements, but assumes no affirmative duties with respect to safety and never directs the method of performance, there is [probably] insufficient control" to render the employer liable for injury caused by unsafe performance of the work.[116] It is extremely important, nevertheless, to understand that if the owner assumes an affirmative duty to implement safety precautions, by contract or conduct, the owner will be liable for injuries caused to others by the unsafe performance of the work, if it negligently allowed the unsafe work to continue.[117]

When sued by an injured worker, a design professional firm should consider whether it can cross-claim against the contractor and subcontractor. Alternatively, it may argue to the jury that the contractor or subcontractor was 100 percent at fault (even if the employer is not a party to the suit—the empty chair defense). There are, however, legal obstacles to either option. First, the design firm typically will not have contractual privity with either the contractor or any of the subcontractors. This means that the only legal basis for a cross-claim (in the absence of contractual indemnity) would be a tort claim for negligence or a statutory claim for contribution.

Second, a design professional firm may not be able to assert a claim of contribution or seek an allocation of fault against the injured worker's employer (whether

---

114. *Werdehausen v. Union Elec. Co.*, 801 S.W.2d 358, 364 (Mo. 1990) (if the employer retained only the right to inspect the construction work to see that contract specifications were met, while the independent contractor controlled how and when the work was to be done, there was not sufficient retained control to subject the employer to liability).

115. *Id., citing* Moloso v. State, 644 P.2d 205, 211 (Alaska 1982); *see also* DeVille v. Shell Oil Co., 366 F.2d 123, 125–26 (9th Cir. 1966).

116. *Werdehausen*, 801 S.W.2d at 364.

117. *Id.*; *see, e.g.,* Jones v. Chevron U.S.A., Inc., 718 P.2d 890, 896 (Wyo. 1986) (because owner of work site retained the right to direct the manner of the independent contractor's performance and assumed affirmative duties with respect to safety, owner owed a duty of reasonable care to the independent contractor's employee, even when the employee was injured doing the very work the contractor was hired to perform); Hammond v. Bechtel Inc., 606 P.2d 1269, 1276 (Alaska 1980) (employers contended that they had no duty to inspect, repair, or maintain the independent contractor's equipment; court held that genuine issues of material fact remained regarding whether the employers were responsible for exercising control over work site and turning over site free of safety hazards).

that employer is the contractor or a subcontractor). For example, Washington's Tort Reform Act[118] was amended to exclude any assessment of fault against an employer with immunity under Washington's Industrial Insurance Act.[119] In other words, the design firm may not be able to sue the immune employer or use the empty chair defense at trial. Thus, it could be held liable for the injured worker's employer's share of fault.[120]

C. *Indemnity for the Design Professional*

A design firm may be able to protect itself from the risk of suit by an injured worker by requiring the owner to indemnify it from personal injury lawsuits brought by contractors' or subcontractors' employees. Subject to any anti-indemnity statute (discussed below), this indemnification clause would allow the design professional firm to tender defense of the claim to the owner. An alternative approach is to prepare contract documents requiring the contractor and subcontractors to indemnify the design professional and owner from worker injury claims and to waive workers' compensation immunity to that extent.

In Washington, these waivers are valid and enforceable. In *Gilbert H. Moen Co. v. Island Steel Erectors, Inc.*,[121] a subcontractor's employee sued the general contractor, claiming that the it had been negligent in failing to require its subcontractors to comply with safety regulations. The contractor settled with the worker and then sued the subcontractor for indemnification. The court held that an indemnification clause, coupled with an express waiver of workers' compensation immunity, was a valid and permissible method for protecting against subcontractor negligence in managing site safety for its own workers.

Anti-indemnity statutes, however, may limit the scope of any such protection. These laws prohibit clauses that purport to indemnify a party for injury caused by the indemnitee's own negligence or willful misconduct, or the negligence or willful misconduct of those under the indemnitee's control.[122] Anti-indemnity statutes result from public policy governing contracts purporting to indemnify parties against their own negligence.[123] "Such contracts may remove or reduce the incentives to protect workers and others from personal injury and the anti-indemnity statute would help to insure that persons responsible for the work were properly motivated to provide safe working conditions."[124]

---

118. WASH. REV. CODE § 4.22.070(1) (2005).

119. WASH. REV. CODE Title 51.

120. *Gilbert H. Moen*, 128 Wash. 2d 745, 912 P.2d 472.

121. *Id.* 128 Wash. 2d at 760–61, 912 P.2d at 480–81.

122. Aetna Cas. & Sur. Co. v. Marion Equip. Co., 894 P.2d 664, 670 (Alaska 1995) (court refused to enforce indemnification provision in lease agreement between construction equipment company and contractor due to anti-indemnity statute).

123. Union Pac. Res. Co. v. Dolenc, 2004 Wy. 36, 86 P.3d 1287 (2004) (anti-indemnity statute applies only to contracts for work dealing directly with the mining of oil and gas).

124. *Id.*

Under an anti-indemnity statute, a design professional firm cannot seek indemnification from the owner for injuries caused by the design firm's sole negligence. It can seek indemnification for such claims when caused by the concurrent negligence of the owner, contractor or subcontractor, but only to the extent of that party's negligence. However, "concurrent negligence" language must be present in the indemnification clause to be valid and enforceable.[125] For example, the Washington law states:

> A covenant, promise, agreement or understanding in, or in connection with or collateral to, a contract ... relative to the construction ... of, any building, ... purporting to indemnify against liability for damages arising out of bodily injury to persons or damage to property:
> (1) Caused by or resulting from the sole negligence of the indemnitee, his agents or employees is against public policy and is void and unenforceable;
> (2) Caused by or resulting from the concurrent negligence of (a) the indemnitee or the indemnitee's agents or employees, and (b) the indemnitor or the indemnitor's agents or employees, is valid and enforceable only to the extent of the indemnitor's negligence and only if the agreement specifically and expressly provides therefor, and may waive the indemnitor's immunity under industrial insurance, Title 51 RCW, only if the agreement specifically and expressly provides therefor and the waiver was mutually negotiated by the parties.[126]

Therefore, to determine the validity of and requirements for an indemnification provision, the design professional must examine the applicable state law, including any anti-indemnity statute.

D. *Theories of Liability against the Design Professional*

Injured workers typically allege one or more of three theories: negligent design; negligent construction management or supervision; and negligent failure to warn of dangerous site conditions.[127] Design professionals are obliged to exercise such

---

125. *Gilbert H. Moen*, 128 Wash. 2d 745, 912 P.2d at 472.

126. WASH. REV. CODE § 4.24.115 (2005); *see also* Hoffman Constr. Co. v. United States Fabrication & Erection, 32 P.3d 346, 359 (Alaska 2001) (indemnity agreements that shift liability from the indemnitee to the indemnitor are valid, subject to the restrictions of ALASKA STAT. § 45.45.900 (1986); and *see* ARIZ. REV. STAT. ANN. § 34-226 (2006); CAL. CIV. CODE §§ 2782, 2782.5 (2006); CONN. GEN. STAT. § 52-572k (2004); HAWAII REV. STAT. § 431:10-222 (2005); IDAHO CODE § 29-114 (2006); IND. CODE ANN. § 26-2-5-1 (2005); MICH. COMP. LAWS ANN. § 691.991 (2005); N.M. STAT. ANN. § 56-7-1 (2006); S.C. CODE ANN. § 32-2-10 (2005); TENN. CODE ANN. § 62-6-123 (2005); UTAH CODE ANN. § 13-8-1 (2005); VA. CODE ANN. § 11-4.1 (2006); W. VA. CODE § 55-8-14 (2006).

127. *See* Le May v. USH Props., Inc, 338 So. 2d 1143, 1144–45 (Fla. 1976).

reasonable care, technical skill and ability, and diligence as are ordinarily required of design professionals in the course of preparing their plans or performing inspections and supervision during construction for the protection of any person who foreseeably and with reasonable certainty might be injured by their failure to do so.[128]

However, a design professional is not liable for injury caused to a third party if the structural failure resulting in the injury is caused when those responsible for construction or supervising the construction deviate from the plans and specifications.[129] This is true even if the owner impliedly warrants the adequacy and sufficiency of the plans.[130]

E. *The Design Professional's Role on Site*

Most jurisdictions refuse to impose liability on design professionals for site safety, unless there has been a clear assumption of the duty to monitor site safety.[131]

---

128. Le May v. USH Props., Inc, 338 So. 2d 1143, 1144–1145 (Fla. 1976).
129. *Goette,* 413 N.W.2d at 856.
130. *Id.*; *see also* United States v. Spearin, 248 U.S. 132, 136 (1918) (if a contractor is bound to build according to plans and specifications prepared by the owner, the contractor will not be responsible for the consequences of defects in the plans and specifications).
131. *Hanna,* 662 P.2d 243 (architect not liable for injury caused by contractor error, where contract established the duty to provide and enforce safe working conditions was with the general contractor); *contrast* Balagna v. Shawnee County, 233 Kan. 1068, 668 P.2d 157 (1983) (summary judgment was improper where engineer had actual knowledge of an unsafe condition); *see also* Krieger v. J.E. Greiner Co., 382 A.2d 1069 (Md. 1978); Walters v. Kellam & Foley, 360 N.E.2d 199 (Ind. Ct. App. 1977) (neither general contractor nor architect had a duty to supervise the construction site or insure plaintiff's safety; but mechanical engineer, mechanical contractor, and HVAC manufacturer all had a legal duty to insure the plaintiff's safety while he worked on the construction project); Brown v. Gamble Constr. Co., 537 S.W.2d 685 (Mo. Ct. App. 1976) (worker was killed when he fell through a hole in the roof of a partially constructed structure; court held that architects were under no duty to supervise construction, unless they expressly agreed to do so); Seeney v. Dover Country Club Apartments, Inc., 318 A.2d 619 (Del. Super. Ct. 1974) (because engineering firm did not retain or voluntarily exercise any control over the manner or methods used by the contractor in its work, engineering firm not liable for injuries sustained by worker); *Reber,* 474 P.2d 852 (where workers were injured in a structural collapse, court held that statute did not impose a duty upon an architect to directly control the contractor's methods, but only to assure that the results complied with design specifications); Day v. National United States Radiator Corp., 128 So. 2d 660 (La. 1961) (architect liable for injuries arising from boiler explosion during installation when architect's plan were not defective); Jones v. James Reeves Contractors, 701 So. 2d 774, 784 (Miss. 1997) (architect had no duty to supervise the site and, therefore, not liable for work site injuries); Hobson v. Waggoner Eng'g, Inc., 878 So. 2d 68, 72 (Miss. Ct. App. 2003), *cert. denied,* 878 So. 2d 66 (2004) (general contractor, not engineer, had duty for supervision of the work and project safety).

General contractors have a nondelegable and specific duty to comply with state safety regulations.[132]

In an early court decision interpreting the duty of design professionals to "supervise" or "inspect" a contractor's work, the Minnesota Supreme Court drew a line between inspection for general conformance versus a duty to ensure that plans and specifications are carried out in every detail.[133] In that case, a contractor did not properly fasten a roof to the building and a claim was brought against the architect for failing to ensure that the contractor followed the plans and specifications. The court rejected the claim, noting that the architect's contractual duty to make periodic site visits to review the quality and progress of the contractor's work (including whether the contractor was performing in accordance with the plans and specifications) did not rise to the level of imposing responsibility to *ensure* that the contractor carried out his work in accordance with the contract documents.

This ruling was cited favorably in *Mayor and City Council of City of Columbus, Mississippi v. Clark-Dietz & Associates-Engineers, Inc.*[134] The court noted that general inspection services do not impose a duty to make "continuous and exhaustive" inspections of the contractor's work.

Contract documents usually contain a limitation on responsibility assigned to the engineer. The engineer typically avoids any assumption of control over or direction of the contractor's means and methods, techniques, sequences, or procedures, including safety precautions or programs. For example, the Engineering Joint Contracts Design Committee (EJCDC) Owner-Engineer Agreement provides:

> [The] engineer shall not, during such visits or as a result of such observations of contractor's work of progress, supervise, direct or have control over contractor(s) work....[135]

The American Institute of Architects (AIA) Owner-Architect Agreement expresses a similar limitation of responsibility:

> The purpose of the engineer's [site] visits ... will be to enable engineer to better carry out the duties and responsibilities assigned to and undertaken by engineer during the construction phase and ... to provide the owner a greater degree of confidence that the completed work of contractor(s) will conform *generally* to the contract documents....[136]

---

132. Kamla v. Space Needle Corp., 52 P.3d 472 (Wash. 2002).

133. Moundsview Indep. Sch. Dist. No. 621 v. Buetow & Assocs., Inc., 253 N.W.2d 836, 837 (Minn. 1977).

134. Mayor of Columbus, Miss. v. Clark-Dietz, 550 F. Supp. 610, 627 (Miss. 1982).

135. EJCDC Document No. 1910-1, Standard Form Agreement between Owner and Engineer for Professional Services, ¶ 1.6.2.3.

136. AIA Document B151-1997, ¶ 2.6.5 (emphasis added).

The design professional's role at the site, as defined by its contractual responsibilities, plays an important part in identifying the degree to which a duty for site safety may be imposed. Courts may look to several general factors in determining whether a design professional affirmatively undertook responsibility for site safety (as discussed above, some of these same elements might also affect whether any statutory immunity applies). These factors include: actual supervision and control of the work; retention of the right to supervise and control; constant participation in ongoing activities at the construction site; supervision and coordination of subcontractors; assumption of responsibilities for safety practices; authority to issue change orders; and the right to stop the work.[137]

Whether an architect or engineer has assumed site safety responsibilities by contract has been extensively litigated. Below are clauses the courts have pointed to in holding design firms liable:

- Engineering firm contracted to develop and execute safety program and had authority to stop contractor operations in case of noncompliance.[138]
- Engineering firm contracted to ensure overall responsibility to maintain safe construction area, including responsibility to take "every precaution against injuries to persons."[139]
- Architect assumed all duties to manage construction, to perform all managerial functions and to "protect Owner's interest in safety."[140]

In contrast, the following provisions were relied upon by the courts in finding that the design professional was not liable under the common law or the terms of an immunity statute:

- Contract with owner precluded on-site engineer from "advising on or issue directions as to safety precautions and programs in connection with the work."[141]
- Engineering firm only undertook "to assure that the projects are constructed in reasonable conformity with plans, specifications, and contract provisions" without any further reference to safety.[142]
- Under AIA standard contract, architect's ability to reject work did not give rise to any duty to the contractor, subcontractors, or their employees.[143]

---

137. *Hobson*, 878 So. 2d at 72; *Hanna*, 662 P.2d at 243.

138. Riggins v. Bechtel Power Corp., 722 P.2d 819 (Wash. Ct. App. 1986).

139. Piccirillo v. Beltrone-Turner, 284 A.D.2d 854, 727 N.Y.S.2d 721 (Ct. App. Div. 2001) (construction manager held liable to pedestrian injured in fall at airport).

140. Simon v. Omaha Public Power Dist., 202 N.W.2d 157 (Neb. 1972) (architect liable for injury to worker who fell through opening in floor deck).

141. Williams, Hatfield & Stoner, Inc. v. Malcolm, 687 So. 2d 295 (Fla. Dist. Ct. App. 1997); Black v. Gorman-Rupp, 791 So. 2d 793 (La. Ct. App. 2001).

142. Estate of Reyes v. Parsons Brinckerhoff Constr. Servs., Inc., 784 So. 2d 514 (Fla. 2001).

143. *Hanna*, 662 P.2d at 243.

- Contract stated that engineer had not been retained to design or review contractor's safety practices or the means and methods of work.[144]
- Architect's contract disclaimed control over, or charge of, construction means and methods or for safety precautions or programs.[145]
- Engineer's contract with the Department of Transportation did not give it responsibility to supervise or control activity resulting in worker's death.[146]

Regardless of the contract language, many courts have imposed liability on design professionals (based upon common law negligence or an exception to an immunity statute) for on-site injuries when the evidence established that the architect or engineer affirmatively acted in a manner to assume responsibility for worker safety.[147]

F. *Design Professional Immunity Statutes*

Many states have addressed this problem through the enactment of design professional immunity statutes, which exempt design professionals from liability for injuries to third persons at construction sites.[148] Under the typical formulation, a design professional is immune from liability to third persons for such injuries unless it specifically assumed responsibility for safety practices in its contract, or it "actually exercised control over the portion of the premises where the worker was injured."[149]

---

144. *Hobson*, 878 So. 2d at 68.

145. Dillard v. Shaughnessy, Fickel & Scott Architects, 864 S.W.2d 368 (Mo. Ct. App. 1993).

146. Hernandez v. Yonkers Contracting Co., 306 A.D. 2d 379, 760 N.Y.S.2d 865 (N.Y. Ct. App. Div. 2003).

147. Bauer v. Howard S. Wright Constr. Co., 101 Wash. Ct. App. 1046 (2000), No. 44817-0-I, 2000 Wash. Ct. App. LEXIS 1689 (July 17, 2000) (summary judgment for engineer reversed based on exceptions in immunity statute for exerting actual control over the site and negligent preparation of plans and specifications; on-site engineer; engineer hired to perform construction monitoring with authority to stop work if unsafe working conditions observed had in fact previously stopped work for that reason; engineer approved changes in the methods for constructing the temporary retaining wall, including the elimination of safety features originally included in the plans); Jones v. Parsons Transp. Group, Inc., 2004 WL 1254029 (D. Md. May 20, 2004) (no immunity statute involved; court held that while the engineer was retained to perform inspection services, review the contractor's site safety plans, and assure compliance with safety standards, there was no evidence that the engineer undertook these services for the benefit of the contractor or its workers); *see also* Frank Wagner, *Liability to One Injured in Course of Construction Based upon Architect's Alleged Failure to Carry Out Supervisory Responsibilities*, 59 A.L.R.3d 869 (1974 and supp.).

148. *E.g.,* CONN. GEN. STAT. § 31-293 (2004); FLA. STAT. ch. 440.09 (2005); GA. CODE ANN. § 34-9-11 (2005); WASH. REV. CODE § 51.24.035 (2005); KAN. STAT. ANN. § 44-501 (2005).

149. *Bauer,* 2000 Wash. Ct. App. LEXIS 1689 (issue of fact existed as to whether consulting engineer exercised actual control during installation of soil nail wall such that engineer could be

The New York design professional immunity statute omits the requirement that the architect or engineer assume responsibility by a *written* contract:

> No liability for non-compliance with any provision of [site safety responsibility] shall be imposed on professional engineers ... architects ... or landscape architects ... who do not direct or control the work for activities other than planning and design.[150]

Thus, in New York, absent a written contractual assignment for site safety, exemption still can be waived if the design professional directs or controls the contractor's work.

Some design professional immunity statutes explicitly decline to extend protection to claims for injuries caused by defective plans and specifications.[151]

G. *Control over the Work and OSHA Fines*

Exercising control over the work also may expose the design professional to regulatory fines. Congress enacted the Occupational Safety and Health Act (OSHA) in 1970 to help reduce employment-related injuries and illness.[152] The Secretary of Labor is vested with the authority to enforce OSHA. Pursuant to that authority, the Secretary has issued general industry standards[153] and industry-specific standards.[154] Under certain circumstances, design professional firms can face civil liability for a contractor's failure to comply with these regulations.[155]

In *CH2M HILL, Inc. v. Herman*,[156] an OSHA citation was issued to the project engineer following a catastrophic work site accident that resulted in three deaths. The engineer's contract provided that it was not responsible for site safety; the contractor bore that duty. Likewise, the engineer did not have the ability to direct the contractor to make corrections, nor did it have stop-work authority; those duties rested with

---

held liable for physical injuries sustained when wall collapsed). *See also* Pelletier v. Sordoni/Skanska Constr. Co., 264 Conn. 509, 526, 825 A.2d 72, 83 (2003) (general contractor not liable by operation of immunity statute for injuries to subcontractor's employee struck by fallen beam); Agner v. APAC-Florida, Inc., 821 So. 2d 336, 340 (Fla. 2002) (design professional immune from liability based on design professional immunity statute concerning motorist killed due to roadside construction).

150. N.Y. LAB. LAW ART. 10, § 241-a (McKinney 2002).

151. *Pelletier*, 264 Conn. at 526, 825 A.2d at 83; *Agner*, 821 So. 2d at 340; Cowart v. Crown Am. Props., 258 Ga. Ct. App. 21, 22, 572 S.E.2d 706, 708–709 (2002).

152. *See* 29 U.S.C. § 651 (1970).

153. *See* 29 C.F.R. Part 1910.

154. The construction standards are set forth in 29 C.F.R. Part 1926. *See also* George v. Myers, 169 Ore. Ct. App. 472, 480, 10 P.3d 265, 270–271 (2000).

155. CH2M HILL, Inc. v. Herman, 192 F.3d 711 (7th Cir. 1999).

156. *Id.*

the municipality/owner. Notwithstanding these limitations to the engineer's scope of service, the OSHA citation was issued because the engineer responded when a differing site condition arose (detection of methane gas) and, at the owner's direction, provided clarification as to what type of electrical equipment could be used below ground where there was potential that the equipment could ignite the gas. OSHA found that by engaging in this advisory activity, the engineer had "engaged in construction work." The court rejected this formulation, however, turning to the effective contract path to find that CH2M HILL's work did not constitute "engaging in construction work." The court also found that because CH2M HILL had limited authority (approving types of equipment) that was subject to the owner's ultimate approval, it lacked sufficient control to be classified as engaged in providing construction services, rather than merely consulting services. The court noted that OSHA's own "substantial supervision" test compelled this finding. The engineer had no authority over site safety or construction means and methods, and it *did not* function in a manner that was "inextricably intertwined with the actual physical labor of construction."

Contrast *CH2M HILL* with *Carvalho v. Toll Bros. & Developers*,[157] which involved a wrongful death action, rather than OSHA compliance. There, the engineer was tasked with on-site construction observation duties, but had no authority for site safety (this duty rested with the contractor). The engineer was obliged to monitor the progress of construction on a time-sensitive project, which included the proper handling of underground utilities crossing through a portion of the project area where trenching was to occur. The engineer also had to ensure that the contractor protected utility lines crossing the trench. The engineer's contract further provided that the condition of the trench was relevant to determining construction procedures. While the engineer was present and observing, a worker was killed when an unreinforced trench collapsed on him.

In this instance, the engineer knew that the contractor chose not to use a trench box. To do so would have required cutting through two utility pipes, which later would have to be repaired, and thereby slow the work. As a result, the court denied the engineer's motion for summary judgment. The court found that the circumstances of the engineer's responsibility for the rate-of-progress of the work, combined with being present when the choice was made *not* to use a trench box, "demonstrate[d] the interrelationship between safety and progress." The court focused on the forseeability of the risk, noting that in other areas where utilities were not a factor, the contractor had recognized the potential danger and used a trench box.

> The connection between the engineer's responsibilities over the progress of work and safety measures at the job site is relevant in determining whether it is fair to impose a duty of care addressed to work site safety conditions.[158]

---

157. 675 A.2d 209 (N.J. 1996).
158. *Id.* at 214.

The court found that the engineer owed a duty of care to workers based upon his responsibility for the work progress and his knowledge, despite the lack of any contractual responsibility for safety.

H. *Contract Provisions*

To ensure that the engineer does not assume responsibility for site safety, the contract should expressly disclaim that obligation. These provisions are important to avoid liability for negligence, and may affect the immunity of the design professional where states grant that protection.

The most common manner in which a design professional will be implicated following an accident arises from fieldwork performed at the site. Design engineers are routinely charged with performing observation services during construction to verify that the contractor is constructing the project according to the plans and specifications. The degree of involvement at the construction site may range from part-time, intermittent observation, to providing a full-time on-site observer(s), to actual construction management. Each of these on-site services represents an escalating degree of involvement for the design professional at the project, and each brings with it varying degrees of scrutiny that will be applied when considering whether the design professional bears responsibility for an accident (and, in certain jurisdictions, whether the design professional will benefit from statutory immunity).

The design professional should guard against inconsistent and potentially conflicting contract terms. For example, the engineer/owner agreement may provide for only "periodic inspections" for conformance with the design intent, yet the general conditions of the construction contract also may require the engineer to "ensure" that the project is being completed according to design parameters. In the event of an accident attributed to the contractor's failure to perform according to the plans and specifications, the engineer may face a claim for failing to catch the error. Relying on a defense of a limited site inspection duty has no guarantee of success. At the time of contracting, the design professional should be alert to potentially conflicting language.

It is also common for there to be a contractual disconnect between the broad expression of the engineer's duties in the master agreement and a severe limitation on those duties in specific task orders that follow. The contractual expression of duties for site services should be consistent and uniform throughout the contract documents.

I. *Duty to Keep Own Employees Safe*

Site safety obligations must be distinguished from responsibility for the safety of one's own employees. Typically, a design firm owes its employees a nondelegable duty to ensure their physical safety in any location to which those employees are sent to work.[159]

---

159. ALA. CODE § 25-1-1 (2005) (employer must make workplace reasonably safe for employees, including use of proper safety devices and adoption of safety methods); Division of Occupational Safety & Health of the Indus. Comm'n v. Chuck Westenburg Concrete

However, an employer normally is immune from civil lawsuits by its own employees for unintentional job site injuries under workers' compensation acts.[160]

Owners sometimes ask design firms to waive workers' compensation immunity and indemnify them from claims for injuries sustained by the design firm's employees. Again, each state's laws must be reviewed to determine the enforceability of that indemnification.

J. *Hazardous Waste Sites*

There are special federal regulations relating to worker safety on sites defined as hazardous waste sites. Under 29 C.F.R. § 1910.120, any "employer" with workers performing remediation or emergency response work on such a site must maintain a site health and safety plan. It sometimes can be unclear in traditional owner/contractor, owner/engineer relationships who is the "site manager." In addition, contractors often do not have the expertise to develop such plans and will ask the engineer to prepare one. Participation in the preparation of a site safety plan could raise questions regarding the engineer's immunity from suit in the event of a worker injury or exposure.

In addition, the regulations require any "employer" who retains contractors or subcontractors to work in hazardous waste operations to inform those entities of any

---

Contractors, Inc., 193 Ariz. 260, 269–70, 972 P.2d 244, 253–54 (1998) (each employer at a construction site is responsible for assuming that its conduct does not create hazards to any employees at the site, is responsible for violations it reasonably could have been expected to prevent or abate, and must make reasonable efforts to detect violations of standards and to exert reasonable efforts to have them abated); D.C. CODE ANN. § 36-228 (2005) (every employer must furnish a reasonably safe place of employment, use safety devices and safeguards, and adopt practices that are reasonably safe and adequate to render such employment and place of employment reasonably safe); OHIO REV. CODE ANN. § 4101.11 (2006) (every employer shall furnish employment that is safe for the employees and shall do every other thing reasonably necessary to protect the life, health, safety, and welfare of such employees and frequenters); WIS. STAT. § 101.06 (2006) (every employer must furnish a safe place of employment for employees and shall do every other thing reasonably necessary to protect the life, health, safety, and welfare of such employees).

160. Flanigan v. Department of Labor & Indus., 869 P.2d 14 (Wash. 1994) (employers who intentionally cause worker injuries are not immune); Wood v. Safeway, Inc., 121 P.3d 1026 (Nev. 2005) (Nevada Industrial Insurance Act, NEV. REV. STAT. 616A–616D (1999), provides the exclusive remedy for employees injured on the job; employer is immune from suit by an employee for injuries arising out of and in the course of the employment). *But see* Granite Constr. Co. v. Mendoza, 816 S.W.2d 756 (Tex. Ct. App. 1991) (contractor's refusal to provide orange safety vests and to close off nearest lane of traffic to protect flagger constituted gross negligence, rendering employer immunity statute inapplicable).

potential safety hazards and to take steps to ensure adequate training of the contractors' and subcontractors' workers. This requirement may extend to the engineering or environmental consultant acting as an owner's representative at one of these sites. The bottom line is that responsibility for site safety at a hazardous waste site can be more complicated and raise different risks than a traditional construction site.

K. *The Safety Dilemma*

When the field representative observes an unsafe condition at the construction site, the question posed is always, "How to respond?" The dilemma is that, as seen above, some courts find liability based upon the failure to act, but taking affirmative steps and becoming involved in site safety could be construed as an assumption of that responsibility. There is authority that holds that architects and engineers, as professionals, cannot idly stand by with actual knowledge of an unsafe condition or unsafe practices on the job site without taking steps to advise or warn the owner or the contractor. While courts recognize that an architect or engineer does not necessarily assume responsibility for implementing workplace safety measures by reporting observed safety violations to the owner or to the contractor,[161] the professional may incur liability if he or she witnesses an unsafe condition, recognizes it as such, and does not warn the workers of the danger.[162]

Most would advise the field representative to avoid the appearance of becoming involved in safety, but respond in some manner when there is an imminent danger of injury or death. The appropriate steps will vary based upon the circumstances, but they should always be followed with written confirmation that the design professional is not assuming any control or responsibility for safety.

Absent an imminent threat of immediate injury or death, field personnel should be trained to follow a notification protocol consistent with the contract's allocation of responsibility. Thus, the field personnel's first step may be to notify his or her own supervisor, an owner representative, or the contractor's site safety representative. However, it will be important for the field personnel to understand that any notification to either owner or, especially, the contractor should be framed as advisory only and not construed as a direction or a stop-work order. Further steps to be taken should include contemporaneous documentation of the situation observed and how and to whom notification information was provided. In the event that the contractor fails to respond, the personnel should leave the site and notify appropriate parties of this action.

---

161. Rafferty v. Century Eng'g, Inc., 2002 WL 480958 (Del. Mar. 22, 2002).

162. *But see* Herczeg v. Hampton Township Mun. Auth., 766 A.2d 866 (Pa. 2001) (court rejects claim that duty arises based solely upon an engineer's actual knowledge of dangerous conditions); Vonasek v. Hirsch & Stevens, Inc., 221 N.W.2d 815 (Wis. 1974) (architect had no duty to warn workers as to obvious hazard).

CHAPTER 7

Design Professional Work Product—Ownership and Protection

BEN PATRICK, ESQ.

I. Introduction

A design professional is hired to create designs, details, specifications, bid sheets, and any of a hundred other things, depending on the needs of the project. Besides construction observation services, the primary value a design professional brings to a project is the value of his or her creativity and experience. Both are used to create a package which will (in theory) accomplish what the owner wants to achieve within a budget the owner finds acceptable. Because of the work that goes into their creation, the fruits of a design professional's creativity can have significant economic value. Innovative connection details, efficient floor plans, and inviting curbside elevations are all valuable design elements. Design professionals who create these designs are understandably interested in protecting them.

Copyrights, patents, and trade secret laws afford design professionals protection for the designs they create. Each system of laws operates differently, affording different levels of protection and requiring different levels of proof from the design professional in order to qualify for protection. This chapter discusses the various methods through which design professionals can secure protection for their designs. Currently, copyright law affords the best protection for design professionals' designs. Courts have recently seen an explosion in design copyright litigation, so this chapter focuses primarily on copyright law as a means of protecting designs. However, this chapter will also discuss patents and the ways in which patents differ from copyrights. Finally, it will discuss trade secret laws, which can be used to ensure that a design professional's former employees do not use innovative techniques learned during their employment to compete with their employer at a later time.

II. What Types of Works Are Subject to Copyright Protection?

A. *Generally*

Virtually any work of authorship can be protected by a copyright: architectural plans, novels, plays, songs, sculptures, paintings, computer programs, and many other works.[1] Within the category of architectural works,[2] authors can copyright plans, specifications, elevations, models, built structures, and even individual details. A work must meet only two requirements in order to be copyright protected: it must fall into one of the categories of works that are subject to copyright protection and it must be sufficiently original to trigger copyright protection. Section IV of this chapter discusses the protection that architectural works are afforded. For purposes of this section, it is sufficient to note that architectural works like those described above fall into one of the categories of works subject to copyright protection. Therefore, the only issue that authors of architectural works must consider is whether their works are sufficiently original to trigger copyright protection.

In the context of copyright eligibility, there is a very low threshold for originality. The United States Supreme Court has said that a work is sufficiently original to qualify for copyright protection so long as the work was independently created by its author (as opposed to being copied from another work) and possesses "some minimal degree of creativity."[3] The degree of creativity required is extremely low—even a slight amount of creativity will suffice. Thus, the vast majority of works will qualify, no matter how crude, humble, or obvious the work may be.[4]

B. *Copyrights Distinguished from Patents*

Patent law is a highly detailed area of law with many of its own unique rules and procedures. It is beyond the scope of this chapter to present the reader with even a fundamental understanding of patent law. However, it is important to distinguish between copyrights and patents, and to understand how patents can play a role in the protection of architectural designs. As noted above, copyrights are relatively easy to obtain, and almost every author is entitled to copyright protection for his or her creations. In contrast, patents are incredibly difficult to obtain, and many inventions do not qualify for patent protection. This is because the process for obtaining a

---

1. 17 U.S.C. § 102 (1990) generally describes works that are subject to copyright protection.

2. The Copyright Act uses the phrase "architectural work." Therefore, this Chapter will also use that phrase. However, engineering designs are equally protected under the Copyright Act; the use of the word "architectural" is not intended by the Act to be exclusionary.

3. Feist Publ., Inc. v. Rural Tel. Serv. Co., 449 U.S. 340, 345 (1991).

4. *Id. See also* Alfred Bell & Co. v. Catalda Fine Arts, 191 F.2d 99, 103 (2d Cir. 1951) ("Originality in this context means little more than a prohibition of actual copying. No matter how artistically poor the author's addition, it is enough if it be his own.").

patent is essentially adversarial: the inventor will need to prove to the U.S. Patent and Trademark Office that the particular invention is unique, original, and not merely an extension of existing technology. However, the heightened burden of obtaining a patent comes with a greater reward: patent holders can prohibit others from using, in any way, the protected invention, as well as any product that impermissibly duplicates the workings of the protected invention. As discussed below, this protection is more extensive than the protection offered by copyrights.

Architectural designs are seldom eligible for patent protection. It is hard to envision a scenario in which an entire architectural plan could receive a patent. However, some elements of architectural designs, particularly technical construction details, can be protected by patents. Designers who have created particularly innovative construction details and wish to obtain the maximum available protection for them should consider obtaining a patent regarding those details.

Library References

C.J.S. *Copyrights and Intellectual Property* §§ 9–19, 92, 101; C.J.S. *Patents* §§ 1–5, 10–12, 15. West's Key No. Digests, Copyrights and Intellectual Property <KEY>3–17; Patents <KEY>1.

III. Who Owns the Copyright?

The threshold question in any copyright dispute is determining who owns the copyright. Ownership of architectural plans is now a common question in contract negotiations between architects and owners. Therefore, it is important to understand who usually owns architectural copyrights and how the nature of that ownership can be altered.

A. *The Default Rule: Plans Are Owned by the Designer*

For all works that are subject to copyright protection, the default rule is that the author of the work owns the copyright.[5] This principle is also reflected in standard industry contracts.[6] While courts in the past have struggled with the question of whether the designer was the author, this question has now been resolved in the designer's favor.

1. *The Work Made for Hire Doctrine*

In the past, project owners who wanted copyright ownership for architectural plans claimed that right under the work made for hire doctrine. A work made for hire is one that is "prepared by an employee within the scope of his or her employment;

---

5. 17 U.S.C. § 201(a).

6. AIA A201-1997, § 1.6.1: "unless otherwise indicated the Architect and the Architect's consultants shall be deemed the authors … and will retain all common law, statutory and other reserved rights, in addition to the copyrights." EJCDC E-500, 2002, § 6.03(A): "All Documents are instruments of service in respect to this Project, and Engineer shall retain an ownership and property interest therein (including the copyright and the right of reuse at the discretion of the Engineer)...."

or ... specially ordered or commissioned for use as a contribution to a collective work ... [or] a supplementary work ... [or] a compilation ... if the parties expressly agree in a written instrument signed by them that the work shall be considered a work made for hire."[7] This doctrine means that the employer is the author of any works made for hire and owns all copyrights in that work.

Owners have previously argued that the nature of hiring a designer makes the works produced by that designer in the course of his or her work for the owner a work made for hire. Courts in recent years have rejected that contention for several reasons. First, the doctrine applies only to nine specifically enumerated categories of work.[8] Of those categories, the only ones that could be said to apply to architectural plans are collective works, supplementary works, and compilations. In truth, none of these categories is appropriate for architectural works' and courts have recognized this fact; moreover, designers are not the owner's employees, they are independent contractors.[9] Because the doctrine applies only to employees, it is inapplicable to the owner/designer relationship.

2. Co-Authorship

Owners have also attempted to claim copyright ownership of architectural plans based on a co-authorship theory. Owners have argued that because their input was vital to the finished plans (program requirements, aesthetic comments, redline changes, etc.), they are co-authors of the plans with the designer. In recent years, courts have rejected this notion.[10]

B. *Conveying Copyright Interests*

In the absence of an agreement to the contrary, the designer is the author of the plans and the copyright owner of the plans. In recent years, owners have become increasingly interested in owning all or part of the copyright. The transfer of a copyright or an interest therein must be done in writing.[11] Owners typically enter into contract negotiations for exclusive copyright ownership of the plans to be created by the designer. There are several reasons why designers should be reluctant to give owners absolute ownership

---

7. 17 U.S.C. § 101.

8. *Id.*

9. Courts have recognized this fact. *See, e.g.,* M.G.B. Homes, Inc. v. Ameron Homes, Inc., 903 F.2d 1486 (11th Cir. 1990); Kirk v. Harter, 188 F.3d 1005 (8th Cir. 1999); Zitz v. Pereira, 119 F. Supp. 2d 133 (E.D.N.Y. 1999), *aff'd,* 225 F.3d 646 (2d Cir. 2000); Bryce & Palazzola Architects and Assocs., Inc. v. A.M.E. Group, 865 F. Supp. 401 (E.D. Mich. 1994); Joseph J. Legat Architects v. U.S. Dev. Corp., 625 F. Supp. 293 (N.D. Ill. 1985). Kirk v. Harter, 188 F.3d 1005 (8th Cir. 1999).

10. *See, e.g., M.G.B. Homes, supra* note 9, 903 F.2d at 1492–93; *Zitz, supra* note 9, 119 F. Supp. 2d at 143–45, 225 F.3d at 672; *Bryce & Palazzola, supra* note 9, 865 F. Supp. at 404; *Joseph J. Legat Architects, supra* note 9, 625 F. Supp. at 297–98.

11. 17 U.S.C. § 204. *See also* Staggers v. Real Authentic Sound, 77 F. Supp. 2d 57, 68-69 (D.D.C. 1994).

of the plans, not the least of which is the designer's desire to reuse parts of the design in later work for other clients. There are also several alternatives to exclusive ownership that will satisfy the needs of most owners. These alternatives are discussed below in descending order of preference from the standpoint of the designer.

1. *Ownership by the Designer with a Non-Exclusive License for the Owner*

Owners are increasingly unwilling to allow the designer to retain absolute ownership over the work. While there are a number of reasons for this, and every owner will have unique needs depending on the nature of the project, the needs of most owners can be satisfied by granting them a nonexclusive license to the work. Under this scenario, the designer retains ownership of the work, but passes a license to the owner which allows the owner expanded use of the work itself. A typical contract provision in such an instance would be:

> The Work is an instrument of the Designer's service. The Designer is the author of the Work and retains all common law, statutory, and other reserved rights in the Work, including the copyright. The Designer hereby grants to the Owner a nonexclusive license to reproduce, modify, and make derivative products from the Work. In return, the Owner agrees, to the fullest extent permitted by law, to indemnify, defend and hold Designer harmless from any claim, liability or cost (including reasonable attorney's fees and defense costs) arising or allegedly arising out of the use of the Work by the Owner if such use has not been explicitly approved in writing by the Designer.

This provision accomplishes several important goals. First and foremost, the designer still retains all rights to the work, and is recognized as the author of the work. Second, the owner is placated by the license grant which will allow it to modify the work and use the work in the future for other projects. This satisfies the needs of most owners. Third, the license granted to the owner is nonexclusive, which permits the designer to grant others a license to the work if the need arises. Fourth, in the event the owner uses the work without consulting the designer, the owner is obligated to indemnify and defend the designer from any damages resulting from this use. It is not hard to foresee instances in which the owner would improperly modify the work and expose the design professional to a claim of some type. The designer needs to be protected in these circumstances. However, it is important to note that many states have anti-indemnity statutes that apply to contracts for design work and that could affect the designer's ability to obtain full protection from the owner. Designers pursuing projects in those states should ensure that the indemnity provision complies with the applicable anti-indemnification statute.

2. *A Reverse License*

As noted above, the grant of a nonexclusive license should satisfy the needs of most owners. However, if the owner insists that it own the copyright for the work, the designer needs to address several key concerns. First, the designer should not grant the owner the rights to details that appear in the work but also appear in other materials prepared by the designer (e.g., a standard wall section). Second, the designer should retain the right to create designs which are derivative of the work. This avoids conflict between the owner and the designer if a future project created by the designer ends up looking too much like the project. Third, as with the previous provision, the designer needs to be protected from any claims resulting from the owner's future use of the work if the designer is not consulted. Fourth, the designer needs to ensure that it will be paid for its work. Fifth, the designer should explicitly disclaim any warranties of fitness for a particular purpose and merchantability in order to forestall any claim, at a later date, that the designer made any implied warranties regarding the design. These needs can be addressed with contractual language like this:

> The Work is an instrument of the Designer's service. The Designer is the author of the Work, and the Designer hereby assigns to the Owner, without reservation, all of Designer's common law, statutory, and other rights in the Work, including the copyright, upon full payment of all monies due to the Designer under this Contract, including payment for all additional services. In the event a dispute regarding payment arises, this assignment is not effective until after the dispute has been resolved. To the extent that the Work incorporates information and/or designs previously developed by the Designer, the Designer retains all of its rights in such information and/or designs. The Owner, in turn, grants the Designer a nonexclusive license to use the Work in any way, including the creation of derivative products. The Owner agrees, to the fullest extent permitted by law, to indemnify, defend and hold Designer harmless from any claim, liability or cost (including reasonable attorney's fees and defense costs) arising or allegedly arising out of any use of the Work by the Owner if such use has not been explicitly approved in writing by the Designer. Under no circumstances shall this transfer be deemed a sale by the Designer, and the Designer makes no warranties, either express or implied, of merchantability or fitness for any particular purpose.

3. *Ownership of the Work by the Owner*

While rare, owners occasionally insist upon retaining full ownership of the designs and do not want the designer to have the right to create derivative works. Such insistence is typically seen where the design is new or unique and the owner is attempting to ensure that it stays that way. If the owner has been unwilling to accept either of the two licensing provisions suggested above, something akin to the following ownership provision should be insisted on by the designer:

> The Work is an instrument of the Designer's service. The Designer is the author of the Work, and the Designer hereby assigns to the Owner, without reservation, all common law, statutory, and other reserved rights in the Work,

including the copyright, upon full payment of all monies due to the Designer under this Contract, including payment for all additional services. In the event a dispute regarding payment arises, this assignment is not effective until after the dispute has been resolved. To the extent that the Work incorporates information and/or designs previously developed by the Designer, the Designer retains all of its rights in such information and/or designs. The Owner agrees, to the fullest extent permitted by law, to indemnify, defend and hold Designer harmless from any claim, liability or cost (including reasonable attorneys fees and defense costs) arising or allegedly arising out of any use of the Work by the Owner if such use has not been explicitly approved in writing by the Designer. Under no circumstances shall this transfer be deemed a sale by the Designer, and the Designer makes no warranties, either express or implied, of merchantability or fitness for any particular purpose.

This provision is identical to the one suggested in Section III.B.2. above, except that the licensing language has been removed. This clause still accomplishes several important goals from the designer's prospective. First, the designer is protected from any liability arising from the use of the work if the designer is not consulted. Second, the owner does not obtain any ownership interest in the designer's standard details incorporated in the work. Third, the transfer is not complete until the designer is paid in full. Fourth, the designer is protected against unwittingly giving any warranties for the design. Because liability arising from warranties is commonly not covered by the insurance carried by designers, this is a significant concern.

4. *Conveying Copyright Interests at a Later Date*

Ideally, all issues involving ownership of the design should be resolved ahead of time, and that resolution should be embodied in the contract between the designer and the owner. However, sometimes ownership of the design will need to be changed after the design has been created. Most commonly, owners who find that a design sells particularly well may be interested in owning the design in order to prevent others (including the designer) from reusing the design in whole or part. Additionally, one party may want to change the initial ownership agreement sometime after the work is complete (typically, the owner decides it wants either a license or exclusive ownership of the work). All of these things can be accomplished, but several restrictions must be acknowledged.

a. *Licenses*

A license is merely a grant of permission by the owner of the rights in the work to another entity. The license can grant the licensee any of a number of rights: the right to use the work but not modify it; the right to modify the work but not use the original work; or the right to do whatever the licensee wants with the work. The license can be a simple, one-page document, signed by both parties, with language like that found in Section III.B.2. above. The license may also explicitly reference the portion of the original contract which makes the designer the owner of the work.

b. *Transfer of Ownership*

It is not uncommon for owners to develop a unique type of buyer's remorse regarding the work; i.e., if the work is particularly good, the owner wishes it had secured exclusive ownership of the work in the first place. In these instances, the owner will frequently ask the designer to transfer ownership of the work to the owner. The question of whether, in a given situation, a designer ought to agree to do so cannot be answered generically—each situation must be weighed on its own merits. Some designers commonly transfer ownership upon request (although seldom for free); others never do so. However, it can safely be said that there are situations in which the designer will conclude that it is in his or her best interests to strike a deal with the owner and transfer ownership of the work. Once again, such a transfer can be accomplished in a simple, one-page document, with language like that found in Section III.B.3. above. A transfer must be in writing.[12] As with licenses, it may be advisable for the transfer agreement to explicitly reference that portion of the original contract which made the designer the owner of the work.

c. *Transfer of Copyrights*

If the work has been formally registered with the U.S. Copyright Office, transfer of the copyright must accompany transfer of the ownership of the design. However, before the designer agrees to such a transfer, one alternative should be suggested. Owners typically want to hold the copyright in the work in order to prevent anyone else from building similar projects. Owners believe (correctly) that many designers have little interest in pursuing copyright infringement lawsuits against other owners, and for this reason, owners want to hold the copyrights so that they can pursue copyright infringement lawsuits themselves. This particular need on the part of the owner can be fully satisfied without a blanket transfer of all copyrights in the work from the designer to the owner. The designer can transfer to the owner the right to pursue copyright infringement actions while still retaining all other copyrights in the work. The key provision of such an agreement would approximate this language:

> The Work is an instrument of the Designer's service. The Designer is the author of the Work, and the Designer hereby assigns to the Owner the right to bring, in the Owner's name, claims alleging infringement of the copyrights in the Work, including claims which may have accrued prior to the date of this Assignment. The Owner shall bear all of the costs, including all attorney's fees, incurred in pursuing any such claims.

Under such an agreement, the designer loses the ability to bring copyright infringement actions itself, but that result is preferable to losing all rights in the work.

---

12. 17 U.S.C. § 204.

C. *Common-Law Copyrights and Registered Copyrights*

Upon the completion of a work, it is automatically protected by common-law copyright. No formal process or procedure is necessary. However, there are several procedures that authors should follow to protect and extend these common-law copyrights.

1. *Copyright Notice*

A formal copyright notice identifies the author and date of publication, and places the public on notice that the work is copyrighted. The most common formal notice is simply a line which states "Copyright 2006 by John Author." Alternatively, some authors use the copyright symbol: "© 2006 by John Author." Under prior versions of the Copyright Act, an author lost copyright protection if the work was published without a formal copyright notice attached to the work.[13] For works first published on or after March 1, 1989, this is no longer the case.[14] Under the current version of the Copyright Act, it is no longer necessary to include a formal copyright notice. However, inclusion of the formal notice is still the best practice, and the author can gain advantages in copyright infringement lawsuits if a notice was placed on the work.[15]

2. *Formal Registration*

The U.S. Copyright Office[16] provides a method for authors to formally register their copyrighted works. Formal registration of a copyright provides two advantages, both related to copyright infringement litigation. First, it provides the author with some proof of the date of creation of the copyrighted work, because the Copyright Office notes the date of receipt of the formal registration. Second, except for limited situations that are not generally applicable to architectural copyrights, an author cannot commence a copyright infringement lawsuit unless the author has first filed a formal registration for the allegedly infringed work.[17]

Library References

C.J.S. *Copyrights and Intellectual Property* §§ 22, 26–35, 39, 71, 86–91, 93, 95–96, 102–103.

West's Key No. Digests, Copyrights and Intellectual Property <KEY>41–41.3, 43–49, 50.1, 50.15, 101.

---

13. *See* H.R. REP. No. 94-1476 (1976), *reprinted in* 1976 U.S.C.C.A.N. 5659, regarding history of 17 U.S.C. § 401.

14. 17 U.S.C. § 401 currently provides that "a notice of copyright as provided by this section *may* be placed on" copies of the work. (emphasis added) This change in the law was part of the Berne Convention amendments to the Copyright Act. *See also* United States Copyright Office Circular 3, *available at* www.copyright.gov/circs/circ03.pdf.

15. 17 U.S.C. § 401.

16. The U.S. Copyright Office's Web site is located at www.copyright.gov. The Web site contains a wealth of information and is generally user friendly.

17. 17 U.S.C. § 411.

IV. What Protection Does the Copyright Afford?

The scope and nature of copyright protection for architectural works have undergone significant evolution in the last twenty years. In order to understand the current state of the law, it is necessary to briefly analyze the foundations of copyright law and its evolution.

A. *A Brief History of Copyright Protection for Architectural Plans*

The right of an author to protect his or her creations is grounded in the Constitution itself, which provides that "Congress shall have Power to ... promote the Progress of Science and useful Arts, by securing for limited Times to Authors and Inventors the exclusive Right to their respective Writings and Discoveries[.]"[18] Protection for architectural drawings was first extended by the Copyright Act of 1909, which provided copyright protection to "[d]rawings of plastic works of a scientific or technical nature."[19] Architectural plans were considered technical drawings under the 1909 Copyright Act. The next major evolution of copyright law came with the Copyright Act of 1976. The 1976 Act continued to afford protection to "technical drawings, diagrams, and models."[20] The House of Representatives Report on the 1976 Act recognized that "[a]n architect's plans and drawings would, of course, be protected by copyright[.]"[21]

The Berne Convention Implementation Act of 1988, which was intended to make copyright protection in the United States coextensive with copyright protection in other countries who were signatories of the Berne Convention, amended the 1976 Act by substituting the phrase "diagrams, models, and technical drawings, including architectural plans," for the phrase "technical drawings, diagrams, and models." Thus, for the first time, the Copyright Act explicitly included architectural plans among the writings subject to copyright protection.

B. *The Useful Article Exception*

The 1976 Act also gave rise to an exception, which to this day poses problems with architectural copyrights. The 1976 Act defined a useful article as an article having an intrinsic utilitarian function that is not merely to portray the appearance of the article or to convey information.[22] Because the intrinsic function of an architectural plan is to convey information (i.e., the manner in which a structure is to be constructed), an architectural plan is not considered a useful article. However,

---

18. U.S. CONST. art. I, § 8.
19. 17 U.S.C. § 5(i) (1909 Act).
20. 17 U.S.C. § 101 (1976 Act).
21. H. Rep., pg. 53-55. H.R. Rep. No. 94-1476, at 55 (1976), reprinted in 1976 U.S.C.C.A.N. 5659.
22. 17 U.S.C. § 101 (1976 Act).

architectural plans do *portray* a useful article: the structure itself. The 1976 Act indicates that the Copyright Act "does not afford, to the owner of a copyright in a work that portrays a useful article ... any greater or lesser rights with respect to the making, distribution, or display of the useful article so portrayed than those afforded to such works under the law, whether title 17 or the common law or statutes of a State, in effect on December 31, 1977 ...."[23] Under the law in effect on December 31, 1977, copyrights for works of utility (including architectural plans) protected only against the copying of such works for purposes of explanation, *not* against the copying of such works for purposes of use.[24]

Thus, the effect of the useful article exception is that an author who owns a copyright in a set of architectural plans can prevent the copying of those plans, but *cannot* prevent someone from building a structure identical to the structure depicted on those plans. Copyright extended only to the *plans themselves*, not to the structure depicted on those plans, even though the design concepts embodied in the plans may never have been known or used before.[25]

A recent court case in West Virginia[26] demonstrated how problematic the useful article exception can be. In that case, the copyright holder (author) demonstrated that an architect and builder were in possession of the author's copyrighted architectural plans and had used those plans to construct a building which, the author contended, infringed on its copyright. The court noted that "[o]ther courts have found that a technical drawings copyright simply does not give the owner the exclusive right to build the structure depicted in the plan, and a copyright owner has no claim against another who builds the structure from an infringing plan."[27] The court concluded that a building "cannot be an infringing copy of a technical drawing."[28]

C. *The Architectural Works Copyright Protection Act*

The most recent evolution of copyright law relating to architectural plans was the 1990 Architectural Works Copyright Protection Act (AWCPA). The AWCPA was intended to provide "previously lacking copyright protection to physical architectural works, not drawings of such works," which already enjoyed copyright protection.[29] Thus, the AWCPA was intended to close the useful article loophole by permitting designers to copyright the physical structure itself, not just the plans that depicted the structure. The author of architectural plans can now hold two separate

---

23. 17 U.S.C. § 113 (1976 Act).
24. Baker v. Seldon, 101 U.S. 99, 105 (1879).
25. Demetriades v. Kaufmann, 680 F. Supp. 658, 664 (S.D.N.Y. 1988).
26. National Medical Care, Inc. v. Espiritu, 284 F. Supp. 2d 424 (S.D. W.Va. 2003).
27. *Id.* at 435.
28. *Id.* at 424.
29. Guillot-Vogt Assocs., Inc. v. Holly & Smith, 848 F. Supp. 682, 686–87 (E.D. La. 1994).

copyrights: one in the technical drawings; and one in the physical structure.[30] However, the AWCPA applies only to structures built after January 1, 1990.[31]

D. *Rights Protected by Copyright*

A copyright confers three essential rights for authors of architectural plans: the right to reproduce the plans themselves; the right to prepare derivative works based on the plans; and the right to build the structure depicted in the plans.[32]

1. *The Right to Reproduce the Plans Themselves*

The author of architectural plans can prohibit others from copying those plans. Thus, a copyright infringement can occur simply by making a photocopy of copyrighted plans, even if no further action is taken. For this reason, most industry contracts contain specific provisions enabling various parties (most commonly, the owner and the prime contractor) to make copies of the plans for the purposes of construction. The owner typically is also permitted to make copies of the plans for the structure maintenance purposes once it is built.

2. *The Right to Prepare Derivative Works Based on the Plans*

From the designer's point of view, perhaps the most important right secured by copyright is the right to prepare derivative works based on the plans. In the field of residential mass-production architecture (which has the most well-developed body of law regarding architectural copyrights), popular architectural plans can have tremendous economic value. Therefore, the authors of those plans have a strong incentive to protect the plans from being knocked off by architects who develop plans derived from the author's original, successful plan.

3. *The Right to Build the Structure Depicted in the Plans*

As discussed above, the Architectural Works Copyright Protection Act granted authors of architectural plans the exclusive right to build the structure depicted.

E. *Scope of Protection for Derivative Works*

Authors frequently base new works on works they have previously created. When a new design is based on a work the author previously created, copyright protection for the derivative design is limited to those elements that are unique to the derivative design.[33] Thus, elements in the derivative design that are also incorporated in the

---

30. *Id.* at 687 (quoting H.R. REP. NO. 101-735, *reprinted in* 1990 U.S.C.C.A.N. 6935, 6950).

31. While no court has yet considered the issue, it seems likely that a building that was substantially completed after Jan. 1, 1990, would be protected, even though construction started in 1989.

32. 17 U.S.C. § 106.

33. 17 U.S.C. § 103.

original design are not protected. The derivative design is afforded protection only for the new elements. The scope of protection for derivative works is lower than for original works. This is logical: derivative works demonstrate less originality and, therefore, are afforded less protection.

F. *Scope of Protection for Works Consisting of Generic Features*

It has been said that there is nothing new under the sun.[34] This sentiment certainly applies to architecture and engineering, as designers constantly rely on standard, generic, and tried and true designs and design elements, and incorporate those elements into new designs on a daily basis. New and unique designs are few and far between, particularly in residential construction. When concepts such as bathroom, closet, and kitchen layouts are considered, the simple fact is that all functional layouts have been used at some time in the past.

Designs that gather together generic or standard features into a functional plan are eligible for copyright protection as a compilation work. However, because the level of originality expressed in compilation works is limited, the level of copyright protection afforded to these plans is similarly limited. The author can copyright the particular choices he or she made when selecting which standard features to include in the design and can copyright the particular way in which he or she arranged those features within the design.[35] This is often referred to as thin copyright protection. When considering whether a design infringes on a thinly protected design, courts have found that even modest dissimilarities between the two designs will support a finding that no infringement has occurred.[36] Again, this is logical: a design that merely combines standard features in a particular way shows little originality and, therefore, is entitled to little (but not a complete lack of) protection.

G. *The Idea/Expression Dichotomy*

One of the fundamental principles of the Copyright Act is that no one can copyright an idea. An author may only copyright the particular manner in which the author expressed that idea. Thus, for example, no author could copyright the idea of

---

34. *Ecclesiastes* 1:9.
35. 17 U.S.C. § 103. *See also* Tufenkian Import/Export Ventures, Inc. v. Einstein Moomjy, Inc., 338 F.3d 127, 136 (2d Cir. 2003); Ale House Mgmt., Inc. v. Raleigh Ale House, Inc., 205 F.3d 137, 143 (4th Cir. 2000); Domingo Cambeiro Prof. Corp. v. Advent, 2000 WL 262597 (9th Cir. 2000); Howard v. Sterchi, 974 F.2d 1272, 1276 (11th Cir. 1992); John Alden Homes, Inc. v. Kangas, 142 F. Supp. 2d 1338, 1344-45 (M.D. Fla. 2001), *aff'd.*, 37 Fed. Appx. 979 (11th Cir. 2001); LaJoie v. Pavcon, Inc., 146 F. Supp. 2d 1240 (M.D. Fla. 2000).
36. *Supra* note 35.

two young lovers kept apart by feuding families. However, the particular manner in which that idea is expressed in *Romeo and Juliet* and *West Side Story* is copyrightable. This distinction between an idea and the manner in which that idea is expressed is referred to as the idea/expression dichotomy.

The idea/expression dichotomy plays an important role in determining the scope of protection afforded to architectural works. One court astutely observed that elements such as the placement of functional elements, routing of traffic through a home, and generic methods of construction are ideas that cannot be copyrighted.[37] Therefore, no designer could copyright the idea of placing a closet in or adjacent to a bedroom, nor could any designer copyright the idea of a walk-through closet placed between the master bedroom and master bathroom. However, if a designer were to develop a unique layout for that closet, that layout would be copyrightable.

A corollary to the idea/expression dichotomy is that if a particular idea can be expressed in only a limited number of ways, no author can copyright his or her particular expression of that idea, because any such copyright would give the author a partial monopoly on the idea itself. In architectural design, a powder room provides an excellent example of an idea that can be expressed in only a limited number of ways. Many homes feature a two-fixture powder room on the main level for use by guests during entertaining, while preserving the privacy of the master bath. There are very few ways to express the idea of a two-fixture powder room: placing both fixtures on one wall with the door at the end or on the other wall, or placing the fixtures facing each other with the door in the middle. Certainly, within those two forms of expression, details such as finishes, dimensions, and the swing of the door can vary. However, the expression remains fundamentally the same. If any designer was given a copyright on one of the expressions of a two-fixture powder room, the architectural industry as a whole would suffer immensely, as no other architect could incorporate that particular two-fixture powder room in their design. Copyright law is designed to promote the progress of science and the useful arts, not to stifle innovation.[38] It is for this reason that copyright law will not permit an author to copyright the expression of an idea if it can be expressed only in a limited number of ways.

Library References

C.J.S. *Copyrights and Intellectual Property* § § 9–10, 16, 19, 40–41, 45, 47, 57, 59, 92, 97.
West's Key No. Digests, Copyrights and Intellectual Property <KEY>4, 6, 12(3), 35, 64.

---

37. Kootenia Homes, Inc. v. Reliable Homes, Inc., 2002 WL 15594 (D. Minn. 2002). *See also* Attia v. Society of New York Hosp., 201 F.3d 50 (2d Cir. 1999); J. R. Lazaro Builders, Inc. v. R.E. Ripberger Builders, Inc., 883 F. Supp. 336, 343 (S.D. Ind. 1995).

38. U.S. CONST. art. I, § 8.

V. What Is Copyright Infringement?

A copyright is both a source of affirmative rights (the owner has the right to reproduce the work and prepare derivative works based on the work) and negative rights (no one else is permitted to reproduce the work or prepare derivative works based on the work). Copyright infringement occurs when someone other than the author or a license holder reproduces the work, prepares a derivative work based on the work, or prepares a work that is impermissibly similar to the copyrighted work. The precise elements of copyright infringement vary in different federal jurisdictions; however, the core elements remain essentially the same. This discussion will focus on the basic elements common to all jurisdictions. Authors should take care to carefully research the law in their own jurisdiction before commencing (or defending) an infringement action.

Generally, an author must prove three things in order to demonstrate copyright infringement: the author is the owner of a valid copyright in the allegedly infringed work; the defendant had access to the copyrighted work; and the allegedly infringing work is substantially similar to the copyrighted work.

A. *Ownership of a Valid Copyright*

The first showing an author must make is that he or she is the owner of a valid copyright in the work. This showing has two essential parts: showing that the author is the owner of the copyright and that the copyright is valid. Demonstrating ownership is typically a simple matter—the plaintiff must either show that he or she is the author of the work or the holder of a valid copyright assignment or a license that gives the plaintiff the right to sue for infringement.

The second showing, that the copyright is valid, is frequently a point of contention in copyright infringement lawsuits. Generally, a copyright may be shown to be invalid in two ways: by demonstrating that the copyrighted work is not subject to copyright protection or the copyright registration is fraudulent. For the reasons discussed in Section II (eligibility for copyright protection), defendants seldom convince a court to invalidate a copyright on the basis that the allegedly infringed work is not subject to copyright protection.

Defendants in copyright infringement lawsuits frequently challenge the validity of a copyright by attacking the contents of the copyright registration. Courts will invalidate a registered copyright when the author obtains that registration through the use of a materially misleading registration. Courts have invalidated copyright registrations based on the author's failure to identify prior works from which the registered work was derived[39] and based on the failure to appropriately attribute

---

39. *See, e.g.,* R. Ready Prods., Inc. v. Cantrell, 85 F. Supp. 2d 672, 691–92 (S.D. Tex. 2000); Russ Berrie & Co., Inc. v. Jerry Elsner Co., 482 F. Supp. 980, 988–89 (S.D.N.Y. 1980); Vogue Ring Creations, Inc. v. Hardman, 410 F. Supp. 609 (D.R.I. 1976); International Biotical Corp. v. Associated Mills, Inc., 239 F. Supp. 511 (N.D. Ill. 1964).

authorship to the true author of the work.[40] Courts typically consider whether the copyright registration misstated information which, if correctly stated, may have caused the Copyright Office to reject the registration. If so, the resulting copyright is invalid.[41] If the Copyright Office would have accepted the registration had the information been correctly stated, the resulting copyright is valid. Courts will typically invalidate copyright registrations only if the allegedly incorrect information has been deliberately misstated. However, due to the possibility that a copyright may be held invalid due to errors in the registration process, authors should take care in the registration process.

B. *Access to the Copyrighted Work by the Alleged Infringer*

A person cannot copy something they have never seen. Therefore, in order to prove that the defendant has infringed on a copyrighted work, the author must prove that the defendant saw the work. This element of proof is called access. Because it is often impossible to prove that the defendant actually saw the copyrighted work, courts have held that access is established as long as the defendant had an opportunity to view the copyrighted material.[42] Furthermore, access can be inferred between parties on the construction team. Thus, if the builder had access to the author's copyrighted material, courts will find that the designer employed by that builder also had access.[43]

C. *Substantial Similarity between the Copyrighted Work and the Allegedly Infringing Work*

The defendant's work infringes on the author's copyrighted work only if it is *substantially similar* to the author's copyrighted work.[44] In the context of copyright

---

40. *See, e.g.,* M.G.B. Homes, Inc. v. Ameron Homes, Inc., 903 F.2d 1486, 1488–93 (11th Cir. 1990); Zitz v. Pereria, 119 F. Supp. 2d 133, 143–45 (E.D.N.Y. 1999), *aff'd*, 225 F.3d 646, 672 (2d Cir. 2000); Fred Riley Home Bldg. Corp. v. Cosgrove, 864 F. Supp. 1034, 1042–45 (D. Kan. 1994).

41. For a detailed discussion, *see* Foamation, Inc. v. Wedeward Enters., Inc., 947 F. Supp. 1287, 1296 (E.D. Wis. 1996). *See also* MELVELLE B. NIMMER & DAVID NIMMER, NIMMER ON COPYRIGHT § 7.20 (1978).

42. *See, e.g.,* Herzog v. Castle Rock Entm't, 193 F.3d 1241, 1249 (11th Cir. 1999); Moore v. Columbia Pictures Indus., Inc., 972 F.2d 939, 942 (8th Cir. 1992); Robert R. Jones Assoc., Inc. v. Nino Homes, 858 F.2d 274, 277 (6th Cir. 1988).

43. Kamar Int'l, Inc. v. Russ Berrie and Co., 657 F.2d 1059, 1062 (9th Cir. 1981); Arthur Rutenberg Corp. v. Parrino, 664 F. Supp. 479, 481 (M.D. Fla. 1987); MELVELLE B. NIMMER & DAVID NIMMER, NIMMER ON COPYRIGHT § 13.02[A] (1978).

44. Courts commonly hold that copyright infringement can be proved either by direct evidence of copying or by demonstrating access and substantial similarity. *See, e.g.*, Nelson v. PRN Prod., Inc., 873 F.2d 1141, 1142 (8th Cir. 1989). However, no court has ever been called on to find infringement via direct copying when the resulting work is *not* substantially similar to the copyrighted work. Therefore, the best view of the law is that substantial similarity will be

infringement, substantial similarity is a term of art. The exact meaning of substantial similarity, and the process for determining whether one work is substantially similar to another, varies from jurisdiction to jurisdiction. However, in general, the determination of substantial similarity is a two-step process.

1. *The Extrinsic Test*

The first step is an extrinsic test whereby the court analyzes the similarity of the two works, focusing on objective similarities in the details of the works. In conducting this test, the court separates the details of each of the works into two categories: generic details which are not subject to copyright protection (including standard features and ideas that can only be expressed in a limited number of ways); and original details, which are subject to copyright protection. The court then compares the original elements of the copyrighted work to the original elements of the allegedly infringing work. If the two works are found to be extrinsically, substantially similar, the court will submit the issue to the jury for the intrinsic test.

2. *The Intrinsic Test*

The jury has the job of conducting the second test: an intrinsic test which consists of evaluating the response of an ordinary, reasonable person to the two works. At this stage, the works as a whole are compared to each other. If the allegedly infringing work is so similar to the copyrighted work that an ordinary, reasonable person would conclude that the defendant appropriated the plaintiff author's protectable expression by taking material of substance and value, the works are intrinsically, substantially similar and the defendant's work infringes on the author's copyrighted work (subject to the defenses outlined below).[45]

D. *Defenses*

There are two primary defenses to a copyright infringement claim: independent creation and fair use.

1. *Independent Creation*

The most important defense a designer can assert to a copyright infringement claim is to argue that the designer independently created the work accused of infringement. If an architectural design is the result of the designer's own creativity and ingenuity, that design does not infringe on any other design, even if that design

---

required even if liability is proved through direct copying. Indeed, the notion that a direct copy can be anything *but* substantially similar to the work from which it was copied is, at best, a philosophical abstraction.

45. *See, e.g.*, Palmer v. Braun, 287 F.3d 1325, 1330 (11th Cir. 2002); Sturdza v. United Arab Emirates, 281 F.3d 1287, 1297 (D.C. Cir. 2002); Yurman Design, Inc. v. Paj, Inc., 262 F.3d 101 (2d Cir. 2001).

is identical to another design.[46] The critical factor in asserting this defense is the defendant designer's ability to prove that he or she, in fact, independently created the design in question. The more alike the two designs are, the more persuasive the plaintiff designer's accusation of copying will be. At a certain point, the plaintiff will be able to convincingly place the two designs next to each other and assert that the level of similarity between the designs means that the defendant designer must have copied the earlier work.

Designers should view the independent creation defense as their best defense against a charge of copyright infringement. Therefore, designers should take affirmative steps to implement interoffice policies designed to place the designer in the best position to assert such a defense. In a typical design process, dozens of alternative designs are considered and either adopted in whole or in part or discarded entirely. All of these designs should be saved, as they will help the designer rebut a charge of copying. The designs that were adopted in whole or in part are particularly critical, because they demonstrate the design process and can be used to help explain certain similarities. For example, a plaintiff designer may assert that the defendant's kitchen layout is so similar to the plaintiff's design that it must have resulted from copying. If the defendant designer can produce his or her earlier concept drawings which show the layout of the kitchen evolving to its final format, the defendant designer will be able to persuasively rebut this charge.

2. *Fair Use*

The fair use defense is seldom applicable in architectural design copyright infringement lawsuits. Therefore, it is not exhaustively analyzed here. The fair use doctrine holds that it is not infringement to use a copyrighted work for purposes such as criticism, comment, news reporting, teaching (including multiple copies for classroom use), scholarship, or research. In determining whether the use of a work in any particular case is a fair use, the court will consider factors such as: (1) the purpose and character of the use, including whether such use is of a commercial nature or is for nonprofit educational purposes; (2) the nature of the copyrighted work; (3) the amount and substantiality of the portion used in relation to the copyrighted work as a whole; and (4) the effect of the use upon the potential market for or value of the copyrighted work.[47] When the copyrighted work is used for commercial purposes, the fair use doctrine seldom applies. For this reason alone, the fair use doctrine is of limited use in architectural copyright infringement lawsuits.

---

46. *See, e.g.,* Repp v. Webber, 132 F.3d 882, 889 (2d Cir. 1997); CMM Cable Rep., Inc. v. Ocean Coast Props., 97 F.3d 1504, 1513, 1525 (1st Cir. 1990); Kootenia Homes, Inc. v. Reliable Homes, Inc., 2002 WL 15594 (D. Minn. 2002).

47. 17 U.S.C. § 107.

3. *Statute of Limitations*

The statute of limitations is very short for copyright infringement. Actions must be commenced within three years after the claim accrued.[48] The statute begins to run when the alleged infringement took place, regardless of whether the author had notice of the accrual of the claim.

E. *Damages*

If the plaintiff designer prevails, and the defendant is found to have infringed on the plaintiff's copyrighted design, the plaintiff is entitled to damages. There are several methods of calculating damages in copyright infringement lawsuits. The plaintiff must chose between one of the following types of damages.

1. *Plaintiff's Actual Damages*

The plaintiff can recover the profits it would have made but for the defendant's infringement on the plaintiff's copyright.[49]

> *Example:* Apartment Builder has worked extensively with Architects A and B and wants to hire one of them to design a new apartment building. Architect B knows that Apartment Builder really liked Architect A's last apartment design. So, Architect B borrows Architect A's design, slightly modifies it, and presents it to Apartment Builder as his proposed design for the new apartment building. Apartment Builder likes the modified design so much, he hires Architect B to design the new apartment building. Apartment Builder would have hired Architect A if not for Architect B's modified design. Architect A, in this instance, will be entitled to recover from Architect B the profits Architect A would have earned designing the apartment building for Apartment Builder, because Architect A would have been awarded the job if not for Architect B's infringement.

2. *Defendant's Profits Resulting from the Infringement*

The plaintiff can also recover any profits earned by the defendant as a direct result of the infringement. Using the example above, Architect A would be able to collect all of the profits Architect B earned from designing the new apartment building. A plaintiff may elect to recover the defendant's profits if the plaintiff's profit margin is smaller than the defendant's profit margin, or in cases where the plaintiff cannot prove that it would have received the work but for the infringement. For

---

48. 17 U.S.C. § 507(b). *See also* Zitz v. Pereira, 119 F. Supp. 2d 133 (E.D.N.Y. 1999), *aff'd.*, 225 F.3d 646 (2d Cir. 2000); Wood v. Santa Barbara Chamber of Commerce, 507 F. Supp. 1128 (D. Nev. 1980).

49. 17 U.S.C. § 504(b).

example, if a one-man architectural firm that operates exclusively in Arizona proves that an architect in Massachusetts has infringed on one of its designs for a single-family custom home by designing a custom home for a client in Massachusetts, that Arizona architect will likely have suffered no actual damages, as it would not have been hired to design the Massachusetts house.

3. *Statutory Damages*

The Copyright Act provides for statutory damages in appropriate cases. The court, in its discretion, decides what the appropriate amount of statutory damages is, within the limitations set forth in the Copyright Act. Typically, statutory damages are not less than $750, nor more than $30,000.[50] However, if the plaintiff proves that the infringement was intentional (i.e., the defendant set out to copy the plaintiff's design), the upper limit for statutory damages is changed from $30,000 to $150,000.[51] A plaintiff typically seeks statutory damages only when its actual damages (as discussed above) are very small, or where the plaintiff believes it cannot prove it suffered any damages. Using the Arizona architect example above, if the Massachusetts architect made no profit on the job, there are no infringer's profits to recover. Therefore, the Arizona architect would likely elect to receive statutory damages.

4. *Injunctive Relief*

In addition to the damages awards discussed above, the plaintiff can (and likely will) seek injunctive relief, a court order compelling the defendants to cease producing the infringing architectural designs and buildings.[52] The plaintiff also can seek an initial injunction, in the form of a temporary restraining order, compelling the defendants to cease producing the infringing architectural designs and buildings during the course of the lawsuit.[53]

5. *Attorneys' Fees and Costs*

Regardless of the measure of damages the plaintiff elects, a plaintiff can recover the reasonable attorneys' fees and costs it incurred pursuing the lawsuit if the plaintiff's copyright was registered before the infringement occurred.[54] However, this provision is mutual. If the defendant wins, the defendant can also request that the court award it the reasonable attorneys' fees and costs it incurred defending the lawsuit.[55] The court has the discretion to decide whether or not to award attorneys' fees and costs in all cases and to determine the size of any award.

---

50. 17 U.S.C. § 504(c)(1).
51. 17 U.S.C. § 504(c)(2).
52. 17 U.S.C. § 502.
53. *Id.*
54. 17 U.S.C. § 505. *See also* Johnson v. Jones, 149 F.3d 494, 506 (6th Cir. 1998); Mason v. Montgomery Data, Inc., 967 F.2d 135, 143 (5th Cir. 1992); Cornerstone Home Builders, Inc. v. McAllister, 311 F. Supp. 2d 1351, 1352 (M.D. Fla. 2004).

Library References

C.J.S. *Copyrights and Intellectual Property* § § 8, 10, 41, 43–46, 48, 57, 61–75, 77–82, 98–100.

West's Key No. Digests, Copyrights and Intellectual Property <KEY>51–53.2, 64, 72–90.

6. *What Information Is Protected by Trade Secret Laws?*

State trade secret laws can provide protection for certain types of information. Trade secrets are protected under state laws, which vary from state to state. However, a trade secret is commonly defined as information which:

1. is not generally known;
2. derives independent economic value from not being generally known; and
3. is the subject of reasonable efforts to protect its secrecy.

A client database is one example of potential trade secret information. During the course of a client relationship, a designer may discover information like the range of fees the client is willing to pay, the number of projects the client undertakes every year, contract clauses the client is willing to negotiate (or not), and marketing information like names of the client's family, and the client's birthday. All of this information may be bundled into a client database which allows the designer to retain the client's business and maximize profitability on projects for that client. This information is not generally known and part of its value to the designer is that none of the designer's competitors know this information. As long as the designer takes reasonable steps to protect this information from discovery, the information is likely subject to trade secret protection. Password-protecting the database is usually sufficient protection to secure the information's secrecy.

Library References

West's Key No. Digests, Antitrust and Trade Regulation <KEY>413, 421.

VI. Conclusion

Most design firms rent their office space. The office is full of computers, plotters, telephones, and furniture, all of which is commonly leased or financed. The only significant asset most design firms possess is the firm's intellectual property, embodied in the designs the firm develops for their clients. As a design firm grows, it typically builds on its past in a literal fashion by developing new designs based, in whole or in part, on previous successful designs. For this reason, designers need to view intellectual property as a critical part of their business and need to take concrete steps to safeguard that property. The first step is to develop a corporate vision—the

---

55. 17 U.S.C. § 505.

designer must determine what should be protected and how it should be protected. The second step involves establishing interoffice policies aimed at protecting the firm's intellectual property. At the same time, the firm can adopt interoffice policies that protect the firm from claims of copyright infringement from other designers. The final step is an ongoing process to follow through on the vision and interoffice policies the firm has established and to regularly audit those policies to see if adjustments are needed. No policy, no matter how well written, is of any value if it is not routinely followed by the employees responsible for executing it. This fact must be kept in mind at the time the policies are developed—policies that are cumbersome, or interfere with key processes, will not be followed, and therefore, are useless.

The purpose of this chapter has been to summarize the law governing protection of designers' intellectual property, particularly copyright law. Designers who want to establish protection for their intellectual property, and who want to protect themselves from claims made by other designers, should give serious consideration to the issues discussed here and begin the process of establishing policies to protect themselves and their designs.

CHAPTER 8

State Regulation of the Construction Manager

KEVIN L. KOLTON, ESQ.
VIRGIL R. MONTGOMERY, ESQ.

I. Introduction

The profession of construction manager has developed in response to owners' needs to manage project time, cost, and quality. In reality, few owners have the staff or expertise necessary to oversee all details of a project, and architects, engineers, and contractors are frequently unwilling or unable to provide such services. As a result, construction managers have been used increasingly to provide project oversight, acting as an extension of staff for the owner. However, there is no universal definition for construction managers because services may vary depending upon the type of project and the contractual relationships created to facilitate its delivery. Construction management is still an evolving professional service. Typically, the responsibilities of a construction manager are defined contractually and contextually in relation to the duties of other parties in a project, namely architects, engineers, and contractors.

Not surprisingly, state regulation of the construction manager is also evolving and multifaceted. This chapter will identify the different statutory schemes for construction management, including the licensure required of those who practice construction management, the procedures state agencies must follow to procure such services, and regulations pertaining to ethics issues in the relationship between the owner and construction manager. It also identifies how the industry has attempted to regulate itself, through its professional organizations and associations and their codes of ethics, and by offering professional certification and accreditation of postsecondary construction education programs.

II. State Licensing and Project Delivery Requirements

Every state has enacted statutes licensing the practice of architecture and engineering. Similarly, many states have enacted statutes governing the licensing and/or registration of general contractors. However, the regulatory model for construction managers is less uniform from state to state. Only two states have enacted laws specifically regulating construction managers as a separate profession. Many states impose few, if any, regulations on the construction manager. Other states regulate construction managers as general contractors and/or design professionals. A discussion of each of the regulatory schemes follows.

A. *States That Specifically Regulate Construction Managers*

Only two states have enacted laws specifically regulating entities as construction managers, as distinguished from design professionals or contractors. Idaho and Oklahoma have regulations and procedures in place for construction management entities that want to work on public projects and for state agencies who hire construction managers for public projects.

1. *Idaho*

The State of Idaho enacted the Construction Management Licensing Act (the Act), an occupational licensing statute that specifically applies to construction management entities that work on public projects. In order for construction management entities to provide services to public works projects in Idaho, construction managers must be licensed by the Public Works Contractors State License Board (the Board). The Board will license individuals, but a construction management firm may provide services under the direct supervision of a licensed construction manager. Licensed architects, registered landscape architects, and registered professional engineers are exempt from obtaining a construction management license.[1]

To qualify for a license, a construction manager must submit a written application to the Board. The Board requires either of the following qualifications: (1) a bachelor's degree in architecture, engineering, or construction management and four years experience in managing construction projects; or (2) a minimum of five years experience in construction management. If the Board finds that the applicant meets these experience requirements to perform construction management services, the construction manager must sit for an examination administered by the Board.[2] A one-year interim license may be issued to construction managers who meet state qualifications but have yet to sit for the Board's examination.[3] Once a construction manager passes the examination, the Board will issue a license authorizing the individual to provide construction management services. Licenses are valid for one year from the date of issue and must be renewed annually.[4]

A firm may neither provide nor hold itself out as providing construction management services without holding a Certificate of Authority (Certificate). Firms

---

1. IDAHO CODE ANN. § 54-4504 (2005).
2. *See id.* § 54-4505.
3. *Id.* § 54-4506.
4. *Id.* § 54-4507.

with one or more principals or employees who are licensed construction managers may apply for a Certificate from the Board. An application for a Certificate must: (1) identify the designated licensed construction manager (designee) or managers; and (2) include a statement signed by the designee accepting responsibility for the construction management services provided by the firm. If the designee ceases to be an employee of the firm, the firm must notify the Board. In this scenario, the Board will suspend the firm's license but allow a firm to continue construction management services under the supervision of another licensed construction manager in order to complete a project that is already under way.[5]

No public entity may enter into a contract for construction management services with an entity that is not licensed by the Board. A licensed construction manager must comply with all notice and bidding laws when soliciting bids or awarding contracts for public works construction. If a construction manager (and that person's employer) has provided design or other construction services on a public works project, that construction manager (and his other firm) may not provide construction management services for that same project.[6]

The Board holds the authority to deny, renew, suspend, or revoke the license of a construction management entity. The Board may discipline or take legal action against entities upon the following grounds: (1) fraud or deception in obtaining a license; (2) incompetence in performance; (3) holding oneself (or one's firm) out as a construction manager without proper licensure; (4) fraud or deception in performance; (5) willful violation of the Act or the rules imposed by the Board.[7]

Entities who are convicted of violating provisions of the Act will be guilty of misdemeanor and may be punished by a fine of five thousand dollars, imprisonment up to one year, or both, at the discretion of the court.[8] At any time, the Board may seek a temporary restraining order on an entity violating provisions of the Act.[9]

2. Oklahoma

In Oklahoma, a qualified individual or business entity that wishes to provide construction management services on public projects must register with the State Construction Administrator (the Administrator) of the Construction and Properties Division (the Division) of the Department of Central Services. The entity must submit an application to be considered for registration. If an application is found complete and the individual or business entity is deemed qualified, that applicant will be added to the official list of registered construction managers maintained by

---

5. *Id.* § 54-4509.
6. *Id.* § 54-4511.
7. *Id.* § 54-4508.
8. *Id.* § 54-4513.
9. *Id.* § 54-4514.

the Division. Registration is valid for one calendar year and renewed annually on or before January 31 of each succeeding year.[10]

Only registered construction managers are qualified to provide construction management services to the state. The State of Oklahoma provides a list of preferred certifications from construction industry organizations and associations. The Administrator has the authority to determine an entity's qualification. However, it is unlikely that an applicant without any certification will be approved for registration.[11] State agencies seeking construction management services develop a project description which includes a narrative of the work being planned, an estimate of the project schedule and costs, and the source of funding for the project. The state agency then requests a list of qualified construction management entities from the Division and must send a Request for Qualifications (RFQ) to each entity on the list. In special circumstances, the Division may precede the RFQ with a request for Letters of Intent (LOI). Upon receiving LOI, the Division will deliver any information on file for each of these entities. Whether a RFQ or an LOI process is used, a screening committee will create a short list of three to five entities best qualified for the project.[12]

An interview committee comprised of a representative of the state agency seeking services and employees of the Division will establish a clear list of criteria in order to interview each entity. The firms are ranked based on the criteria established by the interview committee. These rankings and the committee's recommendations are forwarded to the state agency for review.[13] Upon receipt of the Division's ranking and recommendation, the state agency selects a construction manager. The state agency must submit a report including its selection criteria used to justify its selection.[14]

Before a state construction project begins, the Director of the Department of Central Services (the Director) determines whether a project merits the construction management delivery system. In making this determination, the Director must consider the following factors: time constraints, state agency capabilities, state agency funding, and project components. The Director will consider whether the Division staff can manage the project or if a construction management delivery system is more feasible. The aforementioned factors are outlined in a written response to the state agency requesting construction management delivery system services. An official written response, outlining the determination and justification of the Director, is sent to the requesting agency upon the decision of the Director regarding the use of the construction management delivery system.[15]

---

10. OKLA. ADMIN. CODE § 580:20-17-4 (2005).
11. *See id.* § 580:20-17-3.
12. *Id.* § 580:20-21-4.
13. *Id.* § 580:20-21-6, -7.
14. *Id.* § 580:20-21-8.
15. *Id.* § 580:20-17-5, -6, -7.

Once the Director has designated the project delivery method as construction management, the Administrator determines the construction management type as either *construction management agency* or *construction management at-risk*. If the Director agrees with the Administrator's designation, the prescribed project designation is sought. If the Director disagrees, the Administrator acquires design consultant and construction services in accordance with Oklahoma Administrative Code § 580:20-1 and § 580:20-3. Upon receiving written notice of project designation, the state agency shall complete a requisition form and submit it to the Administrator. The final project designation shall be filed and maintained in the Division project file.

The highest ranking firm, based on prior interviews, will be asked to submit a fee proposal to the Division. The Division will negotiate, write, and process the contract.[16] If the Division is unable to negotiate a satisfactory contract with the firm considered to be the most qualified, negotiations with that firm are terminated. The Division then will undertake negotiations with the next most qualified firm. This process shall continue until a contract is signed with a qualified firm or the procurement process is terminated and a new request for qualifications is initiated.[17]

A two-phase process is used when negotiating a contract for a construction manager at-risk. Phase 1 begins after negotiating fees for pre-construction services. The construction manager at-risk must then establish a Guaranteed Maximum Price (GMP), which should include the construction manager's fees for administering the construction contract and the fee to cover the general conditions for construction. If the GMP is not approved, the Division terminates the construction manager's services or renegotiates. Phase 2 commences on approval of the GMP when the Division produces a Notice to Proceed. The construction manager at-risk is henceforth required to furnish 100 percent performance, payment, and defect bonds, in an amount equal to the GMP or lump sum. When the design is complete and bid documents have been prepared for issuance of work packages to subcontractor trades, the construction manager at-risk becomes essentially a general contractor for the project.[18]

Subcontracts will be awarded to the lowest responsible bidder who meets the minimum pre-qualification requirements. If the GMP is exceeded after subcontractor bids are received, the construction manager at-risk and the Division must determine whether work packages need to be revised or the GMP needs to be adjusted. The construction manager at-risk must adhere to Oklahoma public bidding laws in advertising and packaging awards. If the construction manager at-risk competitively bids work as a lump sum like other bidders, it is permitted to self-perform portions of the work. Intent to self-perform must be declared during the pre-bid meeting held by the Division.[19]

---

16. OKLA. ADMIN. CODE § 580:20-17-9.
17. *Id.*
18. *Id.*
19. *Id.*

B. *Jurisdictions with Little or No Regulation*

Most states impose no obvious regulations directly affecting the construction management profession. In those states, practitioners are advised to scrutinize closely the state's contractor, architect, and engineering licensing laws for possible application, depending on the specific services to be provided by the construction manager. (*See infra* Section II.D.)

In other states, construction managers may face regulations requiring little more than filling out an application and paying a fee. For instance, the State of Delaware regulates contractors, both resident and nonresident. Resident contractor is defined to include any general contractor, prime contractor, construction manager, subcontractor, or other type of construction or construction transportation contractor.[20] This license is little more than a formality requiring an application and fee.

C. *States Regulating Construction Managers as Contractors or Design Professionals*

Of the states that regulate construction managers, most tend to view them as contractors, design professionals, or a hybrid of the two. The following case illustrates the hybrid resulting from the selective application of laws regulating contractors and design professionals to the unique situation of the construction manager.

In *McMaster Construction v. Board of Regents of Oklahoma Colleges*,[21] unsuccessful bidders for a construction management services contract with a public university brought an action against the university and board of regents seeking to void the contract awarded. Plaintiffs argued that the construction management contracts were subject to the Oklahoma Competitive Bidding Act because they were contracts for the improvement or construction of a public building, and that the Act was not followed. Plaintiffs alleged that, if the Bidding Act did not apply, then the State Consultants Act, which governed the hiring of architects and engineers, applied. Plaintiffs further argued that the awarded contracts were void because they required the construction management firm to practice architecture and engineering without a license.

The Oklahoma Supreme Court ruled that the construction manager was exempt from design professional licensing and competitive bidding requirements. The Bidding Act did not apply because the construction management contracts involved professional judgment. Like contracts for architectural and engineering services, they were not subject to contractors' bidding statutes because the services were not subject to uniform specifications. The court further noted that an overwhelming majority of jurisdictions had concluded that construction management contracts which do not call for the furnishing of equipment, labor, or materials were not subject to competitive bidding statutes. With respect to the application of the State Consultants Act, the court noted that it specifically applied only to registered architects, engineers, and land surveyors, but not construction managers. Rejecting the

---

20. DEL. CODE ANN. tit. 30, § 2501 (2005).
21. 934 P.2d 335 (Okla. 1997).

plaintiff's argument that the construction management contracts called for the unlicensed practice of architecture and engineering, the court stated, "Merely because a service could be performed by an architect or engineer does not mean that one must be licensed to perform the service . . . . [M]any of the services performed by the [construction management] companies are generally performed by a general contractor. To accept plaintiffs' argument would mean that all general contractors are engaged in the unlicensed practice of architecture and engineering."

Here, the construction manager is defined both by its similarities to, and differences from, contractors and design professionals, depending on the law considered. From the standpoint of the competitive bidding statute, the construction manager is more like a design professional than a contractor, and from the standpoint of design professional licensing laws, the construction manager is more like a general contractor than a design professional. As discussed below, state statutes regulating construction managers largely define them in terms of their construction counterparts—contractors and design professionals.

1. *Regulation as Contractor*

Some states that regulate construction managers categorize them as a type of contractor and place them within the licensing scheme regulating general contractors. Of particular concern in those states is that the unlicensed performance of general contracting services can have severe repercussions, including waiver of lien rights, fines, and even criminal penalties, among other things. Lack of licensure in a state where licensure is required could also result in a successful bid protest to overturn a contract award under the theory that an unlicensed contractor cannot be a *responsible* bidder. In states licensing contractors, regulation as a contractor usually is accomplished by explicitly including *construction manager* in the statutory definition of *contractor* and by including the activity of *construction management* within the activities of a contractor regulated by the state's contractor licensing statute. Some states do so expressly (e.g., Georgia, Idaho, New Mexico, North Dakota, Oregon, Tennessee, Utah, and West Virginia), while others implicitly include construction managers within their contractor licensing statutes (e.g., Arizona, Arkansas, North Carolina, and Virginia). In the case of the latter, those state statutes strongly imply that construction managers are contractors, most often by referring to the activities of *manage* and *construction* within the definition of contractor or contractor activities.

a. *Express Regulation of Construction Managers as Contractors*

The following states expressly include *construction manager* or *construction management* within their definition of *contractor* and, thereby, regulate construction managers in the same way as contractors.

(1) *Georgia*

The State of Georgia regulates both residential and general contractors. The statutory definition of *contractor* includes *construction management* services where the person performing such construction management services is at risk contractually to the owner for the performance and cost of the construction.[22] Georgia's contractor licensing law establishes a state licensing board for residential and general contractors and a process for obtaining licensure, including examination, and other qualifications, such as education and work experience.[23]

(2) *Idaho*

Similarly, Idaho's Contractor Registration Act includes *construction manager* within its definition of *contractor*.[24] The Act establishes a board to enforce the Act[25] and requires each applicant to, among other things, pay a fee and submit insurance certificates as proof that it carries workers' compensation and general liability insurance. An applicant must never have had a contractor license denied or revoked.[26] Among the penalties imposed for being unregistered, the statute deems an unregistered contractor to have waived its lien rights.[27]

(3) *New Mexico*

In the State of New Mexico, a *contractor* is broadly defined as "any person who undertakes, offers to undertake by bid or other means or purports to have the capacity to undertake, by himself or through others, contracting" and includes construction managers who coordinate and/or manage a construction project.[28] New Mexico requires all persons acting as a contractor to be licensed by the Construction Industries Division of the Regulation and Licensing Department (the CID).[29] Applications are furnished by and submitted to the division along with a fee.[30] No bid on a contract shall be submitted unless the contractor has a valid license.[31] A contractor applying for license with the CID should have four years experience within the ten years prior to application of related trade experience. The CID may make exceptions to this provision if the requirement is deemed too excessive for a particular craft.[32] A party desiring to be deemed "qualified" by the division must pass an examination administered by the CID.[33]

The statute establishes a construction industries commission (the commission) within the division. The commission's primary duty is to establish policy for the

---

22. GA. CODE ANN. § 43-41-2 (West 2005).
23. *See id.* § 43-41-6.
24. IDAHO CODE ANN. § 54-5203 (2005).
25. *See id.* § 54-5206.
26. *Id.* § 54-5210.
27. *Id.* § 54-5208.
28. N.M. STAT. ANN. § 60-13-3 (West 2005).
29. *See id.* § 60-13-12.
30. *Id.* § 60-13-13.
31. *Id.* § 60-13-12.
32. *Id.* § 60-13-14B.8.
33. *Id.* § 60-13-16.

division.[34] The commission will regularly review all licenses issued by the division.[35] Performing contracting work without a license may be settled by payment of an administrative fee if it is a first offense.[36] However, any person acting as a contractor without a proper license is guilty of a misdemeanor. The penalty for such an infraction may include a fine of 10 percent of the dollar value of the contracted work, imprisonment of up to six months, or both.[37]

(4) *North Dakota*

In North Dakota, if a governing body uses a construction manager on a public improvement, the construction manager must be a licensed contractor.[38] Contractors may obtain a license through an application process with the secretary of the state of North Dakota (the registrar). A certificate of liability insurance must be filed with the contractor's application. "A statement from North Dakota Workforce Safety and Insurance that the contractor has secured workforce safety and insurance coverage" must also accompany the application.[39] North Dakota issues four classes of licenses.

Complaints may be filed against a licensee for various acts or omissions, including, but not limited to, abandonment, failure to commence work, incomplete work, diversion of funds, fraudulent practice, misleading statements in application, or engaging in work without proper licensure.[40] The registrar may "suspend or revoke the contractor's license, order a civil penalty of not more than one thousand dollars, order restitution in an amount of not more than five thousand dollars, or impose some lesser sanction or remedy" if a licensee is guilty of an infraction of North Dakota Century Code § 43-07. "Any person acting in the capacity of a contractor without a license is guilty of a class A misdemeanor."[41]

(5) *Oregon*

Oregon regulates the licensing of general contractors in its Occupations and Professions title, Chapter 701 Construction Contractors.[42] The regulations define a general contractor as "a contractor whose business operations require the use of more than two unrelated building trades or crafts that the contractor supervises or performs in whole or part, whenever the sum of all contracts on any single property, including materials and labor, exceeds an amount established by rule by the [Construction

---

34. N.M. Stat. Ann. § 60-13-6.
35. *Id.* § 60-13-15.
36. *Id.* § 60-13-14E.
37. *Id.* § 60-13-23.1.
38. N.D. Cent. Code § 48-01 (2005).
39. *See id.* § 43-07-04.
40. *Id.* § 43-07-14.
41. *Id.* § 43-07-18.
42. Or. Rev. Stat. § 701 (West 2005).

Contractors Board]," excluding specialty contractors or limited contractors.[43] This definition covers the work of a contractor as:

> [A] person who, for compensation or with the intent to sell, arranges or undertakes or offers to undertake or submits a bid to construct, alter, repair, add to, subtract from, improve, inspect, move, wreck or demolish, for another, by any building, highway, road, railroad, excavation or other structure, project, development or improvement attached to real estate or to do any part thereof.[44]

Oregon's administrative code expands the definition of contractor to include construction management. A contractor may not undertake or submit a bid to do work unless the contractor has a current, valid license issued by the Construction Contractors Board (Board).[45] To obtain a license from the Board, the applicant must qualify as an independent contractor,[46] or "an individual or business entity that performs labor or services for remuneration."[47] An applicant for a construction contractor license must submit an application provided by the Board, pass an examination based on the licensure education requirements, submit proof that the applicant has the legal capacity to contract, pay a licensing fee, and obtain a surety bond in the amount of $15,000.[48] Furthermore, a licensed general contractor must obtain insurance covering the work of the contractor for no less than $500,000.[49]

The failure of a contractor to comply with these regulations constitutes a basis for the suspension, revocation, refusal to issue or reissue a contractor's license, and an assessment of a civil penalty as set forth by the regulations.[50] The Board may suspend or refuse to license a contractor who owes an amount pursuant to a final order or arbitration award of the Board, owes an amount under final judgment of a court or civil penalty arising from the construction business, or has had a license to operate as a contractor revoked.[51] Other grounds for discipline include practicing without a license, violating Board rules or orders, knowingly assisting an unlicensed person to violate the regulations, knowingly providing false information to the Board, and violating permit regulations.[52] Practicing without a license, intentionally using

---

43. *See id.* § 701.005.
44. *Id.* § 701.005.
45. *Id.* § 701.055.
46. *Id.* § 701.035.
47. *Id.* § 670.600.
48. *Id.* §§ 701.075, 701.085, 701.125.
49. *Id.* § 701.105.
50. *Id.* § 701.100.
51. *Id.* §§ 701.102, 701.992.
52. *Id.* § 701.135.

a contractor's license number without authorization, and doing so with the intent to deceive the public are Class A misdemeanors.[53]

(6) *Tennessee*

Under Tennessee's Contractors Licensing Act of 1994, a contractor will include a construction manager of any kind on projects exceeding $25,000. In the state of Tennessee, it is unlawful to engage in contracting without a license. Any contractor (person, firm, or corporation) in this state must be licensed.[54]

The state may revoke an entity's certificate of license if a contracting entity is brought to suit and a judgment is rendered against that entity. An entity who contracts without a license is guilty of a Class A misdemeanor. This or other infractions of the Tennessee Contractors Licensing Act of 1994 may be punished by a fine up to $25,000, and imprisonment up to six months after determination by the board that a violation has occurred.[55] Additionally, contracting without a license is subject to citation from the director of the board. The board may impose a civil penalty of not less than $50 nor more than $1,000.[56]

(7) *Utah*

The Utah Construction Trades Licensing Act (Utah Act) includes a construction manager in its definition of a contractor.[57] The Utah Act mandates that any person engaged as a contractor shall become licensed before engaging in the trade of contracting.[58] To obtain a license, a contractor/construction manager must submit an application with the appropriate application fee as well as meet the requirements of an examination. Applicants for a contractor's license must also present evidence of financial responsibility, knowledge, and experience in the construction industry.[59]

(8) *West Virginia*

Under its contractor licensing statute, West Virginia also defines contractor to include a construction manager "who performs management and counseling services for a construction project for a professional fee."[60] West Virginia further defines construction manager to be "a person who enters into an agreement to employ, direct, coordinate or manage design professionals and contractors who are hired and paid directly by the owner or the construction manager."[61]

---

53. OR. REV. STAT. § 701.990.
54. TENN. CODE ANN. § 62-6-102 (West 2005).
55. *See id.* § 62-6-120.
56. *Id.* § 62-6-201.
57. UTAH CODE ANN. § 58-55-102 (West 2005).
58. *Id.* § 58-55-301.
59. *Id.* § 58-55-302.
60. W. VA. CODE ANN. § 21-11-3 (West 2005).
61. *Id.* § 11-15-2.

b. *Implicit Regulation as Contractors*

The following states appear to regulate construction managers as contractors.

(1) *Arizona*

In Arizona, contractors are licensed by the Registrar of Contractors. Contractors include subcontractors, specialty contractors, floor covering contractors, and landscape contractors other than gardeners and consultants who supervise or manage a construction project for the benefit of the property owner.[62] Architects or engineers who hire the services of a contractor for pre-construction activities are not required to be licensed as contractors.[63] Applicants for a new or renewed license are required to submit to the Registrar a verified application, proof of compliance with workers' compensation insurance statutes, and the required bond and fee.[64]

Applicants for a contractor's license are required to have a minimum of four years practical or management trade experience and at least two years experience within the last ten years in the relevant type of construction. No more than two of the required four years of practical or management trade experience may be substituted by technical training in an accredited college, university, or manufacturer's accredited training program.[65] The four-year requirement may be reduced if industry custom and usage conclusively show the requirement is excessive. Work experience or documentation and verification of examination may be waived if the applicant is currently or has been a qualifying party in the state for the same classification within the preceding five years.[66]

The applicant must demonstrate each of the following by written examination: (1) no more than two years before application, if required, qualification in the relevant field of construction; (2) general knowledge of state building, safety, health, and lien laws; (3) knowledge of administrative principles of the contracting business and the rules adopted by the registrar; (4) knowledge of construction plans and specifications applicable to the particular industry or craft; and (5) a general understanding of other related construction trades and any other matters the registrar deems appropriate.[67]

Contracting without a license is prohibited, unless the contracting entity is exempt.[68] The Registrar may on written complaint or the Registrar's own motion investigate the acts of any contractor within the state and may temporarily suspend or permanently revoke any or all licenses if the holder is guilty of or commits any acts or omissions set forth in the Arizona Revised Statutes § 32-1154A(1)–(24).[69]

---

62. ARIZ. REV. STAT. ANN. § 32-1101(B) (2005).
63. *See id.* § 32-1121(A)(7).
64. *Id.* § 32-1122(B) (1)–(2).
65. *Id.* § 32-1122(F)(1).
66. *Id.* § 32-1122(F).
67. *Id.* § 32-1122(F)(2).
68. *Id.* § 32-1154(B).
69. *Id.* § 32-1154(D).

The Registrar may impose a civil penalty not to exceed $500 for each violation of subsection § 32-1154A(23).[70] Failure to pay any civil penalty imposed under this subsection results in the automatic revocation of the license thirty days after the effective date of the order providing for the civil penalty. No future license may be issued to an entity unless the entity pays any outstanding civil penalty.[71]

Violation of any act specified in § 32-1154A(7) or (10) or acting in the capacity of a contractor without a license is a Class 1 misdemeanor.[72] The contractor must pay a fine of not less than $1,000 for the first offense and not less than $2,000 for any subsequent offense. The Registrar may also issue injunctive relief in addition to all other civil or criminal remedies.[73]

(2) Arkansas

Arkansas defines a contractor as:

> [A]ny person, firm, partnership, copartnership, association, corporation, or other organization, or any combination thereof, who ... attempts to or submits a bid to construct, or contracts or undertakes to construct, or assumes charge, in a supervisory capacity or otherwise, or manages the construction, erection, alteration, or repair, or has or have constructed erected, altered, or repaired, under his or her, their, or its direction ... when the cost of the work to be done, or done, in the State of Arkansas by the contractor, including but not limited to, labor and materials, is twenty thousand dollars ($20,000) or more.[74]

Any person wanting to be licensed as a contractor in the State of Arkansas must file with the Contractors Licensing Board thirty days prior to any regular or special meeting a written application accompanied by a payment in an amount determined by the Board but not to exceed $100. The thirty-day requirement may be waived if the contractor has an original application on file and successfully completed any required examination. In order for a contractor to be licensed or renew its license, it must submit to the Board a financial statement audited by a certified public accountant or registered public accountant. When deciding whether to license a particular contractor, the Board will consider the contractor's (1) experience; (2) ability; (3) character; (4) the manner of performance of previous contracts; (5) financial condition; (6) equipment; (7) any other fact tending to show ability and willingness to conserve the public health and safety; and (8) compliance with the provisions of this chapter or any other law of the state.

---

70. *Id.* § 32-1154(D).
71. *Id.* § 32-1154(D).
72. *Id.* § 32-1164(A)–(B).
73. *Id.* § 32-1166(A)–(C).
74. Ark. Code Ann. § 17-25-101(a)(1) (West 2005).

A contractor will violate the contractors licensing law if it bids or contracts to construct a project worth $20,000 or more without a license. If a contractor presents or files the license certificate of another, gives false or forged information to the Contractors Licensing Board, impersonates another, or uses an expired or revoked certificate of license, the contractor shall be liable to a fine of not less than $100 and no more than $200 for each offense.

(3) *North Carolina*

Under its contractor licensing statute, North Carolina defines contractor as:

[A]ny person or firm or corporation who for a fixed price, commission, fee, or wage, undertakes to bid upon or to construct or who undertakes to superintend or manage, on his own behalf or for any person, firm or corporation that is not licensed as a general contractor pursuant to this Article, the construction of any building, highway, public utilities, grading or any improvement or structure where the cost of the undertaking is thirty thousand dollars ($30,000) or more, or undertakes to erect a North Carolina labeled manufactured modular building meeting the North Carolina State Building Code....[75]

(4) *Virginia*

Virginia regulates contractors under statutory code title 54.1. Professions and Occupations. By definition, a contractor includes any person who contracts for managing or superintending services of construction projects.[76] Contractors are regulated by the state's Board of Contractors (the Board).[77] No entity may contract for construction services without being licensed through the Board.[78] Generally, licensed engineers and architects are exempt from licensure.[79]

Virginia offers three classes (A, B, and C) of contractor licenses corresponding to the cost of the construction undertaken.[80] The procedure for obtaining a Class A and B license is identical. A written application must be submitted to the Board along with an application fee. The Board will review each applicant's performance record, and if the Board deems the entity qualified for licensure, that entity is subject to a Board examination. An entity that satisfies the Board's examination will receive the license of the appropriate class.[81] A Class C license applicant is not subject

---

75. N.C. GEN. STAT. § 87-1 (2005).
76. VA. CODE ANN. § 54.1-1100 (West 2005).
77. *See id.* § 54.1-1102.
78. VA. CODE ANN. § 54.1-1103.
79. *Id.* § 54.1-1106.
80. *Id.* § 54.1-1103.
81. *Id.* § 54.1-1106, -1108.

to examination.[82] The Board determines the expiration and/or renewal term of a license.[83] The Board holds the authority to suspend, revoke, or deny a license.[84] Contracting without a license is a Class 1 misdemeanor and the Board may impose a fine of up to five hundred dollars per day in violation.

2. Regulation as Design Professional

Few states expressly license or regulate construction managers as design professionals. However, those that do usually allow contractors to act as construction managers without first obtaining an architect or engineering license. In effect, these states allow construction managers to be licensed either as design professionals or general contractors. Other states specifically require construction managers to be licensed as design professionals or contractors. The public procurement laws in a number of states treat construction managers as design professionals. In those states, policies and procedures for letting construction contracts, and for hiring contractors, architects, and engineers, usually treat construction managers as professionals like architects and engineers, rather than contractors.

a. Express Regulation as Design Professional

A number of states expressly regulate construction managers as design professionals. As discussed in greater detail below, these states include Arkansas, California, Florida, Illinois, Mississippi, Nevada, New York, Pennsylvania, South Carolina, and Wisconsin.

Arkansas and Mississippi require those who provide construction management in connection with the development of land areas to be licensed as landscape architects, where the dominant purpose of such services is landscape development, preservation, and enhancement, or determination of land uses, natural land features, functional and aesthetic values, among other things.[85]

In California, construction managers on state and local construction projects should have design professional licensure. The California Attorney General issued an opinion in 1995 concluding that a state or local agency may not contract with a private firm for construction project management services if all or part of such services are to be performed other than under the direction and control of a registered engineer or licensed architect.[86]

Florida also requires construction managers on public projects to have professional licensure. In the construction of educational facilities, Florida authorizes the use of construction managers who are responsible for all scheduling and coordination in both the design and construction phases and are generally responsible for

---

82. *Id.* § 54.1-1108.2.
83. *Id.* § 54.1-1109.
84. *Id.* § 54.1-1110.
85. Ark. Code Ann. § 17-36-102 (West 2005); Miss. Code Ann. § 73-2-3 (West 2005).
86. 78 Op. Cal. Att'y Gen. 48 (1995) (Opinion No. 94-819).

the successful, timely, and economical completion of the construction project. The construction management entity must consist of or contract with licensed or registered professionals for the specific fields or areas of construction to be performed, as required by law.[87]

Illinois, New York, and Pennsylvania include construction management in the activities defined as the practice of architecture.[88] However, all three states also provide exemptions for certain others to provide construction management services. The Illinois Architecture Practice Act of 1989 provides:

> Nothing in this Act shall be deemed or construed to prevent . . . the offering or preparation of environmental analysis, feasibility studies, programming or construction management services by persons other than those licensed in accordance with this Act, the Structural Engineering Practice Act of 1989 or the Professional Engineering Practice Act of 1989.[89]

Similarly, New York's architecture practice act provides: "This article shall not be construed to affect or prevent . . . [c]ontractors or builders from engaging in construction management and administration of construction contracts."[90] A New York case, *Charlebois v. J.M. Weller Associates, Inc.*, noted "certain activities . . . such as construction supervision and management, are specifically exempt from the [architecture] licensing requirements."[91] Finally, Pennsylvania's architecture licensing act provides: "Nothing contained in this act shall be construed to prohibit . . . [t]he preparation of any shop drawings or the performance of construction management services by persons customarily engaged in construction work."[92]

Nevada regulates construction managers as contractors,[93] but the Nevada Supreme Court held that a licensed architect's lack of a contractor's license did not preclude contractual recovery for construction management services. The court found that a licensed architect's scope of practice under the licensing statutes can include construction management and an architect does not need a contractor's license to recover compensation for construction management services if the architect's contract with the project owner specifies such services.[94] Thus, Nevada licenses construction managers as design professionals or contractors.

---

87. FLA. STAT. ANN. § 1013.45 (West 2005).
88. 225 ILL. COMP. STAT. 305/5 (West 2005); N.Y. EDUC. LAW § 7301 (McKinney 2005); 63 PA. CONS. STAT. ANN. § 34.3 (West 2005).
89. 225 ILL. COMP. STAT. 305/5 (West 2005).
90. N.Y. EDUC. LAW § 7301 (McKinney 2005).
91. 136 A.D.2d 214, 526 N.Y.S.2d 648 (N.Y. App. Div. 3rd Dept. 1988).
92. 63 PA. CONS. STAT. ANN. § 34.15 (West 2005).
93. NEV. REV. STAT. ANN. 624.020 (West 2005).
94. Kourafas v. Basic Food Flavors, Inc., 120 Nev. 195, 88 P.3d 822 (2004).

South Carolina regulates construction managers by requiring them to obtain either design professional or contractor licensure in one or more of the following professional classifications: (1) a licensed general or mechanical contractor; (2) a registered engineer; or (3) an architect.[95] An architect or engineer licensed in South Carolina who is monitoring the execution of design plans or is performing as an on-site representative for construction quality or quality assurance, or both, for a project owner is not a construction manager for the purpose of this law.[96]

In at least one state, a court holding indicates that a construction manager's professional classification may depend on the type of insurance it carries. A construction manager in Wisconsin will be viewed as a design professional if it carries professional liability insurance. In *1325 North Van Buren LLC v. T-3 Group Ltd.*,[97] the court ruled that a contract for construction management services was a contract for services; hence, the economic loss doctrine did not apply to bar the owner's negligence claims against the general contractor.[98] The general contractor performed no construction and had no design authority. The fact that the construction manager/general contractor carried professional liability insurance convinced the court that the entity was providing professional services and owed a duty to adhere to professional standards.

b. *Regulation under State Procurement Laws*

(1) *State Laws Exempting Design Professionals and Construction Managers from Competitive Bidding Requirements*

One way in which state laws treat construction managers as a kind of design professional is from the perspective of competitive bidding laws. Typically, contracts for construction are awarded to contractors through a competitive bidding process. However, contracts for architecture, engineering, and construction management services are frequently exempt from a competitive bid process because they are contracts for professional services.

For instance, Pennsylvania's County Code,[99] the Philadelphia Regional Port Authority Act,[100] and the Port of Pittsburgh Commission Act[101] all exempt construction managers from the competitive bidding process. Similarly in 1978, the Montana Attorney General opined that a construction manager's employment was exempt from the competitive bidding process for professional services.[102] In Alabama as

---

95. S.C. CODE ANN. § 40-11-320 (2005).

96. *Id.*

97. 701 N.W.2d 13, 284 Wis. 2d 387 (Wis. Ct. App. 2005).

98. *Id.* This interpretation of the economic loss doctrine is unique to Wisconsin courts, and is not necessarily shared by courts in other jurisdictions.

99. 16 PA. CONS. STAT. ANN. § 1802 (West 2005).

100. 55 PA. CONS. STAT. ANN. § 697.11 (West 2005).

101. *See id.* § 698.31.

102. 37 MONT. OP. ATTY. GEN. 735 (1978).

well, awarding authorities are exempted from advertisement or open bidding requirements when hiring for construction management services.[103]

On public works projects, Illinois, Indiana, and Pennsylvania courts have ruled that construction managers are exempt from statutes governing the competitive bidding for construction contracts. In *Attlin Construction, Inc. v. Muncie Community Schools*,[104] the court of appeals held that the Muncie Community School Board was exempt from utilizing the competitive bidding process when seeking to hire a construction manager who performed services similar to those of architects or engineers. Similarly, in *Shively v. Bellevue Township High School District No. 201*, an Illinois contractor challenged the township's award of a construction management contract without competitive bidding.[105] The court of appeals determined that a contract for construction management services is exempt from the competitive bidding process because it required professional skills provided by entities, such as architects or engineers, exempt from competitive bidding. In a taxpayer's action against the Boyertown Area School Board,[106] the Supreme Court of Pennsylvania also held that a contract for construction management services was exempt from competitive bidding requirements because it required skills provided by entities, such as architects or engineers, not subject to the competitive bidding statute.

> (2) *State Laws Mandating That State Agencies Hire Design Professionals and Construction Managers by Weighing Qualifications and Experience, over Price, Unlike Contractors*

Another way in which state laws treat construction managers as a kind of design professional is in the criteria the state must use in hiring construction managers. Rather than price being the preeminent criteria—as with construction contracts awarded to the lowest bidder—professional qualifications and experience are the primary criteria used in many states to hire design professionals and construction managers.

> (a) *California*

In California, a state or local agency selects construction management firms on the basis of demonstrated competence and the professional qualifications necessary for the services required. Construction management firms interested in working on state or local projects are encouraged to submit a statement of qualifications annually. State and local agencies can evaluate the current statements of qualifications and then discuss with no less than three firms the required services being sought. The agencies rank the firms based on qualifications criteria and attempt to negotiate a contract with the best qualified firm. If no reasonable and satisfactory contract is

---

103. ALA. CODE § 39-2-2 (2005).
104. 413 N.E.2d 281 (Ind. Ct. App. 1980).
105. 329 Ill. App. 3d 1156, 769 N.E.2d 1062 (5th Dist. 2002).
106. Malloy v. Boyertown Area Sch. Bd., 540 Pa. 308, 657 A.2d 915 (1995).

negotiated, agencies then may undertake negotiations with the next highest ranked firm until an agreement is reached.[107]

(b) *Florida*

Florida maintains a selection process based on qualifications similar to California's. Construction management entities wishing to perform services for a Florida state agency are encouraged to submit annual statements of qualifications. For each proposed project, the agency seeking construction management services evaluates current statements of qualifications on file with the state. The agency conducts discussions with no fewer than three firms regarding their qualifications to furnish the required services. The agency then ranks at least three firms in order of preference and qualifications to be selected for the required services. The agency negotiates a contract for compensation, which the agency deems to be fair and reasonable, with the most qualified firm. If a negotiated contract is not reached with the most qualified firm, the agency undertakes negotiations with the next most qualified firm. This process continues until an agreeable contract is reached.[108]

(c) *Kansas*

Firms performing construction management at-risk services in Kansas are encouraged to submit annually to the Secretary of Administration and to the State Building Advisory Commission a statement of qualifications and performance data. When the construction management delivery method is utilized for a project, the Commission prepares a list of three to five firms that are qualified to serve on the project. A negotiation committee proceeds to negotiate a contract with the most qualified firm. If the committee is unable to negotiate a contract with the most qualified firm, negotiations are terminated and the committee negotiates with the next ranked firm. The process continues until a contract is satisfactorily negotiated.[109]

(d) *Massachusetts*

The Public Construction Alternative Delivery Methods Code of Massachusetts utilizes a two-phase selection process for the selection of a construction management at-risk firm. A pre-qualification committee is created to review construction managers' responses to the request for qualifications (RFQ) issued by the public agency seeking construction management services. The first of the two-phase selection process begins once the public agency gives public notice of the building project and solicits responses to an RFQ. Interested construction management firms are invited to submit proposals, along with a statement of qualifications, in response to the RFQ. Upon receipt of the responses, the committee will rank the three most qualified construction management firms, and commence the second phase of the

---

107. CAL. GOV'T CODE § 4526 (West 2005).
108. FLA. STAT. ANN. § 287.055 (West 2005).
109. KAN. STAT. ANN. § 76-786 (2005).

selection process. Contracts are negotiated with the most qualified firm until a reasonable contract is reached. All of the pre-qualification committee's decisions are final and not subject to appeal except on the grounds of fraud or collusion.[110]

(e) *Missouri*

When a public works project in Missouri utilizes construction management services, the public owner advertises and solicits proposals from qualified construction managers.[111] Construction management firms interested in supplying their services to the public project submit a proposal to the public owner. For each construction manager who has submitted a proposal, public owners are to consider the following:

- fees for overhead and profit;
- reimbursable costs for reimbursable items as defined in the public owner's request for proposal;
- qualifications;
- demonstration of ability to perform projects comparable in design, scope, and complexity;
- demonstration of good faith efforts to achieve compliance with federal, state, and local affirmative action requirements;
- references by owners for whom construction management has been performed;
- financial strength;
- qualifications of in-house personnel who will manage the project; and
- demonstration of successful management systems which have been employed for the purposes of estimating, scheduling and cost control.

The public owner may negotiate a contract with any construction manager deemed qualified. If a contract is not negotiated, the public owner may again advertise and solicit proposals.[112]

(f) *North Carolina*

North Carolina publicly solicits architectural, engineering, surveying, and construction management at-risk services. Qualified firms are selected on the basis of demonstrated competence and qualification without regard to fee other than unit price information. Contracts for those services are to be negotiated at a fair and reasonable fee with the best qualified firm. If a contract cannot be negotiated with the best qualified firm, negotiations with that firm are terminated and initiated with the next best qualified firm. Selection of a firm includes the use of good faith efforts by the public entity to notify minority firms of the opportunity to submit qualifications for consideration.[113]

---

110. Mass. Gen. Law Sch. 149A, § 5 (2005).
111. Mo. Rev. Stat. § 8.679 (2005).
112. *See id.* § 8.681.
113. N.C. Gen. Stat. § 143-64.31.

Public entities that contract with a construction manager in the State of North Carolina report to the Secretary of Administration the following information on all projects where a construction manager at-risk is utilized: a detailed explanation of the reason why the particular construction manager at-risk was selected; the terms of the contract with the construction manager at-risk; and a report on the form of bidding utilized by the construction manager at-risk.[114]

(g) *South Carolina*

In South Carolina, a state agency seeking construction management services establishes its own architect-engineer, construction management, and land surveying services selection committee. This selection committee is responsible for developing a description of the proposed project, enumerating all required professional services for that project, and preparing a formal invitation to firms for submitting information. The invitation may include the general scope of work, a description of all professional services required for that project, the submission deadline, and how interested firms may apply for consideration. Interested firms are required to respond to the invitation by submitting a current and accurate services questionnaire and any other information that may be required. The agency selection committee interviews at least five persons or firms who have responded and are deemed most qualified. The agency selection committee's determination for which person or firm will be interviewed is in writing and based on its review and evaluation of all submitted material. The agency selection committee evaluates each of the persons or firms interviewed in view of the following: past performance; ability of professional personnel; demonstrated ability to meet time and budget requirements; location; recent, current, and projected workloads; creativity and insight related to the project; and related experience on similar projects. A contract is negotiated with the most qualified person or firm at a compensation that is fair and reasonable to the state. Negotiations commence in the same manner with the second, third, fourth, and fifth most qualified until a satisfactory contract is negotiated. If no agreement is reached with one of the five, additional firms or persons are selected in order of their competence and qualifications.[115]

D. *Other Issues of Concern in Architect, Engineering, and Contractor Licensing*

As mentioned previously, in those states that do not regulate construction managers, the practitioner is well advised to consult the state's laws regarding architect and engineer licensing, as well as the state's laws regarding licensing or registering of general contractors. Because the activities performed by a construction manager on a project may fall within the scope of another licensed profession, they may be subject to those licensing laws. As such, laws are typically enacted to ensure

---

114. *Id.*
115. S.C. CODE ANN. § 11-35-3220 (2005).

professional competence and protect the health, safety, and welfare of the public; the penalties for the unlicensed or unregistered practice of these professions can be severe. As an example of how these laws operate, California's architecture, engineering, and general contracting licensing laws are discussed below.

1. *Design Professional Licensing*

The practice of architecture in California is defined as "offering or performing, or being in responsible control of, professional services which require the skills of an architect in the planning or site selection, and the design, in whole or in part, of buildings, or groups of buildings and structures."[116] Professional services are defined in the statute to include, among other things, investigation, evaluation, planning, creating schematic and preliminary drawings, coordination of consultants, contract administration, and construction observation.[117] The last two services, which commonly fall within the practice of architecture in California and many other states, are of potential concern for the construction manager.

To obtain a license to practice architecture in California, an applicant must have eight years of training and educational experience in architectural work and must pass an examination.[118] A five-year degree from an accredited school is considered the equivalent of five years of training and educational experience.[119]

The licensing statute also specifies penalties and grounds for discipline. If a person has acted, or is about to act, in violation of the statute, the board that enforces the licensing statute may apply to the state court for an injunction or restraining order.[120] The board also has the power to levy a fine up to $5,000, or to suspend or permanently revoke the license of an architect who violates the act.[121] An architect convicted of a crime relating to his professional duties or functions can also be punished under the licensing statute.[122] An architect can also run afoul of the licensing statute by obtaining a license fraudulently, practicing under an improper name, aiding another in the unauthorized practice of architecture, or committing fraud, negligence, or incompetence in architectural practice.[123] A person who practices architecture without a license is subject to a fine between $100 and $5,000 and a one-year prison term in county jail.[124]

---

116. CAL. BUS. & PROF. CODE § 5500.1(a) (West 2005).
117. *See id.* § 5500.1(b).
118. *Id.* §§ 5551, 5552(b).
119. *Id.* §§ 5551, 5552(b).
120. *Id.* § 5527.
121. *Id.* §§ 5560, 5565(d).
122. *Id.* § 5577.
123. *Id.* §§ 5579–5585.
124. *Id.* § 5536(a).

In California, an engineer is defined as:

> [A] person engaged in the professional practice of rendering service or creative work requiring education, training, and experience in engineering sciences and the application of special knowledge of the mathematical, physical and engineering sciences in such professional or creative work as consultation, investigation, evaluation, planning or design of public or private utilities, structures, machines, processes, circuits, buildings, equipment or projects, and supervision of construction for the purpose of securing compliance with specifications and design for any such work.[125]

To obtain a professional engineer's license, an applicant must possess six or more years of approved engineering experience and pass an examination administered by the board that enforces the requirements of the licensing statute.[126]

The engineering licensing statute also specifies grounds for discipline and pen-alties. The board that enforces the licensing statute has the power to "reprove, suspend for a period not to exceed two years, or revoke the certificate" of a professional engineer who commits certain crimes or acts enumerated in the statute.[127] An engineer convicted of a crime related to an engineer's qualifications or duties and found guilty by the board of practice-related fraud, negligence, or incompetence may be disciplined.[128] The board may also discipline an engineer found guilty of breaching a contract to perform professional services.[129] Any person who practices in the state without a license is guilty of a misdemeanor and subject to the powers of the board.[130]

2. Contractor Licensing

California regulates the licensing of general contractors in its Business and Professions Code, Division 3, Chapter 9 Contractors.[131] The code defines a contractor, including subcontractors and specialty contracts, synonymously with builders.[132] The broad definition includes:

> [A]ny person who undertakes to or offers to undertake to, or purports to have the capacity to undertake to, or submits a bid to, or does himself or herself or

---

125. *Id.* § 6701.
126. *Id.* § 6751(b).
127. *Id.* § 6775
128. *Id.* §§ 6775(a)–(c).
129. *Id.* § 6775(d).
130. *Id.* § 6787(a).
131. *Id.* § 7000.
132. *Id.* § 7026.

by or through others, construct, alter, repair, add to, subtract from, improve, move, wreck or demolish any building, highway, road, parking facility, railroad, excavation or other structure, project, development or improvement, or to do any part thereof, including the erection of scaffolding or other structures or works in connection therewith, or the cleaning of grounds or structures in connection therewith, or the preparation and removal of roadway construction zones, lane closures, flagging, or traffic diversions, or the installation, repair, maintenance, or calibration of monitoring equipment for underground storage tanks, and whether or not the performance or work herein described involves the addition to, or fabrication into, any structure, project, development or improvement herein described of any material or article of merchandise.[133]

Persons who provide air conditioning, heating, or refrigeration services; any person or entity that undertakes, offers or purports to have the capacity to undertake to construct any building or home improvement project; a temporary labor service agency that provides employees to perform contracting work; and any person who performs tree removal or pruning services are also considered contractors.[134] In California, as in many states, the statutes governing general contractors are broad enough to be of potential concern for the construction manager.

Contractors' licenses may be issued to individual owners, partnerships, and corporations.[135] California's licensing procedure for contractors includes completing an application providing information required by the registrar of contractors, payment of the application fee, and passing a written examination that tests the applicant's knowledge of California laws and the contracting business and trade.[136] No examination is required for the limited specialty license classification, and some requirements may be relaxed for applicants already licensed in another state through reciprocity.[137] Finally, the applicant must submit evidence of financial solvency; meaning, the applicant's operating capital exceeds $2,500.[138]

Practicing without a contractor's license is a misdemeanor unless the person has been previously convicted of such offense.[139] The punishment for a repeat offense is a court imposed fine of twenty percent of the price of the contract under which the unlicensed person performed the work, or $4,500, whichever is greater.[140] Additionally,

---

133. *Id.* § 7026.
134. *Id.* § 7026.1.
135. *Id.* § 7065.
136. *Id.* §§ 7065, 7072.
137. *Id.* §§ 7065.1, 7065.2, 7065.4.
138. *Id.* § 7067.5.
139. *Id.* § 7028.
140. *Id.* § 7028.

the offender shall serve ninety days in county jail.[141] Where a licensed or unlicensed contractor violates the regulations to cause substantial injury to another person, the registrar may apply to the superior court of the county in which the violator maintains a place of business or resides for an injunction.[142] It is also unlawful for any person who is a member or officer of a licensed contracting entity to act as an individually licensed contractor without having a license.[143]

The registrar of contractors may on its own motion, and will, upon receiving a complaint, investigate the action of a contractor, and may suspend, revoke, or deny the renewal of a license for actions constituting cause for disciplinary action.[144] Grounds for contractor disciplinary action include:

- failing to comply with contractors laws;
- failing to comply with a final citation;
- advertising for construction work without being properly licensed;
- being disciplined in another state for an act that constitutes grounds for disciplinary action in California;
- the abandonment of any construction project or operation without a legal excuse;
- diverting funds or property received for a specific construction project;
- withholding payment to subcontractors;
- departing from accepted trade standards or the plans and specifications of the project;
- violating safety provisions of the labor code resulting in the death or serious injury to an employee;
- disregarding and violating the state building laws;
- requiring the execution of a release in violation of the California Labor Code;
- failing to keep records and failing to comply with requests for information or records;
- failing to cooperate in the investigation of a complaint;
- omitting or misrepresenting a material fact in obtaining or renewing a license;
- breaching a project or operation contract;
- acting under an inactive license;
- avoiding or settling for less than the full amount of lawful obligations incurred;
- aiding or abetting the evasion of contractors laws;
- signing a false certificate of contracting experience;
- willfully or fraudulently acting as a contractor causing substantial injury to another;

---

141. *Id.* § 7028.
142. *Id.* § 7028.3.
143. *Id.* § 7028.5.
144. *Id.* § 7090.

- subverting an investigation;
- acting in a capacity that varies from the license held;
- suspended or expired license;
- contracting with an unlicensed contractor;
- violating special asbestos-related contracting laws;
- willfully refusing to prosecute a construction project with reasonable diligence causing material injury to another;
- failing to pay and falsely denying the claimed amount due;
- employing or serving as an unlicensed executive; and
- being convicted of a crime substantially related to the duties of a contractor.[145]

Penalties for contractor misconduct include denying the issuance or renewal of a license, suspending or revoking a license, imposing on the licensee compliance with specific conditions, and civil penalties not exceeding $5,000.[146]

Library References

C.J.S. *Architects* §§ 1–3, 6–7, 10, 12–14; C.J.S. *Landlord and Tenant* §§ 337–339; C.J.S. *Public Contracts* §§ 6, 14–24; C.J.S. *Schools and School Districts* §§ 406–416, 435, 443, 761; C.J.S. *States* §§ 270, 274, 278, 280–288; C.J.S. *Trading Stamps and Coupons* § 6.

West's Key No. Digests, Licenses <KEY>11, 20–25, 38, 40, 40; Public Contracts <KEY>5; Schools <KEY>80; States <KEY>96–98.

III. Ethical Considerations in the Relationship between Construction Manager and Owner

Licensing and procurement requirements comprise much of the regulation of the construction manager. Ethical aspects of the relationship between the construction manager and owner are also subject to other laws and regulations in the public sector and industry self-policing efforts in the private sector.

According to a 2004 survey conducted by Construction Management Association of America (CMAA) and the management consulting firm FMI,[147] more than eighty percent of the construction managers, contractors, specialty contractors, architects, and other construction industry participants responding personally

---

145. *Id.* §§ 7099-7123.

146. *Id.* §§ 7095-7099.6. The civil penalty maximum increases to $15,000 for violations of § 7114 (aiding the evasion of contractor law) and § 7118 (contracting with an unlicensed contractor).

147. Construction Management Association of America, FMI/CMAA Survey of Construction Industry Ethical Practices (2004), *available at* http://cmaanet.org/user_images/ethics_survey.pdf.

experienced, encountered, or observed acts or transactions they considered unethical. About 65 percent of respondents agreed or strongly agreed that "the construction industry is tainted by prevalent acts that are considered unethical," such as misreporting of costs or time and misuse of equipment and supplies. About half agreed or strongly agreed that "the construction industry is tainted by prevalent illegal acts like bribery." The top five critical ethics issues listed were bid shopping, change order games, payment games, unreliable contractors, and claims games.[148] The common factor is the attempt by one party to profit at the expense of another, to obtain something not contractually bargained for without the knowledge or consent of the other party.

Ethics problems have been addressed in various ways in the public and private sectors. In the private sector, a host of organizations have appeared over the past several decades to self-regulate the construction industry. Several, including the CMAA and the Project Management Institute (PMI), offer professional certifications, and have developed professional codes of ethics. Other entities, such as the American Council for Construction Education (ACCE) and the Associated Schools of Construction (ASC), offer program accreditation promoting curricular and degree standards and excellence in construction education and research.

The traditional design-bid-build scheme developed out of the earlier master-builder model from a perceived need to protect the owner's interests by assigning design and construction contracts to separate entities to eliminate potential conflicts of interest in defining the project budget. On public projects, such concerns partly inform separate procurement procedures for contractors and design professionals. Competitive bidding laws requiring that contracts be awarded to the lowest responsible bidder govern most public sector construction contracts. Competitive proposal and negotiation processes requiring contracts for professional services to be awarded to the most qualified firm typically govern the procurement of architecture, engineering, and construction management services in many states. To avoid real or potential conflicts of interest, a number of states statutorily enforce arm's-length relationships between designer, manager, and constructor (and thus, prohibit other project delivery methods, such as design-build).

A. *Public Projects*

Public entities have addressed ethics issues in a variety of ways. Most states prohibit contingent fees because of the conflict of interest it creates for the public servant. Many states enforce the arm's-length relationship between construction entities by prohibiting an entity from performing design or management and construction on the same project because of the conflict of interest it creates for the construction manager.

---

148. *Id.*

1. *Contingent Fees*

Many states have enacted statutes regulating the payment of contingent fees, rebates, and kickbacks. The hiring agencies of these states, as well as employees who participate in the selection process, are commonly prohibited from having a financial relationship with contracted entities. California and Florida, discussed below, are just two examples of states with such regulations.

a. *California*

In California, the selection process of architects, engineers, land surveyors, and construction project managers for public works projects prohibits practices that might result in unlawful activity. These activities include, but are not limited to, rebates, kickbacks, or other unlawful consideration. Government agency employees are specifically prohibited from participating in the selection process when those employees have a relationship with a person or business entity seeking a contract.[149] No public official at any state or local government level shall make, participate in making, or in any way attempt to use his official position to influence a governmental decision in which he knows or has reason to know he has a financial interest.[150]

b. *Florida*

Every professional contract entered into by a Florida public agency must contain a prohibition against contingent fees as follows:

> The architect (or registered surveyor and mapper or professional engineer, as applicable) warrants that he or she has not employed or retained any company or person, other than a bona fide employee working solely for the architect (or registered surveyor and mapper, or professional engineer, as applicable) to solicit or secure this agreement and that he or she has not paid or agreed to pay any person, company, corporation, individual, or firm, other than a bona fide employee working solely for the architect (or registered surveyor and mapper or professional engineer, as applicable) any fee, commission, percentage, gift, or other consideration contingent upon or resulting from the award or making of this agreement.

If this provision is breached, the agency shall have the right to terminate the agreement without liability and deduct from the contract price the full amount of such fee, commission, percentage, gift, or consideration.[151]

Any entity other than a bona fide employee working solely for an architect, professional engineer, construction manager, or registered land surveyor who solicits contracts for professional services for any other entity to receive consideration, fees, or

---

149. CAL. BUS. & PROF. CODE § 4526 (2005).
150. *See id.* § 87100.
151. FLA. STAT. ANN. § 287.055 (West 2005).

gifts contingent on a contract award for professional services is guilty of a first degree misdemeanor. Any professional services firm or entity that offers consideration, fees, or gifts contingent on a contract award for professional services is guilty of a first degree misdemeanor. State agency officials are subject to these same provisions and penalties.[152]

2. *Separate Contracts/Entities for Design, Management, Construction*

Another ethics concern in the relationship between the construction manager and the owner is aligning the construction manager's financial interest with that of the owner rather than with that of any other party on the project; such as the entity performing the construction. A number of states have addressed this concern by prohibiting or limiting construction managers from actually performing the trade contract work, and requiring that work be performed by another party. Indiana, Louisiana, Michigan, Missouri, and South Carolina are examples of such states.

a. *Indiana*

The State of Indiana limits the employment of and contracts with construction managers. No local government unit may employ the architect or engineer who provided design services on a public construction project or his affiliate to be the construction manager on the project it designed. Additionally, a local government unit may not allow a general contractor, or any separate trade contractor, that performs work on a public construction project, to serve as the construction manager of the project.[153]

b. *Louisiana*

The State of Louisiana advertises public works projects publicly and awards contracts to the lowest responsible bidder. All public work exceeding the contract limit, including labor and materials, to be done by a public entity shall be advertised and let by contract to the lowest responsible bidder who bid according to the contract, plans, and specifications as advertised; no such public work shall be done except as provided. No construction manager, or any other third party consultant, employed by a public entity, may manage a construction project as a general contractor or act in the general contractor role to oversee, direct, or coordinate individual trade contractors on behalf of the public entity, accept bids, or itself bid on the public work or components of the public work, with respect to which the manager or consultant is employed or contracted to manage or consult.[154]

c. *Michigan*

Whether or not construction managers may also perform trade work is less settled in Michigan. In 1994, the Attorney General of Michigan opined that a construction manager employed by the board of education of a local school district may not

---

152. *See id.*
153. IND. CODE ANN. § 5-16-10-2 (West 2005).
154. LA. REV. STAT. ANN. § 2212 (2005).

be the assignee of contractors' construction contacts with the school district.[155] On the other hand, a 1992 Attorney General opinion stated that a local school district board of education may employ a construction manager to contract directly with trade contractors for construction of school facilities only where the board retains and exercises statutory authority to set bid specifications, advertise for bids, accept or reject bids, and require and receive a statutorily mandated security amount.[156]

d. *Missouri*

Upon award of a construction management services contract in Missouri, the successful construction manager contracts with the public owner to furnish his skill and judgment in cooperation with, and reliance on, the services of the project architect or engineer. The construction manager furnishes business administration, management of the construction process, and other specified services to the public owner and performs in an expeditious and economical manner consistent with the interest of the public owner. If the public owner determines it is in the public's best interest, the construction manager may provide or perform basic services for which reimbursement is provided in the general conditions to the construction management services contract. However, the construction manager is not permitted to bid on or perform any of the actual construction on a public works project in which it is acting as construction manager, nor may any construction firm that controls, is controlled by, or shares common ownership or control with, the construction manager be allowed to bid on or perform work on such project. The actual construction work on the project is awarded by competitive bidding as provided by law. All successful bidders contract directly with the public owner, but perform at the direction of the construction manager unless otherwise provided in the construction manager's contract with the public owner. All successful bidders must provide payment and performance bonds to the public owner. Pertaining to the payment of prevailing wages, all successful bidders must meet the obligations of a prime contractor to whom a contract is awarded. In addition, all nonresident employers are to meet the bonding and registration requirements.[157]

No construction management services contract may be awarded by a public owner on a negotiated basis as provided herein if the construction manager, or a firm that controls, is controlled by, or shares common ownership or control with the construction manager, guarantees, warrants, or otherwise assumes financial responsibility for the work of others on the project, or provides the public owner with a guaranteed maximum price for the work of others on the project, or furnishes or guarantees a performance or payment bond for other contractors on the project.

---

155. Op. Mich. Att'y Gen. 116 (1993–1994), 1994 Mich. OAG No. 6789, 1994 WL 580234 (Mich. A.G.).

156. Op. Mich. Att'y Gen. 185 (1991–1992), 1991-1992 Mich. OAG No. 6734, 1992 WL 573169 (Mich. A.G.).

157. Mo. Ann. Stat. § 8.683 (West 2005).

In any such case, the contract for construction management services shall be let by competitive bidding as in the case of contracts for construction work.[158]

e. *South Carolina*

No architect or engineer performing design services in South Carolina, or construction manager performing construction management services, pursuant to a contract awarded, may perform other work on that project as a contractor or subcontractor either directly or through a business in which that person or his or her architectural, engineering, or construction management firm has greater than a five percent interest. Safety compliance and other incidental construction support activities performed by the construction manager are not considered work performed as a contractor or subcontractor. If the construction manager performs or becomes responsible for safety compliance and other incidental construction support activities, and these support activities are not in compliance with the provisions of South Carolina's Occupational Health and Safety laws, then the construction management firm is subject to all applicable fines and penalties.[159]

3. *Exceptions to Prohibitions against Single Entities*

In other states, the construction manager is permitted to perform construction through competitively bid construction trade contracts. This process is followed in Massachusetts, Minnesota, North Carolina, Oklahoma, and Texas.

a. *Massachusetts*

The statutes governing the Massachusetts State College Building Authority require trade contracts to be prequalified and competitively bid. However, a construction management at-risk firm may submit its qualifications to bid on trade contract or subcontract work provided that the construction management at-risk firm customarily performs the work for which it submits its qualifications, performs the work with employees on its own payroll, and meets all the requirements of the selection process.[160]

b. *Minnesota*

In Minnesota, the construction manager at-risk shall competitively bid all trade contract work for the project from a list of qualified firms, subject to the availability of such qualified firms for the specific work. The list of qualified firms shall be based on an open, competitive, and objective pre-qualification process which takes into consideration the firm's experience as a constructor; the capacity of key personnel; technical competence; capability to perform; the past performance of the firm and its employees, including its safety record and compliance with state and federal law; availability to and familiarity with the project locale; and other considerations as defined by the

---

158. *See id.* § 8.685.
159. S.C. CODE ANN. § 11-35-3245 (2005).
160. MASS. GEN. LAWS ANN. ch. 73, § 1-21G (2005).

construction manager at-risk and the commissioner. The construction manager at-risk and the commissioner shall jointly determine the composition of the list of qualified firms. The criteria shall not impose unnecessary conditions beyond reasonable requirements to ensure maximum participation of qualified contractors. With the commissioner's approval or request, the construction manager at-risk may also submit bids for trade contract work. Additionally, the construction manager at-risk and the commissioner shall enter into a guaranteed maximum price contract for the project.[161]

c. *North Carolina*

In North Carolina, the construction manager at-risk shall contract directly with the public entity for all construction. The public entity shall publicly advertise, pre-qualify, and accept bids from subcontractors for all construction work. The pre-qualification criteria shall be determined by the public entity and the construction manager at-risk to address quality, performance, the time specified in the bids for performance of the contract, the cost of construction oversight, time for completion, capacity to perform, and other factors deemed appropriate by the public entity. The public entity shall require the construction manager at-risk to submit its plan for approval by the public entity prior to soliciting bids for the project's first-tier subcontractors. A construction manager at-risk and subcontractors shall make a good faith effort to recruit and select minority businesses for participation in contracts. A construction manager at-risk may perform a portion of the work only if bidding produces no responsible, responsive bidder for that portion of the work, the lowest responsible, responsive bidder will not execute a contract for the bid portion of the work, or the subcontractor defaults and a pre-qualified replacement cannot be obtained in a timely manner. In addition, the public entity must approve the construction manager at-risk's performance of the work. All bids shall be opened publicly, and once they are opened, shall be public records. The construction manager at-risk shall act as the fiduciary of the public entity in handling and opening bids. The construction manager at-risk shall award the contract to the lowest responsible, responsive bidder, taking into consideration quality, performance, the time specified in the bids for performance of the contract, the cost of construction oversight, time for completion, and other factors deemed appropriate by the public entity and advertised as part of the bid solicitation. The public entity may require the selection of a different first-tier subcontractor for any portion of the work, provided that the construction manager at-risk is compensated for any additional cost incurred.[162]

d. *Oklahoma*

All construction contracts or subcontracts for work to be performed for any school district pursuant to a construction management project delivery method shall

---

161. MINN. STAT. ANN. § 16C.34 (West 2005).
162. N.C. GEN. STAT. § 143.128.1 (West 2005).

be awarded in accordance with the provisions of the Public Competitive Bidding Act of 1974. If a construction manager at-risk wishes to self-perform portions of the construction work to be performed, the construction manager at-risk may self-perform portions of the work provided the construction manager at-risk competitively bids the work under the same terms and conditions as the other bidders and the construction manager at-risk is the lowest responsible bidder for the construction subcontract. No work shall commence until the school district executes a written contract and the contractor and subcontractors submit bonds and proof of insurance as required by the appropriate contract.[163]

e. *Texas*

When selecting a construction manager at-risk in Texas, the agency shall select or designate an engineer or architect who shall prepare the construction documents for the project. If the engineer or architect is not a full-time employee of the agency, the agency shall select the engineer or architect on the basis of demonstrated competence and qualifications. The agency's engineer, architect, or construction manager-agent for a project may not serve, alone or in combination with another, as the construction manager at-risk unless the engineer or architect is hired to serve as the construction manager at-risk under a separate or concurrent procurement. These provisions do not prohibit the agency's engineer or architect from providing customary construction phase services under the engineer's or architect's original professional service agreement.[164]

A construction manager at-risk shall publicly advertise and receive bids or proposals from trade contractors or subcontractors for the performance of all major elements of the work other than the minor work that may be included in the general conditions. A construction manager at-risk may seek to perform portions of the work itself if the construction manager at-risk submits its bid or proposal for those portions of the work in the same manner as all other trade contractors or subcontractors and if the agency determines that the construction manager at-risk's bid or proposal provides the best value for the agency.[165]

B. *Private Projects*

In the past decade, a number of national organizations have formed in the private sector to promote self-regulation of the construction management profession. Several offer professional certifications to qualified individuals. Other entities offer program accreditation and encourage construction education through curriculum and degree standards. A number of these national organizations have also adopted professional codes of ethics to improve the standard of practice for construction management professionals.

---

163. OKLA. STAT. ANN. tit. 61, § 220 (West 2005).
164. TEX. EDUC. CODE ANN. § 44.038 (Vernon 2005).
165. *See id.*

1. *Certification*

Certification attests to an individual's professional qualifications and assures employers that a person has been tested and found qualified to meet industry standards. Professional associations have established certification programs to set standards for professionals and administer programs that lead to certification. Certification entitles an individual to use the acronym associated with the certification on documentation and correspondence.

The Construction Management Association of America (CMAA) is dedicated to promoting "professionalism and excellence in the management of the construction process." The Construction Manager Certification Program is administered by the Construction Manager Certification Institute (CMCI), an independent administrative body of CMAA. CMCI's mission is "to recognize through certification, individual knowledge and experience that meet the established practices of Construction Management." The Certified Construction Manager (CCM) is someone who has voluntarily met the prescribed criteria of the CCM program with regard to formal education, field experience and demonstrated capability, and understanding of the construction manager body of knowledge. To become a CCM through CMCI, one must have a requisite amount of experience or education, or both. Eligible candidates must have forty-eight months experience as a construction manager and meet a minimum education standard (in construction management, architecture, engineering, or construction science) or a standard term of experience (in addition to the required forty-eight months) in general design/construction. Qualified candidates must complete the CMCI application process by documenting their experience and submitting an application to the Certification Board of Governors for review and evaluation. The final step of the process is the certification exam. It tests general knowledge of design and construction through experience in the industry and material in CMAA's Capstone and CM Standards of Practice. Applicants who meet experience standards and achieve a passing score on the certification exam will be offered CMAA certification through the CMCI.

The American Society for the Advancement of Project Management (ASAPM) advances the project management discipline. ASAPM actively encourages its member to be involved in the following areas:

- developing industry standards;
- publishing articles in magazines and publications;
- enhancing education and learning;
- designing and implementing projects that conduct research;
- involving business leaders in roundtable discussions;
- creating an award program for public recognition of excellence; and
- implementing a competence-based certification program to acknowledge and attest to individuals' professional qualifications.

The ASAPM certification program measures competence for individuals working in different areas of project management (including construction management).

ASAPM is scheduled to offer their PMCert award during the second quarter of 2006. The program will offer four levels of certification: Certified Project Directors (IPMA-A), Certified Senior Project Manager (IPMA-B), Certified Project Manager (IPMA-C), and Certified Project Practitioner (IPMA-D). Full details of the awards, their descriptions, and application process will be announced on the PMCert Web site, http://www.pmcert.org, when the information is available.

The Association for the Advancement of Cost Engineering (AACE) is a professional society for cost estimators, cost engineers, schedulers, project managers, and project control specialists. AACE International offers four certification programs: Interim Cost Consultant Certification, Certified Cost Consultant, Planning & Scheduling Professional Certification, and Earned Value Professional Certification. Each certification program requires a candidate to sit for a multiple-choice examination. Some certifications require a practical application portion, which could include a professional memorandum or solving a complex planning/scheduling problem.

2. Accreditation

The Project Management Institute (PMI) develops and promotes the field of project management. It administers a globally recognized, rigorous education and/or professional experience and examination-based professional credentialing program. PMI offers certification as either a Project Management Professional or a Certified Associate in Project Management, depending on credentials. Earning a professional credential through PMI means that one has demonstrated the appropriate education and/or professional experience, passed a rigorous examination, agreed to abide by a professional code of conduct, and maintains his or her active credential through meeting continuing certification requirements.

In addition to PMI's Certification Program, PMI offers accreditation to universities through The Global Accreditation Center for Project Management (GAC). To earn the GAC Accreditation credential, universities must successfully complete a peer evaluation and demonstrate compliance with the GAC Accreditation Standards. The GAC is considered a specialized accrediting body and holds full membership in the Association of Specialized and Professional Accreditors (ASPA), one of two agencies recognized by the U.S. Department of Education that certify university accrediting bodies.

The American Council for Construction Education (ACCE) is the main organization for accrediting postsecondary schools in construction education. The ACCE is a private not-for-profit corporation whose mission is to be a leading global advocate of quality construction education programs and to promote, support, and accredit quality construction education programs. The ACCE's purpose is to promote and improve construction education and research at the postsecondary level. Additionally, the ACCE engages in accrediting construction education programs offered by colleges and universities. The ACCE provides a list of colleges and universities which have or are seeking accredited programs of study in construction. To be considered for accreditation through ACCE, a program in construction education must be located in an institution of higher learning that is legally authorized to provide a program of education beyond that of the secondary level.

The Associated Schools of Construction (ASC) is an association whose membership is comprised of both individuals and organizations. The ASC is dedicated to the development and advancement of construction education, the sharing of ideas and knowledge, and promoting excellence in curricula, teaching, research, and service. There are four classes of organizational membership: Institutional, Associate, Industry, and Government. Individuals who are not directly associated with organizational members and have an active interest in construction or construction education may apply for individual membership. Membership in the ASC provides access to a variety of programs and services designed to help better represent the interests of both academic and industry professionals interested in institutional construction education.

3. *Codes of Ethics*

All of these construction management organizations share a common commitment to their members' ethics responsibilities. Each organization is dedicated to establishing and maintaining a high standard of professional conduct by instructing members to engage in certain behaviors/activities and refrain from others. Most organizations formally adopt these policies as a Code of Ethics, but not all are so definitively established. The following standards are commonly expressed by many of the organizations: a commitment to continue developing professional skills, a concern and dedication to the public welfare, a pledge to practice with honesty and fairness, and a promise to accept only assignments that one is qualified to perform. Additionally, most ethics codes share common restrictions on certain behaviors including: avoiding conflicts of interest which may impair objectivity or integrity of service, rejection of bribery, and refraining from discrimination.

While most ethics codes function as a self-policing mechanism, some organizations outline a process for punishing violations of these standards. For example, PMI regulates member ethics through a formal procedure. PMI designates an Ethics Review Committee and an Ethics Appeal Committee, each composed of at least seven PMI members. Ethics violations and/or complaints are heard and decided by these committees. Disciplinary action may include termination of PMI membership, public reprimand, membership probation, and membership suspension for a designated period of time.

Fifty-State Matrix of Construction Manager Licensing Laws		
State/Jurisdiction	**Specific Regulations**	**Regulated as Contractors**	**Regulated as Architect/Engineer**
Alabama	no	no	no
Alaska	no	no	no
Arizona	no	**YES**	no
Arkansas	no	**YES**	**YES**
California	no	no	**YES**
Colorado	no	no	no
Connecticut	no	no	no
Delaware	no	**YES**	no
Florida	no	no	**YES**
Georgia	no	**YES**	no
Hawaii	no	no	no
Idaho	**YES**	**YES**	no
Illinois	no	no	**YES**
Indiana	no	no	no
Iowa	no	no	no
Kansas	no	no	no
Kentucky	no	no	no
Louisiana	no	**YES**	no
Maine	no	no	no
Maryland	no	no	no
Massachusetts	no	no	no
Michigan	no	no	no
Minnesota	no	no	no
Mississippi	no	no	**YES**
Missouri	no	no	no
Montana	no	no	no
Nebraska	no	no	no
Nevada	no	**YES**	no
New Hampshire	no	no	no
New Jersey	no	no	no
New Mexico	no	**YES**	no
New York	no	no	**YES**

Fifty-State Matrix *Continued*

State/Jurisdiction	Specific Regulations	Regulated as Contractors	Regulated as Architect/Engineer
North Carolina	no	**YES**	no
North Dakota	no	**YES**	no
Ohio	no	no	no
Oklahoma	**YES**	no	no
Oregon	no	**YES**	no
Pennsylvania	no	no	**YES**
Puerto Rico	no	no	no
Rhode Island	no	no	no
South Carolina	no	**YES**	**YES**
South Dakota	no	no	no
Tennessee	no	**YES**	no
Texas	no	no	no
Utah	no	**YES**	no
Vermont	no	No	no
Virginia	no	**YES**	no
Washington	no	No	no
Washington D.C.	no	No	no
West Virginia	no	**YES**	no
Wisconsin	no	No	no
Wyoming	no	No	no

IV. Practice Aids

A. *Checklist*

☐ Does the state license construction managers as an independent profession?
☐ Does the state license construction contractors? If so, study the activities defined as "contractor" and "contracting" to see if construction management, construction manager, or manage is specifically mentioned or implied as a regulated profession.
☐ Alternatively, do the activities described as contracting under the statute potentially or actually overlap with the activities to be performed by the construction manager?
☐ Study the state's architecture and engineering licensing laws to see if construction management falls within the regulated activities performed by

architects or engineers. If it is, see if any exceptions exist to allow construction managers to perform such services.
- ☐ Alternatively, do the activities described as the practice of architecture or engineering potentially or actually overlap with the activities to be performed by the construction manager? If so, check the architecture/engineering licensing statute(s) to see if the licensing requirement can be satisfied if the construction manager has a licensed architect or engineer on staff in charge, or whether the firm itself has to qualify as a licensed architecture or engineering firm.
- ☐ For state entities wishing to hire construction managers, check procurement statutes to see whether or not construction managers are exempt from competitive bidding statutes. Determine whether they are hired using a qualifications-based selection procedure.
- ☐ For state entities and construction management firms, determine whether state statutes prohibit the same entity from performing design, management, and construction.
- ☐ For owners hiring construction managers (public and private), consider pre-qualifying firms based on personnel possessing industry recognized certifications, and/or degrees from industry accredited construction education programs.

B. *Research and Resource Links*

American Council for Construction Education, http://www.acce-hq.org
American Society for the Advancement of Project Management, http://asapm.org
Associated Schools of Construction, http://www.ascweb.org
Association for the Advancement of Cost Engineering, http://www.aacei.org
Construction Management Association of America, http://cmaanet.org
Project Management Institute, http://www.pmi.org

References

STATE-BY-STATE GUIDE TO ARCHITECT, ENGINEER, AND CONTRACTOR LICENSING (Stephen G. Walker et al. eds., Aspen 1999). This two-volume reference discusses federal and state licensing requirements, with each chapter dedicated to a separate jurisdiction.

KENNETH M. ROBERTS, *Effective Use of Design-Build on State and Local Government Projects*, in DESIGN-BUILD FOR THE PUBLIC SECTOR 155 (2003). This chapter describes state procurement laws affecting architects, engineers, contractors, design-builders, and construction managers.

DESIGN-BUILD CONTRACTING HANDBOOK (Robert F. Cushman & Michael C. Loulakis eds., 2nd ed. 2001). This book covers legal and regulatory issues relevant not only to design builders but also to construction managers.

Philip L. Bruner & Patrick J. O'Connor, Jr., Bruner & O'Connor on Construction Law (2002). This treatise addresses major issues in construction law, including construction management.

Construction Management Association of America, Capstone: The History of Construction Management Practice and Procedures (2003). This publication covers topics such as the construction manager as professional, project delivery, construction management contract documents, construction management legal issues, construction manager risk allocation, and management.

CHAPTER 9

Construction Management Overview

RICHARD D. CONNER, ESQ.
SHIV GHUMAN O'NEILL, ESQ.

I. Introduction

Today the roles of architect, contractor, and construction manager are typically defined on a project-by-project basis. Architectural firms furnish construction management services and construction management firms furnish services traditionally performed by architects and contractors. The roles of the contractor, architect, and construction manager are often blended, and the scope of work for each must be defined and coordinated within the project contracts on a project-by-project contract basis.

The term construction management, only a concept in the 1960s, is well-defined today by prominent construction industry associations.[1] These definitions have in

---

1. The Construction Management Association of America (CMAA) defines construction management as professional management practices applied to construction projects from inception to completion for the purpose of controlling time, scope, cost, and quality. CMAA CONSTRUCTION MANAGEMENT STANDARDS OF PRACTICE (2003). The American Society of Civil Engineers (ASCE) indicates that construction management consists of that group of professional management activities that is distinct from normal architectural and engineering services and is related to a construction program. *CM Responsibilities during Design*, 113 ASCE J. OF CONSTR. ENG'G & MGMT. 21341 (George Stukhardt, ed., Mar. 1987). The Associated General Contractors of America (AGC) defines construction management as one effective method of satisfying an owner's building needs. It treats the project planning, design, and construction phases as integrated tasks within a construction system. The tasks are assigned to a construction team consisting of the owner, the construction manager, and the architect-engineer. Members of the construction team ideally work together from project inception to project completion, with the common objective of best serving the owner's interests. Interactions between construction cost, quality, and completion schedule are carefully examined by the team so that a project of maximum value to the owner is

common the central theme of applying professional management to the design and construction of a project. However, definitions are merely starting points in shaping a construction manager's role in a particular project. The construction manager's role in a particular project cannot be fully realized until the:

- scope of construction management services is clearly defined;
- type of owner-construction manager contract is chosen; and
- responsibilities among the construction manager and other project participants are clearly allocated.

Every team requires a leader, especially the construction project team. Over the years, diverse opinions have developed as to who is best suited to lead the team—the architect, the contractor, or an independent management firm employed by the owner. At project inception, the owner is in complete control. Gradually, the owner contracts away its control to managers, designers, and contractors until, in the end, the owner is only responsible for two things—providing the construction site and paying the various entities. Popular names have developed to describe the various contractual bundles utilized by the owner to procure design, construction, and professional management services for a project. The contractor-led team is described as the design–build system. The architect/engineer-led team is described as the traditional system, or design-bid-build. With the traditional system, the owner may employ one or more prime contractors.

The owner may elect to employ an independent design professional and utilize the services of a single general contractor who provides construction input during the design phase. This last project organization is referred to as the construction management at-risk delivery system (CM At-Risk). Generally, a CM At-Risk is responsible for the means and methods of construction and a contractually specified maximum project cost. The construction manager does not act as the owner's agent. Instead the construction manager performs a variety of front-end management services (e.g., value engineering of design) and takes financial risk in the project's outcome. The CM At-Risk is placed in a legal position similar to that of a general contractor entering into a traditional construction agreement. The CM At-Risk system is discussed in more detail later in this chapter.

---

realized in the most economic time frame. ASSOCIATED GENERAL CONTRACTORS OF AMERICA, CONSTRUCTION MANAGEMENT GUIDELINES NO. 2, DOCUMENT 540 (1979). The American Institute of Architects (AIA) notes that construction management consists of management services provided to the owner of a project during the design phase, construction phase, or both by a person or entity possessing requisite training and experience. Such management services may include advice on the time and cost consequences of design and construction decisions, scheduling, cost control, coordination of contract negotiations and awards, timely purchasing of critical materials and long-lead items, and coordination of construction activities. AIA, ARCHITECT'S HANDBOOK OF PROFESSIONAL PRACTICE, GLOSSARY OF CONSTRUCTION INDUSTRY TERMS 8 (1991).

Under any of the alternative systems mentioned above, the owner may elect to employ an independent manager for project oversight and to schedule and coordinate the services of its design professionals and contractors. While it is not a distinct delivery system, this type of management assistance is typically referred to as Agency CM. Agency CM contemplates a contractual framework where an independent manager leads the project team by coordinating the various design and construction services in the name of the owner (discussed in more detail later in this chapter). The construction manager has no vested financial interest in the project—in either its design or construction—and maintains a fiduciary responsibility to act on the owner's behalf and to provide impartial advice concerning the construction project.

Both contractual arrangements Agency CM and CM At-Risk are widely accepted and utilized.

A hybrid of design–build and the traditional system in the United States is called bridging.[2] In the bridging process, there are two design teams. The first design team is employed under contract with the owner. This design team develops the project's functional and aesthetic requirements, which are set forth in a conceptual design and performance-type specifications. These are the bridging documents. Final construction drawings are developed from the bridging documents by a design–build contractor who employs as consultants the architect and engineers of record. Agency CM services may be utilized by the owner on a complex project to manage the bridging process.

Whatever combination is used, the owner's overriding goal is the same—to have its construction project completed with acceptable quality, within budget, and on time. In the United States, the largest construction firms furnish services under all of the systems mentioned above. The largest firms listed in surveys conducted by *Engineering News Record* each year provide services in the traditional and design–build methods and some form of construction management. Construction management services have become an important part of the revenues earned by the largest firms. By combining project management services traditionally furnished by design professionals and the construction skills of a general contractor in a single entity, those furnishing construction management services have created effective project controls for their clients and powerful marketing tools for their firms.

During the development years of construction management, there was no uniformity in the scope of those services or the standards for performing them. Industry

---

2. When bridging is used, the bid documents leave considerable latitude for contractors to look for economies in construction technology. Typically, construction does not begin until the final construction drawings are complete, and it is clear that there are no misunderstandings about what was intended by the bid documents. The biggest problem with bridging is that it is fairly new in the United States. As one commentator notes, the United States "construction industry is large," and replete with many contractors, architects, engineers, construction managers, manufacturers, and suppliers. Tradition is the great facilitator. When you change the process you must manage well. ANDY PRESSMAN, PROFESSIONAL PRACTICE 101, 283 (1997).

associations responded to the need. Standard form agreements from the American Institute of Architects (AIA) and Associated General Contractors of America (AGC) began to appear in the late 1970s and into the 1980s, each with its own industry-specific description of what a construction manager does. The Construction Management Association of American (CMAA) promulgated its standard form agreements for construction management in 1990.[3] While these form documents have similarities in content and process, each association takes a different approach to the construction manager's function and tends to reflect the perspectives of its members. All three associations publish complete sets of ready-to-use companion documents. Some owners and construction managers have developed their own contract documents, which incorporate important elements of the standard forms, but are more self-protective.

When an owner engages the service of a construction manager, it is essential that the contract documents clearly define the role of each party, the responsibilities they assume, and more importantly, the responsibilities and liabilities assigned to other contracting parties. Otherwise, some of the services may conflict or become duplicative.[4]

This chapter overviews the development of construction management as a professional service and compares industry contract forms that implement the use of construction management services.

II. Historical Evolution of Construction Management

By whatever definition, construction management is a fundamental realignment of traditional construction relationships. It severs management functions from both the contractor and the design professional and consolidates them in a new entity.[5] The notion of having management services provided by an entity independent of the architect and contractor appears to arise from the architect's withdrawal from its former role as master builder and, more recently, the cost of financing complex projects, unpredictable labor demands, and rapidly developing technology. These developments have encouraged the use of professional management at various phases of the project.

By way of background, the separation of architectural design from the construction process began long ago. As far back as 1450, Leon Battista Alberti "gave popularity to

---

3. Representatives of 37 firms met in Indianapolis on Oct. 28, 1981, to explore the need and interest in forming a national construction management association. The group represented a balance between general contracting, architectural, engineering, and construction management firms from across the nation. The organization eventually became CMAA.

4. Richard D. Conner, *Contracting for Construction Management Services*, 46 LAW & CONTEMP. PROBS. 5–19 (1983).

5. C. Allen Foster, *Construction Management and Design-Build/Fast Track Construction: A Solution Which Uncovers a Problem for the Surety*, 46 LAW & CONTEMP. PROBS. 95 (1983).

the notion that the architect as a scholar and an artist was separate and far apart from the mason and other craftsmen that actually carried out the architect's design."[6]

The emergence and development of a distinct architectural profession in the United States are described in *The Architect's Handbook of Professional Practice* § 1.11 (1994), and in *Construction Law* by Philip L. Bruner and Patrick J. O'Connor, Jr., Vol. 5, Chapter 17 (2002). Citing *The Architect's Handbook of Professional Practice*, Bruner and O'Connor wrote that from the time of the ancient Greeks until the nineteenth and twentieth centuries the architect reigned supreme as master builder, and that this role changed dramatically as a result of developments in technology, specialization, building delivery systems, governmental regulations, and the "litigiousness" of modern society.[7]

Carl M. Sapers, a widely recognized authority on legal aspects of architectural practice and its history, writes that architecture established itself as a profession in the United States during the nineteenth century when professional organizations, licensing laws, codes of ethics, and specialized schools were established. In 1857, the American Institute of Architects was founded in New York City, and in 1865, the first U.S. architecture school opened at the Massachusetts Institute of Technology.[8]

The breakout of architecture in the United States occurred after the Civil War with increases in large-scale public projects. There was a need for assurance of expertise to complete projects in the public interest. Sapers notes that "as private and corporate wealth accumulated in America[,] clientele developed who expected buildings of more elaborate design—often imitating the great mansions of Europe. The design took on new importance and professional architects began operating independently from the construction trades."[9]

During the nineteenth and well into the twentieth centuries, architects performed a distinctively hands-on supervisory function during the construction phase. Well into the 1960s, the debate concerning the extent of the architect's supervisory role on the project can be seen in court opinions from the various states.[10] In 1972, the AIA

---

6. LEON BATTISTA ALBERTI: ON THE ART OF BUILDING IN TEN BOOKS (Joseph Rykwert et al. trans., MIT Press 1988); G. William Quatman & Martin Sell, *Return of the Master Builder: Designer-led Design-Build*, presented at the CNA—Schinnerer 44th Annual Meeting of Invited Attorneys, 2005, *at* http://www.aiamail.aia.org/SiteObjects/files/db_cna_paper.pdf.

7. PHILIP L. BRUNER & PATRICK J. O'CONNOR, JR., BRUNER & O'CONNOR ON CONSTRUCTION LAW § 17.1 (2002).

8. Carl M. Sapers, Legal Cases and Materials for the Construction Professional (1993) (unpublished material used by Professor Carl Sapers at the Harvard University Graduate School of Design).

9. *Id.*

10. Carl M. Sapers, Legal Cases and Materials for the Construction Professional (Carl M. Sapers, copyright 1993) (unpublished material used by Professor Carl Sapers at the Harvard University Graduate School of Design).

modified its A201 general conditions by removing provisions requiring the architect's professional services to include supervision of the work. Prior to the modification, some courts tended to view requirements of general inspection or supervision by the architect as creating a duty or obligation to control the contractor's method and manner of performing the details of the work. Removing the word *supervision* in Article 38 of the AIA General Conditions was controversial. One commentator noted that in 1961, twenty-five of the forty-four general conditions in the AIA's General Conditions included elements of the architect's supervisory function during the process of construction. He posed the question, "Do architects wish to have their supervisory function belittled in order possibly, but by no means certainly, to escape a claim of negligence...."[11]

During the 1960s and 1970s, at a time when project management demands were increasing, the hands-on role of the architect had ceased. The commentator noted, "[a]s projects became larger and technically more complex, the owner's need for accurate information and guidance throughout the project's life increased and a fourth project participant was introduced, the construction manager."[12]

Much has been written about the birth and evolution of architecture. The origins of construction management as a profession (if it is a profession!) are not widely chronicled. Perhaps the best summary appears in a CMAA publication utilized in connection with its certification program, *Capstone: The History of Construction Management Practice and Procedures* (Capstone). According to Capstone, construction management had its beginnings in the early to mid-1960s.[13] Gary S. Berman, P.E., an experienced construction manager and construction management historian, notes in an article that the parents of construction management appear to be the nation's largest general contractors in the early 1960s.[14] Berman points out that at that time, the need for capital facilities and improvements in the marketplace was in great demand. Major urban cities were in need of new or replacement facilities and infrastructure. The private sector had increasing demands for capital expansion, the federal government had major requirements for new and expanded facilities, and the Americans were racing the Russians to the moon.[15]

---

11. Parker, S*upervision, an Argument for the Use of the Word "Supervision," in* AIA General Conditions Article 38 (1961).

12. Gary S. Berman, *Are Foxes Watching the Owner's Henhouse? An Examination of the Architect's and Construction Manager's Roles in Managing and Administering the Design and Construction Process*, CME Journal (2003), *available at* http://cmaanet.org/user_images/berman.pdf.

13. Construction Management Association of America, Capstone: The History of Construction Management Practice and Procedures (2003).

14. Gary S. Berman, *The Morphing of the Architect's Role and How It Is Impacting the CM*, CME Journal (2002), *available at* http://www.greyhawk.com/news/technical/Construction_Management_The_Morphing_of_the_Architect_Role_and_How_it_is_Impacting_the_CM.pdf (presented at the CMAA National Conference).

15. *Id.*

At the same time, major technology and methodology developments were occurring that would play a significant role in the evolution of construction management. First came the advent of the computer and the information explosion which followed. Next was the development of sophisticated scheduling techniques such as the critical path method (CPM). These were followed by other new techniques such as fast-tracking, value engineering, life cycle costing, and constructability reviews. These tools and capabilities significantly advanced the application of management principles in the delivery of capital projects. In 1963, New York City Madison Square Garden was one of the earliest reported projects to use construction management; Teshman Realty and Construction Co., Inc. served as construction manager. Construction management was used later in 1965 on the 100-story John Hancock Center in Chicago and again in 1967 on the Twin Towers of New York City's World Trade Center.[16]

A 1970 General Services Administration (GSA) study concluded that GSA's current methods of contracting caused its buildings to require more than twice the time for construction than that of similar buildings in the private sector.[17] This report recommended that GSA abandon its "out-moded procedures" and use phased construction in conjunction with construction management (pointed out in a commentary).[18] Owners were looking for alternatives to the traditional approach in an effort to reduce the overall time and cost associated with project delivery.

In 1975, the $42 million National Air and Space Museum project located in Washington, D.C. became the first GSA project to be completed using construction management. In the early 1970s, the Department of Health, Education and Welfare employed construction management with guaranteed maximum price contracts for federally assisted hospital, school, and laboratory construction.[19]

In August of 1975, three prominent construction industry associations, AIA, AGC, and American Council of Engineering Companies (ACEC) jointly issued a general position statement recognizing construction management as a separate discipline.[20] In their statement (the Joint Association Statement) the associations pointed out that, from a practical standpoint, an effective construction management organization is likely to be a multidisciplined organization. Construction management is an appropriate function for construction contractors, as well as architectural or engineering firms, or divisions thereof, as long as those organizations or divisions, in fact, have construction management capabilities. The statement went on to express

---

16. CAPSTONE, *supra* note 13, at 3.
17. Berman, *supra* note 14, at 7.
18. *See id.*
19. *See id.*
20. National construction management committees of the Associated General Contractors of America, the American Institute of Architects, and the American Consulting Engineers Council unanimously agreed and issued a statement recognizing the importance of the construction management process and agreeing to work collectively on a national comprehensive construction management program to implement certain principles.

the authors' view that neither architects, nor engineers, nor contractors should take any action in connection with licensing or other laws which would have the effect of reserving to themselves construction management markets. The Joint Association Statement marks the starting point for recognition of construction management as a separate profession. The Joint Association Statement (reproduced in note 21) is useful to anyone needing to explain and justify the distinctions between construction management and other construction professions.[21]

---

21. The full text of the statement is as follows:

Denver, CO
August 15, 1975

The Construction Management Committees of the AIA, AGC, and ACEC, in joint conference, recognize the importance of the Construction Management process and have agreed to work together collectively on a National Comprehensive Construction Management Program to implement the following:

1. To define, develop, and disseminate the standards and levels of quality of Construction Management;
2. To develop guidelines and educational programs for the assistance of Members and Owners engaged in Construction Management;
3. To relate and coordinate with all elements in the industry performing Construction Management services to develop acceptable industry-wide standards for Construction Management;
4. To monitor and to make joint recommendations with respect to legislation and regulations at the federal, state, and local levels affecting Construction Management; and
5. To communicate and maintain liaison with the owners and users of Construction Management services, advising and assisting them on the best and most economical procedures

The following statement is made with respect to Construction Management services provided to the Owner or User/Client, being referred to herein as CM. Construction Management requires a number of skills relating to the knowledge of construction process, and knowledge of the design process, and the fundamentals of general and project management.

A background in construction contracting, architecture contracting and engineering can provide a basis of experience for entering the field of Construction Management. However, the basic minimum capabilities of contractors, architects, and engineers do not necessarily or automatically provide an individual with all of the skills required of a competent CM (Construction Manager).

From a practical standpoint, an effective CM organization is likely to be a multi-discipline organization. However, CM is an appropriate function for construction contractors, as well as architectural or engineering firms or divisions thereof, so long as said organization or division, in fact, has CM capabilities.

It is the view of this group that neither architects, engineers, nor contractors should take any action in connection with the licensing or other laws which would have the affect of reserving to themselves Construction Management markets.

In 1982, the Construction Management Association of America held its inaugural meeting in Denver, Colorado.

In 1986, GSA decided to use Construction Quality Management (CQM) as one of its options to manage projects. Instead of multiple prime trade contracts being coordinated by a construction manager acting as an agent, GSA began awarding a single construction contract to a general contractor and hiring a construction manager to act as GSA's agent under a CQM contract. Requests for proposals for such services issued by the GSA indicate that, while the number of construction contracts changed, the basic principles and guidelines of the original *GSA System for Construction Management* essentially remained the same.[22]

One commentator has observed that, while GSA put the name of construction management on the map, it should not be credited with construction management's development as it is practiced today. That credit belongs to the scores of private sector firms that pioneered this distinct contracting systems until it grew in wide use.[23]

III. Recognition of Construction Management as a Profession

Construction management is referred to as a profession with regularity. The term *professional* includes those who are compensated for their performance, as opposed to an amateur, as well as those who perform with a certain degree of expertise and loyalty to the client. One commentator notes that articulating the meaning of professional is a bit like Justice Potter Stewart's attempt to characterize pornography: "it's difficult to define, but I know it when I see it."[24]

Applying the term *profession* to construction managers and to construction management firms established and operated for the sole purpose of construction management is appropriate. Professions generally are founded on a body of specialized information and discipline, along with education and training. Construction management is composed of a distinct group of services, and those providing the services do so while displaying many of the characteristics of a professional. Indeed, some construction management practitioners are registered architects and licensed contractors.

However, use of the term *profession* in conjunction with construction management is to be distinguished from the term *professional* as applied to a registered or licensed member of a *learned profession*. Since the 1980s, the construction management business has shown signs of becoming a distinct learned profession, but it is not there yet. Rather, construction managers are individuals experienced in the various disciplines involved in the construction industry, and construction management firms are staffed with such individuals. A registered architect furnishing construction management services may be held to a professional standard of care, while a

---

22. Berman, *supra* note 14, at 7.
23. *Id.*, *supra* note 14, at 9.
24. PRESSMAN, *supra* note 2, at 2.

licensed general contractor, or an unlicensed individual or firm may be held to a different standard.

The extent to which construction management is regulated as a profession by governments is beyond the scope of this chapter. However, a few general comments on the subject of regulation are appropriate to sensitize the reader. Even where there may not be legislation governing construction management specifically, the licensing and registration laws of the various states applicable to contractors and architects may include the performance of functions closely associated with construction managers. In some states, a construction manager could be subject to both an architect registration statute and a contractor licensing statute.

Many of those furnishing construction management services have united in fellowship within leading industry associations. One of those groups, the CMAA, has a primary objective to promote professionalism in furnishing construction management services.

In furtherance of professionalism within construction management practice, the CMAA promulgated its *Construction Management Standards of Practice* in 1987.[25] The CMAA standards are intended as a guide to the range of services that constitute professional construction management. The standard guidelines are not intended to include any hard, fast rules of procedure, and they are not considered to be a detail procedure manual. The *Construction Management Standards of Practice* provides standard definitions and addresses seven primary areas of construction management functions including:[26]

- project management;
- time management;
- cost management;
- quality management;
- contract administration;
- safety management; and
- program management.

Each function is further broken down into five phases:

- pre-design;
- design;
- procurement;
- construction; and
- post-construction.

---

25. CONSTRUCTTION MANAGEMENT ASSOCIATION OF AMERICA, CONSTRUCTION MANAGEMENT STANDARDS OF PRACTICE (2003).

26. *Id.*

These phases are consistent with the CMAA-suggested scope of services contained in the CMAA contract documents and with established uses in the construction industry.

The CMAA began a Construction Manager Certification Program and a practitioner who meets all of the certification requirements is designated a Certified Construction Manager (CCM).

The preparation of standard form contracts by leading construction organizations, the promulgation of standards for performing construction management services, degrees in construction management being offered by universities (e.g., East Carolina University, North Carolina State University, and the University of Wisconsin), and the popularity of credentialing of construction managers and project managers evidences a trend toward professionalism in the construction management industry.

Library References

C.J.S. *Architects* §§ 1–3, 7, 19, 21–23; C.J.S. *Negligence* §§ 162–164.
West's Key No. Digests, Licenses <KEY>11; Negligence <KEY>321, 322.

IV. Construction Management and the Professional Standard of Care

As with other relationships, both in and outside the construction industry, the construction manager's performance will be measured by reference to a legal standard of care. The Construction Manager will be held to a standard of care normally applied to persons and professions or trades requiring special skills.[27] As a general rule, when a person holds himself out to the public as a member of a particular profession or trade, there is an implied agreement with those who employ him that he (1) possess that degree of knowledge and skill ordinarily possessed by others in the profession or trade; and (2) will perform the services for which he was engaged with that degree of prudence and care ordinarily possessed and observed by others engaged in the same or like employment.

Performance up to this standard does not require that the work performed be perfect. The construction manager has a duty to exercise ordinary skill and competence as exhibited by other construction managers. A failure to discharge that duty may subject the construction manager to liability. However, poor results alone, without a breach of duty, will not subject the construction manager to liability.

Although the standards for a construction manager's performance are determined by the customs and standards of other construction managers, determining a universal standard of performance for all construction managers is difficult, if not impossible. This difficulty arises not only because the particular tasks undertaken by construction managers vary widely from project to project, but also because of the different disciplines construction managers bring to the project. Some construction

---

27. JAMES J. ADRIAN, CM: THE CONSTRUCTION MANAGEMENT PROCESS 64–65 (1980).

managers come to practice construction management as part of their architectural or engineering practices, while other construction managers rely primarily upon their general contracting skills. The difference in the source of the construction manager's skills is a reflection of the different types of entities that may engage in construction management. In resolving issues of alleged construction manager negligence, courts will likely compare the particular construction manager's duties and obligations to those of construction managers similarly situated by experience, profession, and trade, rather than to those of other construction managers with whom the construction manager has little in common, contractually or professionally.

Library References

C.J.S. *Negligence* §§ 163–164.
West's Key No. Digests, Negligence <KEY>322.

V. Differences from Design and Construction Professions

Construction management includes a comprehensive management and control effort applied to the project. This type of comprehensive management, while routine with construction management, may be missing with other delivery systems. The construction management effort generally includes the following:[28]

- development of a written scope understood by all of the participants;
- development of thorough design criteria for issue to the designer;
- design quality assurance throughout the design process;
- consideration of material, systems, and process alternatives;
- constructability review;
- code compliance review;
- milestone cost estimating—to ensure design complies with budget;
- matching construction spending to funds availability;
- construction specification enforcement; and
- continuous schedule enforcement.

It is intended by construction managers that implementing these management activities turns the planning, design, and construction process into one that maximizes the owner's control over the project scope, quality, time, and cost, and adds predictability of the outcome of the project from start of programming to completion of construction.

The most frequently cited criticism of construction management services is that the construction manager adds another level of cost and bureaucracy to the project. While it can be argued that services associated with such costs may actually reduce the overall project cost, it should be noted that an owner can realize the benefit of the construction management services without necessarily committing to large increases in costs by supplementing its own project management capabilities as necessary and selecting a construction manager on a service-specific basis.

---

28. CAPSTONE, *supra* note 13, at 32.

The owner has the option of tailoring its use of construction management services to its needs in order to provide the best combination of project control and cost. For example, many owners have a large contingent of inspection personnel, but may lack sufficient management experience to enact effective project cost controls. Alternatively, an owner may wish to have more construction knowledge built into the design process by engaging a construction manager to perform a value engineering or constructability review. The owner may also desire enhanced scheduling expertise in coordinating the various designers and contractors for a multiple-phase effort.

Other owners may be very comfortable with their design team, but may need assistance in finding qualified contractors to perform the work. Owners may use a construction manager's construction close-out services to resolve intractable problems on projects that degenerate due to disputes with the contractor over schedule and delay issues.

The services furnished by construction managers can differ markedly from those furnished by a design professional or general contractor. As a part of basic services, architects and engineers do not furnish detailed cost estimating services. Cost estimating is offered as an additional service only or limited to conceptual estimates. Cost estimating is a basic service offered by construction managers. Cost estimating services may be furnished initially and thereafter updated as a basic service by the construction manager during all phases of the project.

Architects and engineers do not furnish scheduling services as a part of their basic service. Scheduling and coordination services are typically furnished by architects and engineers only as an additional service.

As a part of their scheduling and coordination function performed as a basic service, construction managers will assist in preparing an overall project master schedule and will update this schedule periodically during all phases of the project. Frequently, trade contractors are required by the project's master specifications to furnish their own schedules that are integrated with the master schedule for the project. The schedules furnished by the trade contractors are often resource and manpower loaded. Periodically, the construction manager will review the schedules furnished by the contractors and will monitor and report on their progress. The contractor's schedule may be used by the construction manager in connection with the review and certification of the contractor's application for payment.

On larger more complex projects, the construction manager's scheduling and coordination function may include monitoring the progress made by design professionals in completing the project design documents.

Performing and coordinating the process of value engineering and constructability reviews during the various phases of design document preparation is a basic service furnished by construction managers.

VI. Construction Management Services Overlap with Other Professions

Services furnished by the construction manager will, of necessity, overlap with services from the architect and engineer in the areas of programming, design development, major system selection, value engineering, constructability reviews, and construction contract administration. During the process of contracting for professional services for the project, it is advisable that revisions be made in the contracts between owner and construction manager, and architect and engineer that promote cooperation and coordination to avoid duplication of services, i.e., handling the coordination of the various architectural and engineering drawings and specifications, and addressing errors or omissions in the design to the maximum extent possible without waiting well until into construction.

The construction manager's role during various phases of the project differs under the standard forms, and is discussed in detail in other parts of this chapter. For example, under the CMAA agency forms, completion of the various phases of project design falls in the architect's domain. However, each of the design architect's other traditional areas of expertise is performed by a construction manager with input from the architect when requested by the construction manager. This process is in sharp contrast to the construction manager's role as specified in the AIA forms.

A fundamental economic decision for the owner is whether to separate the design review function from the architect. The owner may feel that the architect should be in charge of design optimization on large projects and that a fixed fee does not present the disincentive to reduce construction costs as may be the case with a percentage fee. Others may argue that optimizing the design through value engineering would best be left to others rather than to the project architect, on the basis that there would appear to be little economic incentive toward cost savings in design where the architect is paid on the basis of a percentage of the construction cost.

VII. Major Categories of Service

Under the traditional project delivery system, an owner hires a design professional to develop a complete set of plans and the work is awarded to a contractor who then completes the work either with its own forces or with subcontractors it hires. During the construction phase, the design professional usually provides contract administration services, which can include monitoring the progress of construction, providing interpretations of design documents as needed, inspecting the work, advising the owner when payments should be made to the contractor, and resolving disputes between the contractor and the owner.[29] The contractor is responsible for constructing the project in accordance with the contract documents within the applicable time and budget constraints.

A construction manager can offer an owner expertise in design, construction, and management and provide a broad range of services from the beginning of the

---

29. Alan E. Harris, *Construction Management: Using the Design Professional as the CM*, in THE AMERICAN BAR ASSOCIATION FORUM ON THE CONSTRUCTION INDUSTRY, EIGHTH ANNUAL FORUM, CONSTRUCTION MANAGEMENT: MAKING THE RIGHT CHOICE (1992).

design phase through project completion.[30] Construction management can be used with the traditional project delivery approach described above or with alternative project delivery approaches; such as multiple prime contractors or design-build.[31]

There are two primary contracting arrangements for construction management: Agency CM, also referred to as construction manager advisor, and CM At-Risk. This section describes these arrangements and briefly discusses other categories of service a construction manager can provide.

A. *Agency CM*

The Agent CM essentially serves as a professional consultant to the owner. In this arrangement, the construction manager contracts directly with the owner only and has no contractual relationship with the architect or the contractor.[32] Because the Agent CM is not in privity with the contractor, it is not responsible for construction delays or cost overruns caused by contractor's acts or omissions.[33] The Agent CM is the owner's agent. Because of this relationship, the Agent CM may have fiduciary duties to the owner.[34] Of course, these duties can be contractually modified or eliminated.

The Agent CM is usually compensated in the same manner as the design professional and other professional consultants, i.e., it is paid a fee for its services. Although the methods for calculating the fee may vary, the underlying intent is that the construction manager's compensation is not tied to the profitability of or liability for the construction.[35] Therefore, the construction manager is able to provide unbiased advice because there is no conflict of interest with the owner.[36]

The scope of services provided by the Agent CM can vary depending on the owner's needs. The Agent CM can assist the owner during the pre-construction phase in developing a budget and design criteria, selecting a project delivery method, selecting a design professional, providing advice regarding constructability and value engineering, developing a schedule, and assisting with the bid process or other

---

30. Richard D. Conner, *Contracting for Construction Management Services*, 46 LAW & CONTEMP. PROBS. 6 (1983).

31. MICHAEL C. LOULAKIS, ET AL., CONSTRUCTION MANAGEMENT LAW AND PRACTICE 14 (1995).

32. Linda Chiarelli & Lawrence Chiarelli, *The Role of the Construction Manager on a Design-Build Project*, 15-APR CONSTRUCTION LAW. 58, at 3 (1995).

33. Christi L. Underwood, *Rights and Liabilities of Construction Managers*, FLORIDA CONSTRUCTION LAW AND PRACTICE, Ch. 4, at 5 (2003).

34. *See Atlin Const., Inc. v. Muncie Cmty. Schs.*, 413 N.E.2d 281, 284 n.3 (Ind. Ct. App. 1980).

35. Chiarelli, *supra* note 32, at 3.

36. PHILIP L. BRUNER & PATRICK J. O'CONNOR, JR., 1 BRUNER AND O'CONNOR ON CONSTRUCTION LAW 605 (2002).

methods of contractor selection.[37] During the construction phase, the construction manager can provide project administration services such as monitoring the schedule and budget, processing payments, inspecting the work, processing change orders, and monitoring the project safety program.[38]

An Agent CM is often used in fast-track projects with a single general contractor or multiple prime contractors. In a fast-track project, the initial stages of procurement and construction overlap with the design phase of the project. Such projects require a high level of management and coordination, which an experienced construction manager is well suited to provide. However, this type of construction manager can be valuable to an owner in any project delivery system, especially if the owner lacks construction experience and knowledge.[39]

An important feature of the Agent CM arrangement is that it allows competitive bidding for the construction work and therefore, meets the competitive bidding requirements placed on public bodies.[40] Most jurisdictions do not require Agent CM contracts to be competitively bid.[41]

B. *CM At-Risk*

The CM At-Risk arrangement is also known as the construction manager as constructor or as construction manager/general contractor.[42] The fundamental distinction between an Agent CM and a CM At-Risk is that the CM At-Risk is responsible for the construction, in addition to providing pre-construction services to the owner.[43]

The CM At-Risk not only contracts with the owner but also holds the trade contracts.[44] Therefore, the CM At-Risk bears the risks of late completion, defective performance, and cost overruns.[45]

The CM At-Risk is usually compensated on a fee basis for its pre-construction services.[46] However, at some point, the CM At-Risk assumes liability for the cost of construction and is usually paid on a cost-plus-fee with a guaranteed maximum price basis for its services during the construction phase.[47] Although the CM At-Risk

---

37. Kenneth M. Cushman & Joyce K. Hackenbush, *Construction Project Risk Allocation: The Owner's Perspective*, in HANDLING CONSTRUCTION RISKS: ALLOCATE NOW OR LITIGATE LATER 10 (Practicing Law Institute 2002).

38. *Id.*

39. Cushman, *supra* note 37, at 11.

40. BRUNER & O'CONNOR, *supra* note 36.

41. *See id.*

42. *See id.*

43. Lynn R. Axelroth, *What's An Owner or Lender To Do When Confronted with AIA Documents*, in HANDLING CONSTRUCTION RISKS: ALLOCATE NOW OR LITIGATE LATER 10 (Practicing Law Institute 2003).

44. *Id.*

45. John I. Spangler, III & William M. Hill, *The Evolving Liabilities of Construction Managers*, 19-JAN CONSTRUCTION LAW. 30, at 1 (1999).

46. Cushman, *supra* note 37, at 11.

47. *See id.*

may have fiduciary obligations to the owner during the pre-construction phase, it is an independent contractor once it assumes liability for the cost of construction.[48] In most jurisdictions, CM At-Risk contracts in which the construction manager guarantees a maximum price must be competitively bid.[49]

C. *Program Management Services*

A construction manager can also provide program management services. A program manager is a dedicated person or entity who manages the development and implementation of an entire project program.[50] Program management has traditionally been utilized with public school systems and with large infrastructure projects, but can also be used in the context of private commercial projects, especially multi-project construction for large owners who require detailed engineering design.[51]

A program manager can provide a variety of services, including:

- analyzing the owner's current facilities and program requirements;
- assisting in site evaluation, selection, and acquisition;
- assisting with regulatory and permitting requirements;
- managing site investigation;
- managing a community relations program;
- managing insurance needs;
- assisting with the procurement of design and construction services;
- drafting and negotiating design and construction contracts;
- helping to develop design criteria;
- project administration, including scheduling, budgeting, handling claims and dispute resolution, monitoring safety and quality assurance programs, and administering contract close out activities; and
- facility management services.[52]

Factors to consider in deciding whether a program manager is needed on a project are:

- size of the project;
- complexity of the project;
- uniqueness of the project; and
- the need to coordinate the project with existing facilities.[53]

---

48. Chiarelli, *supra* note 32, at 3.

49. BRUNER & O'CONNOR, *supra* note 36, at 605, 615.

50. David A. Scotti, *Program Management: The Owner's Perspective*, 16-OCT CONSTRUCTION LAW. 15, at 2 (1996).

51. Christopher L. Noble, *Program Management: The Design Professional's Perspective*, 16-OCT CONSTRUCTION LAW. 5, at 1 (1996).

52. James Duffy O'Connor, *On Platypuses and Program Management*, 16-OCT CONSTRUCTION LAW. 11, at 2, 3 (1996).

53. Scotti, *supra* note 50.

Another factor to consider is whether there will be a series of projects. Using a program manager can provide continuity and apply the expertise developed in earlier projects to later projects.[54]

The form of the contractual relationship between the program manager and the owner can vary. The program manager can be retained as a pure consultant with no responsibility for the design or construction work, or it can assume responsibility for some or all of the design or construction work.[55] Of course, the program manager's liability increases along with its responsibilities.[56]

D. *Value Engineering Services*

A construction manager can also provide value engineering services. Value engineering involves reviewing the architectural and engineering plans and evaluating the life-cycle and construction costs of a project to determine whether lower costs can be achieved without compromising functionality.[57] Value engineering should be done by a team different than the one responsible for the original design.[58]

Two methods commonly used in value engineering are life-cycle analyses and cost comparisons.[59] In life-cycle analyses, the value engineering consultant compares the purchase, operating, and maintenance costs of alternative systems (such as HVAC, plumbing, and electrical) and provides the owner with options based on this analysis.[60] Cost comparison involves recommending alternative building components.[61] Along with the cost comparison, the value engineering consultant usually explains the functional and aesthetic differences between alternative components.[62]

E. *Owner's Representative Services*

A construction manager can also provide owner's representative or peer review services. In this role, the construction manager serves as a consultant to the owner, but plays a more limited role in the project than the construction manager advisor.

The owner's representative ensures that the owner's best interests are at the heart of decisions made in the design and construction processes. The construction manager will work with the design professional during the design phase and ensure that the owner's needs and concerns are being addressed. During the construction phase,

---

54. Scotti, *supra* note 50, at 1.
55. *Id.*, at 4.
56. *See id.*
57. Underwood, *supra* note 33, at 13.
58. Loulakis, *supra* note 31, at 249.
59. Underwood, *supra* note 33, at 13.
60. *See id.*
61. *See id.*
62. *See id.*

the construction manager will often be onsite identifying issues and problems and assisting in resolving them. An owner's representative can be of immense value to an owner who is relatively unsophisticated in the construction process or to any owner if the project is large or complicated.

Library References

C.J.S. *Agency* §§ 240–241, 244–247, 253, 310, 312–313, 316–318, 331–335, 338–339; C.J.S. *Architects* §§ 4, 16–17, 19, 21–24, 32; C.J.S. *Contracts* §§ 11, 341, 343, 380, 393–406.

VIII. Standard Form Agreements

The primary sources for standard construction management agreements are: the Construction Management Association of America (CMAA), the American Institute of Architects (AIA), and the Associated General Contractors (AGC). Each of these organizations takes a different approach to the construction manager's role in the construction process. This section examines the responsibilities allocated to the construction manager under the standard form agreement for agency construction management published by each of these organizations.

A. *CMAA*

Under the CMAA agreement, the construction manager is the leader of the construction team from start to finish.[63] The CMAA approach is premised on the belief that "providing services and assuming authority while having clearly defined responsibilities allows the [construction manager] to more effectively gain control over the risks which create exposure to liability."[64]

1. *Pre-Construction Phase*

The construction manager's responsibilities under CMAA Document No. A-1, Standard Form of Agreement Between Owner and Construction Manager (Construction Manager as Owner's Agent), include:

- preparing and revising a construction management plan.[65]

The Construction Management Plan is the hallmark of the CMAA approach. A typical Construction Management Plan includes the following components:

---

63. Harris, *supra* note 29, at 17.

64. Richard D. Conner, *The Development of and Philosophy Behind the CMAA's Construction Management Documents—Agency Series*, in THE AMERICAN BAR ASSOCIATION FORUM ON THE CONSTRUCTION INDUSTRY, EIGHTH ANNUAL FORUM, CONSTRUCTION MANAGEMENT: MAKING THE RIGHT CHOICE (1992).

65. CMAA Document No. A-1, Standard Form of Agreement Between Owner and Construction Manager (Construction Manager as Owner's Agent) (CMAA Doc. No. A-1), ¶¶ 3.2.1.1, 3.3.1.1.

project description; master schedule; milestone schedule; quality management approach; reference to project documents; project organization chart and staffing plan; explanation of roles, responsibilities, and authority of team members; project budget/work breakdown structure; environmental/archeological considerations; reference to project procedures manual; management information system; communications protocol; and packaging and contracting strategy.[66]

- assisting the owner with selecting the design professional and preparing the contract between the owner and the design professional;[67]
- assisting the owner with orienting the design professional with respect to the project schedule, cost and administrative requirements;[68]
- preparing and revising a master schedule and reviewing contractor schedules;[69]
- conducting a construction market survey including information on local construction services, labor, material and equipment cost;[70]
- preparing and revising a project and construction budget;[71]
- preparing a cost analysis regarding the estimated costs of design and construction alternatives;[72]
- establishing a management information system in order to facilitate communication between the various parties involved in the project;[73]
- reviewing the design documents;
- making recommendations regarding constructability, clarity, consistency, and coordination of documentation among contractors;
- scheduling and packaging the work;[74]
- monitoring design professional compliance with the construction management plan and the management information system;[75]
- conducting value analysis studies;[76]
- managing the bid process;[77] and
- preparing the construction contracts.[78]

---

66. Conner, *supra* note 4.
67. CMAA, Doc. No. A-1, ¶¶ 3.2.1.2, 3.2.1.3.
68. CMAA, Doc. No. A-1, ¶ 3.2.1.4.
69. CMAA, Doc. No. A-1, ¶¶ 3.2.2.1, 3.4.2.2, 3.5.2.1, 3.5.2.2.
70. CMAA, Doc. No. A-1, ¶ 3.2.3.1.
71. CMAA, Doc. No. A-1, ¶¶ 3.2.3.2, 3.3.3.2.
72. CMAA, Doc. No. A-1, ¶ 3.2.3.3.
73. CMAA, Doc. No. A-1, ¶ 3.2.4.1.
74. CMAA, Doc. No. A-1, ¶ 3.3.1.5.
75. CMAA, Doc. No. A-1, ¶ 3.3.1.3.
76. CMAA, Doc. No. A-1, ¶ 3.3.3.3.
77. CMAA, Doc. No. A-1, ¶¶ 3.4.1.1—3.4.1.9, 3.4.3.2.
78. CMAA, Doc. No. A-1, ¶ 3.4.1.10.

2. Construction Phase

The construction manager's responsibilities under CMAA Document No. A-1, Standard Form of Agreement Between Owner and Construction Manager (Construction Manager as Owner's Agent) include:

- maintaining an on-site management team to provide contract administration services;[79]
- assisting the design professional with providing interpretations of contract documents;[80]
- reviewing change orders and making recommendations to the owner;[81]
- inspecting the construction;[82]
- reviewing contractor safety programs;[83]
- rendering decisions on disputes between the owner and contractor;[84]
- processing payments;[85]
- monitoring costs;[86]
- coordinating inspection and testing by independent consultants;[87] and
- determining substantial and final completion.[88]

B. *AIA*

Under the AIA agreement, the architect is the leader of the construction team.

1. *Pre-Construction Phase*

The construction manager's responsibilities under the AIA Standard Form Agreement Between Owner and Construction Manager where the Construction Manager is *Not* a Constructor, AIA Document B801/CMa, include:

- preparing and updating cost estimates and monitoring the budget;[89]
- reviewing design documents, advising on material and equipment selection, project delivery methods, and construction methods;

---

79. CMAA, Doc. No. A-1, ¶ 3.5.1.3.
80. CMAA, Doc. No. A-1, ¶¶ 3.5.1.4, 3.5.1.5, 3.5.1.9.
81. CMAA, Doc. No. A-1, ¶ 3.5.1.9.
82. The construction manager has the authority to reject the work. CMAA, Doc. No. A-1, ¶ 3.5.1.11.
83. CMAA, Doc. No. A-1, ¶ 3.5.1.12.
84. CMAA, Doc. No. A-1, ¶ 3.5.1.13.
85. CMAA, Doc. No. A-1, ¶ 3.5.3.6.
86. CMAA, Doc. No. A-1, ¶ 3.5.3.3.
87. CMAA, Doc. No. A-1, ¶ 3.5.1.7.
88. CMAA, Doc. No. A-1, ¶¶ 3.5.1.15, 3.5.1.16.
89. AIA Standard Form Agreement Between Owner and Construction Manager When the Construction Manager Is *Not* a Constructor, AIA Document B801/CMa (AIA Document B801/CMa), ¶¶ 2.2.3, 2.2.6, 2.2.16, 2.3.3.

- providing cost evaluations of alternative materials and systems;[90]
- preparing a project schedule and construction schedule;[91]
- providing advice regarding constructability;[92]
- providing advice on packaging of the work;[93]
- providing advice regarding allocation of responsibility for safety programs among contractors;[94]
- assisting the owner with selecting and retaining consultants;[95]
- managing the bid process;[96]
- assisting the owner with preparing construction contracts;[97] and
- assisting the owner with permitting.[98]

2. *Construction Phase*

The construction manager's responsibilities under the AIA Standard Form Agreement Between Owner and Construction Manager where the Construction Manager is *Not* a Constructor, AIA Document B801/CMa include:

- coordinating the work;[99]
- monitoring the schedule;[100]
- monitoring costs;[101]
- developing procedures for processing payments;[102]
- reviewing contractors' safety programs for the purpose of coordinating them;[103]
- inspecting the work/monitoring performance;[104]
- assisting the architect in responding to requests for interpretations of the contract documents;[105]
- assisting the architect in reviewing change requests and processing change orders;[106]

---

90. AIA Document B801/CMa, ¶¶ 2.2.3, 2.2.4.
91. The architect has to approve the part of the project schedule related to the architect's services. AIA Document B801/CMa, ¶¶ 2.2.5, 2.2.16.
92. AIA Document B801/CMa, ¶ 2.2.7.
93. AIA Document B801/CMa, ¶ 2.2.10.
94. AIA Document B801/CMa, ¶ 2.2.9.
95. AIA Document B801/CMa, ¶ 2.2.13.
96. AIA Document B801/CMa, ¶¶ 2.2.17, 2.2.19.
97. AIA Document B801/CMa, ¶ 2.2.20.
98. AIA Document B801/CMa, ¶ 2.2.21.
99. AIA Document B801/CMa, ¶¶ 2.3.3, 2.3.6, 2.3.7, 2.3.14.
100. AIA Document B801/CMa, ¶¶ 2.3.5, 2.3.14.
101. AIA Document B801/CMa, ¶¶ 2.3.8, 2.3.9.
102. AIA Document B801/CMa, ¶ 2.3.11.
103. AIA Document B801/CMa, ¶ 2.3.12.
104. AIA Document B801/CMa, ¶ 2.3.13.
105. AIA Document B801/CMa, ¶ 2.3.16.
106. AIA Document B801/CMa, ¶ 2.3.17.

- assisting the architect in reviewing and evaluating claims;[107] and
- assisting the architect in determining substantial completion.[108]

C. *AGC*

The AGC also has its own version of an agency construction management agreement.

1. *Pre-Construction Phase*

The construction manager's responsibilities under AGC 510, Standard Form of Construction Management Agreement Between Owner and Construction Manager (Where the Construction Manager is the Owner's Agent and the Owner Enters into Contractor Agreements), include:

- consulting with the owner and architect regarding site use and selection of materials and equipment, and providing recommendations on construction feasibility and factors related to construction cost, including estimates of alternative designs or materials;[109]
- preparing and updating the project schedule;[110]
- preparing and updating the project budget and construction budget;[111]
- reviewing construction documents for constructability issues;[112]
- making recommendations regarding packaging the work;[113]
- managing the bid process;[114]
- assisting the owner in selecting and retaining consultants;[115] and
- assisting the owner with permitting.[116]

2. *Construction Phase*

The construction manager's responsibilities under AGC 510, Standard Form of Construction Management Agreement Between Owner and Construction Manager (Where the Construction Manager is the Owner's Agent and the Owner Enters into Contractor Agreements) include:

---

107. AIA Document B801/CMa, ¶ 2.3.18.
108. AIA Document B801/CMa, ¶ 2.3.25.
109. AGC 510, Standard Form of Construction Management Agreement Between Owner and Construction Manager (Where the Construction Manager Is the Owner's Agent and the Owner Enters into Contractor Agreements) (AGC 510), ¶ 3.1.2.
110. AGC 510, ¶ 2.1.3.
111. AGC 510, ¶¶ 3.1.4.1–3.1.4.8.
112. AGC 510, ¶ 3.1.5.
113. AGC 510, ¶ 3.1.8.
114. AGC 510, ¶ 3.1.10.
115. AGC 510, ¶ 3.1.12.
116. AGC 510, ¶ 3.1.13.

- assisting owner with preparing and negotiating construction contracts;[117]
- providing on-site project management;[118]
- assisting the owner and architect in reviewing and evaluating claims;[119]
- monitoring and coordinating the work;[120]
- monitoring and updating the project schedule;[121]
- monitoring and updating the project budget;[122]
- processing payments;[123]
- assisting with processing change orders;[124]
- inspecting the work;[125]
- reviewing the safety programs of contractors for purpose of coordinating with the safety programs of other contractors;[126]
- coordinating the work;[127] and
- determining substantial completion.[128]

D. *Comparison of the Standard Form Agreements*

There are differences in the nature and scope of the responsibilities allocated to the construction manager in each of the standard form agreements discussed above. This section highlights some similarities and differences. The chart at the end of this section summarizes this information.

All three of the standard form agreements allocate responsibility for scheduling, cost estimating, value engineering, contract preparation, safety, and processing payments to the construction manager. In addition, all three agreements contain language attempting to limit the construction manager's responsibility for project safety and provide that the construction manager is not responsible for the contractor's means and methods and is not responsible for the contractor's failure to carry out the work in accordance with the contract documents.

All three agreements also allocate responsibility for reviewing the construction documents, processing change orders, and inspecting the work to the construction manager, but the scope of these responsibilities varies between the agreements. With

---

117. AGC 510, ¶ 3.2.2.
118. AGC 510, ¶ 3.2.2.2.
119. AGC 510, ¶ 3.2.2.5.
120. AGC 510, ¶ 3.2.2.6.
121. AGC 510, ¶ 3.2.3.
122. AGC 510, ¶ 3.2.4.
123. AGC 510, ¶ 3.2.4.3.
124. AGC 510, ¶ 3.2.4.5.
125. AGC 510, ¶ 3.2.5.1.
126. AGC 510, ¶ 3.2.5.3.
127. AGC 510, ¶ 3.2.2.6.
128. AGC 510, ¶ 3.2.7.1.

respect to reviewing the construction documents, the CMAA agreement explicitly provides that nothing in the agreement should be construed to mean that the construction manager assumes the duties of the design professional and that the design professional is solely responsible for the design requirements and criteria. The AGC agreement also explicitly provides that the design professional is responsible for "all the design and engineering necessary for the Project, including the preparation of the construction documents."[129] In contrast, the AIA agreement does not contain any such provisions limiting the construction manager's responsibility for design and/or engineering.

With respect to change orders, the CMAA agreement places more responsibility on the construction manager than the AIA and the AGC agreements. Under the CMAA agreement, the construction manager makes recommendations to the owner regarding all proposed change orders (although the construction manager is required to consider the design professional's comments).[130] In contrast, both the AIA and AGC agreements provide the construction manager with recommendations to the owner *and* architect regarding change orders.[131]

Similarly, the CMAA agreement gives the construction manager more responsibility with respect to inspecting the work. Under the CMAA agreement, the construction manager has the authority to reject work on its own.[132] Under both the AIA and AGC agreements, the construction manager can only reject work in consultation with the architect and/or owner.[133]

The agreements also differ in the allocation of the responsibility to interpret the contract documents. Under the AGC agreement, the construction manager has no responsibility for interpreting the contract documents, whereas under the CMAA and AGC agreements, the construction manager assists the design professional with this responsibility.

Another significant area of difference between the agreements is the duty to coordinate. The AIA and AGC agreements both provide that the construction manager shall "coordinate the work."[134] In contrast, the CMAA agreement does not place any such duty to coordinate the work on the construction manager.

The following analysis demonstrates that there are differences in the responsibilities allocated to the construction manager under the three primary standard form construction management agreements. The construction manager must carefully consider these responsibilities and the attendant liability.

---

129. AGC 510, ¶ 2.2.2.
130. AGC 510, ¶¶ 3.5.1.9.2, 3.5.1.9.3.
131. AGC 510, ¶¶ 2.3.17/3.2.4.5.
132. AGC 510, ¶ 3.5.1.11.
133. AIA Document B801/CMa, ¶ 2.3.13; AGC 510, ¶ 3.2.5.1.
134. AIA Document B801/CMa, ¶¶ 2.3.3, 2.3.6, 2.3.7, 2.3.14; AGC 510, ¶ 3.2.2.5.

Construction Manager Responsibilities under Standard Form Agreements

Responsibility	CMAA	AIA	AGC
Scheduling	Y	Y	Y
Cost Estimating	Y	Y	Y
Value Engineering	Y	Y	Y
Contract Preparation	Y	Y	Y
Safety	Limited	Limited	Limited
Processing Payments	Y	Y	Y
Reviewing Construction Documents	Limited	Y	Limited
Processing Change Orders	Y	Limited	Limited
Inspecting Work	Y	Limited	Limited
Interpreting Contract Documents	Limited	Limited	N
Duty to Coordinate	N	Y	Y

IX. Compensation

Construction manager compensation can be structured in a variety of ways. This section discusses several different methods of construction manager compensation.

A. *Time-Based Fee*

Under a time-based fee compensation structure, the construction manager is generally paid an hourly or daily rate plus reimbursable expenses. The hourly or daily rate is usually calculated by multiplying direct personnel expense (the base rate) by a multiplier. The base rate can be the employee's actual hourly salary, the payroll cost of the employee including payroll taxes and insurance, or a predetermined rate not related to the employee's salary.[135] The multiplier includes the construction manager's profit and overhead.[136] The following factors should be considered in determining the multiplier:

- size of the project;
- local labor rates;
- whether nonunion or union workers are used;
- extent of field supervision required; and
- extent of main office support required.[137]

---

135. Loulakis, *supra* note 31, at 179.
136. *See id.*
137. *Id.*, at 180.

Reimbursable expenses normally include reasonable actual expenditures made by the construction manager, its employees, or its professional consultants. The construction manager normally develops its cost estimate for reimbursable expenses based on the understanding it has with the owner, services to be provided, its project plan, and staffing requirements.

Some contracts allow a separate fee in the range of five percent to ten percent to be applied to the cost of reimbursable expenses in order to cover the construction manager's cost of handling the administration of such expenses.

The construction manager's compensation schedule can cover the entire project period, or may be divided into two separate phases, pre-construction and construction. The objective in separating the costs is to enable the parties to better control the construction manager's costs and account for differences in the overhead expenses associated with the construction manager's permanent base of operation versus its field activities. During the pre-construction phase, the construction manager's staff resources are usually corporate, branch, or home office supplied. The multiplier allowed for such staff is often greater than it is for staff permanently assigned to the field.

Examples of common reimbursable expenses recognized during the construction phase include:

- travel and living expenses;
- relocation costs;
- subsistence to key employees;
- communications;
- computer equipment and software;
- vehicles;
- fees;
- permits;
- reproduction and mailing;
- professional liability insurance and other insurance costs for coverage beyond the normal limits carried by the construction manager;
- general conditions (including labor costs); and
- other reasonable field office related expenses.

Certain field construction office costs associated with an assignment are usually recognized as reimbursable expenses. In some cases, part or all of these costs may be avoided by the construction manager when the owner provides field office space, equipment, utilities, supplies, etc., directly and at no cost to the construction manager. Occasionally, the owner may require that one of the trade contractors provide field office support services for the construction manager's field operations as part of a general conditions requirement. Such requirements would obviously reduce the construction manager's normal expected field overhead cost and reduce any multiplier applied to the construction manager's field labor cost for the construction phase.

Payments under this method of construction manager compensation are usually based on the actual labor hours and direct expenses expended by the construction manager's staff and as authorized by the contract terms and conditions. A reasonable retention of five percent to ten percent is normally withheld by the owner until substantial completion of the project is achieved. Thereafter, the retention is substantially reduced and paid in full when the project is considered to have reached final completion, as defined by the contract documents.

A time-based fee compensation structure requires detailed cost records setting forth the exact amount of time spent, the nature of the work, and who did the work. Therefore, it is helpful for the parties to agree at the beginning of the project as to what type of records must be kept.

B. *Lump Sum/Fixed Fee*

Under a lump sum or fixed fee compensation structure, the construction manager agrees to provide services in exchange for a predetermined dollar amount (the fixed fee). The fee includes direct costs, overhead, and profit. When this method is employed, a fixed fee is usually established for services during the pre-construction phase and a separate fixed fee is usually established for services during construction. Although this type of compensation structure provides certainty to the owner, it is highly possible that services initially not contemplated by the parties may be required. Therefore, a fixed fee agreement should include procedures for pricing such additional services.[138] Increases to the fixed fee are normally recognized for services that extend beyond the anticipated completion date, for circumstances or delays beyond the control of the construction manager, and for increases in the scope of services that were not anticipated when the contract was signed.

When a fixed fee compensation structure is used in the context of agency construction management, it is important to carefully define the scope of services to be provided by the construction manager in order to avoid a conflict of interest. If a fixed fee compensation structure is used in the context of a construction manager at-risk, the construction manager bears all the risk of cost overruns. Therefore, it is important that the design documents are as complete and accurate as possible before the fee is determined. However, although the construction manager bears the risk of cost overruns, it also will enjoy extra profit if the work costs less than anticipated.

The fixed fee approach can adversely affect the parties involved. For example, unexpected problems can require a construction manager to expend more cost than budgeted for staff to perform certain tasks or to resolve problems. The construction manager could also be required to provide a greater level of effort than originally anticipated. Such factors can have a serious impact on the construction manager's fixed fee budget. Conversely, some owners believe that a temptation always exists for the construction manager to put forth less effort than anticipated or actually required in order to maximize the construction manager's profit.

Progress payments to the construction manager under the fixed fee method of compensation are usually paid in accordance with a predetermined schedule of

---

138. Loulakis, *supra* note 31, at 181.

monthly progress payments for the pre-construction phase, and the reported aggregate percentage of construction work actually completed by the trade contractors during construction. A retention of five percent to ten percent is considered reasonable through substantial completion of the project. Thereafter, the retention is usually reduced and paid in full when the project is considered to have reached final completion as defined by the contract documents.

C. *Cost of Work Plus a Fixed Fee*

Under this compensation structure, the construction manager is reimbursed for defined costs and receives a fee in addition to reimbursed costs.

The fee generally includes profit and overhead. The fixed fee portion under this method of compensation will generally range from two percent to four percent of construction costs, with most fees closer to two percent. Factors having a bearing on the amount of fixed fee include size, complexity, duration, the services to be provided, and the terms and conditions agreed to by the parties for direct reimbursable costs. In addition, the use of multipliers applied to certain allowable costs can also have a bearing on the amount of fixed fee agreed to by the parties.

The cost of work plus fixed fee method of compensation can also be divided into two separate phases. For the pre-construction phase, a fixed fee and certain direct expenses recognized by the parties as reimbursable may be established. Other compensation options sometimes used for the pre-construction phase are fixed fee or a lump sum. The pre-construction phase usually ends at the time of the award of the first construction contract. During the construction phase, a separate fixed fee is established and all project related costs are usually reimbursed, as per the terms and conditions of the contract.

The fee usually includes the cost of officers, principals, and home office support staff not directly assigned to the construction site, all costs not directly related to site operations, and costs not identified as reimbursable under the terms and conditions of the contract. Examples of costs that are not normally a part of the reimbursable expenses include:

- home office overhead;
- recruitment costs for job site and home office personnel; and
- profit.

Reimbursable expenses generally include:

- the costs of all job site personnel;
- staff relocation costs;
- travel and living expenses;
- subsistence to key field employees;
- consultants;
- the cost of all job site material;
- equipment;
- communications;

- vehicles;
- project-related fees paid;
- reproduction;
- postage;
- computers; and
- any other field office-related expenses.

Other reimbursable expenses may include the use of consultant's home office personnel who may be involved in purchasing, expediting, scheduling and estimating, and any other specific functions, tasks, or services that are usually part-time and project- or field-support related. Further, the cost of general condition items and any field labor associated with the performance of general conditions would be considered reimbursable expenses.

The AGC, AIA, and CMAA standard form contracts all address the issue of reimbursable costs. The parties should clearly identify and agree on what is and is not included as part of the fixed fee, and identify such items in the contract before a fixed fee is established. In addition, the contract should also be clear in its definition of reimbursable costs and how the fixed fee and reimbursable costs will be paid.

Progress payments are usually made based on a percentage of the fee in terms of actual construction progress achieved, plus actual labor costs, and actual allowable reimbursable expenses incurred by the construction manager to date. Retention may range from five percent to ten percent through substantial completion. Thereafter, it may be reduced and paid upon final completion, as defined by contract terms and conditions.

Increases in compensation limits may be allowed when scope changes, additional construction management services, or delays beyond the control of the construction manager are encountered. In some cases, the parties may agree that the fixed fee never increases. In other situations, the construction management contract may be silent on these issues. Such silence can be a source of disagreement when these circumstances arise.

D. *Fee as a Percentage of Construction Cost*

Under this method of compensation, the construction manager usually begins by providing construction management services during the pre-construction phase on a fixed fee or lump sum basis. Construction management services rendered during the construction phase are based on a fixed percentage of the estimated or actual construction costs.

If the estimated or actual construction costs (based on contract award values) are increased due to approved changes, the amount of the construction manager's compensation may be increased based on a predetermined percentage of the costs of such changes. The fee percentage applied to additional work may sometimes be based on a sliding scale (usually decreasing), depending on the size of the increase of construction cost.

Progress payments are usually made to the construction manager based on the reported percentage of actual construction work completed, with reasonable terms for fee retention until final payment is made to the construction manager.

This form of compensation is not recommended, as it is arbitrary and not related to the effort that may be required. For example, a greater effort may be required for a smaller dollar value project due to technical complexity or schedule compression.

E. *Cost of Work Plus a Fee with a GMP*

The most common compensation structure in the construction manager at-risk context is the cost of work plus a fee with a guaranteed maximum price (GMP). Under this compensation structure, the construction manager at-risk assumes the risk of all costs exceeding the GMP. This compensation structure benefits the owner by placing a cap on the construction manager's compensation and shifting the risk of cost contingencies to the construction manager.

In order to calculate a fair and reasonable GMP, the design (working drawings) should be in the range of at least fifty percent to sixty percent complete and the technical specifications should be at least eighty percent complete. The GMP may include the estimated and/or bid cost of certain trade contracts, purchases from suppliers and vendors, general condition items, the construction manager's fee, reimbursable expenses, and sometimes a contingency to protect the construction manager from unforeseen costs at the time the GMP is prepared. It is not unusual for costs of certain specific items to be identified separately or even capped as part of the GMP. For example, the construction manager's fee, the cost of general conditions, and the construction manager's contingency may be identified separately as part of the overall GMP, but any costs exceeding such allowances are usually paid on a cost-reimbursable basis. The use of specific allowances for unknowns should be kept to a minimum.

Under a GMP arrangement, the construction manager's compensation is normally broken into two phases and identified separately for the pre-construction phases. During the pre-construction phase, the construction manager will usually be compensated on a fixed fee, lump sum, or cost-reimbursable basis and paid in monthly portions. Another option is to compensate the construction manager on labor times a multiplier for overhead and profit, with a not-to-exceed cost limitation. The pre-construction phase usually ends upon the award of the first construction contract and when the work begins in the field.

It is important that the phrase *Guaranteed Maximum Price* be clearly defined in the contract as to what it includes and specifically excludes. When an agreement on a GMP is reached with the owner subsequent to signing the initial contract, an amendment to the original contract is usually required. Some believe that a new contract should be established, since the principles, terms, and conditions change. If for some reason the GMP for the cost of construction cannot be agreed upon, the construction manager may continue at the owner's option to perform construction management services during the construction phase as the owner's agent. Usually, the owner also maintains the option to terminate the contract.

Once a GMP for construction cost is established, the construction manager takes on the risk of completing the project. Trade and specialty contracts are bid, negotiated, awarded, and usually held by the construction manager. It is the responsibility of the construction manager to stay within its estimated costs and contingency allowances.

On some GMP projects, the owner may reserve the right to approve all trade and purchase contracts. Under such conditions, the construction management contract should be clear as to which party pays the difference between the estimated trade contract cost and the actual trade contract award value when the award value approved by the owner is greater than the estimated cost element included in the GMP.

Assuming the actual construction costs could be less than the GMP established for construction costs, there may be situations where the owner only wants to pay the actual cost of completing the project. In such cases, the contract should be clear as to who derives the benefit of any cost savings. This is best achieved by a shared incentive savings clause included in the contract.

In most contracts, the owner retains the right to make changes. Accordingly, the contract should be clear as to how the GMP will be equitably adjusted for both added and deleted work. In some contracts, there may be no adjustment in fee for changes. The construction manager is only compensated for the cost of the work as defined under reimbursable costs. If the project is extended beyond the scheduled completion date for reasons beyond the control of the construction manager, a time extension and an increase in the construction manager's compensation may be required.

It is important that the term *construction costs* or *project costs* be clearly defined. The definition of construction costs is usually stated as the cost of all work required to complete the construction as defined in the construction contract documents. Normally, it does not include such items as land costs, professional service costs, or the cost of movable equipment, furniture, etc. In addition, definitions should be provided for reimbursable and nonreimbursable costs.

On some projects, a construction manager may be responsible for providing general condition items. General condition items may include such items as:

- site security;
- scaffolding;
- hoists;
- signs;
- safety barricades;
- water boys;
- cleaning;
- dirt chutes;
- cranes;
- shanties;
- preparation for ceremonies, including minor construction activity in connection therewith;
- temporary toilets;

- fencing;
- sidewalk;
- bridges;
- a first aid station;
- trucking;
- temporary elevators;
- special equipment;
- winter protection;
- temporary heat;
- water and electricity;
- temporary protective enclosures;
- field office and its related costs such as equipment;
- furnishings and office supplies;
- progress photographs;
- messenger service;
- installation of owner-furnished items;
- post and planking;
- general maintenance;
- subsoil exploration;
- refuse disposal;
- field and laboratory tests of concrete, steel, and soils;
- surveys;
- bench marks and monuments;
- storage on-site and off-site of long-lead procurement items; and
- miscellaneous minor construction work.

Payments to the construction manager during the construction phase are usually made on a monthly basis and in proportion to the cost incurred and the construction work actually completed. A retention by the owner of five percent to ten percent of the construction manager's earnings is considered to be reasonable. Two payment issues requiring attention include the potential for overpayment and calculating any sharing of cost savings (if cost savings are made part of the contract requirements).

If for some reason the GMP is exceeded, the construction manager will be responsible for the additional costs unless the owner is determined to be partially or totally responsible. If the owner is found to bear responsibility, the construction manager may be entitled to a cost adjustment and/or a time extension.

There are two points that are sometimes not understood by owners or properly addressed in some GMP agreements. Both points are common causes of compensation disputes between owners and GMP construction managers. The first is: who owns the savings or the contingency that is sometimes created in the actual buy-out of the trade contracts and purchases after the GMP has been agreed to by the parties? The second is: who benefits if the actual construction costs are less than the established GMP? Unless the contract specifically provides otherwise, the answer to both of these questions is the construction manager, since the construction manager is the party taking the risk.

The parties can, of course, contract for alternative savings arrangements. Some alternatives are:

- splitting the savings on an agreed upon percentage between the construction manager and the owner;
- allowing the owner to retain some amount of the savings before the savings are split; and
- allowing the construction manager to retain some amount of the savings before the savings are split.[139]

F. *Comparison of Compensation Arrangements*

Of the alternatives presented above, the first three are consistent with the agency relationship because the calculation of the construction manager's compensation does not conflict with the protection of the owner's interests. However, using the percentage of construction cost to be determined, construction manager compensation presents an inherent conflict. Where the construction manager fails to deliver the project at the lowest possible cost, the construction manager's compensation will actually be greater than if cost containment had been achieved. The construction manager, in essence, is rewarded for the higher construction cost. This alternative is inconsistent with an agency relationship.

A construction management contract with a GMP is also inconsistent with an agency relationship. Because costs exceeding the GMP may be paid out of the construction manager's fee, the construction manager has an incentive to keep costs within the GMP. The construction manager may keep construction costs from overrunning the GMP by performing the work and by cutting corners to save money, time, or costs. Much like a general contractor, the construction manager has some incentive to put its own interests above those of the owner.

In summary, there are a number of factors to be considered for establishing fair and reasonable compensation for construction management services. These include the following:

- The owner and the construction manager should agree on a scope of services and the use of a particular form of construction management agreement prior to the construction manager submitting its cost proposal.
- The owner should consider the cost of construction management services only after the establishment of the technical requirements and after careful consideration of the construction manager's qualifications, related experience, and proposed staffing plan.
- The construction management contract proposed to be used by the parties should clearly define the risks and authority assumed by the construction manager, and distinguished from those retained by the owner.

Keep in mind that the use of any standard form agreement usually requires some changes in order to be tailored to a particular project. Legal counsel should be involved.

---

139. Loulakis, *supra* note 31, at 186.

The owner and the construction manager should establish a clear understanding of what is included and excluded in any fee arrangement and the cost of the work. This includes definitions for such terms as Guaranteed Maximum Price (GMP), cost of the work or cost of construction, project costs, allowances, contingencies, reimbursable costs, nonreimbursable costs, fixed fee, multiplier, etc., as may be appropriate for the particular form of construction management contract to be used. When applicable, the owner and the construction manager should establish a reasonable overhead allowance based on an audit of allowable and reasonable costs.

Contract provisions should address how changes in scope and additional construction management services will be handled.

When a GMP is used, it should be based on the total estimated cost of construction and not on individual line items within the total estimated cost. For GMP contracts, the owner and the construction manager should establish a clear understanding of the type and use of any contingency funds. A construction manager's contingency (ranging from one percent to three percent) for errors in estimating or for covering unforeseen costs relating to construction is considered to be reasonable. Such contingency is identified separately and paid based on documented evidence of expenditures by the construction manager.

Consideration should be given to using a shared savings clause that provides an incentive for the construction manager to achieve performance under the GMP approach. Contract provisions should address extended construction management service costs for any project extension caused by reasons beyond the control of the construction manager. Finally, the construction management contract should establish terms for progress payments.

Library References

C.J.S. *Architects* §§ 28–31; C.J.S. *Contracts* §§ 377–407.

X. Practice Aids

A. *Treatises*

Philip L. Bruner & Patrick J. O'Connor, Jr., Bruner & O'Connor On Construction Law (2002).

Michael C. Loulakis, Jon M. Wickwire, Mark A. Berry & Thomas J. Driscoll, Construction Management Law And Practice (1995).

B. *Web Sites*

Construction Management Association of America, http://www.cmaanet.org
Associated General Contractors, http://www.agc.org
American Institute of Architects, http://www.aia.org
Design-Build Institute of America, http://www.dbia.org

CHAPTER 10

Construction Manager's Responsibilities: Pre-Design, Design, and Pre-Construction Phase

PAUL H. MARTINEZ, ESQ.
RASHIDA Y.V. MacMURRAY, ESQ.

I. Introduction

This chapter summarizes and addresses the nuances of the construction manager's role, responsibilities, and relationship with the design team (owner and its consultants) during the pre-design and pre-construction phases of a construction project. This chapter assumes that the design professional and construction manager are independent entities separately engaged by the owner and does not address projects in which the construction manager is also the design builder.[1]

In the early development of the construction management concept, architects and engineers attempted to expand their basic portfolio of services by venturing into the construction management arena. As with many businesses that move outside their core area of expertise, the results were not favorable. The majority of construction managers in business today have morphed from general contractors into part-time or full-time construction managers. Therefore, this chapter will focus on construction management from the perspective of a construction contractor rather than a design professional.[2] When providing services to the design team, the construction manager should be wary of stepping over the murky line of services that should

---

1. See Chapter 13, *Construction Manager and the Design-Build Project*.
2. Large Architectural, Engineering and Construction (A/E/C) firms may appear on the surface to perform both design and construction management services. However, these firms are made up of multiple separate business entities and the design entity will rarely be the entity performing construction management. The comments related to independent construction managers apply to construction managers which are separate corporate entities that may be owned by a parent A/E/C firm.

(or must, by state licensing statute) be performed by licensed professionals. Many construction managers treat this phase of the project as a marketing tool that may lead to a negotiated agreement to manage the construction phase. The resulting liability exposure can far exceed any good will that might be earned by waiving the fee for profit. This chapter will address several areas of potential liability that construction managers should take into consideration when they enter into agreements to provide pre-design and pre-construction services. Additionally, we have included model agreements and a check list that can be used when drafting or reviewing agreements for pre-construction services.

For each of the construction phases we discuss below, our goal is to outline those tasks that a construction manager should undertake, as well as those tasks that are more properly left to the owner or to consultants.

II. Relationship with Owner

The owner of any construction project is ultimately responsible for ensuring that the project is successfully completed. In most instances, the owner lacks the specialized expertise to coordinate all of the different aspects of the project. Alternatively, the owner does not have sufficient internal resources to self-perform pre-construction services. In other instances, the owner may be averse to accepting risk and wish to transfer some or all of its risk to the construction manager and/or the design professional. Consequently, the owner will solicit proposals from construction managers to perform pre-construction services.[3]

The construction manager's role during the pre-construction phase will vary in accordance with the scope of the services that the owner requests and the construction manager agrees to provide. On most pre-construction projects, the construction manager will (1) prepare a series of cost estimates at predetermined stages of design development; (2) prepare the schedule for the design phase as well as a preliminary schedule for the construction phase; (3) perform value engineering analysis; (4) perform a constructability review; and, perhaps, (5) develop the construction logistics plan.

Later, if the construction manager is to manage the construction phase either as an agent or as an at-risk constructor, the construction manager's duties may also include preparation of bid packages, bid evaluation, and preparation of recommendations to the owner for the award of trade contracts. Where there are construction materials and equipment that have relatively long delivery requirements, the construction manager may be asked to purchase or assist the owner in purchasing long-lead items. If the construction manager is expected to enter into an at-risk agreement for the construction phase,[4] pre-construction services will include developing the

---

3. Irvin E. Richter & Roy S. Mitchell, Handbook of Construction Law and Claims (1982).

4. *See* Chapter 11, *Construction Management Contract Responsibilities Construction and Completion Phases.*

guaranteed maximum price. If the construction manager is brought on board very early in the process, it may also be involved in assisting the owner, architect, and other design professionals with the pre-design acquisition of information and governmental approvals.

Sometimes, disputes develop between the construction manager and other design team members with respect to each party's role and responsibilities. These disputes can be avoided if methods are employed to promote open communication. In particular, an experienced construction manager can facilitate communications between the owner, design professionals, and other stakeholders involved in the project.

Although not licensed to provide design services, most construction managers have significant experience that enables them to assist the design professionals with meeting budgetary requirements imposed by the owner and with evaluating the feasibility and constructability of the proposed design.

Unfortunately, many project team relationships deteriorate because of unrealistic expectations on the part of the owner and/or design professionals. In many instances, specifically with a less sophisticated owner, a lack of understanding of the actual construction costs may result in a grossly underfunded project. Consequently, it is imperative that the owner retain a construction manager who is in tune with actual construction costs to assist in the development of the initial budget. If the owner or the design professional establishes a costs estimate that is too low, the project may be doomed before it ever gets into construction.

III. Pre-Design

For more sophisticated owners, some or all of the pre-design may be performed in-house without the assistance of an independent construction manager. Before establishing the economic feasibility of the project, many owners are reluctant to spend the money necessary to engage a construction manager. However, depending on the sophistication of the owner's in-house staff, the role of the construction manager during the pre-design phase can greatly impact the success of the project. On the other hand, the construction manager must be aware that pre-design contains the word *design*; this is an area where the construction manager may inadvertently take on professional liability or perform services that, by law, are to be performed only by licensed professionals.

A. *Project Scope*

Establishing the program[5] and defining the project scope is a formidable task for any owner. The project scope should not only identify the site, regulatory issues, use(s), goals, milestones, potential opposition, and stakeholders, it should also define the roles to be assumed by the project's team members (i.e., owner, architect, consultants,

---

5. The owner's program will outline the purpose, infrastructure, and facilities required to fulfill that purpose as well as the basic services required to support the planned facilities.

construction manager, etc.). It is imperative that the owner recognize the critical nature of formulating the project scope. Taking the time to plan the project carefully during pre-design is a good investment. It is true that people do not plan to fail, but rather they fail to plan. By recognizing that a project has many discrete parts that must be coordinated, the construction manager can apply its specialized management expertise to pre-design tasks as well as pre-construction.

Many abandoned projects, construction claims, and cost overruns can be eliminated or at least minimized if the owner makes the required investment in time and resources to complete the pre-design phase. Changes made to the scope during pre-design have minimal adverse financial consequences. Overall, when compared to the total cost to construct a project, the funds and resources required to properly complete the pre-design phase are small. The construction manager should recommend to the owner that it is never too early to start coordinating the details of the project and assigning team members to manage these tasks. The construction manager's role is to anticipate the owner's short-term and long-term project needs from the pre-construction phase to project close out. To achieve this, the construction manager should provide multiple options, from which the owner can select, which will assist the owner to focus on the decisions that will eliminate potential problems before they mature into deal killers.

B. *Environmental Impact Statements*

An Environmental Impact Statement (EIS) is a report prepared in accordance with the National Environmental Policy Act (NEPA)[6] for federal projects or the applicable State Environmental Quality Review Act (SEQA). The EIS report is intended to provide a means for owners, governmental agencies, and the public to review the environmental ramifications, if any, associated with the proposed project.[7] Anything that a reasonable person would consider significant should be addressed in the EIS. The EIS should address the total impact on the environment. Further, according to the Council on Environmental Quality (requirements may vary according to applicable SEQA), the EIS should generally address the direct and indirect effects that the project will have on the environment and possible methods for mitigating them. The EIS inserts the balancing of social, economic, and environmental issues into the project planning. If there is opposition to a project, attacking the EIS is the method most frequently used to delay the project and dissuade the owner from building in the opposition's back yard (also known as not in my back yard—NIMBY).[8] There are consultants that specialize in this area, and these consultants should be retained directly by the owner and not the construction manager. Few construction managers have expertise in this area and the risk does not counterbalance the fee for

---

6. 42 U.S.C. § 4321 (1969).

7. MICHAEL L. MCKINNEY, ENVIRONMENTAL SCIENCE SYSTEMS AND SOLUTIONS (2003).

8. This is a common acronym for persons or groups that oppose new real estate development, manufacturing plant, or public works projects that have been proposed for their town or neighborhood.

profit (more than likely, no fee for profit) associated with the construction manager's involvement in pre-design.[9]

1. *Phase I Environmental Assessment*

A Phase I Environmental Assessment (EAS) is intended to satisfy due diligence requirements for a real estate purchaser, a construction lender, or a title insurer. They are performed in accordance with American Society of Testing Materials Standards[10] and usually involve an historical investigation, including a physical survey of the site and surrounding properties, on-site inspection, review of building permits, review of environmental records, review of aerial photographs, record search for evidence of underground storage tanks, land use assessment, review of previous owners and uses, review of local geology and hydrology, and a written report that includes recommendations and conclusions. As indicated above, there are consultants that specialize in preparing the Phase I EAS. However, as opposed to an EAS discussed above, it is good practice for the construction manager who intends to manage on-site construction to review the Phase I EAS. In cases where the site or the area in which the site is located was previously devoted to an industrial use or there is a history of environmental problems in the area, if there is no Phase I available, the construction manager should qualify its proposal on the basis that the owner will provide a Phase I EAS that does not require preparation of a Phase II EAS. The construction manager and its subcontractors are responsible for the health and safety of the labor force working on the site and the construction manager should assure itself, in a reasonable manner, that there are no contaminated or hazardous substances on the site that could endanger its employees or subcontractor labor forces. Unfortunately, if the owner adopts a "don't ask, don't tell" attitude regarding potential environmental issues and refuses to order an EAS, the construction manager will be placed in the difficult position of deciding whether to walk away or order its own Phase I EAS.

2. *Phase II Environmental Assessment*

If questions are raised during the Phase I EAS, conducting a Phase II would be recommended as part of the conclusions and recommendations included in the Phase I EAS. Again, there are consultants who specialize in Phase II studies but in addition to the consultant, testing agencies will probably be required to perform soil borings, soil samplings, ground water surveys, asbestos surveys, lead paint surveys, and other testing and analyses. Negative or questionable findings during the Phase II

---

9. Some Architect/Engineer/Construction manager firms may perform environmental assessments with separate business entities.

10. AMERICAN SOCIETY OF TESTING MATERIALS STANDARDS, STANDARD PRACTICE FOR ENVIRONMENTAL SITE ASSESSMENTS: PHASE I ENVIRONMENTAL SITE ASSESSMENT PROCESS, E1527-05 (Nov. 1, 2005).

investigation may necessitate further testing, notification of appropriate governmental authorities, and the development of a remediation plan. There is no established protocol since each Phase II EAS depends on the potential problems discovered during the Phase I EAS. When health and safety issues are involved, the construction manager should err on the side of caution.

C. *Site and Existing Building Survey Hazardous and Contaminated Substances*

Before construction is performed in, or before the demolition of, buildings known or thought to contain mold, asbestos, lead paint, PCBs, and other hazardous or contaminated substances, an environmental consultant should be retained to survey and test for hazardous materials and develop a remediation plan. An owner either knows or should know that these substances are in its building and should assume the risk associated with abatement or cleanup. It is only fair since the owner is going to obtain the greatest benefit from the construction project. The construction manager, on the other hand, should strenuously resist becoming directly involved in retaining the environmental consultant or the testing agencies. The construction manager's insurance program may prohibit such activity. In addition, many of these consultants are relatively small businesses with levels of errors and omission insurance of $1 million or $2 million and limited assets. If there is a problem, it will usually carry unpleasant publicity and could place the construction manager's assets at risk.

D. *Geotechnical Investigation*

The geotechnical investigations are conducted early in the pre-design phase at a time before the initial budget has been established and when most owners wish to minimize costs. Geotechnical investigations involve high risk, and many architects insist that the owner contract directly with the geotechnical engineer. At a minimum, the geotechnical reports should include soil borings, soil investigation, determination of ground water level, and a recommendation of the foundation system or systems that could be used. At best, this investigation is a snapshot of the conditions observed.

Geotechnical investigations frequently miss changes in soil characteristics, presence of rock on portions of the site, environmental soil conditions, buried foundations, and other abandoned materials and equipment. Construction managers usually will insist on contract language that allows them to submit a Type I or Type II claim for differing subsurface conditions such as the language found in American Institute of Architects (AIA) Document A201-1997, § 4.3.4.[11]

Agreements executed by geotechnical engineers frequently include a limitation of liability that will not go very far in rebuilding a structure that sustains damages as a result of errors or omissions. Even though this is clearly a design function, some owners will ask the construction manager to engage the geotechnical engineer and/or the boring or testing agency. Construction managers are not design professionals and

---

11. Differing Site Conditions, 48 C.F.R. § 52.236-2 (2000).

should be even more insistent than architects in avoiding any involvement in geotechnical investigations. Construction managers also need to be careful not to assume risk for subsurface conditions through the backdoor by value engineering—the recommendations made by the geotechnical engineer. Any such recommendation should be conditioned on written approval by the geotechnical engineer—not the architect.

E. *Funding*

The funding source of a construction project is typically the most time-sensitive and integral component of the project plan. During the pre-design phase, the construction manager can assist the owner with developing a project budget, including identifying creative strategies for financing the construction of the project. Engaging a construction manager who has a good reputation and relationship with the lender(s) is often factored into a decision by the owner to engage the construction manager during pre-design and/or pre-construction. However, construction managers are typically in the construction business and are not developers. During the early 1980s, some construction managers decided to enter into speculative real estate development; generally, they did not obtain the results they expected. Construction managers have different risk tolerance than owners and developers, and except for captive construction managers owned and operated by developers or owners, construction managers should not venture into speculative real estate development.

During the funding process, owners are asked to execute completion and environmental guarantees. Occasionally, the construction manager may be asked by the owner to execute a completion guarantee. On the surface, the risk associated with a completion guarantee appears to be similar to that associated with a performance bond. However, by executing a completion guarantee, the guarantor agrees to complete the project and assume the risk for differing conditions, excusable delays, design defects, and other costs that otherwise would entitle the construction manager to a change order and payment outside the lump sum or guaranteed maximum price. Unless offset by significant fee enhancement or an equity interest in the project, this risk is inappropriate for a construction manager operating under a traditional fee arrangement to assume.

F. *Land Acquisition*

Land acquisition, even though part of the pre-design phase, is usually handled solely by the owner. The construction manager may be asked to prepare budget estimates for a series of potential sites. Further involvement in site acquisition would be limited to situations where the construction manager also operates as the design builder. In most instances, the owner has procured the land for erecting the structure, or purchased the building it desires to renovate. In the event that the construction manager is retained at the point of project creation, the construction manager may be able to introduce the owner to developers and consultants that specialize in alternative or creative solutions for land acquisition through partnerships with government or private entities. For instance, a Fortune 500 company that wants to build a facility may be able to partner with a quasi-public agency to build and lease back the facility with tax incentives. Public-private partnerships can maximize the owner's return on their investment with minimal risk.

G. *Engaging Design Professional*

The relationship between the owner and design professional can be difficult to define because the design professional can serve three different roles. Understanding the dynamics and the contractual relationship between the owner and design professional should be a priority of the construction manager. Therefore, the construction manager will be able to better define its role in the pre-construction phase of the project.

First, the design professional can be retained as an independent contractor to prepare the design. In this capacity, the design professional is responsible for preparing the construction drawings and specifications and limits its duties to managing the design aspects of the project. The design professional remains liable for errors and omissions in performance of its services.

Second, the design professional can be retained as an agent of the owner in performing inspection services and contract administration. In this capacity, the design professional would be responsible, in addition to preparing the construction documents, for observing the progress of construction work, certifying payment, and reporting to the owner. Further, the design professional has express or implied authority to negotiate on the owner's behalf with respect to modifications in the construction agreement.

Finally, the design professional can be retained as an independent arbitrator of disputes between the owner and contractor. In this capacity, the design professional may be asked to interpret the contractual rights and duties of the owner or contractor under the owner-contractor agreement.[12]

In most project scenarios, the design professional's role is a combination of the roles discussed above. Selecting a design professional is one of the owner's most important decisions and has the greatest impact on the project's success. Specifically, the owner will have to rely on the design professional to transfer the owner's vision onto the construction drawings. If the design professional is unable to prepare the drawings in a timely fashion, the success of the project could be in jeopardy.

Depending on the scope and size of the project, the owner may use a "direct selection" method.[13] In this instance, the design professional may be personally referred by a friend or the design professional's reputation in the region may make him or her a viable choice. For large projects or government-funded projects, the

---

12. There is a movement within the American Institute of Architects (AIA) to remove the architect from this responsibility in the 2007 version of the AIA documents and substitute an independent neutral chosen by the parties.

13. IRVIN E. RICHTER & ROY S. MITCHELL, HANDBOOK OF CONSTRUCTION LAW AND CLAIMS 40–43 (1982).

best method of selecting the design professional would be through a Request for Proposal (RFP) method.

In relatively few instances, the design professional and construction manager may be the same entity. In particular, design professionals have traditionally held a high level of expertise in project management, sometimes including project scheduling, quality control, and contractor payment review.[14] Consequently, if the design professional assumes the responsibility for the design as well as the project coordination, the owner-architect agreement should clearly separate the scope of activities between situations when the design professional is acting as just a design professional and situations when the design professional is acting as a construction manager. If the design professional wears the construction manager hat, the relationship between the parties of the project team must be defined through the use of explicit contract language.[15] The inherent problem with combination organizations is that it is difficult to separate the design professional's professional liability from the liability for construction means and methods that would be assumed by the construction manager.

Owners can use project-specific contracts or pre-printed forms provided by the AIA[16] or the Associated General Contractors of America (AGC).[17] The AIA and AGC publish a family of contract documents that can be used as a starting point of contract negotiations. As a standard form, these agreements are designed to establish a relationship between the parties in a standard situation. However, every project is unique and modifications of the documents may be required and, in most cases, recommended.

The construction manager rarely gets deeply involved in the selection of the design professional. Unofficial recommendations may be solicited if the construction manager has been selected or has been engaged. In the vast majority of projects, design professionals are selected before the construction manager is selected. As part of the process, the design professional and owner should discuss the owner's expectations, budgetary requirements, and timing, as well as the impact that budget creep may have on the feasibility of the project. If already engaged, the construction manager should also be involved in this process.

H. *Adjacent Site Investigation*

The design may be dependent on an evaluation of adjacent properties. This is particularly true in a high-density urban environment. The construction manager may be asked to participate in nondestructive testing and recording of existing

---

14. *See generally* AIA B141, Standard Form of Agreement Between Owner and Architect ¶ 2.6.
15. *See* www.aia.org; www.agc.org.
16. www.aia.org.
17. www.agc.org.

conditions. Before performing such services, it should be clear in the construction management agreement that the construction manager is performing these services at the direction and under the supervision of the design professional. The construction manager does not want to share responsibility for errors and omissions related to the interpretation of information provided to the construction manager by the design professional. The information collected should be limited to data and the construction manager's employees should not, even if requested, offer opinions that could be construed as design related.

There is also a separate nondesign-related investigation of adjacent properties that is often made by construction managers. This investigation involves a recording of existing conditions prior to construction activities in an effort to minimize frivolous claims by adjacent property owners of damage caused by the construction manager's operations. This investigation usually is performed as part of the preconstruction services.

I. *Utility Service*

Frequently, construction managers are asked to take part in evaluating the suitability of the utilities that service the site. Again, this is a design issue and the construction manager's responsibilities should be limited to locating existing utilities rather than offering opinions as to whether or not they will support the proposed project. Additionally, construction managers should not take the lead in negotiations with utilities. There are many examples of projects that were substantially completed but were still waiting for the public utilities to complete the service connection to the new construction project. The owner should take the lead and should maintain responsibility for such negotiations.

Library References

C.J.S. *Health and Environment* §§ 105–107, 110, 112–129.
West's Key No. Digests, Environmental Law <KEY>579–611.

IV. Design

A. *Phases*

The first step in any design is developing the owner's program. During the program phase, the owner, and on larger projects the architect and the construction manager, establishes the project objectives to be achieved and the parameters within which those objectives will be satisfied. The owner may develop its program internally or retain a program manager. The program manager may be a consultant that specializes in the owner's planned facilities. Program management is not a delivery system and differs from project management. In particular, a program manager should be an extension of the owner's in-house staff. The program manager's scope of services varies based on the needs of the owner. Typically, there are at least five phases of services that define the program manager's scope of services: program planning, design, construction, occupancy, and facilities management for one or more discrete projects. Depending on the owner's needs, additional services may be needed to accomplish the owner's objectives.

A program manager is particularly useful where the owner is building multiple facilities or a new headquarters. The owner's program should include a description of the owner's short-term and long-term objectives, the owner's financial and budgetary needs, the owner's space requirements, and the owner's scheduling priorities. This is by no means an all-inclusive list. Unlike the construction manager's role on a discrete project, the program manager is responsible for the overall development and coordination of multiple services.

Although the design professional retained by the owner will be responsible for the final design, the construction manager's input is critical to the overall success of the final design. Consequently, the construction manager should have a comprehensive understanding of the project's design requirements.

The construction manager should also appreciate the scope of the design professional's role during pre-construction and design development stages. Specifically, the construction manager can provide material pricing and availability of the proposed building systems. For example, if the project team learns that there may be a steel shortage in the upcoming months of the project, the construction manager may suggest an alternate building material to accomplish the owner's instructions. In the event that the owner does not want to use an alternate building material, the construction manager can assist the owner with revising the long-term project schedule.

The construction manager should incorporate the following list of questions into its initial program discussions with the owner:

- What is the target completion date? (i.e., When does the project need to be completed?)
- Who is responsible for making decisions? (i.e., defining the approval authorities)
- How will project information be disseminated among the parties?
- Who is responsible for maintaining the project schedule?

The construction manager on large or multi-site projects may be called on to fulfill the dual responsibility of program manager. After the program has been established and the design professionals have been engaged by the owner, the design process is customarily broken down by architects, and sometimes engineers, into three phases: schematic design; design development; and construction documents.

As a project team member, the construction manager will, in each phase of the design, be involved in preparing an estimate, preparing or updating the progress schedule, value engineering, and reviewing constructability. The construction manager's responsibility is to assist the design professionals and the owner without assuming or usurping design responsibility and liability. This is an iterative process and should not be conducted in isolation.

1. *Schematic Design Drawings*

Schematic design is the process of transforming the owner's program into drawings which illustrate the scale and relationship between various project components. At the beginning of the schematic design phase, a pre-design conference and planning session is usually conducted between the owner and the design professional. Preferably, if the owner has retained a construction manager, the construction manager will also attend the pre-design conference and planning session. Other attendees may include the program manager (if not the construction manager), the owner's other consultants, and the architect's consultants, as may be appropriate. One of the objectives in the meeting should be to develop the basis on which the schematic design will be developed. Another objective for the meeting should be development and review of the preliminary design schedule.

In this phase, a detailed code and life safety analysis of applicable federal, state, and local jurisdictional building codes should be conducted. Also, it is recommended that the owner and the design professional coordinate introductory meetings with local government officials, including the local zoning board, to confirm that the project will satisfy all legal and project requirements.

At a minimum, the design professional should provide the following documentation for the owner's review: site plan showing building, parking, and grades; exterior concepts and image boards to communicate the design intent; architectural site plan; floor plans of all floors, including the proposed structural grid; summary of applicable building codes; proposed exterior finishes with color concept perspectives; horizontal control dimensions; and proposed finished floor elevations. The purpose of the schematic design phase is to provide the owner with an opportunity to confirm whether the development of the design is consistent with the owner's vision. Additionally, changes at this level of the design have minimal impact on the overall cost of the project.

Throughout the schematic design phase, the design professional should conduct weekly or biweekly formal design review meetings with other members of the project team and with the owner to review the status of the schematic design and design parameters, and to incorporate feedback from the owner and other team members. When the schematic design has reached a point where input is required from other disciplines, the design professional should retain consultants to develop the structural system, the site plan, the mechanical plans, and the electrical plans. These consultants usually include a structural engineer, a civil engineer, and a mechanical engineer to generate the supplemental design drawings.

The construction manager is responsible for coordinating and updating the design schedule. In addition, the construction manager may be asked to perform quick estimates to be used as part of a decision-making process in selecting systems to be incorporated into the design. Value engineering and constructability analysis are invoked on an as-needed basis. After the owner has approved the schematic design, the first formal estimate beyond the preliminary estimate will be prepared along with a more detailed progress schedule for completion of the design and an updated construction schedule for the owner's review and approval.

2. *Design Development Drawings*

During the design development phase, the schematic design is further developed through the use of larger scale plans as well as elevations, sections, and details. The project team should continue to have periodic design review meetings to review design progress and to solicit input from team members. The construction manager will review the overall project schedule and conduct a design scheduling meeting to coordinate the completion of the design with the construction sequencing plan.

The final design development package should include all of the relevant drawings and specifications to describe the design intent and finishes adequately. The construction manager should perform value engineering and constructability review during the design development phase. Any changes to the design, including additions or deletions, resolved during the design development phase, will have a minimal adverse impact to the overall project budget. After the design development drawings and outline specifications have been completed and approved by the owner, another formal estimate will be prepared by the construction manager along with a more detailed design and construction progress schedule for the owner's review and approval. This estimate will be compared with the schematic design phase estimate, and adjustments will be made, if required, to the construction documents. Design development drawings and specifications are, on average, forty percent to fifty percent complete construction documents. The owner should approve the construction manager's estimate and the updated progress schedule at the commencement of the construction document phase.

3. *Construction Drawings and Specifications (Construction Documents)*

The construction documents are intended to convey, in as much detail as necessary, the graphic and quantitative information required by the trade contractors to perform the construction work. Design review meetings should continue during the construction document phase to discuss and resolve any outstanding issues with respect to the design. The construction manager should review the overall project schedule and conduct coordination meetings with the design professionals to make sure the remaining design activities are in compliance with and integrated into the construction schedule.

At the agreed upon point during the construction document phase (which should be in excess of seventy-five percent complete, as determined by the construction manager), a complete set of documents should be delivered to the construction manager. The construction manager will then assemble the bid packages and solicit bids from trade contractors, which will be used to serve as the basis for the guaranteed maximum price, or if the construction manager is to operate as an agent of the owner or the project is to be completed on cost-plus basis, the final formal estimate. The bid documents should include, at a minimum, the design drawings and specifications, instructions to bidders, and other relevant documents necessary to assist the construction manager in obtaining bids from trade contractors.

B. *Estimate*

One of the construction manager's primary tasks during pre-construction is to provide construction cost information so that the owner can complete and update its cost-benefit analysis and have an understanding of the return expected from its investment. Construction managers are closer to the changes that take place in the construction marketplace and they provide the owner with a reality check. Sometimes, sophisticated owners will employ a separate cost consultant to perform a parallel estimate for comparison with the construction manager's cost estimate. Those owners who infrequently perform construction are usually the most out of touch with the actual costs. This is a very high-risk process because the owner's expectations upon which preliminary decisions are made may not be realistic. Owners who have become overly enamored with their project do not react favorably to the reality check provided by a major cost estimate.

An estimate is a rough or approximate calculation based upon the information available. Owners sometimes overlook the fact that the estimate is approximate and that an estimate merely represents the best judgment of the construction manager as to what the construction market will be when the project is eventually released to trade contractors for bid. Although providing estimates is an everyday occurrence for construction managers, they need to be aware of the dilemma faced by the owner's managers who discover they have to return to their senior management with a scaled back plan, or even worse, drop the plan entirely. An owner's desire to build a landmark or win an architectural award may no longer be possible. Most owners will treat this as a business setback that has to be managed and overcome. However, the construction manager needs to protect itself from those few owners who would rather shoot the messenger than accept defeat. Because of the importance of estimates, this section goes into some detail and can be skimmed by those familiar with the estimating process performed by construction managers.

Estimates should contain more than just a number. For example, there should be a narrative description of the project and document list, elements of the project should be quantified, assumptions and qualification used to prepare the estimate should be recorded, construction phasing assumed by estimator should be recorded, effect of means and methods on cost and schedule should be recorded, contract requirements that may affect the contract and schedule should be addressed, and a comparison should be made with the previous estimate including an analysis of scope changes that may be included in the latest estimate. Formal estimates consume the vast majority of costs associated with pre-construction and the construction manager should limit its responsibility to an agreed upon specific number of estimates. For example, the construction manager may agree to deliver one detailed estimate each (1) upon drawings approved by the owner at the completion of schematic design; (2) upon drawings approved by the owner at completion of design development;

(3) at sixty percent complete construction documents; and (4) for a guaranteed maximum price at ninety percent buyout of major trades. Without a defined duration and number of estimates, the construction manager could be liable for providing repeated costly estimates until the owner is satisfied.

Program and schematic estimates are prepared before a great deal of information is available. At this point in the design process, major decisions on the location of the project, systems to be employed, and area required may not be finalized. Preliminary estimates are also known as *order of magnitude*, *square foot*, or *conceptual* estimates. It is common for preliminary estimates to examine several alternatives so that cost can be included as part of decision-making process. It is a good idea before preparing the preliminary estimate to conduct a project charrette with other project team members to vet the project scope and review details that are being discussed among designers and/or the owner but have not yet been incorporated into the design. The results of a charrette can greatly increase the accuracy of the preliminary estimate. Preliminary estimates utilize historical data and may be based upon floor area, the number of cars in a parking garage, the number of rooms in a hotel, the number and type of units in multifamily housing, or the number of beds in a hospital. Preliminary estimates may also be based on historical data for similar projects to which parametric ratios and cost indices (escalation) are applied. If additional systems information is available, unit or assembly prices for the structure, exterior walls, roofing, and major mechanical and electrical equipment may be factored into the estimate. There should be a low expectation of accuracy commensurate with the level of information available. Preliminary estimates based on early design information generally have an average accuracy range within twenty percent.

Design development and early construction document estimates are based on quantity survey or drawing take-offs and review of the outline specifications. Assumptions are made to fill in the missing design information, and the estimate includes narrowed assumptions and qualifications. For buildings, foundation design may be assumed from available data. A preliminary finish schedule reviewed with the owner and architect is a valuable tool for establishing the quality of materials that will be required. Labor availability and other market conditions should be reviewed for cost impact. Design contingencies should be applied to cover further development of the design. A construction contingency should be included (outside the design contingency) for the exclusive use of the construction manager during the construction. Input from major trade contractors may also be included as part of the estimate. Allowances[18] may be included for elements that have not been clearly defined such as millwork, carpet, and light fixtures.

---

18. Allowances are used where there is doubt as to material selection or insufficient information to price the element. An estimate to furnish and/or install certain construction elements is included as part of the construction manager's estimate. The owner has the right to specify the materials, equipment, or system to be installed and is responsible for costs incurred by the construction manager in excess of the allowance.

The final estimate or guaranteed maximum price (GMP) estimate, if there is one, is prepared after the construction documents are at least seventy-five percent complete. The construction manager's risk associated with a GMP is reduced by the percentage of trades that can be bought out before the GMP is prepared. As part of the GMP process, the construction manager is required to make assumptions for the uncompleted portion of the design, which is reasonably inferable from the construction documents. However, the GMP should not include changes in scope, systems, kinds, quality levels, quantities of materials, finishes, or equipment from those shown on construction drawings and specifications, or reasonably inferable from them. Some or all of these would be covered by the design contingency held by the owner outside of the GMP. The construction contingency that would be established by the construction manager in the GMP is intended for unanticipated costs encountered by the construction manager during construction such as unanticipated market conditions, buyout overruns, subcontractor default, schedule recovery for unexcused delay not recoverable from subcontractors, and other events outside the reasonable control of the construction manager. Detailed scope definitions should be prepared for each trade and included in bid packages sent to pre-qualified subcontractors (unless prohibited by law for public projects). Allowances may be established by the architect or suggested by bidders. Except for work that may be self-performed, costs are developed from bids submitted by subcontractors and suppliers. The construction manager's experience may indicate that there are certain risks that may not be included in a subcontractor's bid prices which the construction manager assigns to holds, i.e., temporary heat, de-watering, etc. A hold is part of the estimated direct construction costs, and the construction manager assumes the risk of an overrun. With an allowance, the owner assumes the risk of an overrun. The assumptions and qualifications included in the proposal should be further narrowed and coordinated with assumptions and qualifications received from bidders. The GMP should include the following[19]:

- a list identifying drawings and specifications, addenda, and other documents used in preparation of the GMP;
- a list of allowances and definition of materials and labor included in such allowances;
- the construction manager's assumptions and qualifications;
- a breakdown of the GMP into costs of the work (trade categories, holds, general conditions, insurance) and the construction manager's fee;
- alternates included in the GMP and those available (with cut-off dates) for subsequent acceptance by the owner;
- the date of substantial completion and other milestones (if applicable); and
- proposed incentives (if any) for the construction manager's performance.

---

19. *See* AIA Document A121/CMc-2003, § 2.2.

The GMP should always include an expiration date. If the owner does not accept the proposal by that date, it will be withdrawn by the construction manager. If the GMP is not accepted by the owner in a timely manner, the construction manager's participation in the project will come to an end. On the other hand, some owner-generated agreements provide that a rejection of the GMP will not terminate the contract, and the construction manager shall instead complete the project on a cost-plus and agreed upon fee basis. Securing the owner's acceptance of the GMP usually includes negotiation and, when there are budget pressures, scope reduction or value engineering. Issues that are related to the performance of construction under a GMP are outside the scope of this chapter.

C. *Project Progress Scheduling*

Time is another area of considerable risk to a construction manager. On a private project, the owner's return on investment is dependent on a specific completion date. Public projects such as a school must be open for the beginning of the school year. Most construction phase agreements and some pre-construction agreements contain a *time is of the essence* provision. If time is of the essence, a delay caused by the fault of the construction manager would constitute a breach of contract and could result in the construction manager's being liable for damages incurred by the owner as a result of the delay. If there is no time is of the essence provision, reasonableness will be the guide when measuring the timeliness of the services performed by the construction manager. Scheduling can be defined as planning the order (logic) in which construction will be performed, breaking down major tasks into manageable activities, and assigning a time for each activity. Scheduling is not an exact science, but there is an industry standard for coordinating the individual tasks of any project. The most widely accepted scheduling method is the Critical Path Method (CPM), which has been used in the construction industry for more than thirty years. CPM scheduling is a technique that breaks down the entire work into individual activities and requires an analysis of the number of days required to perform each individual activity. The logic applied to a CPM project schedule depends on the interrelationship between various activities, i.e., which activity or activities must be started or possibly completed before the next activity can start. In addition to assigning a reasonable duration to an activity, resources (labor and materials) may also be applied to each activity. The CPM is a representation of a project plan in a schematic diagram or network that depicts the sequences and the interdependent relationships of all of the activities of the project.[20] Preparing a schedule forces the construction manager and other participants to understand the entire project and to visualize the steps that must be taken to complete the project. The shortest path of interrelated activities found in the critical path network (the critical path), predicts when the project should be substantially complete.

---

20. *See* Continental Consol. Corp., ENGBCA No. 2743 et al., 67-2 BCA, ¶ 6624 (1967); *see also* Dobson v. Rutgers, 157 N.J. Super. 357, 384 A.2d 1121 (1978) (citing to ANTILL WOODHEAD, CRITICAL PATH METHODS IN CONSTRUCTION PRACTICE (2d ed. 2005)).

Substantial completion[21] is the major goal of most construction projects since the start of owner occupancy usually begins at substantial completion. Besides completion of the work, final completion includes delivery of record drawings, maintenance manuals, waivers, and other documentation that takes time to assemble but does not really affect physical completion. Some public owners have instituted onerous final completion provisions that seemingly can take longer than the physical construction period to complete. During the development of a schedule, what-if scenarios are explored in an effort to develop an optimum plan. Float is the difference in time between when an activity should be completed and when the activity must be completed to avoid an extension of the critical path. Activities on the critical path do not have float.

One of the important issues in scheduling is who controls the float. During preconstruction, the CPM schedule usually is used to gauge progress. During construction, the CPM schedule is a forecast of whether or not the construction manager will complete the project on time. Control of float becomes a liability issue if the construction manager has contractual responsibility for completing the project in an agreed upon time period. If the construction manager has assumed this risk, the construction manager should have control over its own destiny. The CPM can also be used to ensure that all of the critical construction items are completed in a timely manner and to identify tasks on the critical path that must be completed within the allotted time and cannot be delayed.[22] However, construction managers should be careful not to assume responsibility for meeting multiple milestone dates which effectively prohibit the use of available float and pass control of float to the owner. After the CPM schedule is in place, it enters the maintenance phase and is used to measure actual against planned performance.

D. *Value Engineering*

Value Engineering (VE) was adopted by construction industry to differentiate the service performed by the construction manager from that of the design professional. It was originally developed by General Electric Corp. during World War II and is defined as analysis of programs, services, and supplies performed by the construction manager that are directed at improving performance, reliability, quality, safety, and life cycle costs. Most owners have a much more simplistic view of VE as a method of reducing costs without unduly changing the scope or intent of the design.

When the construction manager is brought on board early in the design process, the construction manager's senior operations staff can provide valuable insight into the use of alternate systems, products, or equipment without disrupting the design process. The experience level of the staff assigned to VE and the breadth of projects

---

21. Defined in AIA A201-1997, General Conditions of Contract for Construction, § 9.8.1, as "the stage in the progress of the Work when the Work or designated portion thereof is sufficiently complete in accordance with the Contract Documents so that the Owner can occupy or utilize the Work for its intended use."

22. Peter Kiewit Sons Co. v. Iowa S. Utils. Co., 355 F. Supp. 376 (S.D. Iowa 1973).

constructed by the construction manager coupled with the architect's knowledge and experience will have a synergistic effect on the design process. Applying VE when construction documents are nearing completion usually is not as effective because the design is relatively fixed and significant changes are difficult or expensive to accomplish.

If at the time the construction manager joins the project team, the first estimate is above the owner's budget—a very common occurrence—engaging in value engineering may not reduce the cost to the degree required. At this time, the construction manager may be required to mix scope reduction into the VE suggestions. The Department of Defense attempted to go even farther by institutionalizing the VE process into the procurement process through the use of a Value Engineering Change Proposal (VECP) that required contract modifications be tied to VE proposals and provided an incentive to the contractor or construction manager to reduce costs. The VECP process apparently went too far and was not widely used and finally abandoned by the Department of Defense. The incentive to employ VE enjoyed by construction managers is to assist the owner in holding the budget in check so that the construction manager can successfully move from pre-construction to construction.

The construction manager is not the design professional and should be careful to limit its role to making suggestions that can be either approved or rejected by the architect and owner. The construction manager does not want to step over the line and assume design responsibility and liability, and should defer the final decision to the architect. When VE includes a significant modification to the design, the construction manager should insist that these changes be reflected in the architect's construction documents.

E. *Constructability*

The term *constructability* according to the Construction Industry Institute is defined as "the integration of construction knowledge and experience in planning, design, procurement, and construction phases of projects consistent with the overall project objectives." Constructability is achieved through the effective and timely integration of construction input into planning and design as well as field operations. For maximum effect, it should be initiated early in the design process and performed at predefined points during the design process in accordance with a well thought-out plan.

Construction managers may place constructability review milestones on the design schedule to take place in the schematic design phase, the design development phase, and the construction document phase. The reviews may be broken down further by design disciplines; for example, site, architectural, structural, mechanical, and electrical. The construction manager's constructability review is intended to improve coordination among the construction design documents, identify potential construction problems, and mitigate claims that may flow from the problems identified. Constructability should focus on the relationship between the drawings and specifications prepared by the design professional and the means and methods to be employed by the construction manager during construction.

Constructability review usually involves a committee appointed by the construction manager and includes experienced senior level staff that may not be assigned to the project. The review has to be tailored to each design phase. For example, the schematic phase may concentrate on the site, selection of materials, and location of utilities.

During the schematic and design development constructability reviews, the committee should take into consideration that the design is a work in progress and should not focus on spelling and other obvious errors that will be cleaned up as the design progresses. The review for the design development phase may focus on systems to be employed, including the foundation, structure, exterior wall, roof, and major mechanical and electrical systems. The review of the construction documents should be completed before the seventy-five percent completion and should focus on space limitations, tolerances, connections with existing structures, potential leaks, construction difficulty, and coordination of the drawings and specifications.

The construction manager is not the licensed architect or engineer for the project and should avoid comments in the constructability reports that would more properly be included in a peer review by design professionals. Problems noted by the construction manager in the design can instead be geared toward industry standards, previous construction experience with similar designs, and previous experience related to re-work or warranty issues. Unless a construction manager is careful in preparing its comments, constructability review can shift design responsibility to the construction manager. There is published project execution plan data which includes constructability as an element.[23]

F. *Site Logistics*

As part of pre-construction, the construction manager should develop the site logistics plan which relates the construction sequencing envisioned for the project with site safety, security, and the movement of labor and materials. Depending on site constraints, there may be several alternates developed before the final plan is selected. The site logistics plan may include the following: (1) protection of adjacent property, roads, and utilities; (2) location of construction fence and gates; (3) crane locations and crane erection path; (4) location and type of sidewalk bridges required for pedestrian protection; (5) temporary storage of materials and equipment; (6) location of hoists and lifts; and (7) temporary roads, staging areas, parking, and location of temporary offices and storage trailers. The safety of public and on-site construction labor is of paramount concern when developing the site logistics plan.

The site logistics plan should take into consideration site constraints that may affect the plan selected, such as (1) hazardous materials or contaminated soil located

---

23. CONSTRUCTION INDUSTRY INSTITUTE, GUIDELINE FOR IMPLEMENTATION OF CII CONCEPTS: BEST PRACTICES FOR THE CONSTRUCTION INDUSTRY 12–25 (Sept. 1995). For a comprehensive discussion on constructability with respect to architect/engineer professional liability claims see P. Douglas Folk, *Constructability Reviews: An Effective Tool for Improving Construction Documents and Reducing Claims*, CONSTRUCTION BRIEFINGS No. 2006-04, at 1 (2006).

on the site; (2) noise restrictions and work hours; (3) dust protection; (4) storm water run-off; (5) restrictions on deliveries or movement of materials and debris from the site; (6) existing businesses to continue operation during construction; and (7) potential neighborhood opposition or complaints. The care and quality exhibited by the construction manager in developing and selecting the site logistics plan are very important to establish a good basis for project safety as well as reducing the impact that the project will have on the neighborhood.

G. *Green Building and Related Compliance*

There could be liability for the construction manager hidden in the rush by the construction industry toward the laudable goal of sustainable building construction. A good rule of thumb for contract review is never to include (or skip over) unfamiliar terms to you. Therefore, it is important for the person reviewing and measuring the risk in pre-construction agreements to understand the green building certification program. The United States Green Building Council (USGBC) is an organization that promotes environmentally responsive and sustainable design, and has developed the LEED® (Leadership in Energy and Environmental Design), a system for energy conservation, waste reduction, and sustainability in the design and construction of buildings. Owners who will occupy their buildings may be focused on their investment return while developers who will sell the building or seek a triple net lease may be more focused on the marketing enhancement that LEED® will bring to the project.[24] LEED® criteria exist for new construction (NC), renovation of existing buildings (EB), and commercial interior fit-out (CI). Pilot certification programs are being developed for core and shell projects (CS), homes (H), and neighborhood development (ND). Certification levels by USGBC are designated certified, silver, gold, and platinum. The number of LEED® credits required to achieve certification depends on the project classification. For example, with LEED®-NC (public, institutional, and private new construction), simple certification requires 26-32 points, silver requires 33-38 points, gold requires 39-51 points, and platinum requires 52-69 points. Points for LEED®-NC are assigned depending on the project goals achieved for each of the prerequisites:[25]

1. Sustainable Sites (14 points)—develop only appropriate sites, reuse existing buildings and/or site, protect natural and agricultural areas, reduce need for automobile use, and protect and/or restore natural sites.
2. Water Efficiency (5 points)—reduce the quantity used by the building and reduce the burden on the municipal water supply and treatment.

---

24. Scott Lewis, *Succeed at LEED: Focus on LEED Process*, ENVIRONMENTAL DESIGN & CONSTRUCTION, *available at* http://www.edcmag.com/CDA/Articles/Leed/04c937e14c697010Vgn VCM100000f932a8c0 (July 1, 2004) (registration required).

25. U.S. GREEN BUILDING COUNCIL, LEED—NC, *at* www.usgbc.org/file.asp?DocumentID= 1317 (2005).

3. Energy and Atmosphere (17 points)—establish energy efficiency and system performance, optimize energy efficiency, encourage renewable and alternative energy sources, and support ozone protection protocols.
4. Materials and Resources (13 points)—use materials with less environmental impact, reduce and manage waste, reduce the amount of materials needed.
5. Indoor Environmental Quality (15 points)—establish good indoor air quality; eliminate, reduce, and manage the sources of indoor air pollution; ensure thermal comfort and system controllability; and provide for occupant connection to the outdoor environment.
6. Innovation in Design (5 points)—additional points are available for exceptional performance and innovation not otherwise specifically recognized by LEED®-NC.

Construction managers are marketing their commitment to and experience with LEED® as a qualification credential. Like any other risk in the construction arena, owners will seek to pass the risk of falling to achieve certification status to their construction managers and construction managers will either push back (negotiate a level playing field) and/or pass the risk through to their subcontractors. Owners may seek a warranty from the construction manager that certification will be achieved or that the savings predicted for the additional investment will be realized. At least one commentator has even suggested liquidated damages linked to energy performance.[26]

There is no established body of case law to provide guidance and these relatively new risks are difficult for the construction manager and the owner to quantify. If the owner's perceived expectations are not realized, the construction manager may find itself saddled with extensive corrective work or default by a key subcontractor or supplier could leave the construction manager with no cost-effective method to achieve the required certification points.[27] Constructing a more environmentally friendly building is good for everyone and it should not be derailed by overreaching and misplaced risk transfer. Construction managers should only be asked to assume the risk for matters within their control. Care should be taken to assure that the LEED® certification requirements do not amount to a performance specification that transfers both design and construction liability to the construction manager.

Construction managers should avoid warranting or guaranteeing certification since achieving that goal is heavily dependent on actions being taken by the owner and the design professionals. A careful review of the construction manager's marketing materials may avoid an unexpected implied warranty. Construction managers should

---

26. Ujjval K. Vyas, *Making the Case: Including Legal Issues in the Green Building Agenda*, U.S. GREEN BUILDING COUNCIL GREEN BYTES (Chicago Chapter), *available at* http://chapters.usgbc.org/chicago/docs/Resources/GreenBytes/GreenBytes05_05.pdf (May 2005).

27. Ujjval K. Vyas, *A Greener Landscape on the Horizon*, AE/PRONET, *at* http://www.aepronet.org/ge/no32.html (2005).

also seek protection from owners for the implementation of innovative designs (additional points) that may be difficult or impossible to attain.

Library References

C.J.S. *Architects* §§ 16, 19, 21–23; C.J.S. *Contracts* §§ 11, 343, 359, 579–581, 583. West's Key No. Digests, Contracts <KEY>199, 205.25, 205.35, 211, 213(2).

V. Pre-Construction

A. *Long-Lead Materials and Equipment*

The construction manager, with the design professional's assistance, will prepare a list of materials and equipment which may have long-lead times and should be purchased as soon as possible so as not to cause delay. Examples of typical long-lead items are structural steel, electrical switchgear, emergency generators, unique window wall systems, boilers, and chilled water systems. The construction manager may purchase long-lead items either as an agent of the owner or directly. As an agent of the owner, the owner enters into purchase orders directly with the required suppliers and the construction manager manages the procurement and expediting process for the owner. On commencement of construction by the construction manager, the owner's purchase orders may be assigned to the construction manager.

B. *Permits and Approvals*

The construction manager should make sure that it does not have responsibility for zoning, tax abatement, and land use review which should be the design professional's and/or the owner's responsibility. The design professional should submit drawings for review as soon as possible during the construction document phase and earlier if permitted by the local building department. The design professional's interpretation of the building code and other applicable rules and regulations may not be accepted by the building department. These issues should be addressed as soon as possible to avoid the need for a major redesign.

The construction manager should avoid assuming responsibility for the building permit, which contains information available only to the owner. The construction manager should not be placed in the position of making representations concerning the information that is supplied by the owner. However, the construction manager may suggest that the owner retain a permit expediting service to facilitate approval of its construction drawings with the local authority. The construction manager can pull the permits, but the owner should file the application, engage the expeditor (if applicable), and pay the fee which may be substantial. The design professional should file necessary signed and sealed drawings and specifications as required by the building code applicable to the local jurisdiction where the project is located.

The construction manager needs to make sure that any differences between the filed set of signed and sealed drawings and specifications and the construction documents used for trade contractor bid packages are resolved by the design professional and the owner before commencement of construction. Otherwise, the work may be stopped by the building inspector if the work performed does not match that shown on the drawings and specifications filed as part of the building permit application process.

C. *Early Work*

Projects are almost always behind the owner's perceived schedule and, therefore, it is common for construction managers to be asked to perform certain construction work before the design has been completed and during the time when the construction manager is performing pre-construction services. This work may include abatement of hazardous materials, demolition, removal of contaminated soil, excavation, piles, foundations, erection of structural steel, etc., which are collectively referred to as "Early Work."[28]

The construction manager may perform Early Work either as an agent of the owner or at-risk. As an agent of the owner, the owner enters into contracts with required trade contractors and the construction manager manages the construction for the owner. Owners sometimes select the agency relationship for Early Work so that they are free to solicit proposals from other construction managers if negotiations for the construction phase with the pre-construction construction manager reach an impasse. On the other hand, some owners do not feel comfortable directly entering into contracts with trade contractors and ask the construction manager to perform Early Work on a cost-plus-a-fee (where the scope is not defined or a sequence of trade contract bid packages is to be issued) or fixed price basis (scope defined).

D. *GMP Estimate*

Part of pre-construction in many cases involves preparation and acceptance of the GMP estimate (covered in detail above).

Library References

C.J.S. *Contracts* §§ 380, 393–407.
West's Key No. Digests, Contracts <KEY>232, 241, 243.

VI. Compensation

A. *Consulting*

Construction managers are in the business of construction, and the consulting services provided during pre-construction afford the construction manager the opportunity to gain a competitive edge in the selection of the construction manager

---

28. When used herein, the term "Early Work" includes any physical construction work performed by the construction manager before entering into the construction agreement with the owner. Some of the other terms used by owners for this work are "Early Action Work," "Early Construction Work," etc.

for the construction phase. As a result, consulting services during pre-construction frequently are provided at cost without a fee for profit. If the construction manager believes the proposed construction project could be a financially attractive, the direct costs actually incurred by the construction manager may have no relation to the compensation. Preparing a GMP is usually the most costly portion of this process and reimbursement for some or all of these costs may be deferred until after the construction contract is awarded, i.e., included as part of the construction manager's general conditions contained in the GMP. As opposed to other services provided by the construction manager, the compensation for pre-construction services is not balanced against the risk assumed. The construction manager needs to ensure that the zeal to obtain a new project does not outweigh the diligence that should be applied to preparing and reviewing the pre-construction agreement.

The compensation methods discussed below are most commonly used by construction managers for pre-construction services.

1. *Time and Reimbursable Expenses*

If there is no defined scope, the services are usually performed on a timecard basis. The construction manager is reimbursed at agreed all-inclusive billing rates or for the actual salary and wages of the employees assigned to the project plus a multiplier applied to wages and salary covering insurance, benefits, and taxes. In addition to time, the construction manager is reimbursed for costs incurred such as reproduction of drawings and specifications (a major expense when soliciting prices from trade contractors), messenger service, overnight courier, out of town travel, etc.

2. *Not-to-Exceed*

Some owners feel more comfortable if there is a not-to-exceed price set for the construction manager's services. This method does not usually involve a guarantee and the construction manager performs its services on a timecard manner as discussed above but is not authorized to incur expenses beyond the not-to-exceed amount. Reimbursable expenses may or may not be included in the not-to-exceed amount.

3. *Fixed Price*

Fixed price means the construction manager has agreed to perform a predefined scope of services for a guaranteed cost. This is the most prevalent method used by private owners for short-term (three to eight months) pre-construction services. If compensation is less than costs incurred, the construction manager assumes the risk that the project will go forward. The construction manager's salaries and wages make up the largest portion of such services, and fixed prices should be conditioned on a fixed duration and fixed scope of services.

B. *Early Work*

Regardless of whether or not the construction manager is at-risk, the construction manager will almost always apply a fee for profit to Early Work. The construction manager is reimbursed on a cost-plus-a-fee, fixed price, or GMP basis. There is liability that may be assumed by the construction manager if the construction project is awarded to another construction manager. For example, the general conditions costs included as part of the Early Work may not be sufficient, coordination with another construction manager could increase the construction manager's staffing requirements, or the logistics plan adopted by the other construction manager may increase costs for Early Work subcontractors. An exit strategy should be in place and assigning Early Work subcontractors to the new construction manager should be a priority.

VII. Changes

A change is an alteration of the contract that modifies any or all of the following:

- the scope of work;
- the contract price; and/or
- the contract duration.

Changes within a pre-construction agreement usually involve additional services. Changes in such agreements become an issue only if compensation is based on GMP or fixed price. Before the construction manager can claim that the service is additional, there must be a clear definition of the services that are included in the contract.

A. *Consulting Additional Services*

To avoid a dispute with the owner, it is good practice to include in the agreement a listing of events that commonly give rise to additional services by the construction manager:

1. an extension in the duration of services to be provided;
2. a material change in the project, including, but not limited to, size, quality, complexity, the owner's schedule or budget, or procurement method;
3. performance failure by the owner, the owner's design professionals, or other consultants;
4. preparation for and attendance at a dispute resolution proceeding or a legal proceeding, except where the construction manager is a party thereto; and
5. change in services requested by owner beyond the services listed in the agreement.

For pre-construction services, the construction manager is performing tasks through employees who have an established cost. Anything that changes the assumptions as to the time required to perform the designated tasks should be the subject of additional compensation, i.e., change in services.

B. *Change Orders for Early Work*

A changes clause is essential when Early Work is included as part of the pre-construction agreement to control cost and risk allocation. The purpose of the changes clause is to provide a means to alter the scope of the Early Work (or services) or duration. All construction contracts should include provisions for preparing and executing documentation related to changes. In contracts containing a changes clause, the owner has the right to change its mind, add or delete work, or make such changes as it deems appropriate. Some changes are owner-initiated changes such as directives, change orders, or constructive changes. Alternatively, the changes can be initiated by the construction manager or design professionals.

The complexity of the change provision depends on the scope of the Early Work to be performed. For Early Work (as opposed to pure consulting pre-construction agreements), the change process is as follows: first, an information distribution or identification (e.g., Request for Information, Field Work Order, Field Condition Report, etc.) is provided to or initiated by the design professional. Second, the construction manager is requested to submit pricing and scheduling compilation for the proposed scope of work. Third, the submission is reviewed by the design professional and the owner. Fourth, the owner makes a decision and action based on review. Finally, three scenarios may arise: (1) the owner provides approval in the form of a written change order; (2) the owner allows the construction manager to proceed but disputes time or cost; or (3) the construction manager abandons the change request.

As stated above, the owner or design professional can initiate and define the scope of changed work and negotiate with the construction manager for an agreed price. If the parties are able to agree, then the change order will be issued. If the parties are unable to agree, then a Construction Change Directive (CCD) may be issued for the undisputed portion of cost/time proposal. For example, AIA Document A201-1997, General Conditions to the Contract for Construction, provides that the architect is required to prepare change orders and construction change directives. AIA Document A201-1997 also encourages the architect to obtain written agreement from the contractor that the proposed orders for minor changes will not result in claims for additional time or money.

VIII. Insurance

Insurance is an area of pre-construction agreements that is given little attention. Insurance is intended to transfer the risk of accidental loss from the construction manager to an insurance company. If not covered by insurance, any liability assumed by the construction manager is a potential loss. Insurance covers losses that flow from bodily injury or property damage, but does not cover losses resulting from cost overruns or defective work or services. Property insurance covers property that is owned by the insured. Builder's risk insurance covers the work performed prior to completion. The construction manager's personal property insurance covers the construction manager's personal property such as, furniture, tools, and equipment. Occurrence policies cover losses that occur during the policy term regardless of when the claim is made. Claims-made policies generally cover losses only for claims filed during the term of the policy and sometimes for an extended discovery period of one year or two years after the policy expires. The insurance required should be identified in the agreement and depends on the scope of services to be performed. At a minimum, every construction manager should have the following insurance coverage:

- Commercial General Liability (CGL) Insurance including Excess Liability;
- Commercial Automobile Liability Insurance; and
- Workers' Compensation Insurance.

For a more comprehensive discussion of the above insurance policies, see Chapter 14. The following discussion of insurance is limited to issues relating to pre-construction.

A. *Construction Manager's Professional Liability Insurance*

Even though construction managers may not be performing professional services, a major loss will drag the entire pre-construction team into the dispute. There is no clear distinction between pre-construction services performed by the construction manager and the services performed by design professionals. If the loss involves errors and omissions, the design professional's insurance may not be adequate, and the owner will look to the experienced construction manager who should have known that there was an error or omission in the drawings it reviewed. CGL policies do not cover this type of loss. If the construction manager, directly or indirectly, may be found to have negligently provided design services, that risk can be transferred to a claims-made professional liability insurance policy.

In reviewing the pre-construction agreement language, the construction manager should be aware that professional liability insurance typically excludes coverage for liability arising from indemnity clauses, agreements to defend, warranties, guarantees, and certifications. What may appear to be innocuous language could lead to a refusal by the insurance carrier to pay damages and defend the claim. As the line between design and construction services blurs, the risk to the construction manager increases exponentially.[29] That line can easily become transparent during pre-construction. Many in the construction industry are unaware that this coverage is available, but this is a valuable tool that should be considered as an addition to the construction manager's insurance program.

B. *Contractor's Pollution Liability Insurance*

The owner frequently adds abatement of hazardous or contaminated materials on the list of Early Work to be performed by the construction manager. If at all possible, the construction manager should avoid assuming responsibility for hazardous and contaminated substances. Losses resulting from such activities are not covered by

---

29. Fred Muse, *Professional Liability: Are Contractors Adequately Protected?*, IRMI.COM, *at* http://www.irmi.com/Expert/articles/2000/Muse12.aspx (Dec. 2000).

CGL insurance. The owner selects the site that contains hazardous or contaminated substances and the owner should not be allowed to transfer that risk to a construction manager working for a relatively small and short-term fee.

If the construction manager is to perform abatement work, the construction manager should operate as the owner's agent, and the owner should enter directly into contract with the cleanup and abatement trade contractors. In addition, the construction manager should avoid assuming responsibility for environmental consultants hired to perform surveys and testing and to monitor the performance of the cleanup and abatement trade contractors. Such consultants should be engaged directly by the owner.

It is also recommended that the owner defend, indemnify, and hold the construction manager harmless from losses arising from abatement and cleanup of hazardous and contaminated substances. If the construction manager decides to undertake abatement and cleanup activities, it should include the sizeable cost associated with project-specific Contractor's Pollution Liability (CPL) as part of its compensation. Construction managers who engage environmental consultants or provide in-house environmental services should instead include a combined professional and pollution liability policy. The CPL policy will cover pollution incidents that arise from the construction manager's activities as well as the activities of its subcontractors. Both occurrence and claims-made policies are available. The benefit of occurrence policies is that claims occurring during the project are covered regardless, even if made long after the construction manager has left the job site.

C. *Builder's Risk Insurance*

Most construction managers do not link the need for builder's risk insurance with pre-construction. However, if the construction manager is on-site during a major loss, the construction manager could become embroiled in a dispute with the owner's insurance carrier(s). Even though not directly involved in the loss, there is always guilt by association, and the construction manager's reputation may also affect the situation. Therefore, it is good practice to advise and follow up with the owner to make sure that all risk property insurance be in place before any improvements are made to the project site.

Builder's risk insurance, also known as course of construction insurance, covers new structures while they are being built, as well as materials and equipment stored on site (and in some cases, in transit or stored off-site) which are intended to be incorporated into the structure. For renovations, builder's risk insurance covers only the improvements being made. The owner's property insurance covers the unimproved property. Builder's risk policies should be all-risk for the full value of the project on a replacement costs basis and should include, without limitation, insurance against the perils of fire (with extended coverage) and physical loss or damage, as well as coverage for theft, terrorism, mold, glass breakage, vandalism, malicious mischief, collapse, earthquake, flood, windstorm, falsework, testing, and start-up.

There are two types of policies available: (1) reporting form, which involves periodic payments based upon the value of the work in place; and (2) nonreporting policy, which requires an up-front payment of the entire premium and no reporting. The construction manager, subconsultants, and sub-subcontractors should be added as additional insureds. If the owner supplies builder's risk insurance, a waiver of subrogation for losses covered by builder's risk should be a mandatory addition to every agreement. The construction manager or its subcontractors cause many of the builder's risk losses. The construction manager does not want to find itself without a viable defense to litigation by a builder's risk insurance carrier.

The construction manager should review the builder's risk policy of the owner and make sure that the owner is responsible for losses not covered by the deductible. Deductibles are frequently in excess of a quarter of a million dollars and the construction manager should make sure that it does not assume liability for the deductible or for losses which exceed the policy limits. A simple innocuous clause that says the construction manager is responsible for repairing any damage to the work can be used to pass on the responsibility for the deductible to the construction manager. Losses to the work itself are excluded from coverage by the construction manager's CGL insurance and would have to be paid from the profit that was to be earned from the project. For example, if there are multiple tenants performing construction on a shopping center, the construction manager should seek a waiver of a subrogation from all of the owners.

Library References

C.J.S. *Insurance* §§ 54, 950, 970–973.
West's Key No. Digests, Insurance <KEY>2323, 2359, 2383.

IX. Suspension or Termination

The construction manager's agreement for pre-construction services should provide reimbursement for costs incurred up to the date of any subsequent termination, as the owner should not be given the impression that the construction manager is responsible for costs associated with completing the construction manager's pre-construction services, especially when there is no substantial profit associated with those services. This situation changes if the construction manager has agreed to perform Early Work at-risk. The construction manager in this situation may be responsible for direct costs incurred to complete the Early Work. However, consequential damages resulting from such termination for cause should be eliminated if the contract contains a mutual waiver of consequential damages provision.[30] It is important to refer specifically to damages flowing from termination for cause in the mutual waiver of consequential damages provision.

---

30. *See, e.g.,* AIA A201-1997, § 4.3.10.

If the due diligence investigation conducted prior to commencement of services discloses potential problems that cannot be readily resolved, the construction manager may wish to retain the right to terminate for its convenience.

The owner's right to suspend services should be coupled with the right of the construction manager to suspend services for nonpayment.

Library References
C.J.S. *Contracts* §§ 384–385, 387–391, 393–394.
West's Key No. Digests, Contracts <KEY>231, 233.

X. Claims

A. Force Majeure *and Extension of Time*

A *force majeure* event occurs when a part of a contract cannot be performed due to causes that are outside the parties' control and that could not be avoided by the exercise of due care.[31] The construction manager should be granted an extension of the schedule for *force majeure* events (Excusable Delay). To balance the risks fairly, Excusable Delays should also be compensable delays and provide for an equitable adjustment in the construction manager's compensation for costs incurred as a result of the Excusable Delay.

B. *Notice*

To be entitled to an extension of time and an equitable adjustment, the construction manager is usually required to notify the owner and the architect of a claim within a certain number of days. It is important that the construction manager have sufficient time to assemble information and analyze the schedule and cost impact attributable to a delay. Short notice periods imposed by owners are intended to cut off potential claims and avoid responsibility. The construction manager should avoid the requirement to submit a claim on short notice or should include an allowance for reasonable compliance.

Library References
C.J.S. *Contracts* §§ 520–525, 578–579, 585–586.
West's Key No. Digests, Contracts <KEY>309, 298.

XI. Contractual Risk

Just like any other business, if construction managers are busy, they are more inclined to pick and choose the projects they will become involved in, and if business is slow, they are more inclined to take anything that is available. The preconstruction phase of a project does not contain the level of risk associated with

---

31. BRYAN A. GARNER, BLACK'S LAW DICTIONARY (8th ed. 2004).

the construction phase, but there is risk. Once a construction manager is deeply involved in a project, it is very difficult to walk away when the construction project (and an attractive fee for profit) is about to commence. The construction manager should perform a due diligence review of every new project. Some of the questions to ask are:

(1) What is the owner's reputation? For example, has the owner defaulted on a previous project? Is the owner very difficult to work with or is the owner litigious?
(2) What is the owner's experience level? For example, does the owner have a track record for successful projects, or is this project the owner's first or the first project of this type? Does the owner have experienced personnel assigned to the project?
(3) What are the owner's financial resources? Does the construction manager want to invest money in a project that may be abandoned?
(4) Does the owner have title to the real estate to be improved or a lease that allows the owner to perform tenant improvements?
(5) Is this a single entity owner? Single entity owners will have no income to offset unforeseen expenses during construction.
(6) Does the owner's project seem reasonable? If the project does not seem reasonable to the construction manager, there is a greater risk that it will be abandoned.

A. *Scope of Services*

The agreement should clearly define the services that are to be performed by the construction manager. There should be a clear understanding between the owner and the construction manager as to what services are to be provided and what services, if provided, require additional compensation. As discussed above, the number of estimates to be provided should be clearly defined in the construction manager's proposal and agreement. If public hearings are to be required, the number of hearings to be attended by the construction manager should be indicated. The financial responsibility for the building permit should be assigned to the owner or factored into the construction manager's compensation. The financial responsibility for the reproduction of drawings and specifications also should be assigned to the owner or factored into the construction manager's compensation.

B. *Exclusion of Design Responsibility*

The construction manager is performing consulting services and it should be absolutely clear that it is not assuming any responsibility for design. As discussed above, the construction manager's employees assigned to perform value engineering and the constructability analysis must be careful to defer all design decisions to the design professionals. In the early twentieth century, the U.S. Supreme Court held that a contractor is not responsible for design defects (*Spearin* doctrine).[32] Pursuant to

---

32. *United States v. Spearin*, 248 U.S. 132.

one common formulation of this doctrine, the owner is said to impliedly warrant the adequacy of the drawings and specifications, and as long as the work is performed in accordance with such drawings and specifications, the owner (not the construction manager) is liable for damages that may result from design errors and omissions. The *Spearin* doctrine has been followed by state courts, but the precise contours of the doctrine vary from one jurisdiction to another. In addition, the *Spearin* doctrine applies primarily to the construction phase rather than the murky arena of pre-construction work where the construction manager could be accused of taking part in the design process. Specific language in the pre-construction agreement should address this issue and create a bright-line separation of responsibilities.

C. *Waiver of Consequential Damages*

Consequential damages are damages (1) resulting from a breach of contract, including any loss resulting from the owner's general or particular requirements and needs of which the construction manager, at the time of contracting, had reason to know and could not reasonably be prevented; or (2) injury to person or property proximately resulting from any breach of warranty by the construction manager.[33] Consequential damages are foreseeable losses resulting from an inability to use the improvement. Direct damages refer to damages related to cost incurred to complete or correct the work performed by the defaulting contractor.

Before 1997 and the AIA's issuance of the revised document A201-1997, General Conditions of the Contract for Construction, there was little focus on including a waiver of consequential damages provision in construction agreements. Construction managers routinely accepted the risk, and owners expected construction managers to be liable for consequential damages.

A case in Atlantic City, New Jersey, is credited with bringing about this change. In 1981, Perini Corp. entered into a construction management agreement for the renovation of the Sands Hotel and Casino in Atlantic City. The construction manager was to be reimbursed for costs of managing the project and paid a construction management fee of approximately $600,000. During construction, the casino and hotel were in continuous operation. No substantial completion date was inserted into the agreement and there was no time is of the essence provision. The entire project was substantially completed in September 1984. The owner took action to terminate Perini in December 1984, for default arising out of project delays.[34] Perini claimed that the revenue portions of the project were in operation before Memorial Day 1984, and that the only delay was to the glass facade. The court ordered the parties to arbitration. A three-party arbitration panel awarded $14,500,000 in lost profits to the owner. The award was confirmed by New Jersey's highest court in September 1991.[35] The contracting world

---

33. U.C.C. § 7-205 (2005).
34. The court noted that the contract contained language prohibiting termination following substantial completion.
35. Perini Corp. v. Great Bay Hotel & Casino, Inc., 610 A.2d 364 (N.J. 1992).

was shocked by an award of $14,500,000 in consequential damages on a $600,000 fee project. This result influenced the addition of the mutual waiver of consequential damages provision that is contained in § 4.3.10 of A201-1997, General Conditions of the Contract for Construction.[36]

Construction managers should insert a similar mutual waiver into their pre-construction agreements. There could be considerable foreseeable consequential damages associated with delays in commencing construction allegedly attributable to the construction manager's performance. This is especially true if the agreement includes a time is of the essence provision; these provisions, of course, may not be appropriate where consulting services are heavily dependent on third parties and the allocation of delays would be difficult. If the owner insists on such a provision, it should be limited to Early Work. The owner may allege that the construction manager's defective estimates caused the project to be abandoned and left the owner liable for damages to investors and potential users. A waiver of consequential damages should be part of the boilerplate of every construction manager's agreement.

D. *Limitation of Liability*

Some construction managers may include a provision in their agreements that limits their liability to the owner to a designated amount. Obviously, these limits do not apply to third party claims. Some construction managers may argue that this limitation is reasonable because they are not receiving a fee for profit on pre-construction work and they should not be subject to unlimited liability. This is a hard sell to most owners but appears to be coming into wider use. The use of such a provision is presently limited to mega projects where the potential liabilities could put the construction manager out of business.

E. *Guarantee of Estimates*

Each pre-construction agreement should include exculpatory language which provides that the construction manager does not warrant or represent that the bid

---

36. Section 4.3.10 states:

The Contractor and Owner waive Claims against each other for consequential damages arising out of or relating to this Contract. This mutual waiver includes:

1. damages incurred by the Owner for rental expenses, for losses of use, income, profit, financing, business and reputation, and for loss of management or employee productivity or of the services of such persons; and
2. damages incurred by the Contractor for principal office expenses including the compensation of personnel stationed there, for losses of financing, business and reputation, and for loss of profit except anticipated profit arising directly from the Work.

This mutual waiver is applicable, without limitation, to all consequential damages due to either party's termination in accordance with Article 14. Nothing contained in this Section 4.3.10 shall be deemed to preclude an award of liquidated damages, when applicable, in accordance with the requirements of the Contract Documents.

prices eventually received from trade contactors will not vary from the construction manager's estimate and/or the owner's budget.[37] The bids received depend on the competitive market and negotiating conditions in place at the time the bid packages are released. For example, as is common in any market, high demand (busy trade contractors) equals higher bids. If the construction market cools, trade contractors who are actively looking for work tend to sharpen their pencils, assume more risk, and submit tighter (lower) bids. The only guaranteed estimate that should be provided by a construction manager is the GMP.

F. *Indemnification*

Indemnification is a device which usually shifts the risk from owner and architect to the construction manager. There should be no indemnification provision included in pre-construction agreements that do not include Early Work. Owners are so used to requiring insurance and indemnification from contractors that they automatically include such requirements in pre-construction consulting agreements. If the owner does not ask for indemnification and insurance certificates from other nonconstruction-related consultants (e.g., accountants, insurance brokers, attorneys, etc.), why does it need them from the construction manager who is performing similar types of services and who will not be located on the project site?

Indemnification is appropriate if the construction manager is performing Early Work. Indemnification provisions should be narrowly constructed as follows: (1) to mirror the coverage available under the CGL insurance policy—limited to bodily injury and property damages that arise from the performance of the work; (2) they should cover only the negligent act or omission by the construction manager and its subcontractors and sub-subcontractors; (3) the construction work performed by the construction manager which is covered by builder's risk insurance should be excluded from the indemnification provision; (4) indemnification should be limited to losses suffered by third parties; (5) if allowable under applicable state law, comparative negligence should be applied so that each defendant is liable to the extent of its own negligence and for the portion of defense costs associated therewith; (6) indemnification should not include breach of contract which has an adequate remedy; and (7) the design professionals should not be named as indemnitees (Claims brought by injured parties routinely include design issues; the construction manager's insurance carrier should not be required to engage separate attorneys for the defense of design professionals.).

A large number of states have enacted statutes designed to limit or void overbroad indemnification provisions. The earliest statutes carved sole negligence of the indemnitee out of the indemnification provision. Some of these are all-or-nothing statutes that void indemnification provisions if the intent of the language is to indemnify an indemnitee for its sole negligence. Starting in the 1980s, there was a movement in some states to mirror comparative negligence law and void indemnification

---

37. AIA Document B801/CMa-1992, § 5.2.

provisions to the extent they required the indemnitor to indemnify the indemnitee for its own negligence. Other states prohibited indemnification by contractors for services performed by design professionals. The construction manager should be intimately familiar with the statutes relating to indemnification that apply to the states in which it performs work.[38]

Library References

C.J.S. *Contracts* §§ 11, 271, 343, 347, 359.
West's Key No. Digests, Contracts <KEY>114, 197, 205.15; Indemnity <KEY>33(5).

XII. Conclusion

Pre-construction on its surface seems to have considerably lower risk than managing the construction process. With fair balancing of the risks in the pre-design and pre-construction agreement, the construction manager should be able to complete its services without the necessity to set aside money for a potential loss reserve.

XIII. Practice Aids

A. *Stand Alone Pre-Construction Agreement*

AGREEMENT BETWEEN OWNER AND CONSTRUCTION MANAGER FOR PRE-CONSTRUCTION SERVICES

OWNER:

CONSTRUCTION MANAGER:

PROJECT:

---

38. For a more detailed discussion and for cites to state law, *see* JUSTIN SWEET, LEGAL ASPECTS OF ARCHITECTURAL ENGINEERING AND THE CONSTRUCTION PROCESS (7th ed. 2005).

OWNER/CONSTRUCTION MANAGER AGREEMENT FOR PRE-CONSTRUCTION SERVICES

This AGREEMENT made as of first day of <date>, by and between <Owner's name> ("Owner"), a <state> Corporation having its principal place of business at <address> and <Construction Manager's name>, a <state> Corporation having a place of business at <address> ("Construction Manager");

WHEREAS, the Owner plans to construct the <identify project> which consists of <details> ("Project"), as more fully described in the Project Description set forth in Exhibit A attached; and

WHEREAS, the Owner has retained <name of architect>, as the Architect for the Project.

WHEREAS, the Owner desires to retain the Construction Manager to provide pre-construction services in connection with the Project and the Construction Manager is willing to perform such services; and

WHEREAS, the Owner contemplates entering into an agreement for a Guaranteed Maximum Price for the construction phase services but the Owner has no obligation to enter into a further agreement with the Construction Manager; and

WHEREAS, the pre-construction phase services provided under this Agreement shall be performed to the reasonable satisfaction of the Owner;

NOW, THEREFORE, in consideration of the premises and mutual promises herein contained, the Owner and the Construction Manager agree as follows:

SECTION 1: GENERAL

1.1 The Project description and criteria for conceptual design are set forth in Exhibit A attached hereto.

SECTION 2: OBLIGATIONS AND RESPONSIBILITIES OF THE CONSTRUCTION MANAGER

2.1 THE CONSTRUCTION MANAGER

2.1.1 The Construction Manager accepts the relationship of trust and confidence established with the Owner by this Agreement and covenants with the Owner to furnish the Construction Manager's reasonable skill and judgment and to further the interests of the Owner. The Construction Manager shall furnish construction administration and management services and use the Construction Manager's good practices to perform pre-construction phase services in an expeditious and economical manner consistent with the interests of the Owner.

2.2 DUTY OF COOPERATION

2.2.1 The Owner shall retain an Architect, Engineer, and other design consultants to provide professional design services for the Project ("Design Consultants"). The

Owner may retain other consultants to provide services for the Project. The Construction Manager shall cooperate with and coordinate its work with that of the Design Consultants and other consultants retained by the Owner. Nothing in this Agreement shall be deemed to require the Construction Manager to engage in the practice of architecture, professional engineering or other design or construction-related professions regulated under the law of the state in which the Project is located.

2.2.2 If there is a primary tenant(s) or user(s) involved in the Project, the Construction Manager acknowledges that the Owner may consult with such tenant(s) about the design and construction of the Project. The Construction Manager shall anticipate and allow for such consultations when submitting documents for the Owner's review and approval. If the Owner's response is time sensitive, the Construction Manager shall make the Owner aware of the date when such response is required and the anticipated consequences of a late response.

2.3 PROJECT PERSONNEL

2.3.1 The Construction Manager's Principal Project Team is listed on Exhibit B attached. The Owner shall have the right to approve the Construction Manager's Project staff, which approval shall not be unreasonably withheld.

2.3.2 The Construction Manager agrees to furnish a qualified experienced staff for the administration, coordination, and management of the Project. Principals and staff of the Construction Manager shall be available and shall participate as needed. All personnel assigned by the Construction Manager to the Project shall be required to cooperate with personnel assigned by the Owner, by the Design Consultants and by the other consultants to the Project and, in the event the Construction Manager's personnel fail to so cooperate or are found to be lacking in competence, they shall be relieved of their duties in connection with the Project at the request of the Owner.

2.4 EMPLOYMENT PRACTICES

2.4.1 The Construction Manager shall comply with all applicable local, state, and federal employment laws and regulations which apply to means and methods employed by the Construction Manager in the performance of the Work and with applicable trade or collective bargaining agreements. They shall keep and maintain such records as are required by such laws and regulations.

SECTION 3: SCOPE OF CONSTRUCTION MANAGER'S SERVICES

3.1 GENERAL

3.1.1 The Construction Manager shall review the program furnished by the Owner to ascertain the requirements of the Project and shall arrive at a mutual understanding of such requirements with the Owner.

3.1.2 The Construction Manager shall provide a preliminary evaluation of the Owner's program, schedule, and construction budget requirements, each in terms of the other.

3.1.3 The Construction Manager shall expeditiously review design documents during their development and advise on selection of materials, building systems and equipment, and methods of Project delivery relative to the feasibility of construction methods, availability of materials and labor, time requirements for procurement, installation, and construction, and factors related to construction cost including, but not limited to, costs of alternative designs or materials, preliminary budgets, and possible economies.

3.1.4 The Construction Manager shall consult with the Owner and Architect regarding the Construction Documents and make recommendations regarding design details that adversely affect constructability, cost, or schedules.

3.1.5 The Construction Manager shall provide recommendations and information to the Owner and Architect regarding the assignment of responsibilities for temporary Project facilities and equipment, materials, and services for common use of the trade contractors. The Construction Manager shall verify that such requirements and assignment of responsibilities are included in the proposed Contract Documents.

3.1.6 The Construction Manager shall provide recommendations and information to the Owner regarding the allocation of responsibilities for safety programs among the contractors.

3.1.7 The Construction Manager shall advise on the division of the Project into individual contracts for various categories of Work, including the method to be used for selecting contractors and awarding Contracts. If multiple contracts are to be awarded, the Construction Manager shall review the Construction Documents and make recommendations as required to provide that (1) the Work of the contractors is coordinated, (2) all requirements for the Project have been assigned to the appropriate contract, (3) the likelihood of jurisdictional disputes has been minimized, and (4) proper coordination has been provided for phased construction.

3.1.8 Upon request, the Construction Manager shall assist the Owner in selecting, retaining, and coordinating the professional services of surveyors, special consultants, and testing laboratories required for the Project.

3.1.9 The Construction Manager shall provide an analysis of the types and quantities of labor required for the Project and review the availability of appropriate categories of labor required for critical phases. The Construction Manager shall make recommendations for actions designed to minimize adverse effects of labor shortages.

3.1.10 The Construction Manager shall assist the Owner in obtaining information regarding applicable requirements for equal employment opportunity programs for inclusion in the Contract Documents.

3.1.11 The Construction Manager shall develop bidders' interest in the Project and establish bidding schedules. The Construction Manager, with the assistance of the Architect, shall issue bidding documents to bidders and conduct pre-bid conferences with prospective bidders. The Construction Manager shall assist the Architect with regard to questions from bidders and with the issuance of addenda.

3.1.12 The Construction Manager shall assist the Owner in obtaining building permits and special permits for permanent improvements, except for permits required to be obtained directly by the various contractors. The Construction Manager shall assist the Owner and Architect in connection with the Owner's responsibility for filing documents required for the approvals of governmental authorities having jurisdiction over the Project.

3.1.13 The Construction Manager shall assist the design professionals and owner with the discussions and development of the program for off- and on-site utility infrastructure for the various utility companies. Evaluate related information, e.g., load letters, Points of Entry, typical construction details, etc., on basis of constructability and value engineering including cost studies as may be needed to evaluate impact to the project and/or alternatives.

3.2 BUDGET AND COST ESTIMATES

3.2.1 The Construction Manager shall create a comprehensive construction cost estimate with trade breakdowns and unit pricing. Unless otherwise agreed in writing by the Construction Manager, the Construction Manager shall prepare no more than four (4) estimates including the Guaranteed Maximum Price: (1) 100% Schematic Drawings completion; (2) 100% Design Development; (3) at 60% percent complete Construction Documents; and (4) the Guaranteed Maximum Price estimate at 90% complete Construction Documents. Estimates shall carry multiple alternates to reflect the continuing cost studies as the design develops.

3.2.2 Based on schematic designs and other design criteria prepared by the Architect and approved by the Owner, the Construction Manager shall prepare a preliminary cost estimate using area, volume, or similar conceptual estimating techniques. The preliminary cost estimate shall be submitted to the Owner for review by the Owner, if applicable, the primary tenant(s), the Design Consultants, and the other consultants and for approval by the Owner. From time to time, as requested by the Owner, the Construction Manager shall provide budgets and cost estimates for changes made by the Owner's users or primary tenant(s) to the base building scope of work.

3.2.3 When Design Development Documents have been completed by the Design Consultants and approved by the Owner, the Construction Manager shall prepare and submit to the Owner a detailed estimate with supporting data for review by the Owner, the Design Consultants, and the other consultants and for approval by the Owner.

3.2.4 When Construction Documents have been completed by the Design Consultants to a level of 60%, the Construction Manager shall prepare and submit to the Owner a detailed estimate with supporting data for review by the Owner, the Design Consultants, and the other consultants and for approval by the Owner.

3.2.5 If any estimate submitted to the Owner exceeds previously approved estimates or the Owner's budget, the Construction Manager shall make reasonable recommendations to the Owner for scope reduction and/or value engineering to bring the estimated cost of construction within the Owner's budget.

3.2.6 The Construction Manager shall assist the Owner in developing cash flow projections.

3.3 VALUE ENGINEERING

3.3.1 The Construction Manager shall provide value engineering services during the pre-construction phase. The Construction Manager shall submit written value engineering estimates and recommendations throughout the document development phases. The Construction Manager shall provide a sample value engineering report for the Owner's approval and shall provide value engineering services of similar scope and level of effort to that shown in the sample. The Construction Manager recognizes that value engineering is an iterative process and that it may be required to modify its estimates and recommendations in response to review of its reports by the Owner, the Design Consultants, and the other consultants.

3.4 PROJECT SCHEDULE

3.4.1 The Construction Manager is aware that the pre-construction phase is to be completed within a fixed time frame. The Construction Manager shall perform its duties in a timely manner consistent with the Project Schedule.

3.4.2 The Construction Manager shall prepare a master Project Schedule which identifies milestones including but not limited to turnover dates to Owner's user(s), date of Substantial Completion, equipment and material delivery dates, and work necessary to achieve these dates for approval by the Owner. The Project Schedule shall include both design and construction activities. The Construction Manager shall coordinate and integrate the Project Schedule with the services and activities of the Owner, the Design Consultants, the other consultants, and the Construction Manager. The Construction Manager shall obtain the input of the Owner, the Design Consultants, and the other consultants for those portions of the Project Schedule relating to the performance of their services.

3.4.3 As design proceeds, the Project Schedule shall be updated to indicate proposed activity sequences and durations, milestone dates for receipts, and approval of pertinent information, submittal of a Guaranteed Maximum Price proposal, preparation and processing of shop drawings and samples, delivery of materials or equipment requiring long-lead time procurement, the Owner's occupancy requirements showing portions of the Project having occupancy priority, and proposed date of Substantial Completion. If updates to the Project Schedule indicate that previously approved schedules may not be met, the Construction Manager shall advise the Owner of the particular activities causing the schedule to be extended and shall recommend a schedule recovery plan to the Owner.

3.5 ADVICE ON CONSTRUCTABILITY AND LOGISTICS

3.5.1 The Construction Manager shall review design documents and make recommendations on means and methods of construction as well as availability of labor and materials. The Construction Manager shall advise the Owner regarding potential jurisdictional disputes between trades. The Construction Manager shall produce a site safety plan that complies with the requirements of all governing bodies having jurisdiction.

3.5.2 The Construction Manager shall establish a sequence of construction with the Owner, the Design Consultants, and the other consultants with respect to specific site logistics. Construction Manager shall provide site logistics plans and sequence diagrams reflective of the various stages of construction progress (i.e., site work, utility infrastructure, commencement of structures, etc.) and as may be required for proposed construction phasing alternatives.

3.6 MEETINGS

3.6.1 The Construction Manager shall schedule and attend regular meetings with the Owner, the Design Consultants, and the other consultants and shall produce minutes of such meetings when requested by the Owner and distribute them to all attendees and other concerned parties. The Construction Manager has included approximately two (2) meetings per month with the Design Team and the Owner. If so requested by the Owner, the Construction Manager shall attend a reasonable number of other meetings with governmental agencies, community groups, the Owner's primary users or other third parties and shall produce minutes of such meetings for the Owner's use when requested by the Owner.

3.7 [OMITTED]

3.8 PERMITS AND APPROVALS

3.8.1 The Owner shall file, expedite, and pay the fee for the application for the Building Permit and obtain approval of site plan and construction drawings from the Building Department.

3.8.2 The Construction Manager shall obtain Work permits from the governmental agencies having jurisdiction over the Project for the demolition and other preliminary construction Work if directed by the Owner.

3.9 GUARANTEED MAXIMUM PRICE PROPOSAL

3.9.1 The Construction Manager shall propose a Guaranteed Maximum Price ("GMP") when the construction documents are ninety percent (90%) complete. The proposed GMP shall be the sum of the estimated Cost of the Work including General Conditions, Insurance, and the Construction Manager's Fee. The assumptions and qualifications upon which the GMP is based shall be included as part of the proposal and shall have priority over all other Contract Documents.

3.9.2 "General Conditions" shall include Construction Manager's supervisory and administrative staff (on- and off-site); materials, supplies and expenses related to maintenance of the field office and other temporary facilities; clean up labor; blueprinting and copying; rubbish removal; and other miscellaneous costs and expenses as indicated in the GMP.

3.9.3 "Cost of the Work" shall mean subcontractor costs; materials and equipment purchased by Construction Manager; General Conditions; contingency for the sole use of the Construction Manager (upon Owner's written approval); and other costs incurred by the Construction Manager in the performance of the Work.

3.9.4 Insurance may include Subcontractor Default Insurance and Contractor's Controlled Insurance program.

3.10 ACCEPTANCE OF THE GMP PROPOSAL

3.10.1 If the GMP proposal is accepted by the Owner, the parties shall execute a separate agreement for Construction Phase Construction Management services ("GMP Construction Agreement") which shall contain mutually acceptable terms and conditions. Should negotiations between the Owner and Construction Manager concerning the amount of GMP or terms of the GMP Construction Agreement reach an impasse, the Owner shall have the right to award a contract for the services included in the Construction Manager's GMP proposal to an entity other than the Construction Manager.

3.10.2 In the event that the Owner does not award a contract for the services included in Construction Manager's GMP proposal to the Construction Manager, the Owner may, at its sole discretion, (1) retain the Construction Manager under a separate Agreement to complete the Construction Phase on a cost-plus-a-fee basis, (2) retain the Construction Manager to complete the Early Work, or (3) require the Construction Manager to assign such Subcontracts performing Early Work to the Owner.

SECTION 4: EARLY WORK/LONG-LEAD ITEMS

4.1.1 The Construction Manager shall identify long-lead items and early-work contracts ("Early Work"). The Construction Manager shall make recommendations with respect to current market trends that may affect procurement and installation so as to minimize potential delays and/or cost premiums.

4.2.1 When directed in writing by the Owner to perform Early Work including the purchase of long-lead items:

1. The Construction Manager shall pre-qualify several potential subcontractors in each required trade and/or suppliers, for Owner's approval.
2. The Construction Manager shall submit request for bids including Bid Documents prepared by the Design Consultants for the Owner's approval before bids are solicited and shall make such modifications thereto as the Owner deems advisable.

3. The insurance requirements for Subcontractors shall be as set forth in Section 7 of this Agreement.
4. The Construction Manager shall analyze bids and make recommendations for award of Subcontracts and/or purchase orders.
5. The Construction Manager shall prepare written authorization to be executed by the authorizing Construction Manager to enter into Subcontracts for award of Early Work or purchase orders for the purchase of long-lead items.
6. The following representative list of items of Early Work may be supplemented upon written agreement between the Owner and Construction Manager:

   1. Demolition and Abatement (Early Work);
   2. Excavation and Foundations (Early Work);
   3. Structural Steel (long-lead item);
   4. Curtain Wall (long-lead item);
   5. Elevator Equipment (long-lead item);
   6. HVAC Equipment (long-lead item);
   7. Electrical Equipment (long-lead item).

SECTION 5: CHANGE IN SERVICES

5.1 Change in Services of the Construction Manager, including services required of the Construction Manager's sub-consultants (if any), may be accomplished after execution of this Agreement, without invalidating the Agreement, if mutually agreed in writing, if required by circumstances beyond the Construction Manager's control, or if the Construction Manager's services are affected as described in Section 5.2. In the absence of mutual agreement in writing, the Construction Manager shall notify the Owner prior to providing such services. If the Owner deems that all or a part of such Change in Services is not required, the Owner shall give prompt written notice to the Construction Manager, and the Construction Manager shall have no obligation to provide those services. Change in Services of the Construction Manager shall entitle the Construction Manager to compensation pursuant to Section 6.3, and to any Reimbursable Expenses pursuant to Section 6.4.

5.2 If any of the following circumstances affect the Construction Manager's Services for the Project, the Construction Manager shall be entitled to an appropriate adjustment in the Construction Manager's compensation:

1. extension of the duration of Service beyond the term provided for in Section 6.1.1;
2. significant change in the Project including, but not limited to, size, quality, complexity, the Owner's schedule or budget, or procurement method;
3. failure of performance on the part of the Owner or the Owner's Design Consultants or other consultants;

4. preparation for and attendance at a dispute resolution proceeding or a legal proceeding except where the Construction Manager is party thereto;
5. change in services requested by Owner beyond the services included in Article 3.

SECTION 6: PAYMENT

6.1 COMPENSATION FOR SERVICES

6.1.1 The Construction Manager shall be reimbursed for staff costs incurred on the basis of hourly rates listed on Exhibit B. Construction Manager shall notify the Owner promptly should it appear that such costs may exceed _____ and 00/100 dollars ($_____.00) ("NTE"). Without the prior written agreement of Owner, the Construction Manager shall not incur staff costs in excess of the NTE amount. The term of this agreement shall be _____ (___) months unless extended upon the written agreement of the Owner and Construction Manager. Compensation for Reimbursable Expenses is to be as provided in Section 6.4 below; compensation for Early Work is to be as provided in Sections 4 and 6.2 and for Change in Services as provided in Sections 5 and 6.3.

6.1.2 The Owner shall be entitled to direct the Construction Manager to perform Early Work or to perform Change in Services not covered by this Agreement.

6.1.3 If the Construction Manager is requested to perform Early Work or Change in Services, the Construction Manager may present an estimate of costs associated with such work for the Owner's consideration. Early Work and Change in Services shall not be performed by Construction Manager without the prior written approval of the Owner.

6.1.4 The Construction Manager shall maintain separate cost records as to all costs and expenses related to Early Work or Change in Services, which accounts and records with supporting documentation shall be available for inspection and audit by the Owner or its authorized representatives at all reasonable times.

6.2 COMPENSATION FOR EARLY WORK

6.2.1 The costs incurred by the Construction Manager as a result of Subcontracts and/or Purchase Orders entered into for Early Work shall be passed through to the Owner along with Construction Manager's General Conditions costs associated with the supervision and administration of such Subcontracts and a markup of ____ percent (__%) for the Construction Manager's Fee. The monthly fee for Services described in Subparagraph 6.1.1 above shall not include the Construction Manager's costs associated with the bidding, buyout of Early Work. A listing of the items that may be included in Construction Manager's General Conditions associated with such Early Work is set forth in Exhibit C attached hereto.

6.2.2 Based upon Applications for Payment submitted to the Owner by the Construction Manager, the Owner shall make progress payments to the Construction

Manager for work performed and/or materials or equipment delivered. The period covered by each Application for Payment shall be one calendar month ending on the last day of the month.

6.3 COMPENSATION FOR CHANGES IN SERVICES

6.3.1 The Construction Manager shall be reimbursed for Changes in Services performed by Construction Manager at the Staff Billing Rates included in Exhibit B attached hereto and shall be payable monthly as provided in Section 6.5. Such amount will be in addition to any other compensation due or owed to the Construction Manager under this Agreement. Changes in Services include, for example, supplemental estimating support staff requested by Owner between estimate cycles.

6.4 REIMBURSABLE EXPENSES

6.4.1 Payments for Reimbursable Expenses incurred shall be made monthly upon presentation of the Construction Manager's Application for Payment.

6.4.2 Reimbursable Expenses are in addition to compensation for the Construction Manager's Services (excluding Early Work where such expenses are included as part of Construction Manager's General Conditions) and include expenses incurred by the Construction Manager and Construction Manager's employees and subconsultants directly related to the Project, including the following:

1. transportation in connection with the Project, authorized out-of-town travel and subsistence;
2. blueprinting services, reproduction, graphics, postage, messenger and overnight couriers;
3. the expense of Construction Manager's Insurance including Construction Manager's errors and omissions insurance;
4. long distance telephone, facsimile, and data processing charges;
5. computer software and services for project Web site (if any);
6. other out-of-pocket expenses reasonably incurred by the Construction Manager in performance of Services.

6.5 PAYMENT TERMS

6.5.1 The Construction Manager shall submit Applications for Payment by the first of each month for services provided during the previous month. The Application shall separately list the monthly Services payment, Changes in Services performed, and Early Work performed. Provided that the Construction Manager's Application for Payment shall be received by the Owner not later than the first day of the month, the Owner shall make payment to the Construction Manager not later than the last business day of the same month. If the Construction Manager's Application for Payment is received later than the first day of the month, the Owner shall make payment to the Construction Manager no later than thirty days following receipt of the Application for Payment.

6.5.2 In relation only to the compensation for Early Work, the Owner may retain up to ten percent (10%) of a progress payment due to a Subcontractor. At fifty percent (50%) completion by a Subcontractor, no further retainage shall be withheld and at Substantial Completion retainage shall be paid for Subcontractor's Work except for one hundred percent (100%) of the value of unfinished punch list work. No retainage shall be withheld by the Owner on any costs except those to be paid to Subcontractors.

6.5.3 Records of Reimbursable Expenses, of expenses pertaining to a Change in Services, and of services performed on the basis of hourly rates shall be available to the Owner or the Owner's authorized representative at mutually convenient times.

SECTION 7: INSURANCE AND INDEMNIFICATION

7.1 TYPES AND LIMITS OF INSURANCE

7.1.1 The Construction Manager agrees to procure and maintain all insurance provided below with insurance companies satisfactory to the Owner. Except for Workers' Compensation and Employers' Liability, the insurance policies shall name the Owner and any other entities (excluding Design Consultants) as listed in Section 7.1.2, Owner may reasonably request as an additional insured. All policies shall provide that insurance carriers shall endeavor to give Owner not less than thirty (30) days prior written notice of any policy cancellation, non-renewal, or material change. Coverage shall be maintained for the duration of the Pre-construction Services. Before commencing performance of the Services and Early Work, Construction Manager shall provide Owner with original Certificates of Insurance.

    A. Commercial General Liability Insurance

    Occurrence form including premises and operations coverage, products, and completed operations; coverage for independent contractors; personal injury coverage; and blanket contractual liability. Completed Operations shall be maintained for a period of three (3) years following final completion.

Each Occurrence	$2,000,000
Personal & Advertising Injury	$2,000,000
Products & Completed Operations Aggregate	$2,000,000
General Aggregate	$2,000,000

    B. Workers' Compensation

Workers' Compensation	Statutory
Employer's Liability	$2,000,000 Each Accident
	$2,000,000 Policy Limit—Disease
	$2,000,000 Each Employee—Disease

    C. Excess or Umbrella Liability Insurance, no more restrictive than the underlying insurance, with limits of not less than $8,000,000 per occurrence and annual aggregate.

D. Commercial Automobile Liability Insurance with limits of not less than $2,000,000 combined single limit for bodily injury and property damage covering all owned, non-owned and hired vehicles.

E. By so specifying, the Owner may require additional types of insurance. The premiums for such required additional insurance shall be reimbursed by the Owner.

7.1.2 The insurance policies required herein shall name the following entities as additional insureds:

    A. Owner   _____
    B. Lender's Group   _____
    C. Other Entities   _____

7.1.3 PROPERTY INSURANCE

A. Unless Construction Manager is directed in writing to provide such insurance as a reimbursable expense, the Owner shall purchase and maintain, in a company or companies lawfully authorized to do business in the jurisdiction in which the Project is located, property insurance on an "all-risk" policy form, including builder's risk which shall include, without limitation, insurance against the perils of fire (with extended coverage) and physical loss or damage including, without duplication of coverage, theft, terrorism, mold, glass breakage, vandalism, malicious mischief, collapse, earthquake, flood, windstorm, false work, testing and start-up, temporary buildings, and debris removal including demolition occasioned by enforcement of any applicable legal requirements, and shall cover reasonable compensation for Construction Manager's services and expenses required as a result of such insured loss, in the amount of the total value for the entire Project at the site on a replacement cost basis including costs to cover professional fees without optional deductibles. Such property insurance shall be maintained, unless otherwise agreed in writing by all persons and entities who are beneficiaries of such insurance, until no person or entity other than the Owner has an insurable interest in the property required by this Section to be covered. The Owner shall reimburse the Construction Manager for costs associated with reconstruction as a result of a loss which are not covered because of deductibles or sublimits. This insurance shall name Construction Manager, Subcontractors, and Sub-subcontractors, as additional insureds.

B. The Owner shall file a copy of each policy with the Construction Manager prior to commencement of any Work which would or should be covered by such property insurance. Each policy shall contain a provision that the policy will not be canceled or allowed to expire, and that its limits will not be reduced, until at least 30 days' prior written notice has been given to the Construction Manager.

C. The Owner and Construction Manager waive all rights against (1) each other and any of their contractors, subcontractors, sub-subcontractors, agents, and employees, each of the other, and (2) the Architect, Architect's consultants, separate contractors, if any, and any of their subcontractors, sub-subcontractors, agents, and employees for damages caused by fire or other causes of loss to the extent covered by property insurance obtained pursuant to Section 7.1.3 A or other property insurance applicable to the Work, except such rights as they have to proceeds of such insurance held by the Owner as fiduciary. The Owner or Contractor, as appropriate, shall require of the Architect, Architect's consultants, separate contractors, if any, and the contractors, subcontractors, sub-subcontractors, agents, and employees of any of them, by appropriate written agreements, similar waivers each in favor of other parties enumerated herein. The policies shall provide such waivers of subrogation by endorsement or otherwise. A waiver of subrogation shall be effective as to a person or entity even though that person or entity would otherwise have a duty of indemnification, contractual or otherwise, did not pay the insurance premium directly or indirectly, and whether or not the person or entity had an insurable interest in the property damaged.

D. A loss insured under the Owner's property insurance shall be adjusted by the Owner as fiduciary and made payable to the Owner as fiduciary for the insureds subject to requirements of any applicable mortgagee clause. The Construction Manager shall pay Subcontractors their just shares of insurance proceeds received by the Construction Manager, and by appropriate written agreements shall require Subcontractors to make payments to their sub-subcontractors in similar manner.

7.2 INDEMNIFICATION

7.2.1 To the fullest extent permitted by law, the Construction Manager shall indemnify and hold harmless the Owner, and its agents and employees from and against claims, damages, losses, and expenses, including but not limited to attorneys' fees, arising out of or resulting from performance of the Early Work, provided that such claim, damage, loss, or expense is attributable to bodily injury, sickness, disease, or death, or to injury to or destruction of tangible property (other than the Work itself), but only to the extent caused by the negligent acts or omissions of the Construction Manager, a Subcontractor, anyone directly or indirectly employed by them or anyone for whose acts they may be liable.

7.2.2 In claims against any person or entity indemnified under Section 7.2.1 by an employee of the Construction Manager, a Subcontractor, anyone directly or indirectly employed by them or anyone for whose acts they may be liable, the indemnification obligation under Section 7.2.1 shall not be limited by a limitation on amount or type of damages, compensation, or benefits payable by or for the Construction

Manager or a Subcontractor under workers' compensation acts, disability benefit acts or other employee benefit acts.

7.2.3 The Owner acknowledges and agrees that the Construction Manager's indemnification in Section 7.2 applies to and is intended to apply only to the performance of construction work by the Construction Manager and its Subcontractors defined as Early Work under this Agreement.

7.3 RIGHTS OF THIRD PARTIES

7.3.1 Nothing in this Agreement shall create or give to third parties any claim or right of action against the Construction Manager or the Owner.

SECTION 8: ASSIGNMENT, SUSPENSION OF WORK, AND TERMINATION

8.1 SUCCESSORS AND ASSIGNS

8.1.1 The Construction Manager and Owner respectively bind themselves, their partners, successors, assigns, and legal representatives to the other party hereto and to partners, successors, assigns, and legal representatives of such other party in respect to covenants, agreements and obligations contained in this Agreement. Neither party to the Agreement shall assign the Agreement as a whole without the written consent of the other. If either party attempts to make such an assignment without consent, that party shall nevertheless remain legally responsible for all obligations under this Agreement.

8.2 SUSPENSION OR TERMINATION OF WORK

8.2.1 At any time, the Owner may, upon seven (7) calendar days' written notice to the Construction Manager, suspend or delay the Construction Manager's services hereunder. The Construction Manager shall be compensated for expenses necessarily and reasonably incurred in order to suspend its services.

8.2.2 At any time, the Owner may, upon seven (7) calendar days' written notice to the Construction Manager, terminate this Agreement in whole or in part, with or without cause, or abandon all or any part of the Project. The Owner shall pay to the Construction Manager all amounts due in accordance with Section 6 hereof with respect to Services and Early Work (if any) performed prior to the date of termination or abandonment as well as reimbursement for expenses necessarily and reasonably incurred in order to stop the Services and Early Work on the Project.

8.2.3 If the Owner, through no fault of the Construction Manager, does not pay the Construction Manager within seven (7) days after the date established in this Agreement, then the Construction Manager may, upon seven (7) additional days' written notice to the Owner, suspend Services and Early Work until payment of the amount owing has been received. The Project Schedule, and the time for completion of the Construction Manager's Services and/or Early Work shall be extended appropriately. In addition the Construction Manager's Compensation for Services and/or Early Work including the subcontract amount for Subcontractors shall be increased by the amount of the reasonable costs of shutdown, delay and start-up.

SECTION 9: OWNER'S RESPONSIBILITY

9.1 OWNER'S REPRESENTATIVE

9.1.1 The Owner designates as its representative with respect to the Project:

<representative's name, address, telephone number, fax, and email address>

The Construction Manager may rely upon the written approval or decision of the Owner's designated representative or alternate representative as the approval or the decision by the Owner. The Construction Manager shall recognize only written directives from the Owner's designated representative or the alternate representative designated herein.

9.2 STRUCTURAL AND ENVIRONMENTAL TESTS, SURVEYS, AND REPORTS

9.2.1 The Owner shall furnish surveys, reports, drawings, and other information in its possession regarding the existing condition of the Project site. The Construction Manager may rely on the general accuracy of the technical data contained in such surveys, reports, or drawings.

9.2.2 The Owner may retain consultants to conduct additional structural, environmental, or other investigations for the Project. The Construction Manager may rely on the general accuracy of the technical data obtained through the additional investigations and any interpretations or conclusions expressed therein. The Construction Manager shall be responsible for interpreting the results of the investigations to the extent that such results pertain to constructability, accessibility of the site, and general suitability of types of materials or methods of construction.

9.3 OWNER'S ADMINISTRATIVE SERVICES

9.3.1 The Owner shall furnish, at its expense, all legal, accounting and insurance counseling services as may be necessary for the Project to protect the Owner's interests, including such auditing services as the Owner may require to verify the Construction Manager's invoices. Services furnished by the Owner are for the Owner's exclusive benefit.

9.4 CONSTRUCTION MANAGER NOTIFICATION

9.4.1 If the Owner observes or otherwise becomes aware of any fault or defect in the Early Work, prompt written notice thereof shall be given by the Owner to the Construction Manager.

9.5 OWNER'S APPROVAL

9.5.1 The Owner or its designated representative shall examine documents submitted by the Construction Manager and shall render approvals and decisions promptly. The Owner shall furnish required information and services and shall render approvals and decisions as expeditiously as necessary for the orderly progress of the Construction Manager's Services and of the Early Work.

9.6 HAZARDOUS AND CONTAMINATED SUBSTANCES

9.6.1 If reasonable precautions will be inadequate to prevent foreseeable bodily injury or death to persons resulting from a material or substance, including but not limited to asbestos, polychlorinated biphenyl (PCB), mold, and lead paint encountered on the site by the Construction Manager, the Construction Manager shall, upon recognizing the condition, immediately stop Services or Early Work in the affected area and report the condition to the Owner and applicable Design Consultants.

9.6.2 To the fullest extent permitted by law, the Owner shall indemnify and hold harmless the Construction Manager and Subcontractors, and agents and employees of any of them from and against claims, damages, losses, and expenses, including but not limited to attorneys' fees, arising out of or resulting from performance of Services or Early Work in the affected area if in fact the material or substance presents the risk of bodily injury or death as described in Section 9.6.1 and has not been rendered harmless, provided that such claim, damage, loss or expense is attributable to bodily injury, sickness, disease or death, or to injury to or destruction of tangible property (other than the Work itself) but only to the extent that such damage, loss or expense is not due to the negligence of a party seeking indemnity.

SECTION 10: CLAIMS AND DISPUTE RESOLUTION

10.1 EXTENSIONS OF TIME

10.1.1 The Construction Manager shall be entitled to an extension of time for delay in completion of the work caused (1) by the acts or omissions of the Owner, its officers, agents or employees, the Design Consultant or the other consultants; (2) by supervening events or conditions beyond the control of the Construction Manager including, but not limited to, Acts of God or the public enemy, terrorism, inclement weather, war, or other national emergency making performance temporarily impossible or illegal, or strikes or labor disputes. The Construction Manager shall be entitled to an extension of time and for an equitable increase for costs incurred by the Construction Manager and Subcontractors as a result of such delay.

10.2 NOTICE OF CLAIM

10.2.1 If at any time the Construction Manager believes that there is a condition that will entitle it to additional compensation or a change in time ("Claim"), within ten (10) days after the occurrence of the event giving rise to the Claim, the Construction

Manager shall deliver written notice to the Owner stating the general nature of such Claim. Notice of the amount or extent of the Claim with supporting data shall be delivered to the Owner, in writing, within twenty-one (21) days after said written notice has been delivered (unless the Construction Manager reasonably requires additional time to submit additional or more accurate data to support such Claim). If the Owner requires any additional information to evaluate the Claim, the Owner shall request such information from the Construction Manager in writing. After receipt of all information required by the Owner, the Owner shall notify the Construction Manager in writing of its decision within thirty (30) days.

10.2.2 Pending the resolution of any the Claim (including mediation or arbitration) the Construction Manager, as long as the Construction Manager receives payment by Owner for actual costs incurred, shall diligently perform all Services and/or Early Work under this Agreement, including work or services that are the subject of such Claim.

10.3 MEDIATION AND DISPUTE RESOLUTION

10.3.1 Should a dispute arise between Owner and Construction Manager, senior executives of the parties shall first negotiate in good faith to resolve any Claims.

10.3.2 If such negotiations do not resolve said claims, the Owner and Construction Manager shall mediate the same as a condition precedent to the commencement of any arbitration or legal proceeding. Mediation shall be conducted under the Construction Industry Mediation Rules of the American Arbitration Association in effect at the time the dispute arises. Unless the parties mutually agree otherwise, the parties shall share the mediator's fee and any filing fees equally.

10.3.3 Any controversy arising out of or related to this Agreement, or its breach, shall be settled by arbitration, in accordance with the Construction Industry Arbitration rules then obtaining of the American Arbitration Association and award thereunder shall be binding upon the Parties and judgment on the award rendered by arbitrator(s) may be entered in any court having jurisdiction thereof. Each party shall proceed with his obligations under this Agreement pending arbitration proceedings, unless otherwise agreed by the Parties in writing. Written Notice of the demand for arbitration shall be served upon the other party and filed in writing with the American Arbitration Association. The demand for arbitration shall be made within a reasonable time after the controversy arises. Unless otherwise agreed by the parties, arbitration proceedings shall be conducted in the state and county in which the Project is located.

10.3.4 Except for the defense of third party claims, no such action or proceeding shall lie or shall be maintained by the either party unless such action or proceeding shall be commenced within one year after the date final payment is made under this Agreement, or in the event this Agreement is terminated, unless such action or proceeding be commenced within one year after the date of such termination.

10.4 GOVERNING LAW

10.4.1 This Agreement shall be governed and construed in accordance with the substantive law of the State in which the Project is located excluding conflict of law principles.

SECTION 11: MISCELLANEOUS PROVISIONS

11.1 PARTICIPATION IN CLAIMS AND LITIGATION SUPPORT

11.1.1 In the event any claim is made or any action brought in any way relating to the Project or the Construction Manager's services, the Construction Manager shall diligently render to the Owner any and all assistance which the Owner may require. The Construction Manager shall be entitled to compensation for such services as a Change in Service unless the claim is directly related to the Construction Manager's Services or Early Work performed on the Project.

11.2 WAIVER

11.2.1 No delay in enforcing any right, remedy, privilege, or recourse accorded to either party or to which either party may be or become entitled to have or exercise under this Agreement shall diminish, suspend, or exhaust any such right, remedy, privilege, or recourse.

11.3 EXHIBITS

11.3.1 The following Exhibits are attached hereto and made a part of this Agreement:

Exhibit A	Project Description
Exhibit B	Project Billing Rates
Exhibit C	Listing of General Conditions Cost Items Applicable to Early Work

11.4 CONSTRUCTION OF THIS AGREEMENT

11.4.1 This Agreement shall be construed without the aid of any presumption or other rule of law regarding construction against the party drafting the same or any part of it.

11.4.2 Section or paragraph headings in this Agreement are for convenience only and are not to be used in aid of construing the Agreement.

11.5 NOTICES

11.5.1 Any notice or other communication required or permitted to be given under this Agreement shall be sufficient if in writing and shall be considered given when delivered by messenger; by overnight courier; or mailed by certified mail, return receipt requested, postage prepaid to the parties at the following addresses (or at such other address as a party may specify by notice hereunder):

if to the Owner:
<contact name and address>

if to the Construction Manager:
<contact name and address>

11.6 ENTIRE AGREEMENT: AMENDMENT

11.6.1 This Agreement represents the entire and integrated agreement between the Owner and Construction Manager with respect to the provisions contained herein and supersedes all prior negotiations, representations or agreements, either written or oral, between them with respect to the Services to be provided hereunder. This Agreement may be modified only by a writing signed by the Owner and Construction Manager.

11.7 WAIVER OF CONSEQUENTIAL DAMAGES

The Construction Manager and Owner waive claims against each other for consequential damages arising out of or relating to this Agreement. This mutual waiver includes without limitation:

- (a) damages incurred by the Owner for rental expenses and for loss of use, income, profit, financing, business, and reputation and loss of management or employee productivity or the services of such persons; and
- (b) damages incurred by the Construction Manager or any Subcontractor for principal office expenses including compensation of personnel stationed in the principal office, loss of financing, business, and reputation, loss of profit, and loss of management or employee productivity or the services of such persons.

This mutual waiver is applicable, without limitation, to all consequential damages due to either party's termination under this Agreement.

11.8 PARTIAL INVALIDITY

If any term or provision of this Agreement or the application thereof to any person, firm or corporation, or circumstances, shall be invalid or unenforceable, the remainder of this Agreement, or the application or such term or provision to persons, firms or corporation, or circumstances, other than those as to which it is held invalid, shall both be unaffected thereby, and each term or provision of this Agreement shall be valid and be enforced to the fullest extent permitted by law.

11.9 CONSTRUCTION MANAGER'S SERVICES

Evaluations of the Owner's budget, preliminary and detailed estimates prepared by the Construction Manager represent the Construction Manager's best judgment. It is recognized, that neither the Construction Manager nor the Owner has control over the cost of labor, materials, or equipment, over bidding methodology, or over competitive bidding, market, or negotiating conditions. Accordingly, except for the Guaranteed

Maximum Price, the Construction Manager cannot and does not warrant or represent that bids or negotiated prices will not vary from the budget proposed, established or approved by the Owner, or from any cost estimate or evaluation prepared by the Construction Manager. The recommendations and advice of the Construction Manager concerning value engineering suggestions and constructability shall be subject to the review and approval of the Owner and the Owner's Design Consultants. It is not the Construction Manager's responsibility to ascertain that the Drawings and Specifications and other information prepared by the Design Consultants are in accordance with applicable laws, statutes, ordinances, building codes, rules, and regulations. However, if the Construction Manager discovers that portions of such documents are at variance therewith, the Construction Manager shall notify the Architect and Owner in writing. The Construction Manager does not warrant or guarantee schedules except as may be included as part of the Guaranteed Maximum Price.

11.10 INDEPENDENT CONTRACTORS

11.10.1 Nothing in this Part A shall in any way constitute a partnership between, or joint venture by, the Owner and Construction Manager. Neither party shall hold itself out contrary to the terms of this Part A by advertising or otherwise, and neither party shall become liable or bound by any representation, act, or omission whatsoever of the other party contrary to the provisions of this section.

11.10.2 With regard to the activities of Construction Manager pursuant to this Part A, Construction Manager is at all times acting as an independent contractor of, and not as a joint venturer or partner of Owner.

IN WITNESS WHEREOF, the parties hereto have executed this Agreement as of the date first written above.

CONSTRUCTION MANAGER

\<name\>

\<name and title\>

OWNER

\<name\>

\<name and title\>

B. *Part A—Pre-Construction Agreement*

PART A
PRE-CONSTRUCTION AGREEMENT

This AGREEMENT made as of first day of <date>, by and between <Owner's name> ("Owner"), a <state> Corporation having its principal place of business at <address> and <Construction Manager's name>, a <state> Corporation having a place of business at <address> ("Construction Manager");

WHEREAS, the Owner plans to construct the <identify project> which consists of <details> ("Project"), as more fully described in the Project Description set forth in Exhibit A attached; and

WHEREAS, the Owner has retained <name of architect>, as the Architect for the Project.

WHEREAS, the Owner desires to retain the Construction Manager to provide pre-construction services in connection with the Project and the Construction Manager is willing to perform such services; and

WHEREAS, the Owner contemplates entering into Part B Agreement for Construction Management with a Guaranteed Maximum Price for the construction phase ("Part B"); and

WHEREAS, the pre-construction phase services provided under this Agreement shall be performed to the reasonable satisfaction of the Owner;

NOW, THEREFORE, in consideration of the premises and mutual promises herein contained, the Owner and the Construction Manager agree as follows:

SECTION 1: GENERAL

1.1 The Project description and criteria for conceptual design are set forth in Exhibit A attached hereto.

1.2 Construction Manager agrees to provide pre-construction services during the development of the design for the Project including preparation of cost estimates, preparation of design and construction progress schedule, constructability analysis, value engineering analysis and the development of a Guaranteed Maximum Price Proposal. The performance of Early Work, if requested by Owner, shall be in accordance with the terms of Article 7. No field work.

1.3 Construction Manager shall provide pre-construction services for a period of no more than _____ (__) months from date of this Agreement.

1.4 The terms contained in this Part A which are not defined herein, shall have the same meaning as set forth in Part B. Notwithstanding the foregoing, should there be a conflict in the definition of a term used in Part A and definition contained in Part B, for purposes of this Part A, the definition in Part A shall take precedence.

SECTION 2: OBLIGATIONS OF THE CONSTRUCTION MANAGER

2.1 COST ESTIMATES

2.1.1 The Construction Manager shall prepare no more than four (4) estimates including the Guaranteed Maximum Price: (1) 100% Schematic Drawings completion; (2) 100% Design Development; (3) at 60% percent complete Construction Documents; and (4) the Guaranteed Maximum Price estimate at 90% complete Construction Documents. Estimates shall carry multiple alternates to reflect the continuing cost studies as the design develops.

2.1.2 If any estimate submitted to the Owner exceeds previously approved estimates or the Owner's budget, the Construction Manager shall make reasonable recommendations to the Owner for scope reduction and/or value engineering to bring the estimated cost of construction within the Owner's budget.

2.1.3 The Construction Manager shall assist the Owner in developing cash flow projections.

2.2 VALUE ENGINEERING

2.2.1 The Construction Manager shall provide value engineering services during the pre-construction phase and shall provide value engineering services of similar scope and level of effort to that shown in the sample. The Construction Manager recognizes that value engineering is an iterative process and that it may be required to modify its estimates and recommendations in response to review of its reports by the Owner, the Design Consultants, and the other consultants.

2.3 PROJECT SCHEDULE

2.3.1 The Construction Manager shall perform its duties in a timely manner consistent with the Project Schedule.

2.3.2 The Construction Manager shall prepare the Project Schedule which identifies date of Substantial Completion, equipment and material delivery dates, and work necessary to achieve these dates for approval by the Owner. The Project Schedule shall include both design and construction activities. The Construction Manager shall coordinate and integrate the Project Schedule with the schedules provided by the Design Consultants and the other consultants.

2.3.3 As design proceeds, the Project Schedule shall be updated. If updates to the Project Schedule indicate that previously approved schedule may not be met, the Construction Manager shall advise the Owner of the particular activities causing the schedule to be extended and shall recommend a schedule recovery plan to the Owner.

2.4 CONSTRUCTABILITY AND LOGISTICS

2.4.1 The Construction Manager shall review design documents and make recommendations on means and methods of construction as well as availability of labor and materials. The Construction Manager shall advise the Owner regarding potential jurisdictional disputes between trades.

2.4.2 The Construction Manager shall establish specific site logistics. Construction Manager shall provide site logistics plans reflective of the various stages of construction progress (i.e., site work, utility infrastructure, commencement of structures, etc.) and as may be required for proposed construction phasing alternatives.

2.5 PERMITS AND APPROVALS

2.5.1 The Owner shall file, expedite, and pay the fee for the application for the Building Permit and obtain approval of site plan and construction drawings from the Building Department.

2.5.2 The Construction Manager shall obtain Work permits from the governmental agencies having jurisdiction over the Project for the demolition and other preliminary construction Work if directed by the Owner.

2.6 GUARANTEED MAXIMUM PRICE PROPOSAL

2.6.1 The Construction Manager shall propose a Guaranteed Maximum Price ("GMP") when the construction documents are ninety (90%) percent complete. The proposed GMP shall be the sum of the estimated Cost of the Work including General Conditions, Insurance, and the Construction Manager's Fee. The assumptions and qualifications upon which the GMP is based shall be included as part of the proposal and shall have priority over all other Contract Documents.

2.6.2 Unless otherwise stated in the assumptions and qualifications which take precedence, "Cost of the Work," "General Conditions," Insurance and Construction Manager's Fee included in the GMP shall be as defined in Part B (set forth in Exhibit A).

2.6.3 The Construction Manager's Fee for profit to be included in the GMP shall be _____ percent (__%) the Cost of the Work including General Conditions and Insurance.

2.6.4 Owner shall accept or reject the GMP Proposal within _____ (__) days after receipt. If Owner does not accept the GMP within this period, this Part A shall terminate.

2.7 ACCEPTANCE OF THE GMP PROPOSAL

2.7.1 Upon Owner's acceptance of the GMP Proposal described in Section 2.6 above, Construction Manager and Owner shall execute Part B of this Agreement in the form set forth in Exhibit A annexed hereto.

2.7.2 In the event that the Owner does not award a contract for the services included in Construction Manager's GMP Proposal to the Construction Manager, the Owner may, at its sole discretion, (1) retain the Construction Manager under Part B to complete the Construction Phase on a cost-plus-a-fee basis without a GMP, (2) retain the Construction Manager to complete the Early Work (if any), or (3) require the Construction Manager to assign such Subcontracts performing Early Work (if any) to the Owner.

2.8 EVALUATIONS, RECOMMENDATIONS, AND TIME

2.8.1 Evaluations of the Owner's budget, preliminary and detailed estimates prepared by the Construction Manager represent the Construction Manager's best judgment. It is recognized that neither the Construction Manager nor the Owner has control over the cost of labor, materials, or equipment, over bidding methodology, or over competitive bidding, market or negotiating conditions. Accordingly, except for the Guaranteed Maximum Price, the Construction Manager cannot and does not warrant or represent that bids or negotiated prices will not vary from the budget proposed, established or approved by the Owner, or from any cost estimate or evaluation prepared by the Construction Manager.

2.8.2 The recommendations and advice of the Construction Manager concerning value engineering suggestions and constructability shall be subject to the review and approval of the Owner and the Owner's Design Consultants. It is not the Construction Manager's responsibility to ascertain that the Drawings and Specifications and other information prepared by the Design Consultants are in accordance with applicable laws, statutes, ordinances, building codes, rules, and regulations. However, if the Construction Manager discovers that portions of such documents are at variance therewith, the Construction Manager shall notify the Architect and Owner in writing.

2.8.3 The Construction Manager does not warrant or guarantee schedules except as may be included as part of the Guaranteed Maximum Price

SECTION 3: EARLY WORK/LONG-LEAD ITEMS

3.1 The Construction Manager shall identify long-lead items and early-work contracts ("Early Work"). The Construction Manager shall make recommendations with respect to current market trends that may affect procurement and installation so as to minimize potential delays and/or cost premiums.

3.2 When the Construction Manager is directed in writing by the Owner to perform Early Work, all provisions of Part B (Exhibit C) shall be applicable to the Early Work and shall govern the performance of the Early Work as if Exhibit C had been incorporated herein in its entirety, except that the Early Work shall be performed on a cost-plus basis without a guaranteed maximum price.

3.3 The fee shall be _____ percent (__%) of the Cost of the Work including General Conditions and Insurance applicable to Early Work performed by the Construction Manager. The progress schedule for performance of Early Work shall be as agreed between Owner and Construction Manager.

3.4 The proposal for Early Work should be submitted as soon as possible after receipt from subcontractors of bids for performance of such Early Work.

3.5 The subcontracts for Early Action Work shall provide that if Part B is to be performed by entirely unrelated construction manager, the subcontract shall, upon written notice from the Owner and Construction Manager to subcontractor, be assigned by Construction Manager to the designated construction manager as if the contracting parties were construction manager and subcontractor from inception of the subcontractor's work.

SECTION 4: PRE-CONSTRUCTION COMPENSATION

4.1 Owner shall pay Construction Manager, as total compensation for services and expenses described in this Part A, the fixed sum of _____ ($_____.00). The term of this agreement shall be _____ (____) months unless extended upon the written agreement of the Owner and Construction Manager.

4.2 The Construction Manager shall be reimbursed for Changes in Services performed by Construction Manager at the Staff Billing Rates included in Exhibit B attached hereto and shall be payable monthly. Such amount will be in addition to any other compensation due or owed to the Construction Manager under this Agreement.

4.3 Reimbursable Expenses are in addition to compensation for the Construction Manager's Services and include expenses incurred by the Construction Manager and Construction Manager's employees and subconsultants directly related to the Project, including the following:
1. transportation in connection with the Project, authorized out-of-town travel and subsistence;
2. blueprinting services, reproduction, graphics, postage, messenger and overnight couriers;
3. the expense of Construction Manager's Insurance including Construction Manager's errors and omissions insurance;
4. long distance telephone, facsimile, and data processing charges;
5. computer software and services for project Web site (if any);
6. other out-of-pocket expenses reasonably incurred by the Construction Manager in performance of Services.

4.4 The Construction Manager shall submit Applications for Payment by the first of each month for Pre-construction services provided during the previous month. The fixed fee indicated in Section 4.1 shall be invoiced in equal monthly payments over the duration of pre-construction. The Owner shall make payment to the

Construction Manager no later than 30 days following receipt of the Application for Payment. Early Work shall be separately invoiced and payment shall be as provided in Part B.

SECTION 5: INSURANCE

5.1 TYPES AND LIMITS OF INSURANCE

5.1.1 The Construction Manager agrees to procure and maintain all insurance provided below with insurance companies satisfactory to the Owner. Except for Workers' Compensation and Employers' Liability, the insurance policies shall name the Owner and any other entities (excluding Design Consultants) as listed in Section 5.1.2, Owner may reasonably request as an additional insured. All policies shall provide that insurance carriers shall endeavor to give Owner not less than thirty (30) days prior written notice of any policy cancellation, non-renewal or material change. Coverage shall be maintained for the duration of the Pre-construction Services. Before commencing performance of the Services and Early Work, Construction Manager shall provide Owner with original Certificates of Insurance.

    A. Commercial General Liability Insurance

    Occurrence form including premises and operations coverage, products and completed operations, coverage for independent contractors, personal injury coverage, and blanket contractual liability. Completed Operations shall be maintained for a period of three (3) years following final completion.

Each Occurrence	$2,000,000
Personal & Advertising Injury	$2,000,000
Products & Completed Operations Aggregate	$2,000,000
General Aggregate	$2,000,000

    B. Workers' Compensation

Workers' Compensation	Statutory
Employer's Liability	$2,000,000 Each Accident
	$2,000,000 Policy Limit—Disease
	$2,000,000 Each Employee—Disease

    C. Excess or Umbrella Liability Insurance, no more restrictive than the underlying insurance, with limits of not less than $__.00 per occurrence and annual aggregate.

    D. Commercial Automobile Liability Insurance with limits of not less than $2,000,000 combined single limit for bodily injury and property damage covering all owned, non-owned and hired vehicles.

    E. By so specifying, the Owner may require additional types of insurance. The premiums for such required additional insurance shall be reimbursed by the Owner.

5.1.2 The insurance policies required herein shall name the following entities as additional insureds:

    A. Owner   _____

    B. Lender's Group   _____

    C. Other Entities   _____

5.1.3 Property Insurance

The Owner shall file a copy of each "all risk" property insurance (builder's risk) policy with the Construction Manager prior to commencement of any Early Work. This insurance shall be in accordance with the insurance requirements contained in Part B and shall include a mutual waiver of subrogation by Owner and Construction Manager and name Construction Manager as an additional insured.

SECTION 6: SUSPENSION OR TERMINATION OF WORK

6.1 At any time, the Owner may, upon seven (7) calendar days' written notice to the Construction Manager, suspend or delay the Construction Manager's services hereunder. The Construction Manager shall be compensated for expenses necessarily and reasonably incurred in order to suspend its services.

6.2 At any time, the Owner may, upon seven (7) calendar days' written notice to the Construction Manager, terminate this Agreement in whole or in part, with or without cause, or abandon all or any part of the Project. The Owner shall pay to the Construction Manager all amounts due in accordance with Section 4 hereof with respect to Services and Early Work (if any) performed prior to the date of termination or abandonment as well as reimbursement for expenses necessarily and reasonably incurred in order to stop the Services and Early Work on the Project.

6.3 If the Owner, through no fault of the Construction Manager, does not pay the Construction Manager within seven (7) days after the date established in this Agreement, then the Construction Manager may, upon seven (7) additional days' written notice to the Owner, suspend Services and Early Work until payment of the amount owing has been received. The Project Schedule, and the time for completion of the Construction Manager's Services and/or Early Work shall be extended appropriately. In addition the Construction Manager's Compensation for Services and/or Early Work including the subcontract amount for Subcontractors shall be increased by the amount of the reasonable costs of shutdown, delay and start-up.

SECTION 7: MISCELLANEOUS

7.1 PARTICIPATION IN CLAIMS AND LITIGATION SUPPORT

7.1.1 In the event any claim is made or any action brought in any way relating to the Project or the Construction Manager's services, the Construction Manager shall diligently render to the Owner any and all assistance which the Owner may require. The Construction Manager shall be entitled to compensation for such services.

7.2 WAIVER

7.2.1 No delay in enforcing any right, remedy, privilege or recourse accorded to either party or to which either party may be or become entitled to have or exercise under this Agreement shall diminish, suspend or exhaust any such right, remedy, privilege or recourse.

7.3 EXHIBITS

7.3.1 The following Exhibits are attached hereto and made a part of this Agreement:

Exhibit A	Project Description
Exhibit B	Project Billing Rates
Exhibit C	Part B Agreement for Construction Management with a GMP

7.4 CONSTRUCTION OF THIS AGREEMENT

7.4.1 This Agreement shall be construed without the aid of any presumption or other rule of law regarding construction against the party drafting the same or any part of it.

7.4.2 Section or paragraph headings in this Agreement are for convenience only and are not to be used in aid of construing the Agreement.

7.5 NOTICES

7.5.1 Any notice or other communication required or permitted to be given under this Agreement shall be sufficient if in writing and shall be considered given when delivered by messenger; overnight courier; or mailed by certified mail, return receipt requested, postage prepaid to the parties at the following addresses (or at such other address as a party may specify by notice hereunder):

if to the Owner:
<contact name and address>

if to the Construction Manager:
<contact name and address>

7.6 PARTIAL INVALIDITY

7.6.1 If any term or provision of this Agreement or the application thereof to any person, firm or corporation, or circumstances, shall be invalid or unenforceable, the remainder of this Agreement, or the application of such term or provision to persons, firms or corporations, or circumstances, other than those as to which it is held invalid, shall both be unaffected thereby, and each term or provision of this Agreement shall be valid and enforced to the fullest extent permitted by law.

7.7 RIGHTS OF THIRD PARTIES

7.7.1 Nothing in this Agreement shall create or give to third parties any claim or right of action against the Construction Manager or the Owner.

7.8 SIMULTANEOUS EXECUTION

7.8.1 This Agreement may be executed in one or more counterparts, all of which shall be considered one and the same agreement, and shall become effective when one or more counterparts have been executed by each of the parties hereto and delivered to each of them.

7.9 COMPLETE AGREEMENT

7.9.1 This Agreement constitutes the complete agreement and understanding of the parties concerning the subject matter of this Agreement. This Agreement may be modified only by a writing signed by the Owner and Construction Manager.

7.10 EXTENSIONS OF TIME

7.10.1 The Construction Manager shall be entitled to an extension of time for delay in completion of the work caused; (1) by the acts or omissions of the Owner, its officers, agents or employees, the Design Consultant or the other consultants; (2) by supervening events or conditions beyond the control of the Construction Manager including, but not limited to, Acts of God or the public enemy, terrorism, inclement weather, war or other national emergency making performance temporarily impossible or illegal, or strikes or labor disputes. The Construction Manager shall be entitled to an extension of time and for an equitable increase for costs incurred by the Construction Manager and Subcontractors as a result of such delay.

7.11 MEDIATION AND DISPUTE RESOLUTION

7.11.1 Should a dispute arise between Owner and Construction Manager, senior executives of the parties shall first negotiate in good faith to resolve any Claims.

7.11.2 If such negotiations do not resolve said claims, the Owner and Construction Manager shall mediate the same as a condition precedent to the commencement of any arbitration or legal proceeding. Mediation shall be conducted under the Construction Industry Mediation Rules of the American Arbitration Association in effect at the time the dispute arises. Unless the parties mutually agree otherwise, the parties shall share the mediator's fee and any filing fees equally.

7.11.3 Any controversy arising out of or related to this Agreement, or its breach, shall be settled by arbitration, in accordance with the Construction Industry

Arbitration rules then obtaining of the American Arbitration Association and award thereunder shall be binding upon the Parties and judgment on the award rendered by arbitrator(s) may be entered in any court having jurisdiction thereof. Each party shall proceed with his obligations under this Agreement pending arbitration proceedings, unless otherwise agreed by the Parties in writing. Written Notice of the demand for arbitration shall be served upon the other party and filed in writing with the American Arbitration Association. The demand for arbitration shall be made within a reasonable time after the controversy arises. Unless otherwise agreed by the parties, arbitration proceedings shall be conducted in the state and county in which the Project is located.

7.11.4 Except for the defense of third party claims, no such action or proceeding shall lie or shall be maintained by the either party unless such action or proceeding shall be commenced within one year after the date final payment is made under this Agreement, or in the event this Agreement is terminated, unless such action or proceeding be commenced within one year after the date of such termination.

7.12 GOVERNING LAW

7.12.1 This Agreement shall be governed and construed in accordance with the substantive law of the State in which the Project is located excluding conflict of law principles.

7.13 WAIVER

7.13.1 No delay in enforcing any right, remedy, privilege or recourse accorded to either party or to which either party may be or become entitled to have or exercise under this Agreement shall diminish, suspend or exhaust any such right, remedy, privilege, or recourse.

7.14 CONSTRUCTION OF THIS AGREEMENT

7.14.1 This Agreement shall be construed without the aid of any presumption or other rule of law regarding construction against the party drafting the same or any part of it.

7.15 PARTIAL INVALIDITY

7.15.1 If any term or provision of this Agreement or the application thereof to any person, firm or corporation, or circumstances, shall be invalid or unenforceable, the remainder of this Agreement, or the application or such term or provision to persons, firms or corporation, or circumstances, other than those as to which it is held invalid, shall both be unaffected thereby, and each term or provision of this Agreement shall be valid and be enforced to the fullest extent permitted by law.

7.16 INDEPENDENT CONTRACTORS

7.16.1 Nothing in this Agreement shall in any way constitute a partnership between, or joint venture by, the Owner and Construction Manager. Neither party shall hold itself out contrary to the terms of this Agreement by advertising or otherwise, and neither party shall become liable or bound by any representation, act, or omission whatsoever of the other party contrary to the provisions of this section.

7.16.2 With regard to the activities of Construction Manager pursuant to this Agreement, Construction Manager is at all times acting as an independent contractor of, and not as a joint venturer or partner of Owner.

7.17 WAIVER OF CONSEQUENTIAL DAMAGES

7.17.1 The Construction Manager and Owner waive claims against each other for consequential damages arising out of or relating to this Agreement. This mutual waiver includes without limitation:

1. damages incurred by the Owner for rental expenses and for loss of use, income, profit, financing, business and reputation and loss of management or employee productivity or the services of such persons; and

2. damages incurred by the Construction Manager or any Subcontractor for principal office expenses including compensation of personnel stationed in the principal office, loss of financing, business and reputation, loss of profit and loss of management or employee productivity or the services of such persons.

This mutual waiver is applicable, without limitation, to all consequential damages due to either party's termination under this Agreement.

IN WITNESS WHEREOF, the parties hereto have executed this Agreement as of the date first written above.

CONSTRUCTION MANAGER

<name>

_____

<name and title>

OWNER

<name>

_____

<name and title>

C. *Pre-Construction Agreement Checklist*

OWNER/CONSTRUCTION MANAGER PRE-CONSTRUCTION AGREEMENT

Item	Included	Title	Notes
1	❏	Waiver of consequential damages	
2	❏	Indemnification only for Early Work	
3	❏	Insurance requirements	
4	❏	Builder's risk waiver of subrogation	
5	❏	CM added as additional insured to builder's risk	
6	❏	Separate fee where applicable for GMP and cost-plus options	
7	❏	Exclude CM from design responsibility	
8	❏	Limit number of estimates	
9	❏	Limit number of meetings	
10	❏	Scope of pre-construction services	
11	❏	Owner obtains building permit	
12	❏	Method for reimbursement for additional services	
13	❏	Payment for reimbursable expenses	
14	❏	Payment term	
15	❏	Termination provision	
16	❏	Owner indemnification for hazmat where applicable	
17	❏	Right to suspend services for nonpayment	
18	❏	Notice provisions	
19	❏	Dispute resolution provisions	

CHAPTER 11

Construction Management Contract Responsibilities: Construction and Completion Phases

STANLEY A. MARTIN, ESQ.

I. Introduction

The concept of construction management originated with providing a construction perspective during the design process. However, the construction manager's role during construction has evolved far beyond the role of the contractor in the traditional design-bid-build scenario. This chapter examines the construction manager's role during construction and afterward evaluates those tasks and responsibilities construction managers typically undertake, and considers the construction manager's attendant risks.

As noted earlier, construction manager may refer to an advisor, or to a contractor who is not only advising the owner but also undertaking responsibility for performance of the construction work. In this chapter, in the absence of any notation to the contrary, the reference to a construction manager will be in this latter capacity—as one undertaking responsibility for construction cost, schedule, and quality. Throughout the chapter, though, when there are useful comments regarding the advisor construction manager—often referred to as an agent construction manager—they will be noted with reference to the advisory or agent role.

This chapter covers standard or typical practices. The construction manager's professional service aspect is governed largely by contract. The profession is sufficiently young and few common law standards unique to construction management have developed. This chapter does not purport to establish any standards, but rather to report on common practices and issues. Where roles and responsibilities are discussed, they are covered from the standpoint of what many of the industry standard contract forms require of the construction manager. Other publications are identified at the end of this chapter as additional resources.

A. *Construction Manager's Responsibility during Construction*

The construction manager's responsibilities during construction are quite broad. One of the most important functions to fulfill before construction work commences is the procurement of trade contractors. The construction manager at-risk must review the design requirements, create a work breakdown structured into various trades, and then buy out each element of the work. The construction manager advisor typically monitors this effort, weighing in on the issues of subcontractor qualifications, and ensuring that the process of procurement is being conducted in such a way as to advance the overall interests of the owner. For example, if the quality of work is critical for a particular element of the project, such as millwork, then the construction manager advisor will likely recommend standards for qualifying subcontractors, may recommend particular companies, and will also evaluate the qualifications of prospective bidders.

Once construction work is under way, the responsibilities of the construction manager increase significantly. This role takes on that of collaborator with both owner and designer, as well as what would be considered the traditional role of a general contractor. The construction manager at-risk will coordinate the trades at the job site, manage cost controls—through skillful procurement and on-site management of issues and changes—and manage the schedule. The construction manager at-risk will also be responsible for quality control in the completed work.

Controlling construction quality, cost, and schedule also requires planning and anticipation of upcoming issues and events. In that regard, the construction manager must maintain close communication with the owner and designer to advise both of upcoming issues and problems, and recommend a course of action to deal with the same.

As discussed in greater detail below, the construction manager at-risk must also manage various elements of risk. These include not only the cost, schedule, and quality risks noted above, but also the safety of persons and property. This latter risk, as it relates to the work force, is covered by safety plans, training, and enforcement of safety principles on the project. The safety risk concerning third parties and adjacent property is covered in large part by insurance, and the construction manager typically has a role in planning and/or implementing the insurance program for the project.

During construction, the construction manager advisor monitors all of the foregoing activities and reports to the project owner on the work of the design professionals, the general contractor, and the subcontractors. Note that a construction manager agent, or advisor, can be utilized in conjunction with any project delivery method.

B. *Construction Manager's Responsibility during Completion*

As the project nears completion, the construction manager's role shifts from manager of the work force in the field to administrator and coordinator of the transfer of responsibility from the construction crews to the owner or its operations and maintenance staff. This effort includes management of the punch list process to resolve any defects or deficiencies noted, commissioning of building systems and training of maintenance staff, and compiling warranties, guarantees, and certifications from all vendors, subcontractors, and manufacturers.

If the project has been built on a cost-plus-a-fee basis, then the construction manager may also have to undergo an audit by the owner's accountant to verify the costs billed to the project.

II. The Construction Manager's Role, Risks and Responsibilities during Construction

This section covers in detail the various responsibilities of the construction manager during the construction process, and the role played by the construction manager during that aspect of the project. Included in each subsection is a discussion of typical roles, responsibilities, and efforts associated with that component of the project, as well as some discussion about issues that could arise from the construction manager's performance of those services. Where particular liabilities may result from the construction manager's work, those risks are discussed.

In many respects, the role of a construction manager at-risk during construction is quite similar to that of a general contractor. The construction manager's superintendent is on-site coordinating the subcontractors. The construction manager is responsible for maintaining the schedule, coordinating the work of various subcontractors, and generally, acting as a traffic cop. The construction manager conducts weekly or periodic site meetings with the owner and designer for the purpose of addressing and resolving issues during construction. However, in at least one sense, the construction manager is unlike a general contractor when it comes to project costs and budgets. Most construction managers work on a cost-plus-a-fee basis, as opposed to a lump sum basis. Thus, the project budget is managed, monitored, and updated throughout in a manner that is open and transparent to the owner.

A. *Procurement of Trade Contractors*

The construction manager typically supervises the ongoing construction work via site-based and office-based managers. Most construction managers do not perform construction work with their own employees or may provide a small crew of workers to address miscellaneous elements of work not assigned to any particular subcontractor. Accordingly, they must procure, or buy out, in the trade lingo, various trade contractors to perform discrete components of the work.

This is one aspect of the construction manager's responsibility that ties directly into its pre-construction activity. In the traditional design-bid-build scenario, the contractor takes the specifications that have already been prepared, with allocation of work among different specification sections, and divides that work among various trades. One of the perennial issues in that format—an issue that the movement toward construction management was designed to cure—was with gaps and overlaps among the scope of various specification sections. During pre-construction, the construction manager should advise the design professionals as to various ways to allocate the work to avoid both gaps and overlaps. Now, during the procurement aspect of the construction phase, the construction manager must take the overall design and break it down into work scopes for various trades.

1. Work Scope Allocation

If the specifications have been developed with input from the construction manager, then allocation of work scope is more likely to follow along the lines of specification sections. Alternatively, the construction manager will have developed a good understanding of the scope of work to be performed as a result of detailed discussion about the project during the design phase and can use that knowledge to procure various scopes of work by trade.

The construction manager's role in allocating scopes of work calls for it to apply its knowledge of local customs, to avoid potential union jurisdictional disputes between trades, and to consider the impact of any specialized materials or equipment purchases on the proposed scope. This effort should reduce the number of situations where two subcontractors performing work in related trades disclaim responsibility for a particular element of the work that bridges their respective scopes, and it should also minimize the situations where the work of one subcontractor is hampered by the problems encountered by another subcontractor in procuring materials or equipment.

Practice Tip

Allocation or delineation of trade contractor work scope is a typical element of construction manager responsibility that is not covered in many standard form contracts. Consider adding the following to the description of the construction manager's responsibilities:

> *The CM shall review the design documents and make recommendations to the Owner and Architect concerning separation of the work into contracts or specification sections, taking into consideration market conditions, trade jurisdiction, and coordination of procurement of specialized materials or equipment.*

The corollary to this allocation of work scope among subcontractors is that the construction manager may have some liability in the event of gaps in work scope. This liability, in the normal course, will not result in the construction manager's having to reimburse the owner, but rather in the application of contingency funds to cover any gap. Alternatively, the costs to be reimbursed will increase for the added work. Depending on how close the project costs are to the cost cap under the contract (Guaranteed Maximum Price or GMP), the added cost may either cause the GMP to be exceeded or reduce the savings below the GMP.

2. Trade Contractor Qualification

Selection of trade contractors involves reviewing their experience and qualifications, and then assessing whether the bid price and proposal, with all exclusions and qualifications, covers the work scope. Most construction managers lead and are intricately involved in all aspects of this process.

Selection of trade contractors starts with defining the criteria for qualifying them. Most construction managers develop a list of potential subcontractors based on their own experience in the area. Typically, the owner and design professional are also given the opportunity to propose subcontractors from whom proposals will be solicited. Less commonly, the owner or designer may also strike names from the list of prospects for some enunciated reason. The qualification process may be informal and without any solicitation of written submissions from the prospects. Alternatively, prospective subcontractors may be invited to submit information about their company, work history, and personnel. If the process is more structured, the solicitation will be in the form of a written Request for Qualifications (RFQ), and companies responding will need to do so with the information sought in the time and manner provided in the RFQ. For example, § 2.1.6 of the Standard Form of Agreement Between Owner and Construction Manager, AIA Document A121/CMc-2003 (A121/CMc-2003), states:

> The Construction Manager shall seek to develop subcontractor interest in the Project and shall furnish to the Owner and Architect for their information a list of possible subcontractors, including suppliers who are to furnish materials or equipment fabricated to a special design, from whom proposals will be requested for each principal portion of the Work. The Architect will promptly reply in writing to the Construction Manager if the Architect or Owner know of any objection to such subcontractor or supplier. The receipt of such list shall not require the Owner or Architect to investigate the qualifications of proposed subcontractors or suppliers, nor shall it waive the right of the Owner or Architect later to object to or reject any proposed subcontractor or supplier.

The construction manager and owner must have an agreement at the outset of the process for determining qualifications of trade contractors. This agreement should limit the owner's discretion to propose or strike trade contractors, the target number of trade contractors sought for each specialty, and the criteria used to determine which ones are qualified to bid on the project. Further, most projects employ a range of processes for subcontractor qualifications, with the most structured processes being used for the major trades or contracts likely to exceed a certain dollar amount. Minor trades may still be selected on a more informal basis and without need for written submissions on qualifications.

3. Trade Contractor Bidding and Selection

Once a list of qualified trade contractors has been developed, bids are solicited. The construction manager ordinarily reviews the bids and prepares a bid tabulation listing for each prospective subcontractor the various elements of the work, and verifies that the bidder has included them within its price the scope items. For large or critical trades, the construction manager may meet separately with each bidder to review scope items in detail, and verify the bidder's anticipated methodology for performing critical components of its work. A bid tabulation summarizing this detailed information then is presented to the owner, with recommendations by the construction manager for those trade contractors who should be engaged to perform the work.

Final selection of trade contractors depends on the arrangement between the construction manager and owner. If the owner and the construction manager agreed to a guaranteed maximum price for the work, the construction manager may select the lowest qualified bidder. The owner's discretion to object to the lowest bidder, challenge its qualifications, or select one who is not the lowest bidder will be defined in the contract between owner and the construction manager.[1]

Cost overruns are the most significant risk to the construction manager from this aspect of the work. If the construction manager did not properly budget or buy out the project, then the total cost of the work, together with the construction manager's fee and general conditions costs, may exceed the project GMP. Thus, the construction manager may be exposed to losses for which it has no recourse. This cost exposure is no different than what a general contractor risks under a lump sum contract, but in practice, this problem should occur less frequently due to the construction manager's involvement in design phase and pre-construction services, and its ability to influence the specifications to permit efficient buy out of the trades.

B. *Management of On-Site Work*

Daily construction operations are managed by project superintendents. Depending on its size and scope, the project may require more than one superintendent or assistant superintendents. The superintendent coordinates and oversees the work of the various trades at the site. The construction manager's superintendent is also the point of communication regarding on-site activities for any trade contractor's foreman or superintendent.

1. *Coordination of Trades*

The most essential part of a construction manager's on-site work is to coordinate the various subcontractors and trade workers. While each subcontractor is responsible for managing its own work force, the construction manager must see that the activities of the various subcontractors are carried out in an appropriate sequence with regard for construction means and methods and per the project schedule. The construction manager's superintendent must notify each subcontractor as to when it should be on-site, and must also discuss the anticipated sequence of work.

2. *Coordination of Changes and Modifications*

The construction manager is responsible not only for managing the completion of construction but also the *process* of construction. This process invariably includes dealing with changes and modifications in the program and the design during the

---

1. AIA Document A121/CMc-2003, § 2.3.2.2 provides that the GMP may be adjusted if the owner requires the construction manager to select one other than a qualified low bidder recommended by the construction manager.

course of construction. More so than a general contractor, a construction manager is expected to anticipate, plan around, or deal with changes and modifications in the work, and coordinate subcontractors' response to the same. The construction manager is expected to advise the owner and designer of alternative courses of action in the event of a change, with the owner's interests at heart and an eye toward minimizing cost and schedule impact. Common contract language dealing with project costs demonstrates this point:

> The Construction Manager shall develop a system of cost control for the Work, including regular monitoring of actual costs for activities in progress and estimates for uncompleted tasks and proposed changes. The Construction Manager shall identify variances between actual and estimated costs and report the variances to the Owner and Architect at regular intervals.[2]

C. Cost Control

With the usual scenario of a cost-plus guaranteed maximum price contract, the construction manager must take steps to manage the project costs and take corrective action when project costs rise above acceptable levels. In one sense, this role is similar to that of a general contractor performing under a lump sum contract. Where the general contractor must maintain cost control to yield a profit on the project, the construction manager must maintain costs below the GMP to protect its fee. The construction manager's liability for failing to control the costs is ordinarily that it will incur a loss on the project.

Cost control is typically accomplished in a number of ways. First and foremost is the development of a realistic project budget, discussed above. If the budget is sensible and responsive to the design, then the likelihood of surprises during buyout and construction are minimized. Second is the use of contingencies within the budget until the buyout is complete and the highest risk activities completed. Cost control is also carried out via management of the process of changes, and in the construction manager's continued development of value engineering and other cost savings alternatives.

Unlike the construction manager at-risk, a construction manager agent ordinarily has no liability at all for the failure of the prime contractor to achieve the project budget. The construction manager agent's role is that of a reporter to the owner, identifying the current costs and cost projections, and then making recommendations to deal with anticipated cost overruns.

1. Use of Contingency during Construction

The project GMP typically is agreed on before design completion, which is also before buyout completion of subcontractors. Thus, at the time the GMP is agreed

---

2. AIA Document A121/CMc-2003, § 2.3.2.7.

on—particularly in fast track projects—cost risks remain in the buyout process. For that reason, a reasonable construction manager will require the project owner to include a contingency in the construction budget. The amount of that contingency will vary from one project to another, depending on the circumstances, the perceived risk of certain tasks, and the construction manager's comfort level that the anticipated design will be maintained as the design is finalized.[3]

This question often arises: What is a realistic contingency amount? In most cases, the contingency line item ranges from 2.5 percent to 7.5 percent of anticipated project value. Construction of a new auto dealer show room might have a contingency on the lower end of the scale, whereas renovation of an historic building for conversion to an alternative use is an example of a project likely to have a high contingency.

Practice Tip

When establishing the terms of a contingency, it is important for both owner and construction manager to carry contingency line items in their respective budgets. Ideally, the construction manager contingency is used to cover unanticipated costs of buying out the project, or for taking steps to expedite some portion of the work or delivery of equipment or materials. It should not be used to cover anticipated changes in the work. The owner contingency is an overall project contingency, covering both unanticipated soft costs of design and other consultants, as well as construction costs.

Construction managers and owners often negotiate the conditions and circumstances in which the construction manager can tap into the contingency. The construction manager does not want any limits on its ability to use contingency funds to cover unanticipated costs. On the other hand, owners often want the construction manager to seek approval for any expenditure out of the contingency. An appropriate compromise is for the owner to accept that the construction manager can expend funds under a certain limit, without the need to seek prior approval, but with the obligation to report promptly to the owner when such funds have been spent.

The conditions governing the construction manager's right to tap into contingency funds typically are negotiated. Those conditions include:

- when and to what extent the construction manager can use contingency funds to offset cost overruns rather than unanticipated events;
- whether unused contingency funds or allowances for one part of the budget can be used to cover overruns on another part the budget;
- whether the construction manager must obtain the approval of the owner or architect/engineer to use contingency funds; and
- what happens to unused contingency funds at the end of the project.

---

3. Most standard form contracts provide for a contingency. *See, e.g.*, AIA Document A121/CMc-2003, § 2.2.3.

Thus, while the contingency is not the sole method of cost control, it is an essential element when construction work starts before the design is complete. Further, as the price of certain commodities and materials common to construction projects has become more volatile in the past few years, a contingency for unforeseeable price spikes will assist both the construction manager and the owner in getting the best overall cost for the work.

2. Management of Trade Contractor Changes

One critical value that an owner expects from a construction manager is management of the construction process. Nowhere is that management more critical than in dealing with changes in the project. Changes are a fact of life on construction projects. The ability to manage change in order to minimize schedule disruption and the resulting cost impact is one significant reason to employ a construction manager for the project.

Managing changes includes identifying the pertinent trade contractors whose work will be impacted directly or indirectly by the change, getting price and schedule information quickly from those subcontractors, assessing potential alternatives to the proposed change in order to offer cost and schedule alternatives, and then implementing the change.

This is one aspect of the construction manager's work where liability can flow from its failure to carry out its obligations. For example, a proposed change may directly impact the work of one or two trade contractors and also have a potential indirect impact on another contractor. If a trade contractor whose work is indirectly impacted is not consulted for input and pricing for the change order, a situation could arise where that contractor files a claim for the cost impact on its work, even though the owner believes the full cost of the change has already been agreed. In that scenario, the construction manager may have to satisfy the disappointed trade contractor's claim from contingency funds (if its contract allows it to do so), or else cover the cost from its own pocket if the claim cannot be passed through to the owner.

Management of changes also carries an administrative cost. The construction manager must have project staff to communicate with trade contractors, solicit and compile cost information, give timely notice to the owner of pending cost or schedule adjustments, prepare paperwork to document proposed and accepted changes, and then track the trade contractors' charges against the agreed change amount. That is when everything goes smoothly. At times, there are disputes over the true cost of a particular change. On those occasions, the construction manager must shuttle between owner and trade contractor(s) to negotiate a settlement. There are multiple opportunities for errors or disputes in this process, and therefore, increased liability exposure for the construction manager—with greater exposure, typically, to claims by trade contractors that cannot be passed along to the owner.

One final note on changes: experienced construction professionals know that many changes are implemented with nothing more than a handshake agreement. Contractors and subcontractors alike often take certain steps to make changes, or redo small items of work, all without any legal obligation to do so, in order to maintain a good working relationship with others on the project site. There is a lot of horse trading that takes place on many projects, and many potential small claims are resolved without any formal paperwork to accompany them. The construction manager's skill and experience comes to bear in recognizing which matters can be handled informally and which ones require more formal attention and a paper trail.

3. *Continued Consulting on Value Engineering and Substitutions*

As noted above, with the ever-present potential for changes on a project, the construction manager must continue to advise the project owner on cost-cutting measures through the end of the project. Particularly when an owner has exhausted its own contingency, the construction manager will be called on to reduce costs in one area to make up for increased costs in another. The construction manager's failure to control costs does not necessarily result in any increased liability from a legal standpoint, but will clearly have an adverse impact on the construction manager's business relationship with its own customer.

D. *Schedule Control*

In much the same manner that a construction manager must complete the work within a project budget or GMP, the construction manager who is at-risk also must complete the work by a certain deadline. Owners select construction managers based on their perceived ability to forecast scheduling problems and develop the means to resolve or work around them.

An at-risk construction manager who signs a contract with a definite schedule and a *time is of the essence* provision will be liable to the owner in damages for failing to achieve timely completion as the result of its own failures or those of the subcontractors. In this regard, an at-risk construction manager's liability is commensurate with that of a general contractor in the traditional design-bid-build setting.[4]

A construction manager agent will not have any liability for the schedule or the prime contractor's failure to comply with the same. Rather, the construction manager agent typically acts as a consultant to the owner, advising the owner of known or anticipated schedule issues and concerns and recommending methods to deal with them.

1. *Establishing or Maintaining Baseline Schedule*

Just as maintaining cost control starts with a good budget, maintaining schedule control starts with a realistic baseline schedule. Typically, the construction manager develops a prospective schedule during the pre-construction phase of the project. During this time, the construction manager should identify potential long-lead time

---

4. While AIA Document A121/CMc does not include a time of the essence clause, it incorporates by reference AIA Document A201-1997, General Conditions of the Contract for Construction. Subparagraph 8.2.1 in the A201 document provides: "Time limits stated in the Contract Documents are of the essence of the Contract."

items, so that the design for those items can be finalized and the items ordered in sufficient time to maintain the overall schedule. The construction manager should also develop the original or baseline schedule with input from major subcontractors.

Once construction is under way, good practice calls for the construction manager to provide each trade contractor with its portion of the schedule and then seek confirmation that the amount of time allotted to that contractor's work scope is sufficient. The construction manager who fails to do so doubles its risk. First, the project schedule may not be maintained, so the construction manager becomes liable for late completion. Second, the construction manager may lose the cooperation of the trade contractor who has not been given a reasonable time period to perform its work; that relationship may turn into an antagonistic one as well.

In order to achieve the schedule, construction managers also take the owner-approved schedule and pass along the obligation to perform in a timely manner to each trade contractor. As a result, a contractor who fails to perform in a timely manner becomes liable to the construction manager. While some construction managers use this risk management tool (passing schedule risk along to the trade contractor), doing so without having established a reasonable baseline will only invite disputes and ensure that each side keeps churning out the paperwork to demonstrate that any failure to meet the schedule is due to another party.

2. Developing Schedule Updates

Schedules are management tools. A detailed CPM schedule not only shows the sequence and duration of various activities, but also allows the construction manager to monitor and forecast whether the work is keeping pace with the planned sequence or is falling behind. The schedule also allows the construction manager to identify when noncritical items that are falling behind may jeopardize the overall completion.

In order for the schedule to maintain its usefulness as a management tool, it must be updated periodically. The frequency of these updates is typically fixed by contract, but monthly and semimonthly updates are common. Also, the utility of an update requires the construction manager and owner to agree early in the project on what will be the practical meaning of a schedule update that shows a portion of the schedule falling behind. Standard contract terms are that a schedule update, in and of itself, does not modify the contract completion date. Even from the perspective of a construction manager, it is hard to argue otherwise. However, the parties should agree in advance whether the owner may take action if one component of the project falls behind, without any regard for the construction manager's ability to regain the overall schedule.

Practice Tip

The utility of schedule updates can be best understood in conjunction with the construction manager's ability to tap into contingency funds. If one aspect of the project is falling behind, particularly for reasons that are not the fault of a single subcontractor, the construction manager must have the ability to initiate a recovery schedule. If the construction manager is able to use contingency funds to expedite delivery of certain materials, or to pay for premium time so that a particular subcontractor will work its crews for longer hours each week, the construction manager is able to use the schedule updates in order to evaluate what steps must be taken in order to maintain the schedule. A construction manager who is at risk for falling behind in a part of the schedule, and who is at the same time blocked from using contingency funds to address the schedule delay, may take steps just to protect its own flanks and not take steps to regain the schedule. When negotiating the terms of the schedule obligations, it is important to keep in mind terms for the application and use of a project contingency.

---

3. *Control of Trade Contractor Schedules, and Coordination*

Control and coordination of trade contractor schedules are carried out in two ways. First, the trade contractor must be tied contractually to a schedule for its own scope of work that will support the overall project schedule. Second, the construction manager's project supervisors must take steps to enforce and implement the trade contractor's time commitments.

The vast majority of construction managers, much like general contractors, are familiar with the means to contractually enforce schedule requirements. The subcontractor is given a subcontract form with the schedule either attached or incorporated by reference, and the contract terms and conditions impose on the subcontractor an obligation to perform in accordance with that schedule. The issue becomes more complicated on construction manager projects when subcontractors, over the span of the buyout, sign subcontracts incorporating periodic updates or revisions to the schedule. For example, the structural steel subcontract on a project may incorporate Schedule Revision #2 and the curtain wall subcontract may incorporate Schedule Revision #4.

What happens in this instance? This will depend, in part, on other terms in the subcontract. The construction manager must retain the right to make periodic adjustments and revisions to the overall schedule, and each subcontractor must agree *contractually* to abide by those updated contracts. As a result, there still may be cost impacts to sort out, but the obligation to abide by updated schedules should be part of each subcontract, or the construction manager will find itself in a situation of having later-procured subcontractors filing claims due to the failure of earlier-procured subcontractors, or vice versa, to abide by schedule updates.

4. *Identification of Potential Schedule Impacts, and Developing Alternatives*

The term *work-arounds* did not originate with construction management, but it is appropriate for the value a construction manager brings to the table. When the project does not go as planned, and some aspect of the work falls behind, the construction manager must come up with a plan to work around the problem. Often, this involves modifying the sequence of operations, or even the use of alternative materials or components to maintain the overall schedule. The construction manager's ability to anticipate problems is of special importance here. The sooner the potential problem is spotted, the more alternatives there may be to deal with it.

Like the challenge of cost control, an owner expects its construction manager to employ a proactive approach to schedule control. Not only is the construction manager expected to foresee schedule issues, but also to develop plans to work around them so that schedule hiccups do not lead to major indigestion. By failing to achieve timely completion of the work, the construction manager who fails to do so risks being in breach of its contract. The construction manager's liability becomes its exposure to the owner's incremental costs associated with delayed completion, whether expressed as liquidated damages or measured as actual damages.

5. *Float*

Float is the term applied to schedule activities that may slide with no impact on the critical path. In a critical path schedule, most activities have float and only those activities comprising the critical path do not have float. The question that often arises is who owns the float?

The situation is understood best with an example. On a typical high-rise construction project, the critical path moves from the foundation up through erection of the structural steel, followed by installation of the roof before transferring to interior activities. Window installation, while an important aspect of closing in the structure to allow installation of interior finishes and heating of the building, is not typically on the critical path. Suppose the window installation subcontractor, with the construction manager's knowledge, plans to submit alternative windows to those specified because they are less expensive, even though they have a longer lead time due to delays with the manufacturer. The subcontractor makes an early submittal to deal with the longer lead time. However, the owner and designer take much longer than usual to analyze the alternative windows before finally approving them. As a result, the windows are ordered without adequate time to get them on-site and the late window installation impacts the critical path of construction.

Can the construction manager make a claim for extending the schedule because the owner and designer are taking too long to review the submittal? Or, was the owner entitled to take the necessary time and approve the submittal in order to allow the same lead time as would have been adequate with the windows originally specified? The issue of which party *owns* the float is one that many courts have grappled with, and no clear line of cases has evolved. On balance, though, a majority of courts have ruled that the owner owns the float in the absence of a contract clause allocating the same.

All standard form contracts are silent on this point. Thus, allocation of float remains an item of contract negotiation.

E. *Quality Control and Quality Assurance*

Maintaining work quality is one of the three critical elements of the construction manager's duties. If the construction manager fails to comply with quality standards in the plans and specifications, the construction manager will be liable to the owner. In fact, the construction manager's liability, with respect to quality of work, is typically coextensive with that of a general contractor in a traditional design-bid-build scenario with one caveat. The caveat is that many industry standard contract forms allow the construction manager to include, as a cost of the work (to the extent the GMP has not been reached), the cost of remedial work. If defects are found after completion, the construction manager's liability will approach or equal the general contractor's liability in the same situation.

One fundamental element of construction management is that price is not the sole criterion for selection of a construction manager. As a result, most construction management firms advertise and promote their attention to quality in the finished product in an attempt to distinguish themselves from competitors. A truly distinguishing characteristic, though, is not whether a construction management firm can tout its attention to quality assurance, but whether it can demonstrate actual experience in maintaining quality in the project. By signing on to perform the work in accordance with plans and specifications (which typically incorporate other industry standards by reference), the construction manager agrees to see that the work is carried out properly. Its liability for failing to do so will be commensurate with that of a general contractor who does not ensure that the work is performed properly. The cost of correcting, repairing, or replacing defective work will be the construction manager's responsibility.

As with cost and schedule requirements, a construction manager agent will typically have no responsibility for the failure of the prime contractor or subcontractors to perform the work properly.

1. *Responsibilities of the Construction Manager as Design Is Finalized*

During the design process, the construction manager must advise both owner and designer of potential constructability issues presented by the design. Although the construction manager also advises on design elements that will have an impact on the cost and schedule for the project, most of the construction manager's comments are likely to fall into the category of constructability, or the ability of the construction manager and its subcontractors to carry out the design.[5] As issues of constructability have a potential impact on the quality of the completed building, it is incumbent on the construction manager to raise any such issues during the design process.

Since the construction manager is liable to the owner for the failure of the construction to comply with the plans and specifications, the construction manager's failure to address known or suspected issues during the design process will only serve to increase the likelihood of quality issues arising during the construction process.

---

5. Review of the design for constructability, sometimes referred to as *construction feasibility,* is a common element of pre-construction services delineated in industry standard contract forms. *See, e.g.,* AIA Document A121/CMc-2003, § 2.1.2.

2. Responsibility of the Construction Manager Vis-à-Vis Designer

The construction manager is not liable for the design (leaving aside for the moment design/build elements of work). However, the construction manager has an obligation to review the design and to notify the owner of any deficiencies in the design that the construction manager either recognizes or, in the exercise of reasonable judgment, should recognize. One standard contract clause provides, in part:

> The Construction Manager shall consult with the Owner and Architect regarding ... selection of materials, building systems and equipment. The Construction Manager shall provide recommendations on construction feasibility; ... and factors related to construction cost, including estimates of alternative designs or materials ....[6]

Under many standard form contracts, the construction manager's failure to notify the owner of any errors in the design will impose liability on the construction manager for the incremental cost incurred resulting from the construction manager's failure to notify the owner. This does not make the construction manager a guarantor or peer reviewer of the design—as argued by many contractors when attempting to negotiate away the clause—but only for the costs that could have been avoided if a situation was known to the construction manager but it failed to say anything.

F. Risk Management

Risk management is a broad topic with many elements. The focus of this chapter is on the major elements of risk management during construction and the role of the construction manager. The major categories of risk are:

- cost overruns;
- schedule delays;
- quality in performance of the work;
- job site safety; and
- third party claims for property damage or personal injury.

Briefly stated, risk management for cost, schedule, and quality of work is typically accomplished by passing those risks through to subcontractors. Management of job site safety remains with the construction manager, although the construction manager typically requires that its subcontractors also assume responsibility for the safety of their own crews and work areas. Management of potential third party claims is mostly an insurable risk and is handled via an insurance program.

In many respects, the construction manager's handling of risk management issues is quite similar to that of a general contractor. By the same token, the

---

6. *See supra* text accompanying note 5. *See, e.g.*, AIA Document A121/CMc-2003, § 2.1.2.

construction manager's liability for failure to manage and control or delegate risk properly is similar or identical in many respects with that of a general contractor in the traditional design-bid-build setting, with some differences discussed below.

For a construction manager agent, risk management is not a direct concern. Most construction manager agency contracts specifically state that the construction manager agent does not undertake any liability for design services provided by the owner's design professionals, nor for the failure of the contractor and subcontractors to perform the work timely and properly. Since the construction manager agent has little risk as a consultant, such risk can be handled with careful management and with insurance against professional errors and omissions.

1. *Contracts and Delegation of Risk*

For all elements of risk that may arise from performance of work by subcontractors, the construction manager's primary method of risk management is to pass such risks along to the subcontractors via the subcontract. For cost risks, the construction manager typically will enter into subcontracts for lump sum amounts to perform a specified scope of work. Thus, if there are overruns in the cost of performing the subcontractor's scope of the work without any change in that scope, the subcontractor under a lump sum agreement will bear that risk.

In the same manner, schedule and quality of work performance are also passed along to the subcontractors who will be liable to the construction manager if they fail to perform. With this delegation of risk and with each subcontractor responsible for its own failure to perform properly, the construction manager's primary liability on cost, quality, and schedule arises when the construction manager fails to coordinate the work properly.

Contract performance risks may also be covered by the construction manager via payment and performance bonds provided by subcontractors. Thus, the cost, schedule, and/or quality risks presented by a subcontractor going out of business, or failing to pay lower-tier subcontractors and vendors, can be covered in this manner as discussed in more detail below.

Safety of the work site presents a slight variation on the contractual delegation of risk. In some states, statutes make the prime contractor responsible for safety on the work site and do not recognize contractual delegation of this duty. However, delegation of the duty to enforce worker safety and to keep the immediate work area safe may still be permissible even in those jurisdictions.

2. *Insurable Risks*

Insurable risks fall predominantly in the realm of third party claims. Most construction managers carry general liability insurance, which provides protection against claims of personal injury and property damage. Furthermore, on almost all construction projects, either the owner or the construction manager will carry and maintain builder's risk insurance which is property damage insurance covering the work in place. In the event of property damage occurring during construction, the builder's risk policy will likely provide insurance coverage for the costs of repair.

Even though third party risks are typically covered by the construction manager's own general liability insurance, most construction managers pass the requirement of obtaining similar levels of general liability insurance coverage on to each of the subcontractors for two reasons. First, prime construction manager contracts typically require the construction manager to take this step. Second, construction managers utilize insurance risk management tools to lower their own insurance premium costs. Should a construction manager fail to require appropriate insurance from a subcontractor, this failure could constitute a breach of the prime contract. It is not certain that this failure would constitute a material breach giving rise to anything more than nominal liability of the construction manager to the owner. More importantly, in the event of a third party claim arising from that subcontractor's operations, if the construction manager fails to require a subcontractor to obtain the requisite insurance, the construction manager and/or its own insurance carrier could be required to pay the loss that would have been borne by the subcontractor's insurance.

3. Surety Bonds and Performance Risk

The risk to the construction manager that subcontractors will fail to perform is sometimes secured via a performance bond. With such a bond, if the subcontractor defaults in its performance, the construction manager can make a demand on the subcontractor's surety to arrange for completion of the subcontractor's work.

In the construction management project delivery method, it is common that owners do not require any performance bond from the construction manager but the construction manager does require performance bonds from each of the major subcontractors. This is a typical manner for the construction manager to lower its own overall risk of performance to the owner. In the event that a particular subcontractor fails to perform, thereby placing the project cost or schedule in jeopardy, the construction manager can call upon the surety whose liability may be coextensive with that of the subcontractor. This is of particular importance if the subcontractor's failure to perform results from a financial downfall.

While there are innumerable project scenarios involving surety bonds, the construction manager's liability in this area typically arises from two common situations. One is when the construction manager fails to ensure that a subcontractor furnish the requisite bonds. In that scenario, there is no security for the defaulting subcontractor and any project costs incurred in completing the work above the GMP may be at the construction manager's risk. The second scenario is when the construction manager fails to take the proper steps to declare the subcontractor in default and make demand on the surety. The surety in that situation may deny liability on the grounds that an essential condition of its obligation to complete the work (timely and proper notice of default and demand on the surety) has not occurred. The construction manager in that scenario again may find itself liable for excess costs of completing the subcontractor's work that would otherwise have been borne by the surety.

4. *Management of Risk That Is Not Transferred or Insured*

For risks neither transferred by contract nor covered by insurance, the construction manager must manage the risk to avoid or minimize its liability. This aspect of risk management focuses on the word *management*. Following appropriate procedures for verifying subcontractor qualifications, identifying work scope and soliciting bids, maintaining proper supervision of the work in the field, closely monitoring the progress of the work, and taking action to spur progress when it is lagging are all elements of common sense project management, as well as proper risk management tools.

The construction management structure opens lines of communication that historically are more open than between owner and general contractor and project team members that typically act as a team and not in an adversarial manner. The theoretical risk may remain the same—such as, for quality in performance of the work—but by virtue of its involvement in the pre-construction phase, the construction manager typically will identify the primary issues to be faced and, thus have ample opportunity to prepare for such risks. So, as a practical matter, the construction manager's ability to manage the risk is enhanced by the knowledge gained during pre-construction and by the relationship between team members. Thus, the lines of communication developed often result in prompt resolution of issues.

G. *Safety*

Safety is one area where the construction manager at-risk's responsibility and attendant liability is similar or identical to that of a general contractor. Since the construction manager at-risk has overall responsibility for performance of the work, it has liability for safety violations occurring during the work. All of the industry standard form contracts treat the construction manager at-risk's responsibility with respect to safety the same as a general contractor. The construction manager at-risk is not only responsible for safety on the job site, but must also indemnify the owner and designer from any claims arising out of its failure to comply with this obligation.

In contrast, the construction manager agent's liability for safety in performance of the work is remote. It may have a duty to advise the owner of practices that it considers unsafe, but its duty typically is solely to the owner and not to any other person or party. In situations of severe accidents or worker deaths, the U.S. Occupational Safety and Health Administration (OSHA) has charged construction manager agents as supervisors, alleging liability on the part of those construction managers for a violation of OSHA standards. However, in more common situations, construction manager agents and other consultants have not been found liable for violation of OSHA standards. As a result, the construction manager agent has less liability exposure for safety violations.

H. *Dispute Resolution*

Most industry standard form contracts do not assign any formal role in dispute resolution to construction managers. Standard form contracts continue to place the architect, if anyone, in this role.[7] However, the industry practice is quite different. Most construction management firms tout the customer relations aspect of their business and pride themselves on good working relationships with design professionals and, for those having this client base, with institutional owners. As with any service industry not driven by lowest cost, customer satisfaction is critical. As a result, many construction managers find themselves in the role of facilitator or mediator, whether between owner and architect or between owner and subcontractors, or even in dealing with their own claims. In the absence of any contractual requirement for such services, construction managers undertaking this role do not invoke significant practical liability as a result.

I. *Customer Service and Customer Relations*

Construction managers must be responsive to their customers as a matter of survival. This is not an issue of legal responsibility or risk, but one of business judgment. The standard contract forms deal with this not as a customer relations or service issue, but via a list of duties and responsibilities. The concept of service, as that term is traditionally used, is not present in the standard forms. The construction manager who fails to listen to its customer and respond to concerns or inquiries will soon find that it is no longer on proposal lists. As noted before, construction manager selection is not typically via low bid or low price. Customer service remains an essential element of the construction manager's work.

Library References

C.J.S. *Contracts* §§ 508, 531, 557, 565, 573–574, 578–579, 585–586, 592–593; C.J.S. *Counties* §§ 165–170; C.J.S. *Municipal Corporations* §§ 917–933, 946–952; C.J.S. *Negligence* §§ 399, 595–596, 598–601, 618–619; C.J.S. *Principal and Surety* §§ 2–3; C.J.S. *Public Contracts* §§ 6, 14–24, 29–41; C.J.S. *Schools and School Districts* §§ 409–414, 435, 447–450, 761; C.J.S. *United States* §§ 113, 127, 130–136, 139, 143.

West's Key No. Digests, Contracts <KEY>285, 298, 319–321, 285; Counties <KEY>115, 129, 130; Indemnity <KEY>33(5); Municipal Corporations <KEY>234, 254, 255; Negligence <KEY>1205; Principal and Surety <KEY>1; Public Contracts <KEY>5, 23–33; Schools <KEY>80(2), 86(2); United States <KEY>64, 73(19), 74–75.

III. The Construction Manager's Role, Risks, and Responsibilities during Completion

As the project nears completion, the focus of effort shifts. The construction manager starts to come out of production mode and moves into wrap-up mode. The

---

7. *See, e.g.,* AIA Document A201-1997, General Conditions of the Contract for Construction, ¶ 4.4.

completion effort involves a broad range of activities and administrative effort that must bring all outstanding issues and matters to a close. The construction manager must oversee work of multiple trades to complete the punch list and get subcontractors back into the building often for small lists of disparate items to be fixed or completed. At the same time, the construction manager must complete various items of paperwork. All subcontractors will be asked to sign project close-out documents (perhaps, final change orders), provide any manufacturers' warranties, and supply their own warranty documents.

A. *Management of Punch List*

The punch list is often the focus of project completion and turnover to the owner. It is a list compiled by either the construction manager or the designer, or both, on a room-by-room basis, listing items that must be repaired or completed. Scratches in the paint, missing outlet covers, a cabinet door hinge that is not working, a short strip of vinyl base that was never installed, a shrub that has died and must be replaced—all are the type of items that show up on punch lists.

Management of the punch list requires the construction manager to take the lengthy list of items, figure out which subcontractor is responsible for each item, distribute the list to the pertinent subcontractors, and then hound them to get back to the project site to perform the work.

Practice Tip

Most standard form contracts do not deal with the holdback for punch list work in any detail. As a practical matter, there are two points to address in the prime contract which should then be passed along to all subcontractors via the contract general conditions.

First, the amount of the holdback for punch list work should be specified. The value of this holdback varies in practice from 150 percent to 200 percent (meaning, if the cost of repairing a particular item is estimated to be $25, then the punch list holdback will be 150 percent or some other specified percentage of that value). Having this value specified in advance will reduce complaints, confusion, and disputes down the road.

Second, the general conditions should specify who will determine this value and whether it will be a subjective or objective determination. Typically, the architect determines the initial value of the punch list. In theory, however, the construction manager would be in the better position to assess or determine the level of effort required to complete a particular item of work. Since the money is being retained on the subcontractors, via the construction manager, there is no real incentive on the construction manager's part to come up with a number that is too low, as some owners and designers often believe. Take the time in the contract negotiations to identify and respond to each of these points.

As punch list work is performed, the construction manager must confer with the designer and/or owner, verify with them that the work items are now satisfactory, and then pursue release of money relating to the value of completed punch list work. In theory, this process continues on an iterative basis until all punch list items are done and all contract funds released. In practice, the punch list effort sometimes comes to a stalemate when particular subcontractors never get around to completing their work. When that occurs, either the construction manager or the owner must arrange for another contractor to come in and wrap up the incomplete items, or else the construction manager and owner agree to a permanent reduction in the cost of the work to be paid reflecting the value of the unfinished work. Since subcontracts typically are written on a lump sum basis, the permanent reduction results in a change order to the subcontractor, reducing the amount of the subcontract so that the construction manager's costs of performance are likewise reduced.

B. *Commissioning and Turnover*

Commissioning of equipment does not happen in the formal sense on every project, but it does happen whenever the project includes installation of mechanical equipment of any sort. This equipment must be started up and any initial procedures that are recommended or required by the manufacturer must be carried out. "Turnover" is a broader term and encompasses the transfer of control of the facility to the owner's operations personnel. This is an area rife with potential misunderstanding. It is also an area that often is not dealt with adequately during contract negotiations. Standard form contracts do not define the process of commissioning and turnover, or allocate responsibility among the parties for the same. It is important that this point be addressed during the contract negotiations to minimize disputes later.

Practice Tip

Many marketing-savvy construction managers see the turnover and commissioning phase as an opportunity both to minimize risks and to improve customer relations. At the contract negotiations stage consider what elements of project completion could be enhanced by a training session for the owner's operations personnel. Consider how best to organize information about the completed project so that maintenance and operations personnel can readily find critical names, phone numbers, and the like. Although many firms seek to avoid callbacks, others recognize the opportunity to gain the customer's time and attention, and either nip a potential risk in the bud, or else turn a potential problem into another example of a satisfied customer.

For this process to work effectively, both construction manager and owner must cooperate. The construction manager must gather information about operation of any equipment to hand over to the owner and arrange, when appropriate, to get a manufacturer's representative to the project site to work with the owner's operations personnel. The owner must ensure that its own crews are ready and available at the appropriate times for orientation and training on the new equipment.

As a general rule, the construction manager's responsibility and liability will depend on the agreed terms for commissioning and turnover. In the absence of any delineation, the construction manager may have no liability if the owner fails to properly operate the newly installed equipment. As a practical matter, even in the absence of contractual or common law liability, the construction manager will still be called on should equipment fail during the initial months of operation. Thus, it is good practice for the construction manager and owner to agree up front on the effort and coordination both will undertake at the point of transfer of the facility.

1. *Training*

Training is typically conducted by the manufacturer or by an authorized representative of the same. The authorized representative may be the subcontractor who installed the equipment, but usually it is not. Should either owner or construction manager believe early in the project that training by the subcontractor will be appropriate or required for particular items of equipment, the specifications for that equipment should be written to include training within the subcontractor's scope of work. The construction manager and subcontractor will then have a common level of understanding as to what must be done. The subcontractor will be able to take the specification and pass it along to the manufacturer so that the cost of training is included in the price of the equipment.

2. *Start-Up of Equipment*

Some types of equipment are not readily subject to damage from improper start-up while others may be. It is not typically the province of the construction manager to know which is which. As a result, the construction manager, owner, and designer should confer early and then issue specifications that obligate the manufacturer (via a subcontractor) to identify particular start-up procedures or routines, and to monitor or supervise the start-up on-site when appropriate. It is in the best interests of all parties to ensure that this process works smoothly.

While a construction manager may not have liability as a matter of contract for improper start-up, it will undoubtedly be contacted for assistance with a warranty claim if equipment fails for any reason during start-up or shortly thereafter. This issue will be resolved eventually according to the cause of the failure, but efforts taken either at the contract negotiations stage or at the point of initial start-up may go a long way toward avoiding warranty claims later on.

C. *Procuring and Handling Warranties/Guarantees*

One of the critical aspects of project completion is the compiling of warranty documents from all manufacturers and vendors. In keeping with industry practice, the construction manager has a duty under all standard contract forms to compile this documentation.

1. *Responsibility to Obtain Warranty*

All industry form agreements include an obligation to hand over manufacturers' and vendors' warranties as a condition of final payment.[8] This obligation is passed along to subcontractors, but it will eventually fall on the construction manager to complete the task either directly or indirectly.

As a practical matter, the process of obtaining warranties becomes an issue when a lower-tier vendor or a manufacturer has not been paid for its product or equipment. Most vendors in that setting will refuse to furnish a warranty on the assumption that withholding the warranty will serve as leverage to ensure payment. Frequently, this strategy works. Thus, the construction manager may be given the task of seeing that payment disputes are worked out between subcontractor and equipment vendor to obtain all of the warranty and close-out documents.

2. *Responsibility for Performance of Warranty Obligations by Vendors/Subcontractors*

The performance of warranty obligations is an area where theory and practice diverge. In theory, most standard form contracts require the at-risk construction manager to be responsible for performance of warranty work. If that work was performed by a subcontractor, which invariably will be the case because most construction managers do not self-perform work, then the construction manager is responsible for getting the subcontractor back to complete the work. In practice, many construction managers take steps to remove themselves from this role and attempt to get the project owner to deal directly with each subcontractor.

A standard contract modification sought by owners requires the construction manager to guarantee performance of warranty work by the subcontractors. This warranty-within-a-warranty provides more protection for the owner, but is problematic for a construction manager. On balance, the construction manager must expect to remain responsible for getting the subcontractors to return for remedial action. Its failure to do so may result in breach of contract claim and attendant liability for the remedial cost that would or should have been covered by a warranty.

---

8. AIA Document A201-1997, General Conditions of the Contract for Construction, ¶¶ 9.8.4, 9.10.1.

D. *Audit or Cost Verification*

Since the predominant form of compensation on construction management contracts is cost of work plus a fee, many owners will want to audit the construction manager's costs at the end of the project. This is a common requirement, and most construction managers are accustomed to the practice. Final payment sometimes is withheld pending this audit and then adjusted depending on the outcome of the audit. For example, a common contract clause states:

> The Owner's accountants will review and report in writing on the Construction Manager's final accounting within 30 days after delivery of the final accounting to the Architect by the Construction Manager.[9]

A common point of negotiation is whether the construction manager must bear the cost of the audit if the audit uncovers any material billing discrepancies. Since the likelihood of a billing discrepancy on a multimillion dollar project is quite high, most construction managers (and owners) recognize that some discrepancies are to be expected as a matter of practice; these discrepancies can ultimately work out in favor of either owner or the construction manager. Since the owner is liable for the cost of the work, an audit uncovering costs that were not passed along will result in increased liability of the owner to the construction manager (assuming the GMP has not been reached). Conversely, since the construction manager is only entitled to be paid the cost of the work plus its fee, an audit uncovering duplicate billing, for example, will result in the construction manager having to reduce its final payment or even reimburse the owner for the duplicative amount.

This issue is often resolved by having the construction manager be responsible for the cost of an audit or for a portion of the cost relative to the outcome of the audit. For instance, the parties may agree that the construction manager will bear the cost of the audit if any discrepancies uncovered in favor of the owner exceed the cost of the audit itself or the parties may agree that the construction manager will bear the cost of the audit if the discrepancies uncovered are material, leaving to future negotiations between the owner and construction manager when small variations are uncovered.

Library References

C.J.S. *Counties* § 158; C.J.S. *Municipal Corporations* §§ 934–936; C.J.S. *Public Contracts* § 10; C.J.S. *Schools and School Districts* §§ 442, 758; C.J.S. *States* §§ 289–290, 292–293; C.J.S. *United States* §§ 105–106, 117–119.

West's Key No. Digests, Counties <KEY>126; Municipal Corporations <KEY>250; Public Contracts <KEY>16; Schools <KEY>84; States <KEY>104; United States <KEY>70(14), 70(15).

---

9. AIA Document A121/CMc, § 7.2.3.

IV. Coordination with Others—an Independent Duty, or an Adjunct to All Other Tasks?

All standard form contracts call for the construction manager to coordinate its work effort with the designer and owner. The very essence of a construction manager is that of a team player whose interest is in building what the owner wants. The profit motive of the construction manager does not conflict with that essence, at least in theory.

In language that is common to a number of contract forms, the construction manager "accepts the relationship of trust and confidence established with the Owner [by the contract], and covenants with the Owner to furnish the construction manager's reasonable skill and judgment and to cooperate with the Architect in furthering the interests of the Owner."[10] Those interests, in turn, can be advanced only if the construction manager maintains open lines of communication with others on the project team. While coordination with others is an adjunct to the work of a general contractor, for a construction manager it is an essential contract requirement. Any construction management contract containing the language set forth above, or words of similar import, places upon the construction manager an express contractual duty to cooperate with the owner and architect.

Library References

C.J.S. *Contracts* §§ 11, 343.
West's Key No. Digests, Contracts <KEY>198.

V. Construction Manager as Advisor or Agent

Owners sometimes engage a construction manager as an agent to monitor the design and construction process and to act as the owner's eyes and ears on the project. This chapter has been written predominantly from the perspective of the construction manager at-risk, but there are a few points to be made concerning the consulting role of the construction manager agent. For greater discussion about the role of construction manager as agent, with more detailed discussion of practice standards, refer to a publication of the Construction Management Association of American, *2002 CMAA Construction Management Standards of Practice*.

A. *Agency Role as Shield from Third Party Claims*

In a number of jurisdictions, particularly those adopting portions of the Restatement (Second) Agency § 320 (1958), an agent acting entirely within the scope of its agency is immune from third party claims. That is to say, the agent is entitled to indemnity from its principal for such claims. Thus, the construction manager agent not only has no contractual liability for performance of the design or the

---

10. AIA Document A121/CMc, § 1.1.

construction, but also enjoys immunity from independent claims as the result of the agency relationship.[11]

This standard highlights a critical issue: whether the construction manager has truly acted solely within the scope of its agency. It will not take much creativity on the part of a claimant's lawyer to allege that some acts of the construction manager agent were outside the scope of its agency. The reality of construction projects, and the interaction among construction manager agent, owner, designer, and contractor will invariably result in situations where the construction manager agent undertakes a role or responsibility beyond that of its written contract.

So while the construction manager agent is an *agent*, and enjoys some degree of immunity as a result, its primary role ensures that its liability to third parties will be minimal. Of course, this does not have any effect on the construction manager agent's liability to the owner who has engaged it.

B. *Construction Manager's Duty to Owner as Agent*

The construction manager agent's role is that of a consultant. In the absence of a contract setting forth duties and standards of effort, its role will be viewed in the light of a professional providing services. As such, the construction manager agent will be liable to an owner for failing to exercise the standard of care expected of construction manager agents who are similarly situated. Many construction manager agency firms carry professional liability insurance on a claims-made basis providing some degree of protection for owners in the event of such claims.

C. *Imputation of Conduct to Owner*

According to the Restatement (Second) Agency § 320 (1958), the acts of the construction manager agent, acting in its role as agent, are imputed to the owner. Most owners attempt to limit their vicarious liability, at least in part, via the terms of the prime design and construction contracts. Owners typically will provide that the contractor is not entitled to rely upon any direction given by the construction manager, and must receive authorization directly from the owner before incurring any cost or making any change. Of course, such a contractual directive, while shielding the owner in certain respects, has a tendency to impede progress of the project. Recognizing this contradiction, sophisticated owners usually take steps, either in the prime contract or in their relationship with the construction manager agent, to ensure that project decisions are timely made.

Library References

C.J.S. *Agency* §§ 344, 346–349, 352, 369, 384–385, 389, 392, 410.

West's Key No. Digests, Indemnity <KEY>33(5); Principal and Agent <KEY>136, 150(3).

---

11. Rich-Taubman Assocs. v. Comm'r of Revenue Servs., 674 A.2d 805, 808 (Conn. 1996); RESTATEMENT (SECOND) AGENCY § 320 (1958); *see also,* United States Fid. & Guar. Co. v. S.B. Phillips Co., 359 F. Supp. 2d 189, 198–99 (D. Conn. 2005); and Gordon v. Vermont Pure Holdings, Ltd., 2001 WL 1585261, at *2 n.1 (D. Conn. Aug. 14, 2000).

VI. Conclusion: Use of Standard Form Contracts and Documents

Several industry and professional trade associations have published standard form contracts for construction management relationships. To varying degrees, these forms accommodate the issues addressed above. Publications from three of those associations are noteworthy.

A. *CMAA*

The Construction Management Association of America (CMAA), www.cmaanet.org, publishes model contracts both for construction manager at-risk and construction manager agency relationships. The construction manager at-risk contract forms include:

- CMAR-1: Standard Form of Agreement, Owner-CM;
- CMAR-2: Standard Form of Agreement, CM-Contractor;
- CMAR-3: General Conditions, CM-Contractor; and
- CMAR-4: Standard Form of Agreement, CM-Designer.

The CMAA forms for construction manager agent relationships include:

- A-1: Standard Form of Agreement between Owner and CM, CM as Owner's Agent;
- A-2: Standard Form of Agreement between Owner and Contractor;
- A-3: General Conditions between Owner and Contractor; and
- A-4: Standard Form of Agreement between Owner and Designer.

B. *AGC*

The Associated General Contractors of America (AGC), www.agc.org, publishes contract documents for construction manager at-risk relationships. One of the most popular forms, AGC 565/AIA Document A121/CMc, is published by AGC jointly with the American Institute of Architects. The AGC forms include:

- AGC 510: Standard Form of Construction Management Agreement Between Owner and Construction Manager (Where the Construction Manager is the Owner's Agent and The Owner Enters Into All Trade Contractor Agreements);
- AGC 520: Standard Form of Agreement Between Owner and Trade Contractor (Where the Construction Manager Is The Owner's Agent);
- AGC 530: Standard Form of Agreement Between Owner and Architect/Engineer (Where a Construction Manager Acting as an Agent Has Been Retained by the Owner);
- AGC 565: Standard Form of Agreement Between Owner and Construction Manager (Where the Construction Manager Is Also the Constructor) (same as A121/CMc-2003); and
- AGC 566: Standard Form of Agreement Between Owner and Construction Manager (Where the Construction Manager Is Also the Constructor and Where the Basis of Payment Is the Cost of the Work Plus a Fee and There Is No Guarantee of Cost) (same as A131/CMc-2003).

C. *AIA*

The American Institute of Architects (AIA), www.aia.org, is perhaps the most well-known publisher of standard form design and construction contracts. The AIA forms include:

- A121/CMc-2003: Standard Form of Agreement Between Owner and Construction Manager (Where the Construction Manager Is Also the Constructor) (same as AGC 565);
- A131/CMc-2003: Standard Form of Agreement Between Owner and Construction Manager (Where the Construction Manager Is Also the Constructor and Where the Basis of Payment Is the Cost of the Work Plus a Fee and There Is No Guarantee of Cost) (same as AGC 566); and
- A201/CMa-1992: General Conditions of the Contract for Construction, Construction Manager-Advisor Edition.

VII. Treatises and Resources

CONSTRUCTION MANAGEMENT ASSOCIATION OF AMERICA, CAPSTONE: THE HISTORY OF CONSTRUCTION MANAGEMENT PRACTICE AND PROCEDURES (2003).

CONSTRUCTION MANAGEMENT ASSOCIATION OF AMERICA, 2002 CMAA CONSTRUCTION MANAGEMENT STANDARDS OF PRACTICE (2002).

ASSOCIATED GENERAL CONTRACTORS OF AMERICA, CM/GC GUIDELINES FOR BUILDING CONSTRUCTION IN THE PUBLIC SECTOR (1999).

ASSOCIATED GENERAL CONTRACTORS OF AMERICA, PROJECT DELIVERY SYSTEMS FOR CONSTRUCTION (2004).

AGC OF MASSACHUSETTS AND THE BOSTON SOCIETY OF ARCHITECTS, THE CLIENT ADVISOR (2000).

CHAPTER 12

The Construction Manager and Tort Liability

WENDY KENNEDY VENOIT, ESQ.
DAVID A. DELMAN, ESQ.[1]

I. Introduction

This chapter covers the various tort liability claims that a construction manager can face on a project, whether these claims are asserted by the owner, the contractors, or other third parties. This chapter also covers the defenses a construction manager may assert to these various theories of liability.

The *construction manager* label has become a generic term to describe the situation where an owner hires an individual or firm to manage the construction project on the owner's behalf. Since the practice of construction management commenced, the construction manager's role has evolved and can mean different things in different contexts, depending upon the intent of the parties and the terms of the construction management agreement.

The two most common types of construction managers are construction managers at-risk and agency construction managers. Where the construction manager is at-risk, the construction manager assumes, to a greater or lesser extent, some responsibility for cost, schedule, and performance of the structure or facility being built. For example, the construction manager may bear the risk of ensuring the completion of the project for a particular price on a defined schedule, and may even act as the prime contractor for the project who is responsible for the work of the trade contractors. Where the construction manager actually contracts with the trade contractors, its role is more akin to a traditional general contractor and the construction manager will face those liabilities that are customarily imposed on general contractors.

---

1. We would also like to thank our associate, Robert Paterno, for his assistance in drafting this chapter.

In contrast, the agency construction manager operates as the owner's *agent* without assuming any liability for or offering any guarantees regarding the success of the project. An agency construction manager is similar to the traditional owner-architect relationship in that the construction manager acts as the owner's agent, managing the construction project in exchange for a fixed fee or a cost-plus-a-fee. The agency construction manager may also assume a design role on the project, thereby invoking the responsibilities and liabilities traditionally faced by design professionals discussed elsewhere in this handbook.

Of course, there are many variations to the construction management arrangement. Ultimately, the scope and responsibilities of the construction manager will depend on the intent of the parties as reflected in the terms of the agreement executed by the owner and the construction manager. For this reason, it is important for both the owner and construction manager to fully understand their agreement at the outset of the project so that both parties do not develop false expectations of the construction manager's role and responsibilities.

Each type of construction management arrangement offers its own benefits and drawbacks. More importantly, that arrangement will dictate the potential liabilities that the construction manager may face.

This chapter will focus on the tort liability theories that may be asserted against a construction manager by the owner and third parties, as well as the construction manager's defenses to such claims. When possible and appropriate, this chapter will differentiate between the liabilities and defenses available with respect to the two main types of construction management schemes.

II. Tort Liability of the Construction Manager to the Owner

A. *Theories of Liability*

There are many theories of tort liability that an owner may assert against a construction manager. The availability of such theories is largely dependent on the terms of the contract and the specific role of the construction manager. For example, if the construction manager is at-risk, its liabilities to the owner may substantially differ from those of a construction manager acting as agent for the owner. However, the most common tort theories are professional negligence, breach of fiduciary duty, intentional or negligent misrepresentation, unfair trade practices, and fraud. This section will discuss the applicability of each of these theories of liability in the at-risk and agency construction management contracting schemes.

1. *Professional Negligence*

Professional negligence is perhaps the most common tort theory alleged by owners against construction managers. Four essential elements are required to prove professional negligence: (1) the construction manager had a duty to act in accordance with a particular standard of care; (2) the construction manager breached that duty; (3) the breach of the standard of care proximately caused the alleged harm; and

(4) the claimant suffered damages as a result of the construction manager's breach of the standard of care.[2]

a. Duty/Standard of Care

Unless they are spelled out in the contract, the duties owed by the construction manager and the applicable standard of care will be governed by common law. Generally speaking, a construction manager is expected to perform its services with the same degree of care and diligence as a reasonably prudent construction manager performing similar services in the same community.[3] This is the standard of care articulated in many of the standard form contracts.[4] More often than not, the contract under which the construction management services will be performed will establish the duties and standards that the construction manager must meet. The construction manager's duties may also be supplemented by nondischargeable obligations imposed by applicable codes, statutes, and regulations of the state in which the project is being executed.

The application of the standard of care by which the construction manager is judged will vary according to the specific circumstances of the project, such as the location, nature, and complexity of the work. For example, the application of the standard of care to a professional in charge of a large, complex industrial plant will be different than a professional in charge of the construction of a single family residence. Similarly, a professional that holds itself out as a specialist or as possessing some heightened degree of skill or knowledge may be held to a higher standard of care, if a heightened standard is articulated in the parties' agreement.

Unless the construction management contract explicitly states otherwise, the standard by which the construction manager is judged is not one of perfection. In other words, a construction manager does not have a duty to provide a flawless design or to administer the project perfectly. Rather, the construction manager must exercise *reasonable* care and skill in the performance of its work. Negligent

---

2. *See, e.g.*, Garretson v. Miller, 99 Cal.Ct. App. 4th 563, 121 Cal. Rptr. 2d 317 (2002).

3. Constance Frisby Fain, *Architect and Engineer Liability*, 35 WASHBURN L.J. 32, 35 (1995).

4. For example, AGC Doc. 510 (AGC 510) states that the "Construction Manager warrants that it will perform its Services in accordance with the standard of care normally practiced by construction management firms in performing services of a similar nature at the time and place the Services are performed." AGC Document 510, Standard Form of Construction Management Agreement Between Owner and Construction Manager (Where The Construction Manager Is The Owner's Agent and The Owner Enters Into All Trade Contractor Agreements), Art. 5.2.2 (1997). The combined AGC 565/AIA A121 contains a similar statement that the construction manager "covenants with the Owner to furnish the Construction Manager's reasonable skill and judgment." *See* AIA Document A121/CMc and AGC Document 565, Standard Form of Agreement Between Owner and Construction Manager (Where the Construction Manager Is Also The Constructor), Art. 1.1 (1991).

conduct may consist either of an act or a failure to act when there is a duty to do so. In other words, negligence is the failure to do something that a reasonably careful person would do, or doing something that a reasonably careful person would not do, in light of the surrounding circumstances established by the evidence in the case.[5] Applying this general definition of negligence to construction managers, the focus of any inquiry will be on the methods employed by the construction manager, rather than the outcome of those methods. However, more often than not, a bad outcome is a strong indicator of methods that fall below the standard of care. Of course, the parties are always free to negotiate performance guarantees to ensure that the outcome of a project meets the expectations of the owner/client. Performance guarantees are typical when the construction manager is at-risk.

In measuring the construction manager's performance of its contractual obligations to the owner, a comparison is made between the methods actually employed by the construction manager and the methods employed by other construction managers undertaking similar work in the community. What constitutes similar work and the applicable community is grist for the lawyers. If the construction manager does not meet the standard, it will be exposed to liability. Generally, to establish the relevant standard of care, and whether it has been met by the construction manager, it is necessary for the owner alleging professional negligence to present the testimony of an expert.[6] However, this may not be necessary where the professional's negligence is within the comprehension of a layperson.[7]

Generally, it is always advisable to establish the applicable community and similar work standards in the contract. For example, if the project at issue is a petrochemical facility, the contract governing the construction manager's duties will often state that the construction manager will perform its services in accordance with the standards of care and diligence normally practiced in the international oil and gas industry. The words *international oil and gas industry* will likely heighten the standard of performance (i.e., the methods employed) that the construction manager must meet. Additional sources for determining the standard of care and the duties owed by the construction manager include, but are not necessarily limited to, licensing requirements, procurement codes, building codes, ISO certifications, and/or other established industry standards, such as critical path method (CPM) scheduling and cost control/accounting programs.

---

5. *See, e.g.*, PA. SSJI (Civ.) § 3.01 (2003).

6. Fain, *supra* note 3, at 39 ("As a general rule, in cases involving architects and engineers, expert testimony is ordinarily necessary to establish the parameters of the professional standard and the breach of that standard.").

7. Fain, *supra* note 3, at 39–40. *See also* Cipriani v. Sun Pipe Line Co., 574 A.2d 706 (Pa. Super. Ct. 1990); Nelson v. Virginia, 368 S.E.2d 239 (Va. 1988).

A recent case in Connecticut highlights the use and importance of building codes as a measure for an accepted standard of care.[8] In early December 2005, a Connecticut jury delivered a $32 million plaintiff's verdict in favor of a worker who was severely injured after an improperly welded beam fell and struck him while he was working on a construction project. The verdict, which was entered against the general contractor on the project, was premised on the theory that the general contractor had breached the duty of care it owed to the plaintiff, an employee of a subcontractor, by failing to comply with applicable Connecticut building code requirements. Specifically, the jury's verdict in *Pelletier v. Sordoni/Skanska* was based on the trial court's determination that the general contractor had a duty to inspect all welds on the project. In the words of the court: "I agree with the plaintiff that the BOCA National Codes impose a separate and distinct obligation on the permit applicant to conduct special inspections of all steel welds to confirm that they meet contract and code specifications."[9] By the time of the trial court's ruling, it was already undisputed that the general contractor did not inspect all of the welds, much less the weld at issue. Thus, the trial that followed was essentially a hearing on damages.

b. *Failure to Comply with Licensing Requirements*

Although most states do not have separate licensing requirements for construction managers, many states do have licensing requirements for contractors and design professionals, which would also encompass construction managers and/or the activities performed by construction managers.[10] Construction managers should take care to ensure that they are in compliance with any potentially applicable licensing laws.

For example, Colorado defines the practice of architecture to include "observation of construction" and "contract administration."[11] Similarly, Colorado defines the "practice of engineering" to include "observation of construction to evaluate compliance with plans and specifications."[12] Thus, even though Colorado does not have a licensing statute specific to construction managers, a construction manager hired to observe the construction or to administer contracts may be obligated to possess an architecture or engineering license in order to perform these functions.

---

8. Pelletier v. Sordoni/Skanska Constr. Co., No. X06-CV-95-0155184S, 2004 WL 3128800, 38 Conn. L. Rptr. 404 (Conn. Super. Ct. Dec. 16, 2004).

9. *Id.* at *5.

10. For a compilation of the applicable licensing laws in all fifty states, *see* STATE-BY-STATE GUIDE TO ARCHITECT, ENGINEER, AND CONTRACTOR LICENSING (Stephen G. Walker et al. eds., 1999).

11. COLO. REV. STAT. § 12-4-102(5) (2005).

12. COLO. REV. STAT. § 12-25-102(10).

In some states, the definition of the practice of architecture is even broader.[13] For example, in Alabama, the practice of architecture is defined as rendering, or holding oneself out as able to render,

> any service by consultations, investigations, evaluations, preliminary studies, plans, specifications, contract documents and a coordination of all factors concerning the design and observation of construction of buildings or any other service in connection with the design, observation or construction of buildings located within the boundaries of the state, regardless of whether such services are performed in connection with one or all of these duties, or whether they are performed in person or as the directing head of an office or organization performing them.[14]

Thus, a construction manager performing any of these services should, in the abundance of caution, obtain an architect's license before undertaking to perform such services on a project.

Likewise, the definition of engineering has been broadly defined in many states to include "professional observation of construction" and "design coordination," including the review and coordination of technical submissions prepared by others, such as by consulting engineers, architects, landscape architects, land surveyors, and other professionals.[15] In such states, the construction manager who undertakes to coordinate designs prepared by the architects and consulting engineers may also be obligated to maintain an engineer's license.

Lastly, in recent years, a number of states have imposed licensing requirements on contractors generally.[16] The primary purpose of these laws is to protect the public from incompetent contractors and to assure that structures are properly built. Some of these states have specifically identified construction managers as persons who must be licensed as general contractors.[17] This is especially true where the construction manager is at-risk and is acting more like a traditional general contractor.

---

13. *See, e.g.*, ALA. CODE § 34-2-30(2) (2002); ALASKA STAT. § 08.48 (2004); ARK. CODE ANN. §§ 17-15-102, 17-15-301 (2001); CAL. BUS. & PROF. CODE § 5500.1 (1993); NEV. REV. STAT. 623.023 (2003); UTAH CODE ANN. § 58-3A-102(6) (2002).

14. ALA. CODE § 34-2-30(2).

15. ALA. CODE § 34-11-1(7); ALASKA STAT. § 08.48; ARK. CODE ANN. § 17-30-101(3); CAL. BUS. & PROF. CODE §§ 6700–6706.3 (2002); NEV. REV. STAT. 625.050; UTAH CODE ANN. § 58-22-102(9).

16. *See, e.g.*, ALA. CODE §§ 34-8-1 to 34-8-28; ALASKA STAT. § 08.18; ARIZ. REV. STAT. § 32-1151 (2002); ARK. CODE ANN. §§ 17-25-101(a), 17-25-103(a); CAL. BUS. & PROF. CODE §§ 7065–77.

17. ALA. ADMIN. CODE r. 230-X-.10 (2005); ARIZ. REV. STAT. § 32-1103; ARK. CODE ANN. § 17-25-103(a); UTAH CODE ANN. § 58-55-102(12)(a)(v).

Failure to comply with applicable licensing requirements may lead to both civil and criminal penalties depending upon the seriousness of the violation.[18] In some states, a contract to perform services by a person without the necessary license or registration is void and unenforceable.[19] In other states, an unlicensed contractor is prohibited from suing in state court for collection or compensation for contracting work unless duly licensed at the time the contract was entered into and when the alleged cause of action arose.[20] This same prohibition applies to counterclaims and setoffs asserted by the unlicensed contractor. While the lack of license can serve as a shield to a suit by the unlicensed contractor against the owner, it may not necessarily provide an affirmative right of action against the contractor.[21] In other words, the mere lack of a license probably may not entitle an owner to sue the contractor to disgorge monies previously paid. Moreover, in some cases persons who contracted with an unlicensed contractor were estopped from raising this lack of a license as a defense to compensating the contractor.[22]

Generally, violations of the licensing laws do not by themselves create separate civil liability for damages caused by the violation. However, if the act or omission

---

18. *See, e.g.*, ALA. CODE §§ 34-2-34 through § 34-2-36 (architects), 34-8-6(a) (contractors); ALASKA ADMIN. CODE tit. 12, § 36 (2005); ALASKA STAT. §§ 08.48.291 (criminal penalties), 08.48.295 (civil penalties); ARIZ. REV. STAT. §§ 32-145 (design professionals), 32-1164 (contractors); ARK. CODE ANN. §§ 17-30-102(a), 17-25-103; CAL. BUS. & PROF. CODE §§ 5536, 6787(a)–(j), 7028; UTAH CODE ANN. §§ 58-23-3(3), 58-23-18.

19. J&M Indus., Inc. v. Huguley Oil Co., 546 So. 2d 369 (Ala. 1989) (engineer); Twickenham Station, Inc. v. Beddingfield, 404 So. 2d 43 (Ala. 1981); Sarkco v. Edwards, 482 S.W.2d 623 (Ark. 1972); Jones v. Wickstrom, 268 P. 449 (Cal. Ct. App. 1928).

20. ARIZ. REV. STAT. §§ 32-1151 to 32-1153 (applicable to contractors, but not design professionals); ARK. CODE ANN. § 17-25-103(d); CAL. BUS. & PROF. CODE § 7031; Cochran v. Ozark Country Club, Inc., 339 So. 2d 1023 (Ala. 1976); Crow v. Hickman's Egg Ranch, Inc., 41 P.3d 651 (Ariz. Ct. App. 2002); Sanders v. Foley, 945 P.2d 1313 (Ariz. Ct. App. 1997); Bird v. Pan W. Corp., 546 S.W.2d 417 (Ark. 1977); Gorrell v. Fowler, 286 S.E.2d 13 (Ga.), *appeal dismissed*, 457 U.S. 1113 (1982); United Stage Equip., Inc. v. Charles Carter & Co., 342 So. 2d 1153 (La. Ct. App. 1977); Reliable Props., Inc. v. McAllister, 366 S.E.2d 108 (N.C. Ct. App.), *review denied*, 342 S.E.2d (N.C. 1985); George & Son Constr. v. Oren Ltd. & Assoc., 672 P.2d 732 (Utah 1983); Meridian Corp. v. McGlynn/Garmaker Co., 567 P.2d 1110 (Utah 1977).

21. *See, e.g.*, ARIZ. REV. STAT. § 32-1153; Bentivegna v. Powers Steel & Wire Prods., Inc., 81 P.3d 1040 (Ariz. Ct. App. 2004) (holding that an unlicensed contractor cannot be required to return any payments received simply because of the lack of license); *but see* MW Erectors, Inc. v. Niederhauser Ornamental & Metal Works, Co., 36 Cal. 4th 412, 419, 115 P.3d 41 (2005) (holding that, without exception, contractors must be licensed at all times throughout performance of work to receive payment).

22. *See, e.g.*, Herman Chanen Constr. Co. v. Northwest Tile & Terrazzo Co., 433 P.2d 807 (Ariz. Ct. App. 1967); *but see* Hydrotech Sys. Ltd. v. Oasis Waterpark, 52 Cal. 3d 988 (1991).

that constitutes the violation gives rise to an independent cause of action in tort or contract for damages caused by that act or omission, the injured party may proceed with that cause of action to recover damages.

For purposes of this discussion, the licensing statutes are relevant in that they establish the relevant standard of care to which the construction manager will be held. In other words, if a state requires a construction manager to be licensed either as an architect, engineer, or contractor, the construction manager will be held to the same standard of care as any other licensee in that category. The licensing statutes, codes, and regulations will dictate certain duties that the construction manager is obligated to perform. Failure to perform such duties, as in *Pelletier* (discussed above), may subject the construction manager to tort liability. In addition, the construction manager may be subject to discipline or revocation of its license depending upon the severity of the offense.

c. *Failure to Comply with Procurement Codes*

The procurement requirements applicable to the construction manager may also dictate the standard of care. For example, a construction manager who is required to post a bond may be held to the standard of care of a contractor, rather than an architect or engineer. This is because contractors are routinely required to obtain bonds guaranteeing performance of the contract and payment to subcontractors and suppliers, whereas architects and engineers are typically exempt from bonding requirements.

Likewise, where the procurement requirements for a particular project require that the construction manager maintain errors and omissions or professional liability insurance for its work on the project, the construction manager may be viewed as more akin to an architect or engineer and, thus, held to the standard of care of the architect or engineer, rather than as a contractor.

On public contracts, the issue of the construction manager's role and the relevant standard of care can often be determined by reference to the applicable procurement codes of the public agency or organization soliciting the contract. Because these codes often limit the types of services for which a public body can contract, they may assist in defining whether a construction manager is operating either as a design professional or as a contractor. In this way, the procurement requirements can assist the court or the trier of fact in determining whether a construction manager should be held to the standard of care applicable to a contractor or a design professional.

d. *Failure to Comply with Applicable Building Codes*

The relevant standard of care applicable to a construction manager may also be established by reference to building and safety codes. Construction professionals typically are expected to have knowledge of and to comply with the relevant federal, state, and local building codes and safety standards, even if the construction management agreement does not specifically state as much.[23] Thus, a construction

---

23. Construction Litigation: Representing the Contractor § 7.13, at 165 (John D. Carter et al. eds., 2d ed. 1992).

manager who is functioning as a design professional, engineer, or contractor may be obligated to comply with the applicable building and safety codes. Even where a construction manager is serving in a more limited role as agent for the owner without responsibility for design or construction, it may still be held responsible for failures of the designer or contractors to comply with the applicable codes. As the *Pelletier* case makes painfully clear, failure to comply with the applicable building codes may expose a construction manager to huge liability under both the contract and for negligence.[24]

e. Failure to Comply with Industry Standards

In professional negligence cases, the courts may also look to the relevant industry or reference standards in imposing a standard of care upon the construction manager—again, as dictated by the specific function undertaken by the construction manager.[25] These industry standards have been codified and endorsed by industry associations. For example, electrical contractors will be held to standards promulgated or endorsed by the National Fire Protection Association (NFPA), mechanical contractors will be held to standards promulgated or endorsed by the American Society of Mechanical Engineers (ASME), steel contractors may be held to standards set forth by the American Institute of Steel Construction, Inc. (AISC), etc. The American National Standards Institute (ANSI) publishes a list of accredited standards developed in an effort to facilitate the development of American National Standards and a uniform National Building Code.[26]

Failure to comply with these industry standards has supported the imposition of liability for negligence.[27] Whether the standards promulgated by these and other industry associations can or will be used as a benchmark against which the construction manager's conduct will be measured depends on the law being applied, or whether the contract incorporates these standards.

2. Breach of Fiduciary Duty

A construction manager may also face liability for breach of fiduciary duty where the construction manager has expressly assumed the role of a fiduciary vis-à-vis the owner by contract. A fiduciary duty is the highest, most stringent, duty known at law. A fiduciary is obligated to act with utmost good faith and complete loyalty toward its principal, including, but not limited to, the duty to subordinate its own interests to those of its principal.

A fiduciary duty might also be implied if a pre-existing special relationship exists between the owner and construction manager. Being an agent, in and of

---

24. *See also* Mondelli v. Kendel Homes Corp., 631 N.W.2d 846, 852 (Neb. 2001).

25. *See, e.g.*, Baker v. Pidgeon Thomas Co., 422 F.2d 744 (6th Cir. 1970) (Arkansas law).

26. *See* www.ANSI.org. However, noted that some jurisdictions have declined to adopt these standards in determining the applicable standard of care.

27. Mondelli, 631 N.W.2d at 852.

itself, will not automatically foist the added duties of a fiduciary on the construction manager. While all fiduciaries are agents of their principal, all agents are not fiduciaries. Rather, something more must exist between the parties before fiduciary obligations are implied, which the courts universally describe as an "overmastering influence" on one side, and "weakness, dependence and trust justifiably reposed," on the other side.[28] No case has yet to find that an implied fiduciary duty exists between a construction manager and its owner/client. Indeed, the great weight of authority in this country, and throughout the western world, holds that fiduciary obligations do not arise in a purely commercial business context, nor can they absent either an express agreement to serve as a fiduciary or where a pre-existing special relationship exists.[29] Because the obligations of a fiduciary are so onerous and exacting, the construction manager should carefully consider the ramifications of undertaking such a role before agreeing to do so by contract. If the agent construction manager does not intend to undertake the additional duties of a fiduciary, it would be wise to disclaim any such duties in the contract.

3. Intentional/Negligent Misrepresentation

A construction manager may also face liability in tort where the construction manager either intentionally or negligently makes statements which are not true, but which others (whether the owner or third parties) reasonably rely on to their detriment. For example, where the construction manager, to induce an owner to hire him or her, promises that the project will cost a certain amount of money, will be completed within a certain amount of time and/or will be of a certain quality, and those promises turn out to be false, the construction manager may be held liable for misrepresentation.[30]

However, it should be noted that some courts will not impose liability for misrepresentation where the claimed misrepresentation is merely a promise of future

---

28. In re Mid-Island Hosp., Inc., 276 F.3d 123, 130 (2d Cir. 2002); Oursler v. Women's Internet Ctr., 566 N.Y.S.2d 295, 297 (1st Dep't. 1991); MRED General Partner, LLC v. Tower Economics Co., No. 2531, 2005 Phila. Ct. Com. Pl. LEXIS 124 (Pa. Ct. C.P. Apr. 12, 2005).

29. *See generally* E. MCKENDRICK, COMMERCIAL ASPECTS OF TRUSTS AND FIDUCIARY OBLIGATIONS (1992); Hospital Prods. v. United States, Ltd., Surgical Corp. (1994) 156 C.L.R. 4, 55 A.L.R. 417 (Australia); *In re Mid-Island Hosp.*, Inc., 276 F. 3d at 130 (Where the parties deal in an arms-length commercial transaction, "no relation of confidence or trust sufficient to find the existence of a fiduciary relationship will arise absent extraordinary circumstances."); *Oursler*, 566 N.Y.S.2d at 297 (A conventional business relationship "without more" does not create a fiduciary relationship in New York.); *MRED General Partner*, 2005 Phila. Ct. Com. Pl. LEXIS 124, at *5–6 (allegation of agency insufficient to create fiduciary or confidential relationship absent "special relationship" being characterized by "overmastering influence" on one side or "weakness, dependence or trust justifiably reposed" on the other side).

30. Barnett v. Coppell N. Tex. Court, Ltd., 123 S.W.3d 804 (Tex. Ct. App. 2003).

performance (such as, promising to act as a construction manager), rather than a statement of past or existing fact (such as statements regarding prior experience as a construction manager).[31] In those courts, a construction manager's promise to act as a prudent construction manager may not be actionable in tort if it later turns out to be false, while a false statement that the construction manager has extensive prior experience may be actionable as a misrepresentation of past or existing fact.

Moreover, where a construction manager makes false statements of material fact to induce an owner's execution of a contract, the construction manager may also face claims of fraudulent inducement.[32] This generally requires a breach of duty which is different than, or in addition to, the obligations undertaken in the contract.[33] Thus, while a construction manager's failure to perform in accordance with its promises is merely a breach of contract, an action for fraud will lie when a promise was actually made with a preconceived and undisclosed intention of not performing it.[34]

In cases of actionable misrepresentation, liability may be imposed even if the statements at issue are not made directly to the person or entity alleging misrepresentation. For example, in the construction setting, the owner, architect, and trade contractors may receive information generated by the construction manager. These parties may sue for negligent or intentional misrepresentation, even if the construction manager did not directly convey the information, because the owner, architect, and trade contractors are usually within the class of individuals who are intended to receive and use such information.[35] Construction managers have proved to be an attractive target for negligent misrepresentation claims.[36]

4. Unfair Trade Practices

Many states have passed legislation designating certain conduct as "unfair trade practices," rendering the offending party liable for not only damages, but potentially

---

31. Dick Anderson Constr., Inc. v. Monroe Constr. Co., No. DV 01-710, 2004 Mont. Dist. LEXIS 2382, at *14–16 (4th Jud. Dist. Ct. Mont., May 3, 2004) (dismissing misrepresentation claim where construction manager allegedly misrepresented that it would and could handle the duties of construction manager on the project, since alleged misrepresentation related to future acts rather than facts already in existence).

32. *See* Rose v. Simms, No. 95 Civ. 1466, 1995 U.S. Dist. LEXIS 17686, at *21–25 (S.D.N.Y. Nov. 29, 1995).

33. Americana Petroleum Corp. v. Northville Indus. Corp., 606 N.Y.S.2d (Ct. App. Div. 2d Dep't 1994) (the emphasis in this case is that the alleged fraud be independent of the contract).

34. *Rose*, 1995 U.S. Dist. LEXIS 17686, at *23 (citing Cohen v. Koenig, 25 F.3d 1168, 1172 (2d Cir. 1994)).

35. *See* RESTATEMENT (SECOND) OF TORTS § 552 (1977).

36. LESLIE O'NEAL-COBLE ET AL., CONSTRUCTION DAMAGES AND REMEDIES 259 (W. Alexander Mosely ed., 2004).

treble damages and attorneys' fees.[37] These statutes are often relied on in construction cases as an additional cause of action to be asserted against the opposing party.[38] For example, in a case involving a metal refinery in the Australian outback, the owner sought treble damages against the contractor/construction manager based on its representations that the refinery could be built in the manner as designed and that the refinery as designed would produce a certain quantity of metal at a certain grade. When the refinery did not function in the manner as represented, the owner brought suit in arbitration claiming, among other things, breach of contract, professional negligence, and unfair trade practices.

Most state statutes require more than a mere breach of contract to constitute an unfair trade practice. Rather, there must be some deceptive conduct, or the conduct at issue must be offensive to public policy.[39] For example, if the defendant engaged in intentional or negligent misrepresentations, or failed to act fairly and in good faith in its role as the administrator for the construction phase of the project (such as the handling of change orders), it may be held liable for an "unfair trade practices" violation.[40]

In some states, it may be necessary to prove more than a single instance of misconduct to rise to the level of unfair trade practices. Thus, it is often important for the plaintiff alleging unfair trade practices to show that the defendant engaged in similar conduct on other projects and that such conduct was not limited to an isolated instance or the particular contract being sued upon.

In other states, such as Texas, the plaintiff must qualify as a *consumer* to obtain standing under the act. Standing as a consumer usually depends upon the plaintiff's

---

37. *See* CONN. GEN. STAT. § 42-110a to -42-110g (2000) (referred to as the Connecticut Unfair Trade Practices Act or CUTPA); MASS. GEN. LAWS ch. 93A (2005); TEX. BUS. & COM. CODE ANN. § 17.41 to -17.63 (2002).

38. Hoang v. Arbess, 80 P.3d 863, 870 (Colo. Ct. App. 2003) (allowing plaintiff homeowners to pursue claims against construction manager pursuant to the Colorado Consumer Protection Act where plaintiffs presented evidence that construction manager knew or should have known the construction techniques implemented were insufficient to protect against damage from expansive soils and that construction manager should not have directed sales persons to represent otherwise).

39. Williams Ford, Inc. v. Hartford Courant Co., 232 Conn. 559, 591 (1995).

40. *See, e.g.,* Barnett v. Coppell N. Tex. Court, Ltd., 123 S.W.3d 804 (Tex. Ct. App. 2003) (upholding jury verdict finding contractor liable of deceptive trade practices act violations where contractor made misrepresentations regarding the costs and duration of a construction project); Nota Constr. Corp. v. Keyes Assoc., Inc., 694 N.E.2d 401 (Mass. Ct. App. Ct. 1998) (allowing unfair trade practices claim pursuant to MASS. GEN. LAWS ANN. ch. 93A, §§2(a), 11 (2005) to proceed where architect made potentially negligent misrepresentations in the plans and specifications and allegedly failed to act fairly and in good faith in resolving the contractor's disputes under the construction contract relating to change orders and redesign of the project).

relationship to the transaction at issue, rather than the plaintiff's contractual relationship with the defendant.[41] Moreover, businesses with significant assets and/or businesses owned or controlled by a corporation or entity with significant assets may not qualify as "consumers" entitled to maintain an unfair trade practices claim.[42]

B. *Bases of Liability*

1. *Defective Design/Design Errors*

Many times, the construction manager is involved in the pre-design and design phases of the project to provide advice regarding materials, selection of major systems or equipment, and other constructability issues. It is in this early stage of the project that the construction manager can offer suggestions to save time and money. The construction manager may even be involved in the preparation of project specifications and selection of a designer or design-build entity.[43] The construction manager can provide technical and management expertise to prepare the basic configuration in the contract documents. Additionally, the construction manager can help the owner assess not only the capabilities of the designer, but also the contractor, increasing the owner's chances of finding the best design-build team possible.[44]

Nonetheless, the construction manager is generally not the designer and, therefore, is not responsible for the adequacy of the architect's or engineer's plans.[45] However, where the construction manager takes on some responsibility for the design or design coordination, the construction manager may be held responsible for defects in the design.[46] For example, in the context of an Engineering, Procurement and Construction Management (EPCM) Agreement, the construction manager is not only responsible for managing the project, it is also responsible

---

41. D/FW Commercial Roofing Co. v. Mehra, 854 S.W.2d 182 (Tex. Ct. App. 1993).

42. *See* TEX. BUS. & COM. CODE. ANN. § 17.54(4) (excepting from the definition of consumer "a business consumer that has assets of $25 Million or more, or that is owned or controlled by a corporation or entity with assets of $25 Million or more").

43. Linda Chiarelli & Lawrence Chiarelli, *The Role of the Construction Manager on a Design/Build Project*, 15 CONSTRUCTION LAW. 58, 59 (1995).

44. *See id.* at 60.

45. *See* AIA A121/CMc and AGC 565, Arts. 2.1.8 ("It is not the Construction Manager's responsibility to ascertain that the Drawings and Specifications are in accordance with applicable laws, statutes, ordinances, building codes, rules and regulations."), Art. 2.4 ("The Construction Manager shall not be required to provide professional services which constitute the practice of architecture or engineering, unless such services are specifically required by the Contract Documents for a portion of the Work or unless the Construction Manager has specifically agreed in writing to provide such services.").

46. *But see* Braun v. Kelly, 2004 Conn. Super. LEXIS 210 (J.D. Stamford Jan. 29, 2004) (construction manager not negligent where homeowner was unable to establish that construction manager's structural engineer negligently failed to design home to support flooring materials).

for engineering or designing the project, in which case the construction manager will be held to the same standard of care as a design professional and will be held responsible for deficiencies in the design.[47] Even where the construction manager only takes responsibility for coordinating the design of the various design professionals and/or consulting engineers, it may be subject to liability for inadequacies in the design or for negligently failing to detect inadequacies in the design.[48]

Construction managers may also play a part in value engineering, that is, looking at the major cost elements of the project and suggesting less costly alternatives.[49] This generally includes helping the client select contractors for discrete parts of the project and reviewing the plans and bids to find ways to reduce costs.[50] Value engineering is meant to save money by developing more efficient, less costly design alternatives which reduce material and/or equipment needs without sacrificing functionality or operability.[51] Although this can be a difficult task, construction managers may play a vital role in identifying cost-saving strategies that are missed or rejected by the traditional architect/engineer.[52] This service, generally reserved to the architect/engineer, will not generally shift the risk of an inefficient or costly project to the construction manager. Nonetheless, if during the course of design review, the construction manager suggests alternative designs and/or means of construction, materials, or sequences, the construction manager may assume at least some responsibility for the implementation of its suggestions. For this reason, it is

---

47. The undertaking of an architect implies that he possesses skill and ability sufficient to enable him to perform the required services at least ordinarily and reasonably well, and that he will exercise and apply this skill and ability, his judgment and taste, reasonably and without neglect. The architect does not imply or guarantee a perfect plan, warrant his plans and specifications, or guarantee a particular result. This basic standard, requiring reasonable care, skill, and diligence, is the rule in most jurisdictions. *See, e.g.,* Gravely v. Providence P'ship, 549 F.2d 958, 959–60 (4th Cir. 1977); First Nat'l. Bank of Akron v. Cann, 503 F. Supp. 419, 439 (N.D. Ohio 1980), *aff'd*, 669 F.2d 415 (6th Cir. 1982); Seiler v. Levitz, 367 A.2d 999, 1007–08 (Del. 1976); Bay Shore Dev. Co. v. Bondfoey, 78 So. 507, 510 (Fla. 1918); Audlane Lumber & Builders Supply v. D.E. Britt Assoc., 168 So. 2d 333, 335 (Fl. Dist. Ct. App. 1964), *cert. denied*, 173 So. 2d 146 (Fla. 1965); Waggoner v. W&W Steel Co., 657 P.2d 147, 149 (Okla. 1982); Nelson v. Commonwealth, 368 S.E.2d 239, 245 (Va. 1988); Surf Realty Corp. v. Standing, 78 S.E.2d 901, 907 (Va. 1953); 5 AM. JUR. 2D *Architects* § 23 (1962); 6 C.J.S. *Architects* § 27 (1975).

48. *See, e.g.,* American Employers' Ins. v. Maryland Cas. Co., 509 F.2d 128 (1st Cir. 1975).

49. AGC 510, Art. 3.1.2 provides that the construction manager shall consult with the owner and architect regarding site use and improvements and selection of materials, building systems, and equipment, and provide recommendations on construction feasibility and factors related to cost, including estimates of alternative designs or materials.

50. Richard J. Wood, *Capital Improvements: A Guide for the Construction of the Modern Law School*, 27 CAP. U. L. REV. 709, 731 (1999).

51. *Id.*

52. *Id.*

important that the construction management agreement be clear as to which party bears the risk of, and assumes responsibility for, the implementation of any value engineering suggestions.

Even construction managers who play no role in the initial design or coordination of the design may face liability if they fail to detect obvious errors or omissions in the project plans. In the face of a patent defect, the construction manager has a duty to seek clarification from the architect/engineer,[53] and failure to do so may render the construction manager liable to the owner for damages.[54] In such cases, the construction manager is an additional pocket for the owner to sue, particularly where the design professional who developed the inadequate design did not maintain errors and omissions coverage or is otherwise immune from judgment.

2. *Faulty Workmanship and Improper Inspection/Supervision*

A construction manager at-risk is directly responsible for the performance of the trade contractors, just as a general contractor would be. Thus, if the trade contractors render defective or incomplete performance, the construction manager at-risk is responsible to the owner.

In contrast, a construction manager acting as agent for the owner does not have any direct contractual relationship with the trade contractors and, therefore, no control over the trade contractors' means and methods of constructing the project. As a result, the agent construction manager typically does not bear responsibility for the acts and omissions of the trade contractors. This may hold true despite the fact that the owner hired the construction manager for the express purpose of checking the design and monitoring construction for compliance with the plans and specifications.[55] In fact, many of the standard form contracts specifically state that the construction manager is not responsible for the trade contractors' failure to carry out the work in accordance with the contract documents.[56]

While inspection of the construction work traditionally has been left to the architect/engineer,[57] a construction manager is usually given some responsibility to inspect the construction work as part of its day-to-day monitoring of construction progress. In fact, many owners elect to have the construction manager serve as their on-site agent for inspection.

---

53. Beacon Constr. Co. v. United States, 314 F.2d 501 (Ct. Cl. 1963).
54. *See* AIA A121/CMc and AGC 565, Art. 2.1.8 (requires that if construction manager recognizes that portions of the drawings or specifications are at variance with applicable laws, statutes, ordinances, building codes, rules and regulations, it must promptly notify the owner and architect in writing).
55. Aiken County v. BSP Div. of Envirotech Corp., 657 F. Supp. 1339, 1356 (D.S.C. 1986).
56. *See* AGC 510, Art. 3.2.5.4.
57. *See* AIA Document B141-1997, Owner-Architect Agreement, Art. 2.6.2.

However, in most cases, the construction manager *is* obligated to review the trade contractors' work and safety programs to determine whether the work is satisfactory.[58] The construction manager's goal is to guard the owner from defects and work deficiencies, and the construction manager may reject work by the trade contractors that does not conform to the trade contractor agreements.[59] To the extent the construction manager fails to adequately perform these duties, it may face liability to the owner under both the contract and in tort.[60] Moreover, if the construction manager is at-risk and holds the building permit, it may also be liable to injured trade contractors in the event that the injury is the result of a failure to inspect.[61] Even where the construction manager relies on and follows the plans, designs, and specifications of the owner's architects and engineers, it may still face negligence claims from the owner and trade contractors where the construction manager has a duty to inspect the work.[62] Of course, the construction manager's duty will generally be limited to a good faith effort and the exercise of reasonable skill and judgment in the execution of its duties.[63]

Liability has also been imposed on the construction manager for failure to adequately supervise construction activity.[64] This is so even though the construction

---

58. See AGC 510, Art. 3.2.5; AIA A201/CMa, Art. 4.6.5.

59. AGC CONTRACT DOCUMENTS HANDBOOK § 8.03[B], at 749 (J. William Ernstrom & Kevin F. Peartree, eds., 2003).

60. *See* Travelers Indem. Co. of Ill. v. 28 East 70th Street Constr. Co., 296 F. Supp. 2d 476 (S.D.N.Y. 2003) (denying construction manager's motion for summary judgment with respect to owner's breach of contract and negligence claims where there was a genuine issue of material fact as to the construction manager's role on the project and whether construction manager had a duty to inspect and supervise trade contractors so as to guard against conditions which led to pipe burst); Wausau Bus. Ins. Co. v. Turner Constr. Co., 143 F. Supp. 2d 336 (S.D.N.Y. 2001) (denying construction manager's motion for summary judgment and finding an issue of fact as to whether construction manager was negligent in failing to correct safety lapses on the part of the trade contractors which led to a fire); Morse/Diesel, Inc. v. Trinity Indus., Inc., 859 F.2d 242, 2247–48 (2d Cir. 1988) (citing New York cases finding project managers with supervisory obligations over subcontractors liable for negligence).

61. *See* Pelletier v. Sordoni/Skanska Constr. Co., No. X-06-CV-950155184S, 2004 WL 3128800, at *5, 38 Conn. L. Rptr. 404 (Conn. Super. Dec. 16, 2004) (holding that BOCA National Codes imposed distinct obligation on permit holder to conduct certain special inspections).

62. Travelers Indem. Co. of Ill., *supra* note 60 at 483, note 5.

63. Stephen Wichern, *Protecting Design-Build Owners through Design Liability Coverage, Independent Construction Managers, and Quality Control Procedures*, 32 TRANSP. L.J. 35, 53 (2004).

64. Walter G. Wright, Jr. & Stephanie M. Irby, *The Transactional Challenges Posed by Mold: Risk Management and Allocation Issues*, 56 ARK. L. REV. 295 (2003); *see* Centex-Rooney Constr. Co. v. Seaboard Sur. Co., 706 So. 2d 20 (Fla. Dist. Ct. App. 1997).

contractor, rather than the construction manager, is responsible for supervising and directing the work.[65] This liability apparently arises because the construction manager is generally responsible for overall coordination of the project and must exercise reasonable care to prevent material deviations from the plans and specification and to prevent substandard workmanship.

The construction manager may also bear some responsibility for construction close out, such as advising the owner as to the dates of substantial completion and final completion of the work, assisting the architect/engineer in preparing a list of incomplete or unsatisfactory items (the punch list), and collecting and delivering to the owner all close-out documentation from the trade contractors. For example, under AGC Document. 510 (AGC 510), the construction manager bears responsibility for the project close-out process, including checking for and advising the owner of incomplete or unsatisfactory work, as well as testing new utilities and equipment.[66] The equivalent AIA contract does not expressly address close-out tasks, but the construction manager and the architect must advise the owner during the correction period of the contract.[67] Moreover, additional close-out responsibilities may be written into the contract. Failing to adequately perform these services may render the construction manager liable to the owner.

3. *Negligent Cost Estimates/The Over-Budget Project*

One of the biggest perceived benefits to an owner in using a construction manager is the construction manager's knowledge and expertise in estimating labor and material costs for the construction project. Budgeting and estimating services include not only estimates of the hard construction costs, but also many of the soft costs that can add up to a significant portion of the owner's budget (e.g., consultant services, printing, permits, etc.). Often, as part of its basic services, the construction manager assists the owner in preparing a budget that incorporates the owner's needs within a defined project schedule. As the design progresses, and throughout the construction phase, the construction manager is tasked often with updating the owner's budget at agreed intervals.

A key component of this estimating service is providing the owner with the construction manager's estimate of construction costs. In the early stages of design, the construction manager's estimates will be conceptual factored estimates, based on unit prices and assumptions drawn from experience with similar projects. As the design progresses, the estimates will increase in detail, based on material take-offs from approved construction drawings. Additionally, the construction manager will

---

65. *See* AIA Document A201/CMa, General Conditions of the Contract for Construction, Art. 3.3 (1992) ("The Contractor shall supervise and direct the Work using the Contractor's best skill and judgment").

66. *See* AGC 510, Art. 3.2.7.

67. *See* AIA A201/CMa, Art. 4.3.

be tasked often with reviewing design development and making recommendations when the design development increases the project budget.

As the cost estimates progress, or as actual bids for construction services and materials come in from trade contractors and suppliers, it often becomes apparent that the project cannot be constructed within the approved budget prepared by the construction manager. In such cases, owners will often look to the construction manager as a responsible party, seeking recoupment of cost overruns if the budget bust is excessive.

In the traditional lump sum or fixed price contracting scheme, the prime contractor assumes the risk of overall cost control. Likewise, this is the case where the construction manager is at-risk and assumes the financial risks of contracting with the trade contractors.[68]

However, in an agency construction management scheme, the risk of budget overruns is shifted back to the owner if the project is being executed on a cost-reimbursable basis. Nonetheless, even the construction manager acting as agent has certain express and implied obligations to further the owner's interests by furnishing business administration and management services to the project and to perform the work in an economical manner.[69] The construction manager is also obligated to monitor the construction budget—by setting up a system to monitor costs, changes in the work, and the total cost of the project—and to notify the owner whenever projected costs are expected to exceed the budget.[70] A failure to employ a reasonably accepted method of cost control and reporting may subject the construction manager to liability for excessive cost overruns even in a cost-reimbursable contract.

Like a construction manager's other duties, the duty to provide cost control is not a standard of perfection. Just because a project ends up over budget does not necessarily mean that the construction manager has breached its duty of care. Rather, a normal margin of error, given the complexity of the project, must be taken into account in determining whether the construction manager acted reasonably under the circumstances.[71]

---

68. However, note that the construction manager at-risk may not warrant or guarantee estimates outside of the guaranteed maximum price. *See* AIA 121/CMc and AGC 565, Art. 2.1.8.

69. *See* AGC 510, Arts. 2, 3.1.4 (requiring the construction manager to provide an initial budget and a detailed budget as design development progresses), 3.2.4 (requiring construction manager to monitor project costs and advise owner whenever projected costs are to exceed budgeted amounts); AIA B801-1992, Art. 2.

70. *See* AGC 510, Art. 3.2.4.1.

71. Murray H. Wright & David E. Boelzner, *Quantifying Liability under the Architect's Standard of Care*, 29 U. Rich. L. Rev. 1471, 1494 (1995). For example, in Kellogg v. Pizza Oven, Inc., 402 P.2d 633 (Colo. 1965), the architect's cost estimate for the project covered only about thirty-three percent below the final project cost. Evidence indicated that a normal margin of error in cost estimating would permit approximately a ten percent variation. The court, therefore, held that the damages to which the plaintiffs were entitled should be the net of this ten percent normal variation, as well as any items caused by owner-initiated changes. *Kellogg*, 402 P.2d at 636.

The general rule applicable to both construction managers and design professionals is that, absent a specific warranty setting a fixed limit of construction costs in the construction management agreement, the construction manager will not be considered to have breached the reasonable standard of care simply because the actual construction costs/bids exceed the estimate. However, the analysis must continue to determine whether the construction manager did in fact meet the standard of care. Many of the standard contracting forms qualify the construction manager's responsibilities with respect to preparing estimates and budgets, and state that the construction manager does not guarantee the accuracy of any estimates and the estimates reflect only the construction manager's best judgment as a construction professional. These forms also contain disclaimers that the construction manager is not to be held responsible if bids exceed any budget or estimate.[72]

In contrast, for construction managers at-risk where there is a guaranteed maximum price arrangement in the construction management agreement,[73] or the construction manager offers a contractual guarantee that the project price will not exceed the stated sum, the construction manager will face liability exposure if the project costs exceed the guaranteed maximum price. Agency construction managers would be wise to explicitly disclaim or otherwise limit their liability for cost overruns in the contract.

4. Improper Certification of Progress Payments

A construction manager may also have responsibility for reviewing and approving payment requisitions submitted by the contractors and suppliers to the owner, even though this role has traditionally been performed by the architect.[74]

For example, a construction manager acting as the owner's agent is typically responsible for reviewing and processing applications for payment made by the trade contractors, even though the construction manager is not a signatory to the contracts between the owner and the trade contractors.[75] After reviewing the applications for payment, the construction manager must decide how much is due to each trade contractor based upon the construction manager's observations of the progress of the work, and the documentation attached to the pay applications. However, this does not mean that the construction manager must exhaustively review all work performed.[76] Rather, the construction manager must, at the very least, make a reasonably diligent review of the work performed to determine whether the progress claimed in the pay application is consistent with the amount of progress observed in the field. Agent construction managers would be wise to develop an agreed method with the owner by which construction progress in the field will be measured.

---

72. *See* AIA B801-1992, Art. 5.2.
73. *See* AIA A121/CMc and AGC 565, Art. 2.2 (1991).
74. *See* AIA B801-1992, Art. 2.3.11.
75. *See* AGC 510, Art. 3.2.4.3; AIA A201/CMa, Art. 9.4.
76. *See* AGC 510, Art. 3.2.4.4.

The construction manager may incur liability to the owner for failing to properly review payment requisitions or for approving amounts not properly due. While the construction manager's liability to the owner flows primarily from the contract, the construction manager may also be held liable for negligence in performing this function, and even fraud, if the construction manager's conduct rises to the level of intentional misrepresentation or deceit.

The construction manager may also be responsible for assessing change orders from the trade contractors.[77] A construction manager's liability for performing this function largely rests on the terms of the construction management agreement. For example, under AGC 510, while the construction manager must make recommendations to the owner and the design professional and negotiate the change orders with the trade contractors, the final approved decision rests with the owner and the design professional.[78] In contrast, AIA Document A201/CMa states that the owner, architect, and construction manager must all sign change orders, stating their agreement to the change in the work, the amount of the adjustment to the contract price, and any schedule adjustments.[79]

A construction manager at-risk may also be responsible for reviewing payment applications, depending on the terms of the contract. Traditionally, this responsibility remains with the architect who is responsible for reviewing and approving the certificates of payment issued by the construction manager.[80] Even in the traditional structure, the construction manager may still be viewed as having an express or implied obligation to the owner to submit applications for payment that are consistent with the contract payment terms (e.g., containing the correct percentage completion), and failure to do so may subject the construction manager to claims of breach of contract, misrepresentation, or fraud.

5. *Improper Schedule Preparation, Monitoring, and Coordination*

Construction managers are generally responsible to some degree for scheduling, monitoring, and coordinating the trade contractors performing the construction on the project. The degree of this responsibility will vary depending on the specific role of the construction manager (i.e., at-risk or as agent for the owner), and the specific terms of the construction management agreement. For example, AGC Documents require the construction manager to prepare a preliminary project schedule, coordinate and integrate the project schedule with the services and activities of the owner, construction manager, and architect, update the schedule as construction proceeds, and advise the owner if the milestone dates will not be met.[81]

---

77. *See* AGC 510, Art. 3.2.4.5.
78. *See* AGC 510, Art. 3.2.4.5.
79. *See* AIA A201/CMa, Art. 7.2.1.
80. *See* AIA A121/CMc; AGC 565, Art. 7.1.
81. *See* AGC 510, Art. 3.1.3; AIA A121/CMc; AGC 565, Art. 2.1.3.

While effective scheduling and coordination of the work is essential to a successful construction project, the liability that the construction manager may face for improperly performing this function will be commensurate with the degree of responsibility imposed by the construction management agreement. In cases where the owner does not delegate this responsibility to the construction manager, the owner can face tremendous liability to contractors, their subcontractors, and suppliers. Even where the construction management agreement requires the construction manager to schedule and coordinate the sequence of construction, the owner may face liability to trade contractors who are impacted by the construction manager's failure to perform its duties. If the construction manager acts as an agent for the owner, trade contractors would have a direct right of action against the owner for delays and disruptions caused by inadequate scheduling or coordination. Of course, the owner may then seek to recover any liability incurred from the construction manager. All construction managers (at-risk and agency) would be wise to utilize CPM or other recognized scheduling techniques to plan project execution, and to utilize systematic methods to communicate with, and coordinate, the work of trade contractors. Except for the smallest of projects, CPM scheduling is fast becoming the only method that meets the standard of care in the industry.

Procurement can play a major role in managing the project schedule. Generally, the construction manager is not directly responsible for procurement. Rather, the procurement role is delegated to the trade contractors who will install the equipment and materials. Nonetheless, the construction manager will play some role in coordinating procurement, particularly items with long-lead delivery times, to ensure that construction will proceed smoothly. For example, in an EPCM arrangement, the construction manager is also responsible for managing the procurement process. In limited cases, the construction manager may even agree to take on the responsibility for procuring the materials and equipment itself.[82] In such cases, the construction manager will be held responsible for failing to procure materials and equipment in a timely manner, particularly when the construction schedule is impacted.

6. *Improper Processing of Submittals*

Construction managers typically are responsible for reviewing submittals by the project's contractors and suppliers.[83] These submittals must be reviewed and approved or rejected in a timely manner. In an agency construction management context, the construction manager's liability would be limited to the owner, whereas a construction manager at-risk may also face liability to third parties with whom it has contracted.

---

82. For example, in AIA A121/CMc and AGC 565, Art. 2.1.7 (construction manager at-risk with guaranteed maximum price), the contracts for long-lead time items are assigned by the owner to the construction manager, who must accept responsibility for such items as if procured by the construction manager.

83. *See* AIA A121/CMc and AGC 565, Art. 2.3.2.1.

7. *Self-Dealing/Conflict of Interest/Lack of Independence*

The construction manager may, at times, find itself in a position where its interests diverge from those of the owner. This potential conflict of interest situation can arise in many contexts.

For example, where an agency construction manager takes on responsibility for design or procurement, an issue may arise as to whether the relationship of trust and confidence inherent in the agency carries over to the construction manager's design and procurement responsibilities. If the agency relationship extends to these other responsibilities, the construction manager may face liability for a conflict of interest or, worse yet, for a breach of fiduciary duty if it fails to act in the owner's best interests or seeks to profit from the owner with respect to these responsibilities. For example, this can arise where the construction manager charges the owner for the construction manager's cost plus a mark up for the procurement of materials. Later, the owner may claim that the mark up was excessive and breached the construction manager's obligations of loyalty and utmost good faith. Likewise, if the materials are procured at an excessive price and/or through an affiliated company of the construction manager, the construction manager may face liability for self-dealing to the owner.

As a result, a construction manager in an agency role should fully disclose to the project owner in advance the prices that will be charged for the procurement of materials and from whom they will be procured, and obtain the owner's approval of such procurement, to avoid a later claim of conflict of interest.

8. *Collusion with Contractors*

Where the construction manager is at-risk and the project is being executed on a lump sum basis, the exposure for allegations of self-dealing or conflict of interest generally falls away, because the owner has no right to interfere with the means, methods, and techniques employed by the construction manager at-risk. However, if the project is being executed on a cost-reimbursable, guaranteed maximum price basis, the construction manager at-risk should fully disclose the prices, sources, and relationships from which materials and services are procured, to avoid allegations of self-dealing and conflict of interest. Even when the construction manager at-risk undertakes a fixed priced project, its freedom may not be unfettered. For example, the AIA and AGC contract forms require the construction manager at-risk to allow the architect and the owner to review the list of proposed trade contractors for the job and to raise any objections thereto.[84] Moreover, these AIA and AGC contract forms also obligate the construction manager to use its best efforts to perform the project in an expeditious and economical manner consistent with the interests of the owner.[85] To the extent that the work is being performed on a cost reimbursable basis, the construction manager may be further obligated to charge "rates not higher

---

84. *See* AIA A121/CMc; AGC 565, Art. 2.1.6.
85. *See* AIA A121/CMc; AGC 565, Art. 1.1.

than those customarily paid at the place of the project except with the prior consent of the owner."[86] This would apply to the cost of labor, trade contracts, materials and equipment, miscellaneous expenses (such as bonds and insurance premiums, sales taxes, permit and testing fees, arbitration costs incurred by the construction manager with the owner's permission, and some personal relocation costs), and emergency expenses.[87] Thus, if a construction manager at-risk on a cost-reimbursable or guaranteed maximum cost contract charges rates higher than those customarily paid, it may face liability under the contract as well as in tort. The construction manager at-risk may also be liable if it attempts to pass on nonreimbursable costs, such as the cost of repairing damaged or nonconforming work resulting from the construction manager's negligence or failure to fulfill a specific responsibility to the owner. The extent of any such liability will depend on the specific terms of the contract.

However, where the construction manager is acting as agent for the owner and the trade contractor agreements are between the owner and the individual trade contractors, the construction manager has an obligation to act in the interest of the owner. As the owner's agent, the construction manager must act in the owner's best interests and should avoid conflicts of interest that are not fully disclosed. In that case, the owner relies on the construction manager to, among other things, monitor the contractors' performance, monitor the project schedule, verify documentation provided by the trade contractors, and advise the owner of important project developments. If the construction manager fails in these obligations and engages in activities that may be perceived as collusion with the trade contractors—such as approving improper submittals, payments not owed, or improper/inflated change orders—the construction manager can face liability, not only for breach of the construction management agreement but also for claims for professional negligence, breach of agency, misrepresentation and, potentially, fraud.[88]

Library References

C.J.S. *Agency* §§ 240–241, 244–255, 258, 262–271, 292–293, 344, 346–349, 352, 369, 384–385, 389, 392, 410; C.J.S. *Architects* §§ 1–3, 7, 10, 13–14, 19, 21–23; C.J.S. *Contracts* §§ 355, 561–562; C.J.S. *Counties* § 165; C.J.S. *Fraud* §§ 10–17, 23–25, 29–30, 34–35, 59–60, 62–63, 67–75, 77–78, 80; C.J.S. *Municipal Corporations* §§ 917–918, 925; C.J.S. *Negligence* §§ 162–164, 803, 805, 808, 810–811; C.J.S. *Public Contracts* §§ 6, 14–15; C.J.S. *Schools and School Districts* §§ 409–414, 435, 761; C.J.S. *States* §§ 270, 280–288; C.J.S. *Trading Stamps and Coupons* § 6; C.J.S. *United States* § 113.

West's Key No. Digests, Antitrust and Trade Regulation <KEY>259; Contracts <KEY>219, 280(3); Counties <KEY>116; Fraud <KEY>8; Licenses <KEY>11, 26, 38, 40, 41; Municipal Corporations <KEY>235; Negligence <KEY>321, 322, 1657, 1662; Principal and Agent <KEY>48, 69, 136; Public Contracts <KEY>5.1; Schools <KEY>80(2); States <KEY>98; United States <KEY>64.15.

---

86. *See* AIA A121/CMc; AGC 565, Art. 6.1.1.
87. *See* AIA A121/CMc; AGC 565, Arts. 6.1.1–6.1.7.
88. Fain, *supra* note 3, at 44.

III. Construction Manager's Liability to Third Parties

In a construction manager agency situation, the trade contractors contract with the owner and not the construction manager. Since the construction manager has no contract with the trade contractors, the trade contractors cannot bring a breach of contract action against the construction manager for failure to meet its contractual obligations (such as providing incorrect cost estimates or failure to adequately supervise or coordinate the work under the contract). Rather, the trade contractors' only contractual right of action is against the owner with whom the trade contractors are in privity of contract.

This lack of contractual privity between the construction manager and the third party trade contractors may also preclude them from directly holding a construction manager liable in tort due to the economic loss doctrine. In states that follow the economic loss doctrine, trade contractors historically have been precluded from suing the construction manager in tort if the loss is purely economic.[89] If personal injury or property damage (other than damage to the product itself) is involved, the economic loss doctrine does not apply. If applied strictly, the economic loss doctrine would block most, if not all, claims by parties who do not have contracts with one another in the absence of personal injury or damage to other property. While courts in many jurisdictions continue to follow the economic loss doctrine, an equal number of states have either completely abandoned the doctrine or created gaping exceptions which have eroded the protections against tort liability afforded by the doctrine.[90] Given that the economic loss doctrine is applied very differently across the country, few generalities about the doctrine and its application can be made.

A. *Theories of Liability*

1. *Negligence*

Negligence claims against construction managers arise most frequently in the context of personal injury claims brought by workers injured on the construction job overseen by the construction manager. In these cases, the injured third parties may be able to hold the construction manager liable for negligence if they can show that the construction manager contractually assumed or actually exercised responsibility

---

89. The economic loss doctrine is a principle prohibiting the recovery of purely economic damages (as opposed to personal injury or property damages) caused by negligence, when a party lacks privity of contract with the defendant. *See* Engineered Refrigeration Sys., Inc. v. Albertson's Inc., 1996 U.S. Dist. LEXIS 9090 (N.D. Fla. 1996); John Martin Co. v. Morse/Diesel, Inc., 819 S.W.2d 428, 430 (Tenn. 1991); Michael D. Lieder, *Constructing a New Action for Negligent Infliction of Economic Loss: Building on Cardozo and Coase*, 66 WASH. L. REV. 937 (1991).

90. *John Martin Co.,* 819 S.W.2d at 434–35 ("We have examined the competing views. Our conclusion is that there is no clear majority; instead we find a split of authority among the states."); *accord,* Bryant Elec. Co. v. Fredericksburg, 762 F.2d 1192, 1996 (4th Cir. 1985).

for control over workplace safety.[91] This is particularly true if the construction manager takes out a building permit in its own name.

These claims aside, other parties—namely, contractors, subcontractors, and architects—may recover from a construction manager for negligence, in jurisdictions that either do not follow the economic loss doctrine or have eroded its application, if they are able to prove all of the elements of a negligence claim. As with any other negligence claim, the plaintiff must show a duty owed by the construction manager, a breach of that duty, causation, and damages. Negligence cases involving construction managers and third parties often turn on the duty element which can be difficult, but certainly not impossible, to establish.[92]

For example, in *Epicentre Strategic Corp., Michigan v. Perrysburg Exempted Village School District*,[93] a contractor's assignee claimed that the construction manager acted negligently in the scheduling of the project, resulting in increased costs to the contractor. The court held that because there was no contract between the contractor's assignee and the construction manager, the construction manager owed no duty of care to prevent economic losses to the contractor.[94]

However, the court also held that in certain circumstances, a construction manager *could* exercise sufficient control over the contractor to create a "nexus" that would substitute for contractual privity.[95] In determining whether such a "nexus" existed, the court said that it would consider whether the construction manager possessed "excessive control over the contractor through the power to stop the work and give orders about the project," the "degree of control" exerted, and the "amount of interaction" between the construction manager and the third party.[96] In that case, the court held that while the third party had not pleaded that the construction manager exerted excessive control, the existence of a "nexus" creating a duty could only

---

91. *See, e.g.,* Jones v. Parsons Transp. Group, Inc., Docket No. JFM-03-526, 2004 U.S. Dist. LEXIS 10650, at *6–7 (D. Md. May 20, 2004); Ivanov v. Process Design Assocs., 267 Ill. Ct. App. 3d 440, 444–445, 642 N.E.2d 711 (1993); Evans v. Buffington Harbor River Boats, LLC, 799 N.E.2d 1103, 1117 (Ind. 2003).

92. *See, e.g.,* Lewis Jorge Constr. Mgmt., Inc. v. Pomona Unified Sch. Dist., 2002 Cal. Ct. App. Unpub. LEXIS 6624, at *5–7 (2002) (construction manager was not party to owner/contractor contract, had no obligations under that contract, and therefore could not be held liable for negligent performance of that contract); Insurance Co. of the West v. O'Brien-Kreitzberg, Inc., 2002 Cal. Ct. App. LEXIS 6673, at *11 (2002) (unpublished) (finding no duty owed by construction manager to contractor's insurer because contract between construction manager and owner was not intended to benefit that third party; because construction manager owed duty to owner, finding of a duty to contractor's insurer could create a conflict of loyalty).

93. No. 3:04-CV-7467, 2005 U.S. Dist. LEXIS 4320 (N.D. Ohio 2005).

94. *Id.* at *10.

95. *Epicentre,* at *13–14.

96. *Epicentre,* at *15 (internal quotation marks and citations omitted).

be decided upon the presentation of evidence and, therefore, the court denied the construction manager's motion to dismiss.[97]

However, other jurisdictions have been unwilling to allow third party claims for negligence against construction managers. For example, California courts have been largely reluctant to impose liability on a construction manager for negligence claims asserted by third parties such as architects and trade contractors. These courts reason that construction managers generally have no duty to manage their business affairs so as to prevent purely economic loss to third parties.[98]

The economic loss doctrine can also be used to bar claims of negligence between those parties who are in privity of contract. In those jurisdictions that follow the economic loss doctrine, the contract remedies are those that are solely and exclusively available.[99]

2. Negligent Misrepresentation

Third parties who rely on the construction manager's factual representations may also assert the tort of negligent misrepresentation.[100] Most jurisdictions that permit negligence claims against nonprivity construction managers to recover economic loss also permit negligent misrepresentation claims to recover similar loss.[101] Even jurisdictions that bar negligence claims against the construction manager for economic loss, however, sometimes allow actions for negligent misrepresentation as an exception to the economic loss doctrine.[102] Some jurisdictions have adopted Restatement (Second) of Torts § 552[103] and hold that the construction manager owes certain independent duties to the contractor—namely, the duty to exercise reasonable care in the supervision, collection, and distribution of information to the contractor. In

---

97. *Id.* at *15–17.

98. Quelimane Co. v. Stewart Title Guar. Co., 960 P.2d 513 (Cal. 4th 1998).

99. *See, e.g.,* Martin Rispens & Son v. Hall Farms, Inc., 621 N.E.2d 1078, 1089 (Ind. 1993); Moore v. Coachmen Indus., Inc., 129 N.C. Ct. App. 389, 401, 499 S.E.2d 772 (1998); Berschauer/Phillips Constr. Co. v. Seattle Sch. Dist. No. 1., 124 Wash. 2d 816, 826, 881 P.2d 986 (1994); Insurance Co. of N. Am. v. Cease Elec., Inc., 276 Wis. 2d 361, 372, 688 N.W.2d 462 (2004).

100. RESTATEMENT (SECOND) OF TORTS § 552. At least one court has used this provision to hold that a property owner may be liable to a subcontractor for plans negligently prepared by the owner's architect. Gilbane Bldg. Co. v. Nemours Found., 606 F. Supp. 995, 1001–02 (D. Del. 1985).

101. *Gilbane Bldg. Co.*, 606 F. Supp. at 1001; Donnelly Constr. Co. v. Oberg/Hunt/Gilleland, 677 P.2d 1292, 1296–97 (Ariz. 1984); Village of Cross Keys v. United States Gypsum Co., 556 A.2d 1126, 1132 (Md. 1989).

102. *See* AAA Excavating v. Francis Constr., 678 S.W.2d 889, 893 (Mo. Ct. App. 1984); Malta Constr. Co. v. Henningson, Durham & Richardson, Inc., 694 F. Supp. 902, 907 (N.D. Ga. 1988), *aff'd*, 927 F.2d 614 (11th Cir. 1991).

103. The Restatement defines negligent misrepresentation and provides for liability against one who, failing to exercise due care, supplies false information which is relied upon, causing financial loss. *See* RESTATEMENT (SECOND) OF TORTS § 552.

Presnell Construction Managers, Inc. v. EH Construction, LLC,[104] a Kentucky trial court dismissed a contractor's claims against a construction manager citing a lack of privity and attendant duty. The Kentucky Supreme Court reversed, adopting the Restatement view and holding that "the tort of negligent misrepresentation defines an independent duty for which recovery in tort for economic loss is available."[105]

Likewise, in *John Martin Co. v. Morse/Diesel, Inc.*, the Supreme Court of Tennessee held that a trade contractor may recover in tort for negligent misrepresentation against a construction manager acting as agent for the owner.[106] In that case, the construction manager provided incorrect floor elevation measurements to the trade contractor responsible for placing the concrete.[107] Despite the lack of privity between the trade contractor and the construction manager, the court found that the construction manager had a duty to supervise and direct the work properly.[108] The court further held that privity of contract was not necessary in a negligence action to recover economic losses, and that the standard of care applicable to the construction manager was one of "reasonable care and competence."[109]

In yet another case, the highest state appellate court in Pennsylvania held that "an architect in the absence of privity of contract may be sued by a general contractor or the subcontractors working on a construction project for economic loss foreseeably resulting from breach of an architect's common law duty of care in the performance of his contract with the owner."[110] The court took the view that public policy dictated that a design professional should not be exempt from tort consequences, because the design professional intended others to rely on its work.

New York permits neither a negligence claim nor a negligent misrepresentation claim for purely economic loss.[111] Nonetheless, New York courts have suggested that

---

104. 134 S.W.3d 575 (2004).
105. *Id.* at 582.
106. 819 S.W.2d at 429.
107. *Id.*
108. *Id.*
109. *Id.* at 433. *See also* Wesconn Co. v. ACMAT Corp., 2001 Conn. Super. LEXIS 1042, at *12–13 (Apr. 11, 2001) (holding that despite lack of privity, construction manager was in a position to make negligent misrepresentations to subcontractor that could cause damage).
110. Bilt-Rite Contractors, Inc. v. The Architectural Studio, 581 Pa. 454, 480, 866 A.2d 270 (2005) (citation omitted).
111. Widett v. United States Fidelity & Guar. Co., 815 F.2d 885 (2d Cir. 1987); City of New York v. Aetna Cas. & Sur. Co., No. 96 Civ. 6890, 1997 U.S. Dist. LEXIS 9615, at *13 (S.D.N.Y. July 8, 1997) (dismissing negligence cause of action against construction manager because construction manager's responsibilities, including directing, coordinating, and expediting construction of the project, advising and making recommendations to the contractors during the design and construction phases, and providing inspection and coordination of the work of the various contractors on the project, did not show that the trade contractor was a member of a limited class whose reliance on any work product of the construction manager was foreseeable).

negligent supervision claims may be viable where the defendant had a duty to manage, supervise, and inspect the construction because the supervisory duties would have inured to the benefit of the plaintiff subcontractors as well as the project owner, and because the subcontractors' reliance on the construction manager's actions was foreseeable.[112]

3. Interference with Contract

Construction managers at-risk may also face liability for interference with the owner's contracts with other parties.[113] An agent construction manager will not likely face such liability because as an agent, the construction manager is considered to be the owner, and the owner cannot interfere with its own contract. To recover for tortious interference with contract, the plaintiff must prove the following elements: (1) a valid contract; (2) defendant's knowledge of the contract; (3) intentional, tortious interference by the defendant which causes a breach of the contract; and (4) damages.[114]

As with any claim for tortious interference with contract, the plaintiff must show that the construction manager interfered with a contract between the third party and another third party.[115] The plaintiff must also be able to show that the interference with contract was in fact wrongful and without justification, and was done to induce the breach of a contract. As with other tort theories, interference with contract claims against construction managers have not met with much success. For example, in *Badiee v. Brighton Area Schools*,[116] subcontractors sued the construction manager, claiming, *inter alia*, interference with contract on the grounds that the construction manager, at a meeting, used ethnic slurs in an attempt to have the owner cancel its contract with the contractor. The court granted summary judgment in favor of the construction manager, finding that while the construction manager may have acted wrongfully, the contractor had not shown that this action was taken with the intent of having the owner terminate the contractor.[117]

---

112. Morse/Diesel, Inc. v. Trinity Indus., 859 F.2d 242, 247–48 (2d Cir. 1988); James McKinney & Son, Inc. v. Lake Placid 1980 Olympic Games, Inc., 92 A.D.2d 991, 993, 461 N.Y.S.2d 483 (1983).

113. *See, e.g.,* City of New York v. Aetna Cas. & Sur. Co., 1997 U.S. Dist. LEXIS 9615, at *15–16 (denying motion to dismiss tortious interference with contract claim against construction manager).

114. Lama Holding v. Smith Barney, 88 N.Y.2d 413, 424, 646 N.Y.S.2d 76, 668 N.E.2d 1370 (1996); RESTATEMENT (SECOND) OF TORTS § 766.

115. *See, e.g.*, Mideast Systems v. Turner Int'l (Micronesia) Inc., 660 F. Supp. 864, 869–870 (S.D.N.Y. 1987) (dismissing claim for tortious interference with contract against construction manager when plaintiff failed to allege that construction manager interfered with a contract between plaintiff and a third party).

116. 265 Mich. Ct. App. 343, 695 N.W.2d 521 (2005).

117. *Id.* at 367, 695 N.W.2d at 539.

As long as a construction manager acts in the owner's interest (its principal), a third party should not be able to support a claim of tortious interference. In *N.S. Electric Corp. v. Procida Construction Corp.*,[118] the construction manager advised the owner to terminate a contractor. In response, the contractor sued the construction manager for tortious interference with contract. The court granted summary judgment to the construction manager after finding that the construction manager, in so advising the owner, was acting in the owner's interest, just as it was obliged to do.

Third parties may also attempt to recover from construction managers under a similar tort—intentional interference with prospective business advantage. This tort is identical to interference with contract; it concerns prospective, rather than existing, relations. For example, in *Cameo Homes, Inc. v. Kraus-Anderson Construction Co.*,[119] the plaintiff contractor claimed that the construction manager with whom it had worked made derogatory comments about the contractor to owners of various other projects. The plaintiff claimed this induced those owners not to hire the contractor. The defendants responded that they had neither intentionally nor improperly interfered. The court agreed with the defendants, granting summary judgment because the contractor had shown neither an intent to interfere nor any causation between the defendants' actions and the contractor's failure to secure other business.[120]

4. Defamation

Third parties have also been known to assert defamation claims against construction managers. In general, to recover under a theory of defamation, the plaintiff must show that the construction manager either intentionally or negligently made a statement to one or more third parties that adversely affected the plaintiff's reputation.[121]

One scenario in which this might arise is when a construction manager criticizes the performance of a contractor to others. In *Cameo Homes*, the contractor alleged that, after completion of its work repairing city buildings, the construction manager made negative comments about the contractor to local business owners, other contractors, financial institutions, and government entities. In that case, the court granted the construction manager's motion for summary judgment because the contractor failed to plead with specificity the content of the construction manager's allegedly defamatory statements.[122]

---

118. N.Y.L.J., Dec. 13, 1989, at 24 (N.Y. Sup. Ct.).
119. 2003 WL 23100876 (D. Minn. Dec. 3, 2003), *aff'd* 394 F.3d 1084 (8th Cir. 2005).
120. *Id.*
121. *See* RESTATEMENT (SECOND) OF TORTS § 558.
122. 2003 U.S. Dist. LEXIS 21714, at *16–17.

Mere statements of opinion are not actionable. In *Pontos Renovation, Inc. v. Kitano Arms Corp.*, the subcontractor alleged that the construction manager made statements that the subcontractor was "terrible" and "incompetent."[123] In that case, the court held that the contractor could not recover for defamation because such statements were merely the construction manager's opinion.

B. *Bases for Liability*

The liability of a construction manager under one or more of these theories of liability can stem from various wrongful actions taken *vis-à-vis* third parties.

For instance, one job often performed by the construction manager is reviewing bids submitted by potential contractors and making recommendations to the owner as to which bids to accept.[124] If the construction manager, through its actions, causes harm to the bidding contractors, the contractors may claim interference with their bids and have a cause of action for interference with prospective business advantage. For example, the owner in *Pepper-Reed Co. v. McBro Planning & Development Co.*[125] hired a construction company to serve as construction manager of a project. The construction manager rejected the bid of one contractor, despite the fact that this contractor had submitted the lowest bid and had indicated that it could go even lower if necessary. Rather than recommend this subcontractor's bid to the owner, the construction manager offered to perform the work instead, and the owner accepted. The rejected contractor subsequently sued. The court held that the contractor had a viable claim for damages against the construction manager as a result of the construction manager's actions.[126]

A construction manager may also be held liable, in certain circumstances, for defects in the design of the project. While the construction manager does not design the construction project, it may nevertheless be liable to a contractor for defects in such design if the construction manager has some involvement in the design that causes damage to the third party contractor. This issue was addressed in *John Martin Co. v. Morse/Diesel, Inc.*[127] when a trade contractor sued the construction manager for negligent misrepresentation. The trade contractor alleged that the construction manager had negligently approved defective shop drawings and that, in following these shop drawings, the trade contractor had been damaged.[128] The court found that because the construction manager had provided information to the trade contractor, information that the trade contractor had relied on in performing its work, causing loss to the trade contractor, the construction manager could be liable for its

---

123. 226 A.D.2d 191, 640 N.Y.S.2d 525 (1st Dep't 1996).
124. *See* CONSTRUCTION LITIGATION: REPRESENTING THE CONTRACTOR, *supra* note 23, § 7.21 at 175.
125. 564 F. Supp. 569 (D.V.I. 1983).
126. *Id.* at 571.
127. 819 S.W.2d 428 (Tenn. 1991).
128. *Id.* at 430.

actions.[129] Therefore, while the construction manager had not actually produced the defective designs, it was responsible to the trade contractor because it had a duty to exercise reasonable care in reviewing the designs, approving them, and conveying them to the subcontractor.[130]

Similarly, the construction manager is responsible for providing adequate information to the contractors, and the failure to do so may result in liability. For example, the construction manager must provide the contractors with clarifications of drawings, schedules of other contractors, and timely notice of proposed changes to the work.[131] When it fails to provide this information and the contractor is damaged, the construction manager may be liable.[132]

One of the construction manager's primary duties is to schedule and coordinate the work of the contractors hired by the owner.[133] Accordingly, the construction manager is responsible for creating and maintaining a construction schedule and coordinating site access by the various contractors.[134] The construction manager may be liable to those contractors if this coordination is lacking or otherwise causes damage to the contractors. For example, in *John E. Green Plumbing & Heating Co. v. Turner Construction Co.*,[135] the plaintiff plumbing contractor alleged that the defendant construction manager had, in several ways, provided deficient oversight to the construction project. Among other things, the construction manager had allowed the various contractors to do work out of sequence (e.g., a masonry contractor was permitted to conduct its work before the plumbing was completed) leading to increased costs for the plumbing contractor. The Sixth Circuit held that these allegations of damages were sufficient for the contractor to pursue a negligence claim against the construction manager.[136] Several courts have allowed similar claims against engineers acting as construction managers.[137]

---

129. 819 S.W.2d at 431–432.

130. *Id.*

131. *See* CONSTRUCTION LITIGATION: REPRESENTING THE CONTRACTOR, *supra* note 23, § 7.25 at 178.

132. *See, e.g.,* John E. Green Plumbing & Heating Co. v. Turner Constr. Co., 742 F.2d 965, 967 (6th Cir. 1984) (plumbing contractor claimed increased manpower costs as a result of construction manager's project mismanagement, including failure to approve design changes).

133. AIA Document B801-1992, Standard Form of Agreement Between Owner and Construction Manager, Art. 2.3.3; AIA Document A201/CMa-1992, General Conditions of the Contract for Construction—Construction Manager-Adviser Edition, Art. 4.6.3.

134. *See* CONSTRUCTION LITIGATION: REPRESENTING THE CONTRACTOR, *supra* note 23, § 7.27 at 179.

135. 742 F.2d 965 (6th Cir. 1984).

136. *Id.* at 967.

137. *See, e.g.,* Normoyle-Berg Assocs. v. Village of Deer Creek, 39 Ill. Ct. App. 3d 744, 350 N.E.2d 559, 561 (1976) (engineer liable to contractor for improper supervision of construction work because the relationship between the engineer and the contractor created a duty of care); Case Prestressing Corp. v. Chicago Coll. of Osteopathic Med., 118 Ill. Ct. App. 3d 782, 455 N.E. 2d 811 (1983) (recognizing duty to properly supervise owed by engineer to subcontractor).

The construction manager also may be liable to contractors for deficient coordination and scheduling if the contractors are denied timely access to the construction site and are thereby damaged. In *R.S. Noonan, Inc. v. Morrison-Knudsen Co.*,[138] following heavy rains at a construction site, the owner and construction manager permitted several contractors to drain rainwater onto the portion of the property being worked on by the plaintiff contractor. As a result, the plaintiff's work was delayed and it suffered increased costs. The court held that both the owner and the construction manager were liable to the plaintiff contractor for allowing this drainage and, in essence, causing the plaintiff's delays and increased costs.[139]

However, construction managers will not be liable to third parties for deficient supervision in the absence of a duty to such third parties. If the only alleged negligence of the construction manager is a breach of its contract with the owner, this will not be sufficient for a third party to recover.[140] Additionally, when the construction manager acts to fulfill its duty to the owner, a contractor cannot claim negligent supervision.[141]

Another responsibility of the construction manager is to provide construction support services to the contractors working on the construction site. If the construction manager fails to provide these services, including heat, light, and power, it can be held liable to a contractor who is damaged as a result.[142] Similarly, the construction manager may be liable if it undertakes responsibility for job site safety. For example, in *Caldwell v. Bechtel, Inc.*,[143] the construction manager undertook duties to provide "safety engineering services" and to oversee the enforcement of safety provisions and codes and inspect the job site for safety violations.[144] When an employee of the one of the trade contractors contracted silicosis from exposure to silica dust on the job site, the employee alleged that the construction manager had responsibility for "overall direction and supervision of safety measures and regulations in effect, or needed during the course of construction," that construction manager knew or should have known of the danger posed by the silica dust, and that the construction manager negligently failed to rectify the situation.[145] The court found that "once [the construction manager] undertook responsibility for overseeing safety compliance, it assumed a duty of reasonable care in carrying out such duties that extended to workers on the site."[146] In addition, the court explained

---

138. 522 F.Supp. 1186 (E.D. La. 1981).

139. *Id.* at 1191.

140. Guarantee Elec. Co. v. Big Rivers Elec. Co., 669 F. Supp. 1371 (W.D. Ky. 1987).

141. Bagwell Coatings, Inc. v. Middle S. Energy, Inc., 797 F.2d 1298 (5th Cir. 1986).

142. *See, e.g.,* John E. Green Plumbing & Heating Co. v. Turner Constr. Co., 742 F.2d at 967 (6th Cir. 1984) (construction manager could be held liable to contractor for failure to provide temporary heat at construction site).

143. 631 F.2d 989 (D.C. Cir. 1980).

144. *Id.* at 992–993.

145. *Id.* at 993–94.

146. *Id.* at 1001.

that the construction manager's contractual duties to the owner coupled with its superior skills and status as safety engineer created a duty in tort to the employee to take reasonable steps to protect him from the foreseeable risk of harm posed by the silica dust.[147]

Library References

C.J.S. *Agency* §§ 375, 377, 419–421, 423–445; C.J.S. *Architects* §§ 19, 21–23; C.J.S. *Fraud* §§ 10–17, 23–25, 29–30, 34–35, 59–60, 62–63, 67–75, 77–78, 80; C.J.S. *Negligence* §§ 149, 162–164, 180–182, 304, 311, 399, 532–535, 561, 574, 576–577, 595–596, 598–601, 618–619.

West's Key No. Digests, Fraud <KEY>8; Libel and Slander <KEY>6(1), 9(6); Negligence <KEY>321, 322, 480, 1202–1205; Principal and Agent <KEY>159(1); Torts <KEY>241, 242.

IV. Owner's Liability to Third Parties Based on Imputation of Construction Manager's Misconduct

Generally, an owner is liable only for its own acts or omissions. However, there are exceptions to this rule, such as where the owner has an agent acting on its behalf. Where the owner utilizes an agent, the owner is responsible for not only its own conduct, but the conduct of its agent, provided that the agent was acting within the scope of its agency at the time the conduct occurred. This is known as the doctrine of *respondeat superior*.[148] Because the owner's liability is derivative of the liability of the agent, the owner is only liable if the agent is liable. Thus, if the agent was not negligent or has a defense, the owner will not be liable.

In the construction management context, the doctrine of *respondeat superior* comes into play when the construction manager is acting as the owner's agent. In this context, the owner may be held liable for the construction manager's actions as long as the construction manager is acting within the scope of its agency.[149] For example, in *Menard, Inc. v. U.S. Equities Development, Inc.*,[150] the plaintiff contracted to purchase property from the owner. Prior to the sale, the defendant construction manager was supposed to build a retaining wall to be used by the plaintiff after it closed on the property. After the sale, the retaining wall crumbled, and the plaintiff sued the construction manager. However, the plaintiff also sued the construction manager's principal, the previous owner of the property. When the previous owner moved to dismiss this count, the court denied the motion, stating that this owner could be liable under the doctrine of *respondeat superior* for the torts of its agent, the construction manager.[151]

---

147. 631 F.2d at 1001.
148. Taylor v. Phelan, 912 F.2d 429 (10th Cir.), *cert. denied*, 498 U.S. 1068 (1991).
149. AGC CONTRACT DOCUMENTS HANDBOOK § 8.01 (J. William Ernstrom & Kevin F. Peartree eds., 2003).
150. 2002 WL 31050160 (N.D. Ill. Sept. 13, 2002).
151. *Id.* at *7.

In contrast, where the construction manager is at-risk or otherwise acts as an independent contractor (as opposed to as the agent for the owner), the owner typically will not be held responsible for the actions or inactions of the construction manager.[152] This distinction can become murky where the construction manager operates as an independent contractor with respect to certain work and as an agent to the owner as to other work.[153]

V. Construction Manager Defenses

A. *Contract Claims Precluding Tort Claims*

In most cases, an owner will assert both breach of contract and tort claims against the construction manager for breach of duties imposed by the construction management agreement. While some courts will allow both the contract and negligence claims to stand,[154] other courts are more restrictive where the sole claim of negligence is based upon the construction manager's failure to perform under the agreement.[155] This is because, in some states, a claim of negligence is not viable where the contract is the sole source of the legal duty.[156] To justify a tort action, there must be some breach of duty separate and distinct from a breach of contract, even if the origin of the duty is the contract itself.[157] In these jurisdictions, the construction manager can defend against an owner's negligence claims by arguing that the alleged

---

152. Aladdin Constr. Co. v. John Hancock Life Ins. Co., 2005 Miss. LEXIS 717, at *11 (Oct. 27, 2005). However, note there can be cases where there is no vicarious liability even though the construction manager is acting as agent for owner. *See* RESTATEMENT (SECOND) OF AGENCY § 2 cmt. b (2005).

153. 2005 Miss. LEXIS 717, at *20 (citing Kight v. Sheppard Bldg. Supply, Inc., 537 So. 2d 1355, 1359 (Miss. 1989)).

154. *See* Brown Field Aviation Park, LLC v. PB Aviation, Inc., No. D-043154, 2005 Cal. Ct. App. LEXIS 6358 (4th Ct. App. Dist., (unpublished) 2005) (rejecting project manager's argument on appeal that a party cannot recover in tort for breach of a duty that duplicates one created by contract).

155. *Dick Anderson Constr. Co.,* 2004 Mont. Dist. LEXIS 2382, at *23; Saint Patrick's Home for the Aged and Infirm v. Laticrete Int'l, Inc., 700 N.Y.S.2d 28 (Ct. App. Div. 1st Dep't 1999) (property owner could not maintain tort claims against construction manager for breach of duty of care in performance of contract and breach of fiduciary duty, which simply restated cause of action for breach of contract; owner sought only to impose liability in tort upon manager for failure to perform duties allegedly owed to owner pursuant to contract).

156. *Id.*

157. 103 N.Y. JUR *Torts* § 16 (West 2d ed. 2003); Trustees of Columbia Univ. v. Gwathmey Siegel & Assoc. Architects, 601 N.Y.S.2d 116 (Ct. App. Div. 1st Dep't. 1993) (negligence action could be maintained alleging defective design and supervision of construction project, which resulted in partial collapse of building facade, where plaintiff made sufficient showing that defendant construction manager bore and breached a legal duty independent of the parties' contract).

breach of duty is merely a restatement of the owner's breach of contract claim.[158] Once again, the thrust of this defense is the economic loss doctrine which holds that no tort claim made arises from the breach of a contractual obligation.

B. *Contributory/Comparative Negligence*

Because the owner owes certain duties to the construction manager under the construction management agreement, the construction manager may assert contributory and/or comparative negligence as a defense to a professional negligence claim asserted by the owner.

For example, the owner usually owes duties to the construction manager to: (1) provide information in a timely manner, including the owner's program and other project requirements; (2) provide project financing[159]; (3) provide reports, surveys, and tests regarding site conditions; (4) retain and provide the services of an architect/engineer; (5) provide timely review submissions/proposals from the construction manager; and (6) provide the services of other consultants as required.[160]

If the owner fails to fulfill these duties, the owner is generally responsible for the error or omission, not the construction manager. This is significant in that it protects the construction manager from negligence suits by the trade contractors for providing incorrect information and places this responsibility on the owner. In jurisdictions that recognize comparative fault or contributory negligence, the owner's failure to perform these duties may limit the owner's recovery against the construction manager or bar its negligence action against the construction manager altogether.

C. *Indemnity and Contribution*

The construction manager may also be entitled to contribution or indemnity from third parties with respect to owner claims. For example, in *Strategem Development Corp. v. Heron International N.V.*,[161] the project owner brought suit against its construction manager for, among other things, exceeding the project budget and failing to adequately supervise the construction project in which not enough rentable square footage had been provided for in the plans. The construction manager filed suit against the architect for contribution/indemnity on the basis that the architect had resisted its efforts to make necessary changes aimed at reducing project costs, and that the architect had made the incorrect calculations.[162] The court found that

---

158. *Saint Patrick's Home*, 700 N.Y.S.2d at 30–31 ("A tort action may only be asserted in an action for breach of contract where the underlying agreement gives rise to a duty independent of the contract obligation .... Where a party is merely seeking to enforce its bargain, a tort claim will not lie.").

159. If the owner fails to provide evidence of adequate financing, the construction manager may not be obligated to begin or to continue its services. *See* AGC 510, Art. 4.1.2; AIA A201/CMa, Art. 2.2.1.

160. *See* AGC 510, Art. 4 (Owner's Responsibilities); AIA A201/CMa, Art. 2.2; AIA A121/CMc, and AGC 565, Art. 3.

161. 153 F.R.D. 535 (S.D.N.Y. 1994).

162. *Id.* at 540.

the architect owed a duty to the construction manager because their relationship approached privity.[163] Specifically, the court found that the architect was aware from weekly meetings that the construction manager was relying on the architect's figures and used them in calculations presented to the owner.[164] Also, the construction manager was a third party beneficiary of the agreement owner/architect agreement, which stated that with respect to negligence caused by the architect, the architect would indemnify the owner and the owner's representative.[165]

D. *Superior Knowledge*

The construction manager may also be able to defend against an owner's claim with the argument that the owner of the property may have superior knowledge of a defect or deficiency, such as underground conditions, for which the owner, rather than the construction manager, stands responsible. It is well established that an owner may be liable to a party injured on his property if the owner has knowledge of a defect that may subject the party to an unreasonable risk of harm.[166] The owner may escape such liability in situations where full possession and control of the land are surrendered to another party, usually an independent contractor.[167] Therefore, as long as the construction manager can show that it has not taken over complete control of the property from the owner, a defense based on the owner's superior knowledge may be possible if it can be shown that the owner in fact had such superior knowledge.

E. *Limitations of Authority*

A party cannot be held responsible for the actions of third parties unless those third parties are somehow under its control. Thus, while a construction manager is generally responsible for coordinating and scheduling the activities of others (such as architects, engineers, contractors, suppliers, etc.), it is not directly responsible for the performance of those third parties unless the construction manager has some degree of control over them. In other words, the degree of the construction manager's liability to the owner for the mistakes of these third parties will depend upon the construction manager's scope of authority under the construction management agreement. This is particularly true when the construction manager is at-risk and its agreement with the owner places upon the construction manager cost and schedule risk.

F. *Exclusivity of Remedies/Limitation of Liability Provisions*

In some cases, the construction management agreement may contain exclusivity of remedy or limitation of liability provisions that limit the construction manager's

---

163. 153 F.R.D. at 548.
164. 153 F.R.D. at 548.
165. 153 F.R.D. at 548.
166. *See, e.g.*, West v. Briggs & Stratton Corp., 244 Ga. Ct. App. 840, 844, 536 S.E.2d 828, 832 (2000); Frost v. Dayton Power and Light Co., 138 Ohio Ct. App. 3d 182, 199 (2000).
167. *West*, 244 Ga. Ct. App. at 844.

liability under the contract. For example, AGC 510 contains an "exclusive remedy" provision stating that the construction manager's warranty obligations set forth in the agreement constitute the owner's "exclusive remedy with respect to the quality of the services."[168] This same provision further states:

> Construction Manager makes no warranties related to schedules or completion dates, budgets, the cost of the Work or the Project, the Work performed by the Trade Contractors, or any other warranties, express or implied, which are not expressly set forth herein. Construction Manager shall have no liability for any errors or omissions in the Construction Documents or any defects in the Services attributable to Construction Manager's use or and/or good faith reliance upon the Construction Documents or any other information furnished by or on behalf of the Owner or Architect/Engineer.[169]

This same document contains an additional limitation of remedies provision at Article 13.11, stating "[t]he parties' rights, liabilities, responsibilities and remedies with respect to this Agreement, whether in contract, tort, negligence or otherwise, shall be exclusively those expressly set forth in this Agreement."[170]

Such exclusivity and limitation of remedies provisions are often upheld even if the limitation is significant. For example, if the construction manager fails to perform its services adequately and the owner is later required to tear out and replace a trade contractor's work, the owner's recovery against the construction manager should be limited to the value of the construction manager's corrective services; the owner will have to look to the trade contractor for more substantial recovery.[171] These provisions may also preclude the owner from asserting alternative theories of recovery against the construction manager for negligence or other tort theories of liability.

Limitations of liability may be defeated where the construction manager's conduct amounts to gross negligence or willful misconduct. In some, but not all jurisdictions, grossly negligent or intentional misconduct may not be disclaimed by contract. Also, if the construction manager's fault concerns a matter outside the services that the construction manager was obligated to provide under the contract, the limitation of liability clause may be inapplicable.

Library References

C.J.S. *Contracts* §§ 229–230, 238–240, 271, 359; C.J.S. *Negligence* §§ 35, 37, 226–266, 281–291, 300, 302, 313, 317–319, 324–327, 566, 624–646, 648, 654.

West's Key No. Digests, Contracts <KEY>114, 129(1), 206; Indemnity <KEY>33(5); Negligence <KEY>219, 501, 1281.

---

168. *See* AGC 510, Art. 6.2.
169. *See id.*
170. *See id.* Art. 13.11.
171. AGC CONTRACT DOCUMENTS HANDBOOK, § 8.06[B], at 758.

CHAPTER 13

The Construction Manager and the Design-Build Project

JEAN C. ARNOLD, ESQ.
CHRIS SQUADRA

I. Introduction

By 2015, the design-build delivery system is expected to account for 55 percent of the United States domestic construction market, both private and public.[1] In April 2003, Bruner and O'Connor discussed the reasons for the growing popularity of the design-build delivery system by citing statistics from a study of 351 domestic building projects. The study results were published in the November/December 1998 issue of the *Journal of Construction Engineering and Management*:[2]

- Unit cost: at least 6.1% less than design-bid-build projects and 4.5% less than construction management at risk projects;[3]
- Construction speed: at least 12% faster than design-bid-build projects and 7% faster than construction management at risk projects;[4]

---

1. 2 BRUNER & O'CONNOR ON CONSTRUCTION LAW §6:22 (2002), *reprinted in* Phil Bruner & Patrick J. O'Connor, *Comparison of Various Project Delivery Systems—Selecting the Best Delivery System for Your Project*, 491 PLI/REAL 279, 285 (April 2003) (PLI Handling Construction Risks 2003: Allocate Now or Litigate Later Course, Handbook Series, PLI Order No. N0-00BY).

2. *Id. citing* Konchar & Sanvido, *Comparison of U.S. Project Delivery Systems*, 124 J. OF CONSTR. ENG. & MGT. 435 (1998).

3. Phil Bruner & Patrick J. O'Connor, *Comparison of Various Project Delivery Systems—Selecting the Best Delivery System for Your Project*, 491 PLI/REAL 279, at 286 (April 2003) (PLI Handling Construction Risks 2003: Allocate Now or Litigate Later Course, Handbook Series, PLI Order No. N0-00BY).

4. *Id.*

- Delivery speed: at least 33.55% faster than design-bid-build projects and 23.5% faster than construction management at risk projects;[5]
- Cost growth: at least 5.2% less than design-bid-build projects and 12.6% less than construction management at risk projects;[6] and
- Schedule growth: at least 11.37% less than design-bid-build projects and 2.8% less than construction management at risk projects.[7]

With such dramatic results and great press, it is no wonder that the design-build delivery method is becoming the delivery system of choice. However, risk considerations for the construction manager as design-builder in the design-build delivery system are substantial and range from state licensing issues to liability for faulty construction and to vicarious liability for design.[8] This aptly numbered chapter will identify and deal practically with many of the risks the construction manager as design-builder faces as it leads the design-build project.

II. The Construction Manager as Design-Build Team Leader

The focus of this chapter is on the construction manager as design-builder rather than as advisor. The construction manager as design-builder becomes a one-stop shop and leader of the design-build team. From an owner's perspective, this arrangement can be very appealing, as the construction manager as design-builder is the single point of responsibility and liability. For the design-builder, this arrangement is fraught with the highest possible degree of liability as the owner looks to the construction manager as design-builder to manage the design-build process from pre-construction to completion. Such risks include:

- Design—no change orders for design omissions because the construction manager as design-builder is responsible for hiring the architect and providing the design.
- Subcontractor—these are inherent for any general contractor, but now the construction manager as design-builder does not have a general contractor to look to. The risks include prequalification, performance, payment and warranty.[9]
- Insurance and bonding—the construction manager as design-builder is generally responsible for obtaining or at least advising the owner on the insurance program for the project and the requirement for subcontractor bonding.

---

5. *Id.*
6. *Id.*
7. *Id.*
8. A.H. Gaede, Jr., *Risk Management in Design-Build*, 467 PLI/REAL 319 (April 2001) (PLI Handling Construction Risks 2001: Allocate Now or Litigate Later Course, Handbook Series, PLI Order No. N0-0060).
9. St. Paul Cos. v. Construction Mgmt. Co., 96 F. Supp. 2d 1094 (D. Mont. 2000) (holding that in the design-build context, the "independent contractor rule" did not shield the design-build team from vicarious liability for the negligence of the electrical subcontractor).

- Safety—the construction manager as design-builder is generally responsible for the project safety program unless the owner hires a separate safety consultant.
- Quality control—the owner will expect the construction manager as design-builder to develop a quality control and testing program unless the owner hires separate consultants to perform this function.

The buck stops here when the construction manager acts as design-builder rather than as an advisor to the owner. The construction manager as design-builder can look only to its design team and subcontractors for contribution to the broad liability assumed. Therefore, it is of critical importance that the owner buy in to the process, both financially and by active participation and use of an owner's design advisory group and peer review team, along with active use of consultants of the owner's choosing.

In addition, there are three unique liability concerns that the construction manager must address when acting as the design-builder and team leader: (1) conflicts of interest; (2) vicarious liability for design; and (3) state licensing considerations.

A. *Conflicts of Interest*

Where the construction manager (CM) acts as design-builder, "the lines of accountability may be blurred ... especially when the CM acts as contractor. At times it may be difficult to tell whether the CM is acting in the best interest of the owner, or in its own best interest as contractor."[10]

Such inherent conflicts undermine the design-builder's credibility and create unnecessary and counterproductive project disputes. The use of an independent design advisory group and peer review team working for the owner restores the design-builder's credibility and the owner's confidence in the process. The use of such owner teams also improves the process and reduces the design-builder's liability by providing independent checks and balances. These teams and their uses are discussed in more detail in Section V below.

B. *Vicarious Liability for Design*

Where the construction manager acts as design-builder:

the design-build entity assumes greater liability for errors in the specification than does the general contractor under the design-bid-build system. In design-build, the owner usually supplies only performance specifications and maybe some approved conceptual, or schematic, design drawings. The design-build entity then prepares the drawings and specifications for the project, and agrees to deliver a project which meets the performance expectations of the

---

10. Walter J. Sears, III & Arlan D. Lewis, *Risk Management in Design-Build*, 443 PLI/Real 361, at 371 (May 1999) (PLI Handling Construction Risks 1999: Allocate Now or Litigate Later Course, Handbook Series, PLI Order No. N0-002D).

owner. Accordingly, to the extent the plans and specifications contain errors, the design-builder, rather than the owner, may be required to absorb the cost of such mistakes.[11]

C. *State Licensing Requirements*

The key question to ask is: "Can a general contractor, organized and licensed according to the relevant state laws, perform design/build work with its own forces under a direct contract with the owner?"[12] *The Design/Build Deskbook*, Map 2, answers the question for the United States and Canada as follows:[13]

- No, unless "grandfathered," for the state of New York;
- No, for the states of Oregon, North Dakota, Iowa, Wisconsin, Michigan, Illinois, Mississippi, Rhode Island, the District of Columbia and the province of Prince Edward Island;
- Yes for engineering, no for architecture for the states of North Carolina and Idaho and the provinces of Alberta and Ontario;
- Yes for architecture, unclear for engineering for the state of Pennsylvania;
- Yes for architecture, no for engineering for the state of Nevada;
- Unclear for the states of Oklahoma and Louisiana;
- Yes for all other states and provinces not listed above.

Therefore, the construction manager as design-builder must first check state or provincial laws before it can engage in providing the services expected by the owner.

Library References

C.J.S. *Architects* §§ 1–3, 7; C.J.S. *Negligence* §§ 399, 574, 576–577, 595–596.
West's Key No. Digests, Licenses <KEY>11; Negligence <KEY>1202(2).

III. Managing Total Project Costs on the Design-Build Project as a Design-Builder

By definition, the construction manager, when not at-risk, is providing only services. Every member of the team, whether owner, design-builder, or the owner's third party consultants and vendors, expects the construction manager to have a bird's-eye perspective on the project, know every nuance, and prevent every oversight and failure. This is an impossible expectation to meet if there ever were one, but nowhere is this more true than when parties discuss project budget.

---

11. *Id.* at 372.
12. James S. Schenck, IV, *Introduction,* DESIGN/BUILD DESKBOOK § 1–14, 15 (John R. Heisse, II & James S. Schenck, IV eds., 3rd ed. 2004).
13. *Id.*

The design-builder, while having a smaller field of view, shares many of the same concerns in managing the design team's efforts to meet a budget getting through the approvals process with enough design team fees, carrying enough contingency for pre-construction scope changes and escalation, etc.

The liability for both the design-builder and the construction manager is very large here, as it always is when the owner's expectations for team members are very high. The construction manager and design-builder can mitigate these risks by following simple processes:

- *understanding* where the disconnect originates;
- using a good *process* to force communication;
- *documenting* the resulting understanding; and
- attaching the resulting understanding as an exhibit to good, clear, concise, and enforceable *contract language*.

A. *Owner's Budget vs. Raw Construction Cost*

The owner has a very different viewpoint of what constitutes project cost than even the most diligent and experienced design-builder—and many a well-intentioned project has fallen into the abyss between the two. Whether a seasoned developer or a construction neophyte, all owners expect their team members to approach the "what does it cost" question with the same "all costs included" answer they are expecting. Most design-builders would be hard-pressed to provide a comprehensive list of owner costs beyond design, construction, and materials testing fees.

Therefore, the construction manager must lead the way to bridge this lack of communication, fully understanding it is a symptom of a much deeper and more dangerous scope of work and compensation disconnect. While translating the differing definitions of cost, reconciling those differences, and generating clarity from confusion, the construction manager simultaneously reduces the risk of project failure and its own professional liability exposure.

There are two excellent, well-recognized industry tools to get this job done: the responsibility matrix and the design services scope of work. Each forces detailed communication between the parties, and both should become exhibits to the owner/design-builder's contract. The use of these tools provides an orderly procedure for documenting assigned responsibilities for the various tasks, project components, and deliverables. This is especially important for the construction manager acting as the design-builder for the project. Managing the risks and coordinating communication and responsibility are absolutely critical.

The Responsibility Matrix. Customized by the construction manager for each specific project type, good responsibility matrixes are comprehensive and have common elements of columns showing responsibility for design; construction documents; budgeting; furnishing; and installation for discrete portions of the project's scope. These scope items are listed row by row, with "O" (Owner), "DB" (Design Builder), or "Excluded" shown in each column for each row. An example of the typical responsibility matrix is attached to this chapter as Appendix A.

Some sample scope items (i.e., rows in the matrix) of typical points of failure would be data conduit, data junction-boxes, data wiring, data terminations, data wiring testing and tone-out, data wiring turnover, teledata equipment, etc. In this example, the owner may typically employ three or four contractors/vendors outside the general contractor's scope. It is the construction manager's job to understand the scope of work of each vendor and properly coordinate each contract so there are no overlaps or gaps.

The Design Services Scope of Work. Like the responsibility matrix, the design services scope of work uses a rows and columns format, with phases of design and contract administration across the column headings (i.e., Conceptual, Schematic, Design Development, etc.). Down the left side, in a column labeled "Description" is a very comprehensive list of all available design services (e.g., site survey, approvals hearings, architectural design, fire protection system design and engineering, value engineering, etc.). For an excellent example of an accepted industry standard, examine the American Institute of Architects (AIA) Document B163 (both the Excel and Word documents). These documents are available for review and purchase on the AIA's Web site at www.aia.org/docs.

The construction manager's responsibilities are to make sure the owner and the general contractor are on the same page with inclusions and exclusions to the general contractor's proposed services, and to document and memorialize that agreement in the contract. These communication tools can achieve this goal if the construction manager conducts meetings expressly for this purpose with the owner and general contractor before executing the contract.

In an ideal situation, where the construction manager is experienced and proactive, the two tools described above are provided to general contractor candidates as part of the selection process before proposal preparation, and are used to de-scope the candidates before selection. In addition, to avoid liability for scope coordination, this same construction manager uses mirror images of these tools to select all design consultants, contractors, and vendors and to facilitate a seamlessly integrated design.

B. *Escalation:* Force Majeure *or Reasonably Inferable in Today's Economy*

The concept of *force majeure* comes into play most often when a project finishes late: "*Force majeure* delays are those that are beyond either party's control, such as strikes, natural catastrophes, war, civil unrest, and 'acts of God.' The risk of such delays often falls on the owner."[14] However, depending on the contract language the risk may also be shared by the general contractor and construction manager acting as design-builder in the form of a time extension only, with no compensation for the delay damages they may suffer.

If the construction manager acting as design-builder guarantees the completion date or price without exception for a *force majeure* event, the construction manager

---

14. Gregg E. Bundschuh & David Collings, *Insurance and Bonding for a Design/Build Project*, in DESIGN/BUILD DESKBOOK § 4-24 (John R. Heisse, II & James S. Schenck, IV eds., ABA FORUM ON THE CONSTRUCTION INDUSTRY, 3rd ed. 2004).

may assume the risk. Limiting the liability and the risk to the construction manager can be accomplished by including a *force majeure* or other risk retention clause in its contract with the owner that allocates the risk to the owner. The AIA and Associated General Contractors of America (AGC) standard form documents do not include a *force majeure* clause projecting the design-builder from delays beyond the design-builder's control.

This risk may also be managed through the purchase of *force majeure* insurance. In their chapter, *Insurance and Bonding for a Design/Build Project*, Gregg Bundschuh and David Collings discuss the availability of such insurance.[15] They compare and discuss the traditional way of dealing with the risk of delayed completion by purchasing a "soft-cost endorsement" to the builder's risk policy.[16]

The construction manager and owner must analyze their project risks and allocate those risks among them. Where certain risks are unforeseeable or rise to the level of *force majeure* events, the parties may need to turn to specialized insurance products to manage the impact of these events on cost and schedule.

C. *Design Team Fees and Reimbursables vs. Additional Services*

1. *Redesign Requirements for the Design Team as Basic Services*

When the construction manager negotiates the design-builder agreement with the owner, as well as when the design-builder negotiates its subcontract agreement internally with its design team, one of the most important components is the design team's scope of work, its resulting fees, and associated reimbursable expenses. Lack of clarity here in either contract language or detailed exhibits is a harbinger of unpleasant meetings at best, and litigation at worst.

To prevent this, and do a diligent job on the owner's behalf, the construction manager as design-builder should, in turn, ask its design team to prepare a proposed exhibit to the owner/design-builder's agreement, in addition to *The Design Services Scope of Work* referenced above, that identifies a lump sum corresponding to each of the following services and reimbursable expenses (actual text from RFP):

Design and/or Engineering Fees

    Master-planning

    Programming

    Schematic Design

    Design Development

    Construction Documents

---

15. *Id.* §§ 4-24 to 4-26.
16. *Id.*

Bidding and Negotiation

Contract Administration

   Subtotal Design Fees

Reimbursable Expenses

Reproduction and Printing Costs

Postage, Shipping and Long Distance Telephone

Airfare, Hotel, Meals, Mileage and Other Travel Costs

Reimbursable Contingency

   Subtotal Reimbursable Expenses

Total of Fees and Reimbursable Expenses

When requesting this information, to protect the owner and itself from future misunderstandings, the construction manager as design-builder should provide these instructions to its design team, especially with regard to reimbursable expenses. The lack of specificity about printing costs, travel expenses, etc., will almost certainly lead to disputes if not carefully documented before project initiation. The instructions example that accompanies the above services/reimbursables includes the actual text from the RFP and is attached as the Instructions for Using the Fee and Reimbursables Worksheet in Appendix B.

After receiving the schedule of fees and reimbursables, the construction manager as design-builder must perform a fair amount of due diligence validation to lessen its exposure to future claims, and the owner's exposure to future, unexpected costs. The most thorough approach (from the construction manager's perspective) is to:

1. confirm the fee is based on and matches the design services scope of work;
2. coordinate the fees and scope of work with other, similar proposals received from the owner's consulting team;
3. compare the fees to the owner's budget line item for such cost(s); and
4. review the fees with the owner.

Obviously, the above steps can be modified by a design-builder selecting its design team, with similar preventative benefits.

When the construction manager has completed these actions, it should reference the resulting exhibit in the owner-design-builder agreement, and note that reimbursable expenses are to be billed at actual cost against the exhibit amount as a maximum.

In the inflationary economy predicted to be typical for the first and second decades of the twenty-first century, the construction manager must recognize and advise the owner as to reasonably inferable cost increases and their associated redesign costs. With a design-builder, this is an easier task than it would be under other contract formats available to the owner (e.g., stipulated sum, cost-plus-a-fee, etc.), as the design-builder is clearly responsible for providing both the design and the corresponding construction cost estimate to the owner.

Nonetheless, typically the owner and the owner's attorney are relying on the construction manager/design-builder's expertise, as the expert in the process, to provide controls for the risk of redesign cost. Therefore, careful attention must be paid by the construction manager as design-builder when agreeing to contract language covering this concept. One possible solution in the form of contract language that protects the owner (but adds risk to the construction manager as design-builder), is provided by the following example:

> A fixed limit of Construction Cost shall be established as a condition of this Agreement.
>
> In the event the Design-Builder's estimate or a compilation of the lowest bona fide bids or negotiated proposals received by the Design-Builder exceeds the Design-Builder's budget, the modification of Contract Documents shall be the Design-Builder's responsibility, without additional compensation or extension of time from the owner.

Language that would manage the design-builder's risk could read:

> A fixed limit of Construction Cost shall be established as a condition of this Agreement.
>
> In the event the Design-Builder's estimate or a compilation of the lowest bona fide bids or negotiated proposals received by the Design-Builder exceeds the Design-Builder's budget, the modification of Contract Documents shall be the Design-Builder's responsibility, and the Owner agrees to pay other additional expenses and compensation associated with the increased costs and further agrees to an extension of time if requested by the Design-Builder.

2. The Approvals Process: Unknown Hourly Billings or Predictable Lump Sum

Local, state, and federal approvals processes can be a grueling, unpredictable journey into a place where capricious public officials provide subjective interpretations, roadblocks, and seemingly endless reviews, redesign loops, and resubmittals before granting the prized building permit. What is the construction manager as design-builder's responsibility to both fairly and firmly set a fee with the owner that is not subject to ever-increasing additional service requests?

The owner expects that the construction manager as design-builder has the level of approvals experience on its team to understand the specifics required, in general, to meet the approvals process for the project location. It is a fair expectation and, therefore, may be enforced contractually in the design-builder's agreement.

The premise is that the design-builder has sufficient fees to provide the level of effort typically required by similar projects in the locale, and the jurisdictions contemplated. The following language may be incorporated for this purpose:

> As part of Basic Services hereunder, the Design-Builder shall provide all submittals and presentations, attend and participate in all meetings and hearings, whether public or restricted, and provide such professional services required to obtain all project approvals necessary to obtain a building permit, but solely to the extent such services are both reasonable and necessary as part of the normal zoning and approval process for a project of similar size and complexity at the project location.

Less desirable from an owner's perspective, but easier on the construction manager as design-builder, is to specify a list of approvals that are excluded. The design-builder then includes all others in its fee.

Least desirable for the owner, but easiest for the design-builder is the strategy of specifying the specific approval(s) included in the field and corresponding level of effort that will be expected. Contract exhibits with a certain number of meetings, or limited number of pre-construction hours allocated to approvals, and/or Web site URL links to the approvals process flowchart(s) are all good tools to include.

3. *Controlling Owner Changes During Design as a Design-Builder*

The construction manager as design-builder's responsibility is to manage the owner's project input. To the extent this job is done well, the project risk (as well as the construction manager's risk of being found negligent) is reduced substantially. But what does it mean to *manage the owner's project input … well*? Clear, timely information from and unambiguous, irrevocable decisions by the owner are the hallmarks of success. Poorly defined, late information and second-guessed decisions made the previous month lead to compensable project delays and justifiable additional services for the design-builder.

To accomplish this, the construction manager must make it clear from the beginning of the relationship that it is the *design-builder* who knows (or should know) what information its team needs and when it needs it, and what decisions it needs from the owner and when those decisions must be provided. Contractually, that responsibility is defined:

> Before requesting any payment under this Agreement, and as a condition precedent to the enforceability of this Agreement by the Design-Builder, Design-Builder shall have submitted to Owner and Owner shall have accepted in writing a proposed schedule setting out the dates on which Design-Builder plans to complete schematic design phase, design development phase, construction

document phase, and bidding or negotiation phase. This schedule shall also set out the dates by which any actions, decisions, or information are required from the Owner in order to permit the Design-Builder to perform according to this schedule, as well as payments required by the Owner for Basic Services and anticipated reimbursables.

In order to ask the design-builder to provide the list above, and adhere to such language, the owner must have provided a site survey for the proposed location, clear project program, schedule, and construction cost budget. The construction manager as design-builder must evaluate and accept each in terms of the other before executing the owner/design-builder agreement.

4. *Managing a Design Team Schedule: Contractual Responsibility*

A similar concept (and the above language) helps the construction manager as design-builder prepare the overall pre-construction schedule and prevent the liability that may otherwise be created by the owner and design professional unilaterally imposing its design schedule on the project. Design-builder and owner schedule compliance is more easily achieved using a mutually agreed schedule, especially when the construction manager as design-builder monitors this target closely against actual project performance.

The owner's performance against this target schedule must also be carefully discussed, preferably weekly, on most projects. The owner must be made aware of its responsibility early and reminded often to reduce the owner's risk of inadvertently delaying the design-builder and the design and construction processes.

To prevent future additional fee(s) or requests from the design-builder for time extensions during pre-construction from the design-builder based on the owner's failure to provide information when anticipated, the construction manager as design-builder is wise to include the following language in the design-builder agreement:

Additional Services for the Design-Builder are agreed to be compensable by the Owner to the extent Design-Builder is required to make revisions in Drawings, Specifications, or other documents when such revisions are:

1. inconsistent with approvals or instructions previously given by the Owner, including revisions made necessary by adjustments in the Owner's program or project budget;
2. required by the enactment or revision of codes, laws or regulations subsequent to the preparation of such documents; or
3. due to changes required as a result of the Owner's failure to render decisions in a reasonably timely manner solely when such decisions were identified by the Design-Builder in advance and in writing per the requirements elsewhere in this Agreement [*or reference Section number of above language here*].

D. *Predicting Permits, Fees, and Other Governmental Approval Costs*

The construction manager is responsible for making sure the budget is sufficient to cover all fees for government approvals, and that approvals are obtained in a timely manner. While on the face this may seem to be an easy task with little liability, in practice it is fraught with large risk and both delay and damages potential.

For example, water tap fees can range from $20,000 to $1,000,000 for the same building type and size, based on subjective interpretations by officials and project engineering data difficult to ascertain at early budgeting stages. Failure to provide the properly made-out check in the right amount with an approval submittal can drop a project from an approval hearing agenda, resulting in a delay of two weeks, a month, or more.

The proper approach to shed this risk is to have a comprehensive list of fees and permits the owner *is* paying for, and assign responsibility for all other fees and permit costs to the design-builder. For example, if the owner were purchasing the building permit and water tap, the language to accomplish this follows:

> The Owner is responsible for securing the building permit and water tap permit, and paying their associated permitting costs and plan check fees, if any as well as securing all project access, including easements. The Design-Builder shall secure and pay for all other permits and fees; including but not limited to obtaining all other necessary approvals, assessments and charges required for construction, use or occupancy of permanent structures or for permanent changes in existing facilities, and any other permits or assessments required by local, state and federal officials.

A more common allocation of these costs is to have the owner take a larger share, and to use milder language such as this (which still holds the design-builder accountable for everything the owner is not specifically funding):

> The Owner shall secure and pay for the plan checking fee and the building permit. Gas and electrical service applications shall be made and their associated fees paid by the Owner. Assessments against the property, including use tax, developmental excise tax, sewer and water charges (tap fees) for capital improvements and utility extensions shall be paid for by the Owner. The Contractor shall secure and pay for all other permits and governmental fees, licenses, and inspections necessary for the proper execution and completion of the Work which are customarily secured after the execution of the Contract and which are legally required at the time bids are received.

E. *Owner and Design-Builder Contingencies*

1. *Owner Contingency*

The construction manager is in charge of creating and tracking the overall budget for the owner, and will assume the liability if it is not designed with enough flexibility to protect it from all normal impacts. *Force majeure* aside, even some unforeseeable circumstances must be anticipated in these budgets. It is quite a big responsibility, which begs the question: How much contingency is enough to meet the test of reasonable industry standards?

2. *How Much Is Enough?*

The three categories of contingencies are *pre-construction, bidding and escalation,* and *construction*. Their definitions (uses) and percentage ranges are described below:

Pre-construction Contingency: the contingency used to protect the budget against quality and scope creep, as well as construction cost estimating errors, as conceptual drawings and specifications evolve into schematic, design development, and finally, construction documents. The actual percentages for this contingency should be continually reduced as the drawings and specifications solidify, and should be very clearly spelled out in the design/builder agreement. The following language can be used for this purpose:

The Design-Builder shall include the following Pre-construction Contingency in its estimate of Construction Cost presented to the Owner to allow for unforeseeable scope increases in the Contract Documents beyond the Design-Builder's control:

1. at Conceptual Design, 15.0%;
2. at Schematic Design, 10.0%;
3. at Design Development, 5.0%;
4. at ninety percent Construction Documents, 2.0%;
5. at one hundred percent Construction Documents, 1.0%; and
6. at Notice to Proceed with Construction, 0.0%.

Bidding and Escalation Contingency. The volatility the marketplace has exhibited in the price of commodities like concrete, steel, gypsum board, and copper, as well as labor in the last five years, is predicted to continue unabated. For this reason, the owner and construction manager as design-builder should agree, in advance of the execution of the design-builder agreement, on the amount of contingency to be carried for this purpose, on a cost-per-month basis. These contingencies typically have ranged from as low as 0.1 percent per month to a high recently in some markets of 1.5 percent per month.

Referencing a standard to use for adjusting or validating this escalation contingency to protect against disputes with the owner and construction manager is also wise. One well-respected benchmark is McGraw-Hill's long-running weekly periodical, the *Engineering News Record,* whose cost indices are published at http://enr.construction.com/features/conEco/subs/default.asp. Several other reliable sources (e.g., Marshall-Swift, Means Cost Data, D4Cost Data, etc.) are also available to serve as reference standards for material or labor cost escalation.

Construction Contingency. Unanticipated costs incurred after receipt of the notice to proceed with construction are the single most difficult contingency to predict. There are four types of cost impacts covered by this contingency: *unforeseeable conditions*, *errors and omissions in the design documents*, *changes required by approval authorities*, and *owner-initiated changes*.

Unforeseeable conditions include, on most projects, encountering unknown subsurface conditions such as rock or water, weather outside of normal conditions for the site and season, and *force majeure* impacts (see above). Contingencies for this risk during construction range from a low of one percent for a new project on a clean site with no below-grade work, to a high of ten percent for projects where rock, water, weather delays, and strikes are common. The driving factor here is the owner's (*not* construction manager's or design-builder's) tolerance for unbudgeted risk. Be aware that most design-builders or construction managers typically expect the owner to carry this contingency. By preparing an estimate, the owner is required to define its risk tolerance mathematically. In the fairly unusual cases where the design-builder or construction manager discloses this contingency in its budget, the owner is usually asked to define the allowance it will tolerate. This practice forces the owner to carry money outside the budget prepared by the design-builder or construction manager. In the less common cases where the design-builder or construction manager have been held somewhat accountable contractually for this risk, the owner loses accountability by the design-builder's and construction manager's abilities to bury hidden contingency in the estimate line items for site work to offset this assumed risk.

Whether or not the construction manager as design-builder builds this *unforeseeable conditions* contingency into the owner's budget, the design-builder includes it in the team's budget. Alternatively, there may be some combination of the two depending on contractual liability as specified in the agreement.

Errors and omissions in the design documents will drive change orders during construction. The courts have acknowledged two-dimensional drawings and written specifications cannot completely and unambiguously depict very complex, multipart construction projects, especially given the speed with which the drawings and specifications are prepared. Conflicts can and do occur, which, at best, lead to change orders for construction cost during construction, and has contributed to the creation of the design and engineering professional liability industry, among other things.

Design-builders must be very conscious of their sole responsibility for the cost and schedule impacts of these ambiguities certain to be contained in the contract documents created under their purview. These costs, inclusive of the actual value added to the project by the change, run between one and one-half percent and four percent of hard construction costs for new projects on a clean site, given a normal design preparation period. Renovations, atypical projects (heavy industrial, complex laboratories, etc.), and design teams that have been subject to accelerated schedules during pre-construction drive these numbers up.

The rule that *value received should be proportional to compensation provided* applies here. When paying below market rates for design services, the design-builder should expect to carry an errors and omissions contingency toward the higher end of the range; when paying a premium for excellent documents, the opposite is true.

In summation, construction contingency between the owner and the construction manager as design-builder ranges from at least three percent to as much as a ten percent contingency or even more. All of the above factors should be considered when arriving at this figure, as well as the project team's tolerance for risk and ability to fund the project outside the budget.

3. *Owner vs. Design-Builder Contingency*

Access to the owner's contingency by the design-builder is straightforward: the change order process in all standard agreements is well defined. Allowing both the design-builder and the owner to access the design-builder's contingency is, interestingly, a much more difficult discussion during both pre-construction and construction. To prevent future disputes, the construction manager as design-builder must agree on the conditions for contingency use. This sample language helps define these circumstances:

> Under no circumstances is the Design-Builder's Contingency to be used by the Owner for increases in the Scope of Work, nor by the Design-Builder for correcting nonconforming Work, Work items discovered during construction not to be coordinated among the Subcontractor's scope of work due to an oversight of the Design-Builder, or similar Design-Builder errors or omissions during construction.
>
> Design-Builder may use the Design-Builder's Contingency during pre-construction for absorbing cost increases for estimating errors; design document detailing; and verifiable marketplace cost escalation beyond the Design-Builder's control. Design-Builder may use the Design-Builder's Contingency during construction for errors and omissions in the Contract Documents prepared by the Design-Builder.
>
> Access to the Design-Builder's Contingency shall be approved in writing on a per instance basis by the Owner as being in compliance with the above requirements, and such approval shall not be unreasonably withheld. Unused Design-Builder contingency shall be returned to the Owner at Substantial Completion via Change Order.

F. *General Conditions vs. the Weatherman: Controlling Weather Protection*

The owner has the right to expect the construction manager as design-builder to provide enough weather protection during construction to protect the work from all but exceptional weather until it is accepted. Additionally, the design-builder's schedule should anticipate the delays reasonably expected to occur at the project site in a normal season. The weather is an area fraught with liability exposure and should be addressed in the owner/design-builder contract.

The owner's expectations and the design-builder's anticipation of the contract schedule, while reasonable on their face, are difficult to describe with clear, enforceable language. The negotiations on this point should begin during the owner's selection process, while competitive forces are still at work. If the issue is not raised by the owner, the construction manager as design-builder should address the issue to avoid misunderstandings later. The below contract language is illustrative of the type of terms and issues that should be addressed. Note: this language favors the owner and passes most of the risk to the construction manager as design-builder (example reference on fourth line shown for work in Colorado; change as required by location).

> Extensions of the contract completion time will be made for delays due to weather conditions only when such conditions are more severe and extended than those reflected by the ten-year average for the month as evidenced by the Colorado Climate Data maintained by [Colorado State University, Fort Collins, Colorado ... change this reference for project location], or other data as mutually agreed between Owner and Design-Builder, for the project area, and only if a request for such an extension of time is received within ten (10) calendar days of the first date the Design-Builder should reasonably be expected to have calculated the impact of such delay. In requesting and allowing delays for weather, the Owner and Design-Builder will be entitled to consider weather conditions prevailing throughout the entire Contract period. Extensions of time due to weather or other allowable reasons will be granted on the basis of one-and-four-tenths (1.4) calendar days credit for every working day lost, with each separate extension figured to the nearest whole calendar day. The extension of the contract completion time for weather conditions will occur only in the event that the weather in question affected critical activities on the most current Project Construction Schedule, and at least one half of the work force allocated to that item of work was also adversely affected by the same weather conditions.

In Exhibit(s) to the owner/design-builder agreement, the on-site costs for weather protection, snow removal, temporary heat, and similar costs are commonly proposed as allowances by the design-builder, which makes them subject to actual costs in excess of the budgeted amount. While a good risk management tool for the design-builder, it moves risk for overage to the owner and is not in keeping with the above philosophy: the design-builder is the expert on the impacts of average weather on the project's cost and schedule.

Creating a not-to-exceed lump sum (sometimes called a topset) for these items, and billing only actual, properly incurred costs against it allows savings to flow to the owner. The following language is an example of its implementation in an agreement:

> The cost of maintaining access to the Work in spite of, facilitating progress of the Work during, and protecting the Work itself from the impact of, inclement weather, "Weather Protection" shall not exceed the lump sum

"Topset" of amount carried within Exhibit "X—General Condition Costs," given the Work is performed during average weather conditions for the Project site.

The actual costs for this Weather Protection will be billed on an "as incurred" basis, and any unused portion of the Topset amount will revert to the Owner as savings below the Guaranteed Maximum Price.

The Design-Builder will keep such separate and distinct records as are required to easily validate the accuracy of its billing for the costs of Weather Protection.

In the event the Design-Builder believes the above Topset should be raised based on more severe than average conditions, or the Owner believes the above lump sum topset should be lowered based on better than average weather conditions, this request for Change will be processed using the methods identified in the General Conditions for the Contract for Construction and its Supplements.

G. *Value Engineering vs. Value Elimination*

True value engineering (VE) is delivering the same or better project with better value. It is designing the project smarter, not smaller, not with lower quality, and not with less initial cost at the expense of operating costs. Unfortunately, far too often on far too many projects, the construction industry has taken VE to mean *value elimination* instead of its true meaning. Value elimination can result in owner/design-builder conflicts and litigation. The best way to manage this liability is for the construction manager as design-builder to fully understand the owner's program, needs, and vision for the completed project—including future life-cycle operating costs.

1. *Value Engineering—Owner's Life-Cycle Costs Defined and Categories Identified*

The specifics of the value engineering of construction projects, choosing options by advantage, and other complex processes are beyond the scope of this chapter. For a more in-depth study of the subject, and the correct guidelines for the process, a good starting point is the Society of American Value Engineers' Web site, www.value-eng.org.

However, it is important for the construction manager and design-builder to understand that the owner's strong temptation to have its first costs lowered must always be tempered by the project team's proactive analysis of such short-term decisions on long-term, ongoing operational and maintenance expenses—also known as *life-cycle* costs. The tradeoff between the two needs to be evaluated using a present-value of funds calculation, which typically shows first cost increases offset by paybacks through life-cycle costs in less than six to eight years. This analysis drives decision making and helps determine if such tradeoffs are worth pursuing.

The owner's obvious operational expenses for energy costs focus value engineering on mechanical and electrical systems. The life-cycle cost of many other systems with large ongoing maintenance costs also need to be presented by the design-builder and considered by the owner during value engineering. Exterior skin, roof, interior finishes, door hardware, vehicular and pedestrian paving, landscaping, and millwork are the most common candidates.

The design-builder is the team member best equipped and most likely to provide the first cost analysis of various alternatives as well as being able to calculate the long-term savings. Therefore, to assist the owner in making decisions, the construction manager as design-builder, along with the architect hired by the design-builder, should participate in the value engineering process. The following language is one way of making this obligation and the accompanying process clear, along with compensation for the process. (The last sentence should be adjusted by the design-builder, as needed, to reflect the compensation arrangements for these services):

> The Design-Builder shall compile value engineering analyses, advice and recommendations in a written summary to be submitted to the Owner for review and consideration. An initial value engineering summary will be completed within ten (10) business days following the execution of this Agreement and updated thereafter every week until completion of the Final Drawings and Specifications.
>
> Each value engineering proposal submitted by the Design-Builder shall include, without limitation, the following: (1) a detailed description of the difference between the requirements of the Contract Documents and the proposed changes and comparative advantages and disadvantages of each; (2) itemization of aspects of the Contract Documents affected by enactment of the proposal; (3) impact of the proposal upon both cost and the Project Construction Schedule; (4) list of the projects to the extent known, where the proposal or similar proposal was used and the results of that experience; (5) other information reasonably necessary to fully evaluate the proposal; and (6) date by which the Owner must accept the proposal in order for the Design-Builder's cost and time estimates to remain valid.
>
> The Design-Builder shall proceed with performance of the Work as required by the Contract Documents and shall not modify such requirements in accordance with any value engineering recommendations unless such value engineering recommendations are accepted by the Owner in a Change Order or Construction Change Directive. The Design-Builder shall not be entitled to receive any additional fees as a result of formal value engineering recommendations submitted by the Design-Builder.

2. *Managing the Implicit Design-Builder Incentive to Lower Costs at the Expense of Life-Cycle Cost*

When the design-builder has a lump sum contract inclusive of design, engineering, and construction, there is a built-in incentive for the design-builder contractually to lower first cost at the expense of life-cycle cost. Why? Because the owner bears the impacts of life-cycle costs, and under most owner/design-builder agreements, savings below the contract sum flow to the design-builder. In some cases, design-builders share the savings realized through lowered first cost with the design team, regardless of short- or long-term effects of the savings. The design-builder may want to consider the liability implications of this bias and negotiate possible incentives for life-cycle cost savings.

There are two ways to protect the owner and the project from this incentive/bias contractually: the first and most common is to have an owner's peer review team perform both a constructability review (see section below on this process) as well as a life-cycle cost analysis of the drawings and specifications at every stage (i.e., thirty, sixty, ninety, and one hundred percent design completion).

The second method, by far less common but also effective, is to structure the owner/design-builder agreement to loosely follow the guaranteed maximum price approach. Here, the owner pays actual on-site project costs properly expended on behalf of the owner, plus an agreed fee for pre-construction and construction services. This goes a long way in reducing the incentive to lower the quality of the project to the owner at the cost of increasing the operating expenses of the owner at least until the cost reaches GMP. Plus, if the owner is happy with the result, litigation with the construction manager as design-builder will not follow.

These two methods, used in tandem, are what the construction manager as design-builder should endeavor to put in place as checks and balances on the project. The added benefits (and reduction in liability exposure) of the peer/constructability review process are discussed elsewhere in this chapter.

Library References

C.J.S. *Contracts* §§ 11, 343, 513, 515–525, 595.

West's Key No. Digests, Contracts <KEY>197, 303(3), 309.

IV. Ownership and Licensing of the Contract Documents

Ownership of design documents is a crucial and hotly contested portion of the contract language both between the design-builder and the owner as well as between the construction manager as design-builder and architect, within the design-build agreements forming their relationships. The liability issues for the construction manager as design-builder are: (1) nonperformance during pre-construction by the registered design professional; (2) the owner's right to easily terminate the construction manager as design-builder for convenience during pre-construction and the owner's desire, in the event of termination, to practically (and almost simultaneously) engage another firm to continue work, and obtain the tools (CADD files of the design progress to date) to do so; (3) the construction manager as design-builder's responsibility to pay the design teams' fees before being paid by the owner; and (4) the design-builder's and design teams' liabilities for the design when used by others.

In March 2005, the AIA introduced a new family of design-build documents (2004 edition). Currently, both the 2004 and 1996 versions are available for use. However, begin as of August 31, 2006, only the 2004 version will be available. The 2004 version is not merely an upgrade but rather a complete replacement of the 1996 documents.[17] One of the primary changes was in the sections governing the ownership and use of the documents produced through the design-build process. These new documents attempt to address the concerns outlined above and allocate the resulting liability among the parties. The result is an interesting compromise of the competing interests of the architect and design-builder with the owner. Article A.1.6 specifically provides that the design-builder and its architect will retain all common law, statutory, and other reserved rights, including copyright, in the instruments of service (documents and electronic data).[18]

> However, in Article A.1.6.2, the Design-Builder specifically grants to the Owner a non-exclusive license to reproduce and use all of the instruments of service for use in connection with the Project (including any further development of the Project). The above license is expressly conditioned on the Owner complying with all obligations under the contract, including prompt payment of all outstanding amounts due. Article A.1.6.2 further requires the Design-Builder to obtain a similar license from any of its architects or other design professionals providing work on the Project.[19]

Granting a license is not a unique approach to the new AIA documents. Design/Build Institute of America (DBIA) Documents 520, 525, 530, and 535 (1998 versions) provided that if the tail agreement was not executed for the construction of the project by the design/builder, the design/builder must "grant the owner a limited license to use the documents to complete the project, provided the owner compensates and indemnifies the design/builder."[20]

However, the AIA license contemplated under the 2004 documents is broader than the DBIA family of documents. Article A.1.6.4. requires the design-builder and its professionals to convey a nonexclusive license to use the instruments of service for completion, use, and maintenance of the project. This provision only applies if the owner is not in default. The owner must also assume the design-builder's obligations to its design professionals, including payment.

---

17. MATTHEW A. GILLIES, ABA CENTER FOR CLE, A DESIGN-BUILDER'S LOOK AT THE NEW AIA DESIGN-BUILD FAMILY OF DOCUMENTS, accompanying *Building Your Practice Around the AIA's New Design/Build Family of Documents* (ABA audio program broadcast Nov. 8, 2005).

18. *Id.*

19. *Id.*

20. Timothy R. Thornton, *A Comparison of Standard Forms: The Owner-Design/Builder Contract*, in DESIGN/BUILD DESKBOOK § 2-33 (John R. Heisse, II & James S. Schenck, IV eds., 3rd ed. 2004) (comparing the AIA, AGC, DBIA, and EJCDC standard form general conditions to be used in the design-build context).

A. Owner Retains the Right to Own Documents If the Design-Build Contract Is Terminated

The owner is most likely to look to the construction manager as design-builder for compensation if the owner has nothing to show for its money at the conclusion of the design process so that construction can commence. To one extent or another, all the standard forms allow the architect or design-builder to use the ownership of the documents produced as leverage for payment.[21] The forms also protect the architect and design-builder from liability for further use of the documents produced if the design-builder or architect does not continue on the project. The new AIA documents present a much more balanced approach to resolving the tension created by the architect's and design-builder's strong interest in ownership of the design and the owner's interest in building its project.

Additionally, failure to get this ownership language in place makes termination for convenience practically impossible without substantial delay, rework, and unnecessary expenses. If the ownership issue is not addressed in advance, the owner may accuse the construction manager as design-builder with not proactively protecting the owner's interests from a poorly performing design team—one led by the design-builder—at a time when rational discussions are not likely to bear much fruit.

The following owner-oriented language has been successfully used to accomplish the owner's goal of document ownership, addressing the practical issue of transmitting professional documents by electronic media, and using professional documents produced by another designer without compromising coverage under standard insurance policies:

> The Design-Builder shall furnish to the Owner, within three (3) business days of a written request, Electronically Accessible Files and one hard copy of all drawings, renderings or other graphic materials, a copy of the most recently prepared specifications, correspondence, approvals documents, and all other information which the Design-Builder has created in connection with or for the Project (collectively referred to in this Agreement as the "Drawings and Specifications"). Electronically Accessible Files means, in the case of drawings, electronic data files in an industry standard version of Auto CAD compatible data, with all layers fully accessible (not "plot" files); in the case of specifications, fully accessible documents in an MS-Word compatible electronic format (not "password-protected" files); and in the case of correspondence, approvals submittals, documents in their native electronic format as it resides on the electronic storage media in the Design-Builder's office.

---

21. Christopher L. Noble, *A Comparison of Standard Forms: The Design/Builder-Architect/Engineer Contract, in* DESIGN/BUILD DESKBOOK § 3-14 (John R. Heisse, II & James S. Schenck, IV eds., ABA FORUM ON THE CONSTRUCTION INDUSTRY, 3rd ed. 2004) (comparing the AIA, AGC, DBIA, and EJCDC standard form general conditions to be used in the design-build context).

The Project is also the property of the Owner, and the Design-Builder may not use the Drawings and Specifications for any purpose not related to the Project without the Owner's prior written consent. Prints shall be furnished as an additional service at any time required by the Owner. The Drawings and Specifications and any reproductions of them shall be the property of the Owner which may use them without the Design-Builder's permission for any purpose. The Design-Builder may make reproductions of the Drawings and Specifications for the Design-Builder's files. Furnishing the Drawings and Specifications to the Owner is a condition required for final payment of the Design-Builder. The Owner will hold the Design-Builder and design team harmless from all actions or causes of actions arising out of the Owner's reuse of the Drawings and Specifications on other Projects, or to the extent modified by another Design-Builder, to the extent so modified.

Two key points on reuse of documents—both of which the construction manager must anticipate—are addressed in the above language: electronic versions of documents, and reuse of design documents by the owner.

B. *Documents in Electronic Format*

To facilitate construction, and changes, the construction manager as design-builder should attempt to assure that the instruments of service can be modified as the design and construction proceeds. Having the documents in electronic format also facilitates transfer of the documents to another designer or even design-builder.

Electronic versions of documents can be supplied and meet the letter of the contract, but can be worthless for the owner's intended purpose: getting a new design-builder in place and working seamlessly. There are three categories of documents: drawings, specifications, and other documents. Drawings or specifications that have been locked by a disgruntled design-builder have technically been provided electronically, but are unusable. Guard against this risk with the above language. The third category, other documents, is usually for informational use only by the *new* design-builder, and not as critical to receive in accessible files.

C. *Use of Documents by Another Designer*

If documents are to be used by another designer, the architect or design-builder will require indemnification against design errors associated with the use and reuse of the documents. Indemnification against intellectual property infringements is also typically requested by the owner from the design-builder and architect.

Reuse of the design-builder's work product by the owner on other project sites or by a new design-builder typically requires indemnification of the original design-builder by the owner against such reuse. The owner may want to limit such indemnification, when the project site remains constant, to only those modifications made by the new design-builder, as shown above.

Library References

C.J.S. *Copyrights and Intellectual Property* §§ 9, 16, 103–105.
West's Key No. Digests, Copyrights and Intellectual Property <KEY>6, 103.

V. The Owner's Voice: Managing Input, Buy-In, and Approval during Pre-Construction

One of the best ways to manage liability, especially since the source of that liability for the construction manager as design-builder is generally the owner, is to obtain the owner's critical input, vision, and approval throughout the pre-construction process. Section IV briefly addressed the new AIA family of design-build documents with respect to the critical issue of design document ownership produced as a result of the design-build process. Here, we will expand on that brief introduction and discuss managing the resulting liability to the construction manager as design-builder through design-build contracts and frequent buy-in at critical stages in the process by the owner.

AIA Document A191-1996 contained the familiar two-part design-build format, with Part 1 covering program development through preliminary design. "The design/build proposal is submitted at the end of Part 1. If the design/build proposal is accepted, the parties execute Part 2 of the agreement, Part 2 covers completion of design plus construction."[22] The AIA design-build contracts set includes three documents: (1) The Owner and Design-Builder (Parts 1 and 2); (2) The Owner and Design-Builder Construction Contract; and (3) The Owner-Architect Contract. Problems arose with this scheme when price could not be agreed to and design-builders did not want to go through two negotiations—one for the design and one for construction.

Under the new family of design-build documents (2004 edition) introduced by the AIA in March 2005, the design-builder becomes the single point of responsibility. Four documents rather than three are created, including a bridging document in which the owner (with the input of the construction manager, architect, or other consultants) creates its own detailed program. The four documents are: (1) Owner/Consultant Agreement B142 (*new*); (2) Owner/Design-Builder A141 (special terms by exhibits) (*replaces A191 Parts 1 and 2*); (3) Design-Builder/Contractor (special terms by exhibits) (*replaces A491 Part 1*); and Design-Builder/Architect Agreement (*replaces B901 Parts 1 and 2*). The documents also have extensive check-the-box options and a wide-ranging list of available consultant services. The most notable change is the "insertion of a new party, and hence a new contract, into the mix."[23]

AIA Document B142 fills the contracting niches found when an owner contracts with a consultant to assist the owner with any upfront services required when planning

---

22. Thornton, *supra* note 20, § 2-3 (comparing the AIA, AGC, DBIA, and EJCDC standard form general conditions to be used in the design-build context).

23. GILLIES, *supra* note 17.

to use the design-build method. Although loosely based on AIA Document B141 standard Owner-Architect Agreement, B142 "does not limit the consultant to traditional architectural roles and does not exclude other professionals from taking on the Consultant role."[24]

The services called for by the B142 document can be provided by a design professional, construction manager, or other consultant. Where the construction manager acts as design-builder, all services will be covered under one contract with the services to be provided defined by checking the box. Filling the previous gap in services under AIA Document A191-1996 is aimed at improving the design process, adding to the owner's buy-in, increasing owner project input early in the design process, and assuring that the owner owns the results of the pre-construction process—an area laden with disputes and litigation. Exhibit A to the B142 document establishes the owner's design and construction criteria for the project that can range from detailed to simple parameters.[25] Establishing project parameters reduces unexpected and unwelcome surprises. The advantage to the owner is summarized as follows:

> The B142 gives the owner an irrevocable license to use the preliminary design created by the Consultant, regardless of the identity of the Design-Builder and regardless of whether the Owner chooses to hire another architect to finalize the design and perform using traditional design-build methods. Unlike the previous two-part system, if the Consultant creates the preliminary design, then the Owner will no longer be forced to choose between scrapping the preliminary design and paying an exorbitant amount for it if she cannot agree with the Design-Builder regarding the final design or costs.[26]

Speaking from the owner's perspective, another author had this to say about the AIA's new family of design-build documents:

> The good news for owners is that the new forms are much more balanced than were the older A191-based forms, and therefore require fewer modifications. In addition, owners should appreciate: (i) the extensive menu of services available to customize the scope of its "owner's consultant" engagement (which also serves as a useful checklist for structuring the "Project Criteria" to be implemented by the design-builder as provided in A141, Section 8.1.3); (ii) the checkbox approach to determining whether the design-builder's compensation will be a lump sum or based on a "cost plus" method

---

24. *Id.*
25. *Id.*
26. *Id.*

(Article 4); and (iii) the opportunity to select alternative approaches for resolution of disputes (Article 6).[27]

The AGC, DBIA, and EJCDC design-build contract forms provide for extensive development of the owner's program, together with forms of a "tail agreement" for construction.[28] However, none of these documents provided form contracts for the owner's consultant. Savvy construction managers, design-builders, and owners were filling this gap on their own, and the 2004 AIA documents are in line with prevailing practices in the marketplace.

Another method used to manage the issues and liability that can be created through the design-build process is the creation of a design advisory group to assist in the process and a peer review team to assess the design. The critical components of the design advisory group and peer review team can be summarized in the following check list.

The owner's design advisory group and peer review team both must be:

- composed of stakeholders;
- representative of the owner's organization and user groups;
- required to commit time and diligence to the process;
- required to update their user groups;
- ultimately take ownership of the finished project's results (this is critical to the construction manager as design-builder in order to limit liability); and
- formed before request for proposals (RFP) are issued, and dissolved after occupancy.

The owner's peer review team, in addition to the components outlined above, must be:

- composed of designers and engineers not affiliated with the design-builder;
- representative for the owner of each discipline on the design-builder's team.
- qualified, available, and experienced with life-cycle costing;
- able to provide multiple review loops essential to an acceptable project;
- required to stamp constructability and peer review of documents;
- able to instill a sense of ownership in the finished project's design in the owner; and
- formed before RFPs issued, and dissolved after occupancy.

---

27. JOHN E. KOFRON, ABA CENTER FOR CLE, A BRIEF OVERVIEW OF AIA DOCUMENT A141-2004 from THE OWNER'S PERSPECTIVE NEW DESIGN-BUILD FAMILY OF DOCUMENTS, accompanying *Building Your Practice Around the AIA's New Design/Build Family of Documents* (ABA audio program broadcast Nov. 8, 2005).

28. Thornton, *supra* note 22.

The design advisory group and peer review team will then be part of the following approval process:

- preparation and approval of the RFPs;
- selection of the construction manager as design-builder and its design team—sometimes;
- approval of the design-builder's conceptual design;
- approval of the design-builder's fifty percent schematic design submittal;
- approval of the design-builder's one hundred percent schematic design submittal;
- approval of design-builder's fifty percent design development submittal;
- approval of design-builder's final submittal under Part 1 (one hundred percent design development drawings);
- authorization of Part 2, if using AIA 1996 documents; (Note: if using 2004 AIA documents, authorization to proceed with construction, after a constructability analysis (*see* Section VII) is completed);
- approval of design-builder's fifty percent contract documents;
- approval of design-builder's one hundred percent contract documents; and
- notice to proceed with construction to the design-builder.

The use of the team approach, led by the construction manager as design-builder and design professional, reduces the risk of a design that is not constructible or not economically feasible. The one-point-of-responsibility approach of the 2004 AIA documents, while giving the owner the bridging input from consultants of the owner's choosing, manages project risk, liability, and reduces the areas of dispute by improving communication among the stakeholders.

VI. The Construction Cost Estimate

Given the focus of this work, the space limitations and broad extent of this topic, the authors will offer the sequence of the standard construction cost estimates an owner should expect during the design-build process, and some suggested sources for further research.

The programmatic estimate is prepared from the project program, before drawings are available. It is followed by the schematic construction cost estimate that is completed using early drawings and in most cases, outline specifications.

The design development construction cost estimate is created using drawings and specifications that are roughly fifty percent complete, and is usually validated by an estimate from drawings and specifications that are almost ready for bidding (ninety percent construction documents).

For more information on the level of detail, estimating standards of the industry, and managing, reviewing, and discussing construction cost estimates, the American Society of Professional Estimators' *Standards of Estimating Practice Manual* is a great resource. Their home page (http://www.aspenational.com/index.html)

is also helpful in locating other resources to assist with the construction cost estimating process and validation of estimates.

VII. Constructability Review

Constructability problems on a project are those that arise because of poor choices with respect to the means and methods of construction, but constructability may also have a design component to it because "[s]ome designs are more difficult to construct than others. If a design's constructability challenges are not discovered until after construction commences, then the parties' expectation interests may be affected, and the cost of construction can increase."[29]

This risk of design errors under the traditional delivery approach belongs to the owner who "impliedly warrants the adequacy and sufficiency of the plans and specifications."[30] On a design-build project, this risk can shift: "Of course, a more or less wholesale transfer of design responsibility to the contractor can be achieved under a design/build project delivery system. The transfer is, however, never total. Owners, even in a design-build approach, must still communicate their needs."[31]

The key concept is communication. This is the area where the construction manager as design-builder shines by discerning and then communicating the owner's needs and obtaining the input of the stakeholders and design team, with liability mitigation through a peer review process. Constructability reviews are not a one-time review, but rather a review process conducted by the design professional, construction manager as design-builder, and consultants as certain design benchmarks are realized.

Two studies have reported positive findings in support of the benefits and high return on investment of constructability reviews.[32] The authors of *Starting Smart* estimate that dedicating two percent to five percent of the total project cost to pre-project

---

29. 2 BRUNER & O'CONNOR ON CONSTRUCTION LAW § 7:210.

30. 2 BRUNER & O'CONNOR ON CONSTRUCTION LAW § 7:212, at 990, *citing United States v. Spearin*, 54 Ct. Cl. 187, 248 U.S. 132, 136 (1918) (government impliedly warrants the adequacy of design specifications).

31. *Id.* at 991.

32. RALPH S. SPILLINGER IN CONJUNCTION WITH THE FEDERAL FACILITIES COUNCIL STANDING COMMITTEE ON ORGANIZATIONAL PERFORMANCE AND METRICS, ADDING VALUE TO FACILITY ACQUISITION PROCESS: BEST PRACTICES FOR REVIEWING FACILITY DESIGNS (Federation Facilities Council Technical Report #139, National Academies Press 2005) (hereinafter *Adding Value*); G. EDWARD GIBSON & MICHAEL P. PAPPAS, JR., IN CONJUNCTION WITH THE FEDERAL FACILITIES COUNCIL, STANDING COMMITTEE ON ORGANIZATIONAL PERFORMANCE AND MANAGEMENT, STARTING SMART: KEY PRACTICES FOR DEVELOPING SCOPES OF WORK FOR FACILITY PROJECTS (Federal Facilities Council Technical Report #146, National Academies Press 2003) (hereinafter *Starting Smart*).

planning will yield a project that comes in on time and budget. The investment translates to a fifteen percent to twenty-five percent complete design before the design is locked in and the project schedules and budgets are fixed. The study confirmed a 3:1 ratio of construction cost savings to planning dollars.[33]

The authors of *Adding Value* recommend formal constructability reviews at thirty-five percent, sixty percent, ninety percent, and one hundred percent design completion, depending on project size and complexity.[34] *Adding Value* concluded that a comprehensive review of design documents during the design phase should cost between 0.2 percent to 0.5 percent of total project cost. The realized cost savings from constructability reviews approached twenty percent of total project cost or more, when considering the costs of inefficient construction, standby costs, litigation, and poor performance.[35] The authors of *Adding Value* also advised during the constructability process to pay special attention to the civil, structural, electrical, and mechanical interfaces. Historically, thirty percent to fifty percent of all construction change orders result from interference fit problems between trades.[36]

The traditional AIA, AGC, DBIA, and EJCDC documents do not include a procedure for the owner's development of a clearly defined scope of work free of interference problems among trades.[37] However, the new family of AIA design-build documents, with their check-the-box approach, now considers the function and importance of constructability reviews. While not using the wording "constructability review," Exhibit B to AIA Document B143-2004 (Contract Between Owner and Consultant) allows the owner to check the box and choose:

- **Section B.2.2 Multi-Discipline Coordination:** Coordinate services provided by the Architect and the Architect's consultants with those services provided by the Owner, the Owner's consultants, the Design-Builder and the Design-Builder's consultants and contractors.
- **Section B.2.4 Materials, Systems and Equipment Evaluation:** Assist the Design-Builder, with the evaluation of alternative materials, building systems and equipment, together with other considerations based on the Project Criteria, other criteria, the Project budget and aesthetics in developing the design for the Project.

With responsibility comes liability for the Consultant under the new AIA scheme. The construction manager as design-builder will fill this pivotal or bridging role on the design-build project. The benefits to the project and the owner are great, and the likelihood of an on-time, on-budget project with few change orders and a reduction in disputes can be the happy result.

---

33. *Starting Smart, supra* note 32 at 17–18, 43–44.
34. *Adding Value, supra* note 32 at 25–40.
35. *Id.* at 18-19.
36. *Id.* at 43-45.
37. *Starting Smart* at 9–13.

VIII. The Construction Schedule

As with construction cost estimating discussed above, this subject is too extensive to do it justice here, given the limited space available, other than to offer some distinctions for the reader, and sources for further study.

The Master Project Schedule includes all components important to the owner, and typically is broken down into three categories: pre-construction, construction, and owner occupancy and move-in. Each team member (design team, construction team, and owner, respectively) is the lead performer during one of these phases, and is responsible for performing to the milestones in their portion of the schedule.

For more study of the proper scheduling process, terminology, and implementation, it is recommended the reader obtain the *Construction Planning and Scheduling Manual* (2d ed.), which is available for purchase through the Associated General Contractors of America's Web site (www.agc.org).

IX. Application of Payment Review

The construction manager as design-builder's liability is at its highest during the payment review and approval process. The analysis starts with a quick check list for the payment review during the construction process. The list identifies key items that the design-builder must consider in order to minimize both the design-builder's liability and the owner's risk exposure. In brief, the construction manager as design-builder performs five critical tasks when engaged in the payment review process:

1. verify percentage completion and budget;
2. observe the quality of the work, as customarily, visually, and readily accessible;
3. verify schedule with completion and quality;
4. verify timely payment of subcontractors and suppliers; and
5. require and review lien waivers.

In managing liability, the operative verbs for the construction manager as design-builder are to *verify*, *observe*, and *review* the cost and schedule components, and validate the quality control process is in place and being vigorously followed by the design-builder's subcontractors before recommending payment by the owner. The owner's purse strings are the construction manager as design-builder's most powerful tool in project control and management.

The construction manager as design-builder, using a process designed in advance with the full involvement and written approval of the design professional, begins the process by reviewing the schedule of values prepared by the design-builder and each subcontractor to verify that the amounts requested by each subcontractor are consistent with the amount of work completed. When dealing with a project that has a guaranteed maximum price (GMP), the construction manager as design-builder asks the question: "Is the amount requested consistent with the project budget by category?" The design-builder also focuses on schedule of values and budget variances to identify potential discrepancies before they ripen into full-blown disputes.

Next, the construction manager as design-builder, again with the full participation of the design professional and specialty consultants, validates that the design professional and specialty consultants have verified the work quality and compliance with project plans and specifications. The construction manager as design-builder typically establishes an independent testing and quality control (QC) program prior to the commencement of construction. The QC program includes testing and inspection as the project progresses and a system for correcting work before the budget and schedules are severely impacted.

With the prior two steps accomplished, the construction manager as design-builder takes a step back and looks at the big picture to assure that the percentage of completion, or GMP vs. budget ratio, makes sense when considering the design-builder's and subcontractor payment applications. That done, the construction manager as design-builder then considers project schedule and whether the percentage complete or GMP/budget reflect consistent project progress toward timely completion. Because of the dynamic nature of the design-build project, this step is critical to identifying and making course corrections to assure that the project is delivered on time and on budget.

The fourth and fifth steps in the review process are absolutely critical to avoiding one of the biggest liability risks on any construction project: assuring that the owner and design-builder do not have to pay twice for the same work. The liability risk, stated in two words is *mechanics' liens*. Mechanics' liens allow unpaid subcontractors and suppliers to secure payment for the labor and materials provided to a project by recording a nonconsensual lien against the real property benefited by that labor and materials. All fifty states have some form of mechanics' lien remedy. Some require pre-lien notices and others do not. Some states give super priority to mechanic's liens so that the liens relate back to the architect's first work taking priority over construction financing. Other states' lien laws start priority from project commencement. Whatever the statutory scheme, the liability risk can be substantial and should not be overlooked.

While the owner should not expect the design-builder and its subcontractors and suppliers to finance the project, the owner should be able to expect that the project will be properly managed in order to protect the owner from the risk of double payment. The AIA Document A201 standard conditions do not contain a requirement for lien waivers except "if required by the Owner." *See* AIA Document A201-1997. Therefore, it is incumbent upon the construction manager as design-builder to assure that a requirement for signed lien waivers from prior payment applications or draws be submitted to the design-builder showing that the funds paid by the design-builder to the project subcontractors were then paid to sub-subcontractors and suppliers. The following language can be used to minimize the risk to the owner and liability to the construction manager as design-builder:

> Each Application for Payment shall be accompanied by the following, all in form and substance satisfactory to the Owner and in compliance with applicable STATE statutes:

(1) A current Sworn Statement from the Contractor setting forth all subcontractors and materialmen with whom the Contractor has subcontracted, the amount of such subcontract, the amount requested for any subcontractor or materialman in the application for payment and the amount to be paid to the Contractor from such progress payment, together with a current, duly executed waiver of mechanics' and materialmen's liens from the Contractor establishing receipt of payment or satisfaction of the payment requested by the Contractor in the current Application for Payment;

(2) Commencing with the second (2nd) Application for Payment submitted by the Contractor, duly executed so-called "after the fact" waivers of mechanics' and materialmen's liens from all subcontractors, materialmen and, when appropriate, from lower tier subcontractors, establishing receipt of payment or satisfaction of payment of all amounts requested on behalf of such entities and disbursed prior to submittal by the Contractor of the current Application for Payment, plus sworn statements from all subcontractors, materialmen and, where appropriate, from lower tier subcontractors, covering all amounts paid.

(3) The payment schedule is subject to satisfactory progress and completion of the work described in this contract. If Contractor has not completed the percentages of work described as of the date Contractor submits his application for payment, Owner may modify the payment schedule to reflect the percentage of work actually completed. Owner shall hold ten percent retainage from each application for payment.

Final payment shall be made following satisfactory completion of the work. The Contractor shall provide the following items to the Owner before Owner is obligated to make final payment, including, but not limited to, (1) all maintenance and operating manuals; (2) marked sets of field record drawings and specifications reflecting as-built conditions; (3) reproducible mylar drawings reflecting the location of any concealed utilities, mechanical or electrical systems and components; (4) any special guaranties or warranties required by the contract documents; (5) assignments of all guaranties and warranties from subcontractors, vendors, suppliers or manufacturers; and (6) a list of the names, addresses and telephone numbers of all subcontractors and any other persons providing guaranties or warranties.

Retainage shall be paid within one day following full compliance with all provisions of the Contract Documents and Project Completion, including all punch list items and satisfactory walk-through of the project with the Owner, Contractor, and Architect.[38]

---

38. Adapted from ALTERNATIVE CLAUSES TO STANDARD CONSTRUCTION CONTRACTS DISK LIBRARY (James E. Stephenson ed., John Wiley & Sons, Inc. 1993).

This contractual scheme leaves the owner and design-builder exposed to the risk of misuse of funds or double payment for only one month's draw. The construction manager as design-builder and owner also often require conditional lien waivers with each month's draw to assure that mechanic's lien rights are waived as soon as payment is made. If this approach is used, an unconditional waiver must accompany the next month's payment application to assure that the payment condition was met.

The rubber meets the road for the construction manager as design-builder through the detailed review and verification process involving:

- *Who*: knowing the identities of all project subcontractors and suppliers;
- *When*: knowing who worked on the project or delivered materials in the preceding month (the design-builder's daily logs and pre-lien notices are good sources for this information);
- *What*: knowing what work was performed in the prior month; and
- *Verify*: matching the lien waivers submitted with the known information of the who, when and what before recommending payment to the owner.

While a detailed verification process can be tedious, attention to detail truly differentiates the excellent construction manager as design-builder from one who lazily trusts the word of subcontractors and suppliers, some of whom may be lacking in integrity or "robbing from Peter to pay Paul."[39]

Library References

C.J.S. *Garnishment* § 204; C.J.S. *Judgments* §§ 584, 587; C.J.S. *Mechanics' Liens* §§ 2, 220–244, 252–254.

West's Key No. Digests, Mechanics' Liens <KEY>1, 194–201, 207, 208.

X. Use of a Commissioning Agent

A commissioning agent is a firm or individual that specializes in making sure the end result of the design-builder's process meets the owner's expectations. The commissioning agent typically contracts directly to the owner to protect the owner's interests. This is especially helpful for owners who are not fluent in the process or the language of construction, or ones who must answer to large user groups. They

---

39. Many states have statutory schemes to deal with the Peter/Paul scenario. These statutes are often referred to as trust fund statutes that impose a trust on all construction funds requiring that the funds be paid for labor and materials used on a construction project. A misuse of trust funds can result in personal liability for the offending subcontractor and its principals. These statutes also may carry criminal sanctions. For example, *see* COLO REV. STAT. § 38-22-127 (2006); FLA. STAT. ANN. §§ 812.035, 772.11 (1997) (treble damages for theft).

also are another form of total quality management provided by an experienced third party working directly for the owner, and should be strongly encouraged on design-build projects.

One role of the commissioning agent is translating team members' very unique and obtuse professional language into terms understood by all, most importantly, the end user groups. A great example would be a mechanical engineer's term *degree days* being put into layperson's terms and explained by the commissioning agent as the number of days per year the temperature will exceed 74 degrees in your office.

Once the communication has taken place, specifications have been written by the design and engineering team and approved by the commissioning agent, the construction team must be monitored to make sure the specifications are followed. This is the role of the commissioning agent during construction. Primarily focused on mechanical systems before the 1980s, commissioning agents have seen their role grow dramatically in the last ten years to cover all disciplines. They are especially useful on projects with tight owner move-in deadlines, as they shorten punch lists and owner move-in. For more information on building commissioning, please see the Building Commission Association's Web site (http://www.bcxa.org).

XI. Friendly Contract Clauses for the Construction Manager as Design-Builder

A. *Correcting Nonconforming Work*

All standard-form design-build contracts hold the design-builder "responsible for proper execution and completion of the work," and "must correct any work not conforming to the contract documents."[40] However, discovery of nonconforming work is an area fraught with liability for the construction manager as design-builder. The construction manager as design-builder can limit its liability by leaving inspections to the design professional and specialty professionals. The construction manager as design-builder thereby focuses its role to coordination of the inspection process rather than assuming the obligation of discovering nonconforming work. However, the design-builder is still vicariously responsible for its subcontractors' performance and may only be able to share the risk with the design professionals. An example of some of the available clauses used to modify standard contracts, preceded by a description of their uses, follows.

1. Blanket limitation of liability of the construction manager as design-builder:

    The Construction Manager will not be responsible or liable for the acts or omissions of the Architect, Contractor, or any Subcontractor, or any of their agents or employees, or any other person performing any of the Work.

---

40. Thornton, *supra* note 20.

2. Limitation of liability of the construction manager as design-builder and owner for materials testing program where the QC is provided by a separate consultant:

It is the intent of this Agreement to control the quality of the project using a Quality Control/Quality Assurance program. The responsibility for the Quality Control Program is the Consultant's, including but not limited to the services of a certified testing laboratory acceptable to the Owner and construction manager as Design-Builder which will perform the tests as called for in the Specifications produced by the Design-Builder. The conditions that apply to materials testing include, but are not limited to the following:

   i. The frequency of Quality Control testing shall be established by the Consultant and approved by the Design-Builder, and shall be sufficient to insure the delivery of a complete and functional project per the Contract Documents. The amount of testing required by the Specifications shall be seen as the minimum required, and shall be increased, if in the opinion of the Consultant, more testing is needed to meet the above requirement.
   ii. The Consultant shall provide the Construction Manager as Design-Builder and Owner copies of all test results it receives within three business days of receipt of same.
   iii. The Owner shall provide such Quality Assurance testing as it and the Construction Manager as Design-Builder mutually deem adequate for their own needs. The Owner and Construction Manager as Design-Builder shall distribute the results of their own Quality Assurance tests as they, at their sole discretion, deem appropriate. The provision of Quality Assurance testing by the Owner, or lack thereof shall in no manner affect the responsibilities of the Consultant under this Agreement.

3. Limitation of liability of the construction manager as design-builder for observations of the work during construction:

The Construction Manager as Design-Builder will determine in general that the Work is being performed in accordance with the requirements of the Contract Documents, will keep the Owner informed of the progress of the Work, and will endeavor to guard the Owner against defects and deficiencies in the Work. However, the Construction Manager will not be required to make exhaustive or continuous on-site inspections to check the quality or quantity of the Work.

4. Warranty provisions *beware*.

The warranty provisions of standard-form design-build contracts provide added user-friendly protection to the owner, but create liability and obligations for the the construction manager as design-builder. For example, AIA Document A191-1996 (Part 2) § 3.2.9 states:

> The Design/builder warrants to the Owner that materials and equipment furnished under the Contract will be of good quality and new unless otherwise required or permitted by the Contract Documents, that the construction will be free from faults and defects, and that the construction will conform with the requirements of the Contract Documents. Construction not conforming to these requirements, including substitutions not properly approved by the Owner, shall be corrected in accordance with Article 9.

Section 9.2 specifies a corrective period of one year, allows the owner to stop work if corrective work is not timely pursued by the design-builder (§ 9.4), and allows the owner to correct the work at the design-builder's expense if the design-builder does not timely pursue the corrective work (§ 9.5).

In contrast, the AGC version (§ 3.8.2, AGC Documents 410 and 415 (1999)) is user-friendly to the design-builder rather than the owner:

> To the extent products, equipment, systems or materials incorporated in the Work are specified and purchased by the Owner, they shall be covered exclusively by the warranty of the manufacturer. There are no warranties which extend beyond the description on the face of any such warranty. To the extent products, equipment, systems or materials incorporated in the Work are specified by the Owner but purchased by the Design-Builder and are inconsistent with selection Criteria that otherwise would have been followed by the Design-Builder, the Design-Builder shall assist the Owner in pursuing warranty claims. All other warranties expressed or implied including the warranty of merchantability and the warranty of fitness for a particular purpose are expressly disclaimed.

The above clause limits rather than expands the design-builder's liability.

The construction manager as design-builder must weigh its allegiances before selecting the form of contract to be used on the project. To limit its own liability, the construction manager as design-builder would be well advised to use the AGC contract form or alter the AIA forms to limit its liability.

B. *Indemnity*

The most common way for an owner to manage project risk for losses or damages the owner may sustain, in addition to an excellent insurance program on a project, is through an indemnity clause that shifts that risk to the design professionals, consultants and construction manager as design-builder. The theory is that these parties are better able to control and insure the risk than the owner.[41]

The construction manager as design-builder can also approach the indemnity clause so that the clause is written in such a way that the indemnity obligation is

---

41. This assumption is not always true, but most standard contracts include such clauses.

jointly insured and indemnified by the owner (mutual indemnity). AIA Document A191-1996, Part 2, between the owner and the design-builder, is written so that the indemnity obligation is insurable. However, the owner does not have a corresponding duty to indemnify the design-builder. In contrast, AGC Documents 415-1999, § 3.6, and 410-1999, § 3.7, both include mutual indemnity provisions. The construction manager as design-builder can also add limitations on liability; such as, a clause that precludes the recovery of consequential damages. Both the AIA and AGC forms cited above are silent on this point. However, the DBIA form includes a limitation of liability clause precluding recovery of consequential damages.[42]

The new family of design-build documents (2004 edition) introduced in March 2005 addresses indemnity in a way new to the AIA.[43] During the rewrite, the indemnity provisions were written more in line with the AGC forms and include a waiver of consequential damages, loss of use, and loss of profits. This is an advantage to the design-builder but is the clause owners found to be most objectionable:

> Mutual Waiver of Consequential Damages. (Article A.4.1.10) This clause (which mirrors A201 Article 4.3.10), is among the most objectionable to owners, and ranks high on the "Top 10" lists for both A201 & B141. Owners believe the mutual waiver of consequential damages clause provides disproportionate benefits to the other party and inappropriately allocates to the owner substantial risks which the other party is better able to manage, all without corresponding benefit or consideration to the owner. If anything, the objection to the waiver of consequential damages will take a greater meaning in the design-build context, where the owner views the design-builder as the single point of responsibility for design and construction of the entire Project. Look for owners to require elimination of this text (and related provisions, such as those found in Article A.3.17.1 (no indemnity for damage to the work), and Article A.11.4.3 (waiver of consequential damages for loss of use of owner's property)).[44]

In conclusion, the construction manager as design-builder should review the language with its own insurance carrier to assure that its insurance program covers the losses and damages indemnified. To the extent the insurance program does not cover the risk assumed, the construction manager as design-builder must either understand it is assuming the risk or sharing the risk via a mutual indemnity clause with the owner and inclusion of the construction manager as design-builder as an additional insured in the owner's insurance program.

Library References

C.J.S. *Contracts* §§ 509–512, 561, 589–590, 592.
West's Key No. Digests, Contracts <KEY>278(2), 295, 320; Indemnity <KEY>33(5).

---

42. *Id.*
43. GILLIES, *supra* note 17.
44. KOFRON, *supra* note 27.

XII. Appendices

A. *Sample Responsibility Matrix*

Description	Design Responsibility	Budget Responsibility	Furnished By	Installed By
Site				
A. Site Signage				
Building Identification				
Sign Support				
Directional/Traffic and Striping				
B. Site Landscape				
Site Planters				
Irrigation/Drainage				
Planting				
Water Features/Fountains				
Site Furniture				
Moveable Planters/Soil				
Classroom/Office Areas				
A. Furniture				
Moveable				
Fixed				
Artwork				
Blocking for Artwork				
B. Floors				
Carpet				
Carpet Padding				
Tile				
Wall Base				
C. Wall Coverings				
Vinyl				
Paint				
Tile Wall				
D. Lighting				
Plug-in				
Direct Connect/Wall Sconce				
Architectural				
E. Doors				
Finishes				
Hardware				
Millwork				
Peepholes				
Connecting Doors				
F. Window Coverings				
Draperies/Valance				
Blocking for Draperies/Valance				
G. Electrical				
Smoke Detectors				
Life Safety System				

Sample Responsibility Matrix *continued*

Description	Design Responsibility	Budget Responsibility	Furnished By	Installed By
Fan Coil Units and High Pressure Duct				
Medium & Low Pressure Duct				
Grilles, Registers, and Diffusers				
Fire Sprinkler System (performance spec - PNC)				
I. Telephone				
Conduit, Pullwire, Sleeves Boxes				
Equipment				
J. Audio/Visual				
Conduit, Pullwire, Sleeves Boxes				
Television/Remote/Swivel				
CATV and Equipment Head				
VCR's				
Radio/Stereo				
K. Millwork				
Special Trim				
Closet Shelf and Rod				
Millwork				
Blocking for Millwork				
L. Office Amenities				
Microwave/Dishwasher				
Mini-bar/Refrigerators				
Bathroom Amenities				
Bathroom Towels				
Moveable Planters				
Vanity Marble Tops				
Mirror				
Backing for Mirrors				
Public Corridors				
A. Furniture				
Artwork				
Blocking for Artwork				
Ash Urns				
B. Floors				
Carpet				
Carpet Padding				
Tile				
Wall Base				
C. Ceilings				
Millwork and Trims				
D. Wall Coverings				
Vinyl				
Paint				
Tile Wall				

Description	Design Responsibility	Budget Responsibility	Furnished By	Installed By
E. Lighting				
Plug-in				
Direct Connect/Wall Sconce				
Architectural				
F. Doors				
Finishes				
Hardware				
Millwork				
G. Window Treatments				
Draperies/Valance				
Blocking for Draperies/Valance				
H. Electrical				
Smoke Detectors				
Fire Alarm and Annunciator				
Voice Communication System				
I. Mechanical				
Fire Sprinkler System (performance spec - PNC)				
Fire Cabinet and Hose				
Fire Extinguisher				
J. Telephone				
Conduit, Pullwire, Sleeves Boxes				
Equipment				
K. Graphic Signage				
Code Required				
Owner Required				
L. Planters				
Fixed Planters/Soil				
Movable Planters/Soil				
Plants				
M. Ice and Vending				
Ice Machines				
Ice Machine Drain/ Rough-in Connect				
Soda Vending				
Soda Vending Rough-in				
Exhibits and Displays				
A. Furniture/Artwork/Displays/Mirrors				
Movable Furniture				
Fixed Furniture				
Artwork/Displays/Artifacts				
Blocking for Artwork/Displays				
Lighting for Artwork/Displays				
Lighting for Artwork/Display Rough-ins				
Ash Urns				
Mirrors				
B. Millwork				
Special Trim				
Closet Shelf and Rod				
Millwork				
Blocking for Millwork				

Sample Responsibility Matrix *continued*

Description	Design Responsibility	Budget Responsibility	Furnished By	Installed By
C. Floors				
Carpet				
Carpet Padding				
Tile				
Wall Base				
Area Rugs				
D. Ceilings				
Millwork and Trims				
E. Wall Coverings				
Vinyl				
Paint				
Tile Wall				
Fabric				
F. Lighting				
Plug-in				
Direct Connect/Wall Sconce				
Architectural				
Chandeliers				
Structural Support for Chandeliers				
G. Doors				
Finishes				
Hardware				
Millwork				
Storefronts				
H. Window Treatments				
Draperies/Valance				
Blocking for Draperies/Valance				
Millwork				
I. Electrical				
Smoke Detectors				
Fire Alarm and Annunciator				
Voice Communication System				
J. Mechanical				
Fire Sprinkler System (performance spec - PNC)				
Fire Cabinet and Hose				
Fire Extinguisher				
Energy Management (alternate?)				
K. Telephone				
Conduit, Pullwire, Sleeves Boxes				
Equipment				
L. Graphic Signage				
Code Required				
Owner Required				
M. Planters				
Fixed Planters/Soil				
Movable Planters/Soil				
Plants				

Description	Design Responsibility	Budget Responsibility	Furnished By	Installed By
Public Areas				
A. Furniture/Artwork/Mirrors				
Movable Furniture				
Fixed Furniture				
Artwork/Artifacts				
Blocking for Artwork				
Lighting for Artwork				
Lighting for Artwork Rough-ins				
Ash Urns				
Mirrors				
B. Floors				
Carpet				
Carpet padding				
Tile				
Wall Base				
Area Rugs				
C. Ceilings				
Millwork and Trims				
D. Wall Coverings				
Vinyl				
Paint				
Tile Wall				
Fabric				
E. Lighting				
Plug-in				
Direct Connect/Wall Sconce				
Architectural				
Chandeliers				
Structural Support for Chandeliers				
F. Doors				
Finishes				
Hardware				
Millwork				
Storefronts				
G. Window Treatments				
Draperies/Valance				
Blocking for Draperies/Valance				
Millwork				
H. Electrical				
Smoke Detectors				
Fire Alarm and Annunciator				
Voice Communication System				
I. Mechanical				
Fire Sprinkler System (performance spec - PNC)				
Fire Cabinet and Hose				
Fire Extinguisher				
Energy Management (alternate?)				
J. Telephone				
Conduit, Pullwire, Sleeves Boxes				
Equipment				

Sample Responsibility Matrix continued

Description	Design Responsibility	Budget Responsibility	Furnished By	Installed By
K. Graphic Signage				
Code Required				
Owner Required				
L. Planters				
Fixed Planters/Soil				
Movable Planters/Soil				
Plants				
M. Water Features (Exterior)				
Pond Waterproofing				
Water Quality System				
Recirculation Pumping				
Piping System at Above				
System Intakes/Inlets				
Filtration and Chemical Treatment				
Water Supply/Water Level Control System				
Main Electrical Cabinet Panel				
Pond Lighting				
Installation of All Above Equipment				
Conduit, Rough-ins and Connections of All Above				
Testing of All Above Equipment				

Reception and Owner Management Offices

Description	Design Responsibility	Budget Responsibility	Furnished By	Installed By
A. Reception/Front Desk				
1. Furniture				
Movable				
Fixed				
2. Finish Carpentry				
Millwork - Front Desk				
Millwork - Work Counters				
Concierge Desk				
Baggage Shelving				
3. Miscellaneous Amenities				
Telephone Booths				
Key Rack				
Safety Deposit Boxes				
B. Offices				
1. Furniture				
Movable				
Fixed				
2. Office Equipment				
Coffee Machines				
Fax Machines				
Copiers				
Water Coolers				
Installation of All Above Equipment				
Conduit, Rough-ins and Connections of All Above				
Testing of All Above Equipment				
3. Miscellaneous Amenities				
Safe Deposit Boxes				
Drop Safe				

p:\current projects\Littleton Historic Museum\
Responsibility Matrix/Museum Responsibility Matrix 21 OCT O2.xls

© 2002 All rights reserved by
Architectural Resource Consultants, Inc.

Description	Design Responsibility	Budget Responsibility	Furnished By	Installed By
Bulletin Boards				
Decorative Lighting				
Fixed Millwork				
C. Chef's Office				
D. Lounge				
E. Multi-Purpose Room/*Conference Facility*				
1. Audio-Visual				
Compact Disc Players				
Digital Audio Tape Decks				
Cassette Decks				
Tuners				
Amplifiers				
Pre-amplifiers				
Equalizers				
Patching Panels				
Mixing Boards				
Speaker System				
PA System				
TV Monitors				
Audio/Visual Carts				
Audio/Visual Slide/Overhead Projectors				
Mobile Screens				
VCR's				
Video Equipment				
Installation of All Above Equipment				
Conduit, Rough-ins and Connections of All above				
Testing of All Above Equipment				
2. Lighting				
Architectural Lighting Fixtures				
Theatrical Lighting Fixtures				
Theatrical Lighting Structure/Catwalks				
Lighting Track				
Stage Rigging				
Stage Curtains				
Dimming and Control				
Electrical Distribution				
Electrical Fixtures				
Installation of All Above Equipment				
Conduit and Rough-ins Only				
Connections of All Above				
Testing of All Above Equipment				
3. Furniture				
Movable				
Fixed				
Public Restrooms				
A. Finishes				
Wall Tile				
Floor Tile				
Vinyl Wallcovering				
Toilet Partitions				

Sample Responsibility Matrix *continued*

Description	Design Responsibility	Budget Responsibility	Furnished By	Installed By
Vanity Tops				
Plumbing Fixtures				
B. Equipment				
Trash Cans (Built-in)				
Toilet Accessories				
Housekeeping/Maintenance/Receiving				
A. Janitorial				
Cleaning Equipment				
Vacuum Cleaners				
Carpet Cleaners				
B. Furniture and Fixtures				
Furniture				
Files				
Shelving - Prefab				
Shelving - Millwork				
C. Maintenance Equipment				
Furniture				
Files				
Shelving - Prefab				
Shelving - Millwork				
Hand Tools				
Key Cabinet				
Portable Equipment				
Stationary Equipment				
Work Benches				
Spare Stock				
D. Receiving				
Furniture				
Shelving				
Special Systems and Special Construction				
A. Computer Systems (POS/PMS)				
HVAC Management CPU				
POS System (if different than above)				
Terminals				
Dedicated Circuit				
U.P.S.				
Cabling and Terminations				
Installation of All Above Equipment				
Conduit, Rough-ins and Connections of All Above				
Testing of All Above Equipment				
B. Computer Systems (Data Network)				
HVAC Management CPU				
POS System (if different than above)				
Terminals				
Dedicated Circuit				

p:\current projects\Littleton Historic Museum\
Responsibility Matrix/Museum Responsibility Matrix 21 OCT O2.xls

© 2002 All rights reserved by
Architectural Resource Consultants, Inc.

The Construction Manager and the Design-Build Project

Description	Design Responsibility	Budget Responsibility	Furnished By	Installed By
U.P.S.				
Cabling and Terminations				
Installation of All Above Equipment				
Conduit, Rough-ins and Connections of All Above				
Testing of All Above Equipment				
C. Paging				
Paging Equipment				
Antenna				
Paging Cabling and Terminations				
Installation of All Above Equipment				
Conduit and Rough-ins Only				
Connections of All Above				
Testing of All Above Equipment				
D. Acoustics and Sound Systems				
Acoustic Attenuation				
Audio System Equipment				
Audio System Cabling and Terminations				
Installation of All Above Equipment				
Conduit Only				
Rough-ins Only				
Connections of All Above				
Testing of All Above Equipment				
E. Security/Surveillance				
Security Equipment				
Security System Cabling and Terminations				
Installation of All Above Equipment				
Conduit Only				
Rough-ins Only				
Connections of All Above				
Testing of All Above Equipment				
F. Telephone				
PBX Equipment/Wiring				
Instruments				
Telephone Equipment				
Conduit Only				
Rough-ins Only				
Connections of All Above				
Testing of All Above Equipment				
G. Television				
Television Equipment				
Antenna				
Wire Installation/Cover Plates				
Conduit Only				
Rough-ins Only				
Connections of All Above				
Testing of All Above Equipment				
H. Electronic Key				
Key System Equipment				
Keys				
Locks				

Sample Responsibility Matrix continued

Description	Design Responsibility	Budget Responsibility	Furnished By	Installed By
Conduit Only				
Rough-ins Only				
Connections of All Above				
Testing of All Above Equipment				

B. *Instructions for Using the Fee and Reimbursables Worksheet*

1. All fees and reimbursables should be extrapolated through the end of the project as it is currently envisioned.

2. The Construction Manager acknowledges the fees for your firm and its consultants may not split cleanly into the different stages of work requested on this worksheet. Please do your best to anticipate the work and fees required at each stage and allocate the fees accordingly.

3. Reproduction costs will be challenging to accurately anticipate. Assume the printing as required for all your downstream consultants, as well as four copies of each document for the Owner as part of your analysis.

4. All other reimbursables will require similar projection methods (e.g., number of out-of-town meetings, meals, etc.).

CHAPTER 14

Design Professional/Construction Manager Indemnification and Professional Liability Insurance

KATHERINE DAVITT ENOS, ESQ.

I. Introduction

Design professionals and construction managers have complex and challenging roles during any construction project, roles that are further complicated by the contractual relationships such projects require. This chapter defines some of these risks and addresses ways in which they can be managed, controlled, and mitigated.

One of the primary methods of managing these risks is to shift the burden of the risk from one party to another through an indemnification clause. Through indemnification provisions, one party agrees to indemnify the other for certain liabilities resulting from the performance of the contract. A properly drafted indemnity clause will shift the risk of loss to the party best able to manage that exposure. Because of the complex nature of shifting these risks, several issues must be carefully considered. This chapter will address and illustrate some of these considerations, allowing design professionals, construction managers, and their advisors to make informed contractual decisions.

In addition to indemnity clauses, professional liability insurance coverage is another risk-sharing mechanism. Although no federal or state law expressly mandates a design professional to carry professional liability—sometimes known as errors and omissions (E&O) insurance—prudence, professionalism, and often specific contract provisions both dictate and necessitate the importance of purchasing this coverage. This chapter will also shed light on the value of professional liability insurance as a risk management resource. Not all liability policies are the same, nor is the coverage afforded under them. These differences, particularly those between professional liability and general liability policies, warrant exploration of the more significant coverage issues, highlighting some questions that should be addressed when design professionals are confronted with the need to utilize this insurance in the construction context.

II. Standard Indemnification Obligations

One of the most common ways of dealing with risk is to transfer that risk to another party. For design professionals, this is often achieved through indemnification clauses, which are also referred to as hold harmless agreements. Increasingly, owners and clients are asking their design professionals—and even requiring them—to include these clauses in professional services agreements and therefore agree to indemnify their clients against any loss suffered in connection with the project in question.

There are typically three types of indemnification provisions: the broad form, intermediate form, and comparative or limited form. Various courts and attorneys also refer to these clauses as Type I, Type II, or Type III, respectively. Although this chapter will use the more descriptive names, the alternative other designations are included for reference. In each of the indemnification provisions, one party agrees to provide the other with some type of indemnity (to hold them harmless) for loss or damages that were incurred on the project. The difference between each of the various forms of indemnity rests in the scope of the intended indemnification.

Library References
West's Key No. Digests, Indemnity <KEY>33(5), 34.

III. Broad Form (Type I) Indemnification

A broad form indemnity provision transfers the entire risk of the loss from one party to another, regardless of the level of fault of either party. The clause's broad wording applies to any losses that one party may suffer—even those attributable to the indemnified party. The primary beneficiary of this provision is known as the indemnitee and the individual providing the indemnification, the indemnitor.

The following is an example of a broad form indemnification clause:

> Design professional agrees to indemnify and hold harmless the client from all claims, losses, damages, and expenses (including attorneys' fees and other legal expenses) arising out of, caused by, related to, or in any way connected with the Project or the design professional's services.

Design professionals assume a significant amount of risk when they agree to such broad form language. When a client insists on this type of indemnification in the project contract, a design professional should carefully assess the risks involved.

A broad form indemnification provision may actually harm a design professional, and provides little benefit to the client. Where the design professional is concerned, agreeing to broad form indemnification provision means accepting contractual responsibilities beyond their normal insurable legal liability. These provisions ignore the realities of professional practice, the standard of care to which design professionals are held, and the basic fact that most professional service firms do not have sufficient assets to effectively hold the client harmless. This is especially true since these clauses have potentially far-reaching insurance implications as well.

Broad form indemnification provisions *do not* void professional liability insurance coverage. Coverage will exist for those risks that have been accepted through the contract of insurance between the design professional and the insurer. However, an indemnity agreement that goes beyond the damage caused by the insured's negligence in providing professional services would not be covered. The legal standard of care, or negligence standard, for a design professional, absent contractual obligations stating otherwise, requires the design professional to utilize the reasonable judgment and skill that another reasonable prudent professional facing the same or similar facts and circumstances would exercise. Any damages resulting from actions that could not be considered negligence are outside the scope of professional liability insurance.

On the whole, broad form indemnification provisions provide little benefit to a client. A client is not well served by triggering coverage battles resulting from an overbroad indemnity provision. If the intent is to hold the design professional responsible for any claim tangentially associated with the project or the professional's performance, the client is overreaching. Such a client jeopardizes the successful completion of a project and may quickly drive its design professionals out of business.

Library References
West's Key No. Digests, Indemnity <KEY>33(5), 34.

IV. Intermediate Form

The intermediate form of indemnification places the entire burden of a loss on the design professional regardless of the professional's degree of responsibility. Liability for damages is not allocated to the percentage of negligence; even a one percent degree of fault will place the entire burden of the damages on the design professional. The only instance in which the indemnitor would be exempt from the obligation is when the indemnitee is solely at fault for the loss.

Professional liability policies typically do not cover this type of indemnification. When they do, it is only to the extent of the damage caused by the insured's own negligence.

Library References
West's Key No. Digests, Indemnity <KEY>33(5), 34.

V. Comparative (Limited) Form

Under the comparative form of indemnification provision, each party bears only the risk of loss arising out of its own negligence. This is the indemnification available under common law—absent contractual language stating otherwise—and is considered the most equitable allocation of risk. The language of comparative form provisions usually states that indemnity is provided, "but only to the extent" caused by the negligent acts of the indemnitor.

Industry standard contract documents promote the use of the comparative fault indemnification. Under the Engineers Joint Contracts Documents Committee (EJCDC) standard documents between an owner and engineer, reciprocal indemnification provisions address both the design professional's and the owner's obligations. EJCDC E-500, Standard Form of Agreement Between Owner & Engineer for Professional Services (2002) at provision 6.10 Indemnification and Mutual Waiver states:

> A. Indemnification by Engineer. To the fullest extent permitted by law, Engineer shall indemnify and hold harmless Owner, and Owner's officers, directors, partners, agents, consultants, and employees from and against any and all claims, costs, losses, and damages (including but not limited to all fees and charges of engineers, architects, attorneys, and other professionals, and all court, arbitration, or other dispute resolution costs) arising out of or relating to the Project, provided that any such claim, cost, loss, or damage is attributable to bodily injury, sickness, disease, or death, or to injury to or destruction of tangible property (other than the Work itself), including the loss of use resulting therefrom, but only to the extent caused by any negligent act or omission of Engineer or Engineer's officers, directors, partners, employees, or Consultants. The indemnification provision of the preceding sentence is subject to and limited by the provisions agreed to by Owner and Engineer in Exhibit I, "Allocation of Risks," if any.
>
> B. Indemnification by Owner. To the fullest extent permitted by law, Owner shall indemnify and hold harmless Engineer, Engineer's officers, directors, partners, agents, employees, and Consultants from and against any and all claims, costs, losses, and damages (including but not limited to all fees and charges of engineers, architects, attorneys, and other professionals, and all court, arbitration, or other dispute resolution costs) arising out of or relating to the Project, provided that any such claim, cost, loss, or damage is attributable to bodily injury, sickness, disease, or death or to injury to or destruction of tangible property (other than the Work itself), including the loss of use resulting therefrom, but only to the extent caused by any negligent act or omission of Owner or Owner's officers, directors, partners, agents, consultants, or employees, or others retained by or under contract to the Owner with respect to this Agreement or to the Project.

The American Institute of Architects (AIA) also suggests standard indemnification language in the general conditions for construction between the owner and the contractor. The AIA General Conditions, AIA Document A201-1997, Article 3.18.1 Indemnification provides:

> To the fullest extent permitted by law and to the extent claims, losses or expenses are not covered by Project Management Liability insurance purchased by the Contractor in accordance with Paragraph 11.3, the Contractor shall indemnify and hold harmless the owner, Architect, Architect's

consultants, and agents and employees of any of them from and against claims, damages, losses and expenses, including but not limited to attorneys' fees, arising out of or resulting from performance of the Work, provided that such claim, damage, loss or expense is attributable to bodily injury, sickness, disease or death, or to injury to or destruction of tangible property (other than the work itself) but only to the extent caused by negligent acts or omissions of the Contractor, a Sub-contractor, anyone directly or indirectly employed by them or anyone for whose acts they may be liable, regardless of whether or not such claim, damage, loss or expense is caused in part by a party indemnified hereunder. Such obligation shall not be construed to negate, abridge, or reduce other rights or obligations of indemnity which would otherwise exist as to a party or person described in this Paragraph 3.18.

The AIA likely intends a narrow interpretation of this indemnification provision. The clause applies only to the extent that the claim is caused by the negligent act of the contractor or any other responsible party. This indemnity is also limited to claims for property damage (other than the work itself), personal injury, or death. It does not cover economic losses. Similar indemnities are found in the standard construction manager agreements and these indemnities typically get pushed downward to subconsultants. However, the AIA B141 standard contract document between the owner and architect remains silent on indemnification. It assumes that the indemnification rights existing in common law—that each party is responsible for damages that arise from their own negligence—will be applied.

When an owner inserts a broad form indemnification clause in a design or construction contract, a design professional has several options. If possible, the design professional should try to strike the provision in its entirety. While not always successful, striking the clause can be a pivotal risk management tool. More likely than not, the design professional should be able to negotiate a clause with a narrower scope, in which he or she is not obligated to indemnify others for liability arising from their own actions.

At a minimum, the design professional should seek to negotiate a reciprocal indemnity clause in its contracts and make certain that contracts with subconsultants carry comparable indemnities. It is important to remember that an indemnity clause does not extinguish liability to an injured third party. Moreover, regardless of what type of indemnification clause is incorporated in the contract, third parties typically cannot benefit from the indemnification provisions unless they are expressly named. Often in construction contracts, other parties such as construction managers and lenders are expressly made intended beneficiaries of the indemnification. It is important to consider that all the parties intended to be indemnified be specifically identified so that there is little confusion as to the intended beneficiaries. Failure to do so can dramatically increase obligations of the indemnitor.

Library References

West's Key No. Digests, Indemnity <KEY>33(5), 67.

VI. Insurance Impacts

Most standard professional liability insurance policies provide limited form indemnity coverage. However, many indemnification provisions extend their scope beyond this limited coverage by requiring indemnification for damages resulting from the non-negligent performance of the insured. To increase the likelihood that a professional liability insurance policy will respond to a claim, the indemnity should be modified to cover damages *to the extent caused by or arising out of any negligent act, error, or omission.* This clarifies that the indemnification obligation is triggered solely by the design professional's duty not to commit a negligent act or to allow a negligent omission to occur.

Note that a client may ask to transfer defense costs to the design professional when a claim alleges the sole negligence of the design professional or where the damage is caused by the design professional's negligence. The client can also ask for indemnification of reasonable attorneys' fees resulting from a claim alleging damages caused by the negligent act, error, or omission of the design professional once that negligence is adjudicated. This represents a fairly common attempt by owners to shift all of the risk and cost of any claim to design professionals. Professional liability insurance will respond only to the allegation of professional negligence.

When agreeing to this type of indemnification agreement, a design professional assumes these risks as business risks. Naturally, the design professional should be in the best position to decide whether or not these risks are acceptable. To their great detriment, many design professionals fail to fully evaluate these risks or to delegate those same obligations downward to its subconsultants. Moreover, design professionals should also ensure that the general conditions for construction also include the contractor's indemnification obligations as well.

Owners should accept some risk incident to their participation in the project and not attempt to transfer such risks to design professionals through indemnification provisions. In addition to creating an unfair allocation of risks, this practice tends not to be upheld by the courts. Courts will generally enforce the indemnification provision against the drafter of the agreement where ambiguity creates two or more reasonable interpretations of construction.[1] Any indemnification obligation that extends beyond negligence of the insured is likely to draw judicial scrutiny. Moreover, any indemnification provision that asks a design professional to indemnify the client for damage caused by the client's sole negligence is likely to be voided by the courts as against public policy.[2]

Library References

C.J.S. *Insurance* § 54.
West's Key No. Digests, Indemnity <KEY>30(5), 33(5); Insurance <KEY>2383.

---

1. United States v. Seckinger, 396 U.S. 203 (1970).
2. Bethlehem Steel Corp. v. G.C. Zarnas & Co., 304 Md. 183, 498 A.2d 605, 610 (1985).

VII. Restrictions on Indemnifications

Because indemnity clauses have been abused, particularly in the construction industry, many states have imposed statutory restrictions, which vary from state to state. Some states' restrictions hinge on the negligence of the party seeking indemnification prohibiting any transfer of liability either for their sole negligence or negligent acts in which the party was involved. Others prohibit only the transfer of liability for one's sole negligence. Other states' anti-indemnification statutes apply only to design professional contracts while others apply to both construction and design contracts. Still other anti-indemnification statutes distinguish between public or private sector projects.

Several states have yet to enact any anti-indemnification statutes, allowing enforcement of broad form indemnification provisions. However, legislation to prevent the spread of overbroad indemnification clauses in construction is expanding. Qualified legal counsel should review applicable state statutes to determine both what the courts will enforce and what contractual provisions a design professional must negotiate. Some states embrace the legal principle of freedom of contract between parties of perceived equal bargaining position and will enforce any indemnity agreement if the intent is clearly and expressly stated. Many states also struggle with the burdens imposed on the public at large by the broad or even intermediate form of indemnification and often apply rules of interpretation to narrow the scope of indemnification provisions. When courts find an ambiguous indemnification provision, they often apply a strict interpretation that results in no indemnification at all. The impact of this potential judicial determination that no indemnification exists should be contemplated as well.

State laws regarding indemnification vary widely and are frequently changing. Because of the nuances and complexities of anti-indemnification statutes, it is prudent for the design professional to seek legal counsel for their individual contract needs and address their applicable state statutes and provisions before entering into a contract for professional services.

Library References
West's Key No. Digests, Indemnity <KEY>26.

VIII. Understanding Standard Insurance Policies

In addition to indemnity clauses, insurance can also provide another outlet for transfer of risk. There are two primary types of insurance policy forms that exist: claims-made and occurrence. Occurrence forms cover losses that arise during a given period of time. The loss can be reported years later, as long as it occurred during the policy term. In contrast, a claims-made policy only covers claims reported during the policy term—even those that happened many years past. The examples in the box below illustrate the differences of providing coverage between the occurrence and claims-made policies.

> **Claims-Made vs. Occurrence Trigger**
>
> ABC Engineering has had claims-made policies for the last three years. Nation's Insurance Co. provided coverage during 2001, XYZ Insurer provided coverage in 2002, and Main St. Insurance provided coverage in 2003. In 2003, ABC received notice of a claim from a client, even though the event leading to the claim occurred in 2002 (see the timeline below).
>
> ```
> Start of services        Event leading to claim        Claim made
> |────X────────────|────────X────────|────X────────|
>    Nation's - 2001         XYZ - 2002           Main St. - 2003
> ```
>
> **Question:** Assuming that the allegations are within the CGL policy's coverage, which carrier—Nation's, XYZ, or Main St.—would be obligated to provide defense for ABC and possibly pay the client's damages?
>
> **Answer:** *Claims-Made*—The claim occurred in 2003, while Main St. Insurance's policy was in effect. Thus, Main St. would respond to the claim.
>
> **Answer:** *Occurrence*—The event leading to the claim occurred in 2002, while XYZ Insurer's policy was in effect. Thus, XYZ would respond to the claim.
>
> © 2005 by Victor O. Schinnerer & Company, Inc.

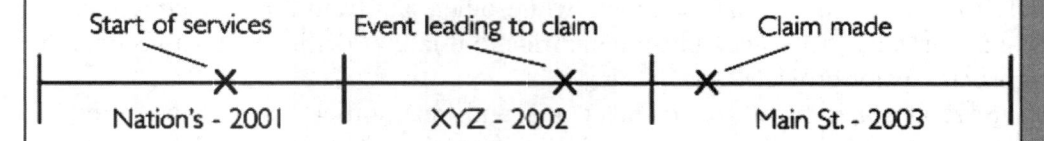

Because professional liability policies are designed for a specific risk—the risk of loss caused by the negligent performance by design professionals—all professional liability policies are written on a claims-made basis. To maintain the protection of a claims-made insurance policy, design professionals need to be aware of several features of this policy form. A claims-made form has no guarantee of continued insurability. It can be canceled by the insured or by the insurance company for various reasons. Additionally, only one limit of liability applies with claims-made policy: the limit in force when the claim is actually brought.

Many design professionals purchase coverage only for a specific project, then drop coverage. Because a professional liability policy does not guarantee coverage for activities performed in the past, it is crucial for a design professional to maintain continuous coverage to get coverage for prior acts. Additionally, when a claims-made policy is allowed to lapse, a design professional runs the risk of being uninsured for a potential claim if the claim for a project is not discovered until after the policy expiration.

Occurrence forms have their own concerns, the primary of which is to make sure that the design professional retains his or her copy of the policy in a safe place, in the event it is needed in the future to show evidence of insurance in effect during a specific time period in which the loss arose. Occurrence forms also have issues that become critical when a design firm merges with another firm, closes its doors, or changes insurance providers. As noted above, most professional liability policies for design professionals are written on a claims-made basis. Most general liability policies are written on an occurrence basis.

A. *Professional Liability Coverage*

Many design professionals recognize their need for professional liability insurance. But what risks does this coverage address? A professional liability policy covers the services a design professional performs for others as an architect, engineer, land surveyor, landscape architect, construction manager, scientist, or technical consultant. It is designed to protect the design professional for any negligent acts, errors, or omissions in their professional services. For coverage to be triggered, the wrong being asserted typically must arise out of the design professional's negligent performance of professional services. Negligence is a civil wrong that is not based on a contract and may be defined by the standard of care. The standard of care for a design professional is defined as the failure to exercise the proper degree of care required by a prudent design professional under similar circumstances. There are four basic requirements that must exist for negligence to be established:

(1) there must be a legal duty of the insured to act or not act;
(2) breach of this duty must be committed;
(3) the act (breach) must be the proximate cause of the injury; and
(4) there must be actual damages.

In today's environment, twenty-six out of every one hundred design firms can anticipate a claim in any given year. Those claims can involve expense payments, indemnity payments, including settlements, judgments, and damages, as well as any deductible obligations the design professional might have. Considering the growing frequency and severity of these claims, carrying adequate insurance coverage is an important consideration for any design professional firm.

B. *Prior Acts Coverage*

Design professionals need to understand how prior acts coverage affects a professional liability policy, especially when changing or merging firms. The primary function of prior acts coverage is to provide for negligent acts or omissions occurring before a claims-made policy period and assumes that the insured did not know about the act or omission precipitating the claim at the start of the policy. Upon the purchase of prior acts coverage, the insurer and insured agree on a retroactive date to determine the cut-off date for claims. Any claims prior to this date are not covered. Even if the claim is first brought after the retroactive date, if the incident that gave rise to the claim occurred before the date, the claim will not be covered.

Some claims-made policies, often the broadest forms available, do not have a retroactive date. Other policies use the policy inception date as their retroactive date, limiting coverage to the date on which the policy was first purchased. When possible, design professional companies often make their retroactive date the same as that on which the firm was formed and began providing professional services.

C. *Tail Coverage—Extended Reporting Periods*

Tail coverage applies to the other end of the policy term: incidents that were reported after the policy's expiration, but occurred while the policy was in effect.

One type of tail coverage is an extended reporting period (ERP). Because it can often take an undetermined period of time, anywhere from one, five, or ten years after a design professional has rendered professional services for a claim to arise, it is important for a design professional to carefully consider purchasing an extended reporting period. With the purchase of an extended reporting period, a design professional is not extending the policy to provide coverage for professional services completed after the policy expires but rather the design professional is extending the time in which they can report a claim that arises out of the professional services they provided prior to the policy expiration date.

The length of that period will often depend on the applicable statute of limitations in the jurisdictions in which a firm practices. Most professional liability policies also provide an automatic sixty-day extended reporting period if the policy is canceled or nonrenewed by either the insurance carrier or the insured. In addition, providers often offer an optional reporting period that the insured can exercise within sixty days of termination. This optional extended reporting period can last up to five years beyond the policy termination date. The limits available to pay a loss reported during an extended reporting period are not renewable at each year. These limits form part of, and are not in addition to, the limits of liability applicable to the final policy year of the practice policy.

Design professionals ceasing business for a while might consider a different kind of post-expiration coverage: a retired professionals' policy. Under this policy, coverage remains in effect for some time after the closing of the firm. This policy typically carries the same limits and deductible as the primary practice policy. The retired professionals' policy can be renewed for a period of time, typically in five-year increments, and may be renewed but is subject to underwriting review to determine whether is should be canceled or is renewable. Although this type of coverage and its subsequent renewal options is not a widely available option, an insurance provider or broker can help obtain this type of coverage.

D. *General Liability Coverage*

Most prudent design and construction professionals also obtain a commercial general liability (CGL) policy. The CGL policy is intended to provide coverage primarily for liability arising out of nonprofessional acts (violations of the personal, business, or property interests of private citizens) that result in bodily injury, property damage, or personal and advertising injury. CGL insurance is designed to cover an insured's liability arising out of incidents on the insured's premises or from the nonprofessional aspects of the insured's practice. In other words, the CGL coverage covers nonprofessional negligent acts. CGL coverage is written on an occurrence basis. As discussed previously, occurrence coverage is triggered when the event giving rise to the claim actually happened. The CGL policy in effect at the time the event or occurrence happened is responsible for the claim even if the claim is not made until years after the policy expires.

General liability coverage typically works on the principle that everything is covered except what is excluded. Therefore, most general liability policies include a significant number of exclusions from coverage—most of which can be insured through other types of insurance. The CGL policy's most significant exclusion is coverage for professional liability claims.

CGL forms also do not usually cover the following risks:

- liability arising from aircraft, automobiles and watercraft;
- liquor liability;
- pollution liability;
- mobile equipment exposures (covered under a property policy); and
- damage to property owned by the insured and damage to property in the insured's care, custody or control.

The following sections of this chapter review the three basic coverage areas of a CGL policy and provide some examples of how they might apply in the design professional arena. However, the complexity of general liability insurance coverage warrants consultation with a qualified insurance broker or agent to assess the liabilities associated with operating a design or construction management firm.

There are three basic coverage areas that comprise a CGL policy: (1) bodily injury and property damage (BI/PD); (2) personal and advertising injury; and (3) medical payments coverage.

E. *Property Damage*

Bodily injury/property damage provides coverage for the legal liability of an insured for bodily injury or property damage to others arising out of nonprofessional negligent acts or for liability arising out of their premises or business operations. The CGL policy defines property damage as "physical injury to tangible property, including all resulting loss of use of that property."

F. *Bodily Injury*

Bodily injury is defined as "bodily injury, sickness or disease sustained by a person, including death resulting from any of these at any time" in Section VIII(F) of ISO's Form CG0001/204. Death that results after a period of time from an earlier injury will be covered by the policy in effect at the time the injury was sustained. The meaning of bodily injury is open to judicial interpretation. Some courts have held that mental injuries and emotional distress can be considered bodily injuries, even in the absence of physical bodily harm.

The following is an example of a BI/PD incident involving a surveyor working on an airport runway:

> The surveyor left his tripod and prism standing upright on the side of the runway when he left for lunch. A small Falcon 900 airplane came in for landing, and the leading edge of the right wing, which extended over the edge of the runway, clipped the equipment. The surveying equipment created a dent in the wing's leading edge, but there were no injuries. However, the repairs to the plane and other expenses, such as downtime, parts, and labor, totaled more than $114,000.

This claim possibly could have been avoided if the surveyor had not left the tripod unattended on the job site. Any equipment used on a job site should be secured before leaving the site to avoid causing possible bodily injury or property damage.

The claim illustrates an interesting point. A design firm's professional liability (PL) policy provides coverage for that firm's professional negligence; the design firm's CGL policy provides coverage for the firm's nonprofessional negligence. In the surveyor's claim example, a compelling argument could have been made that the damage arose out of the surveyor's professional services and, therefore, was expressly excluded and should have been covered by the PL policy.

Unfortunately, insurance providers will occasionally challenge whether a particular claim is covered by their policy or by a policy from another carrier. The surveyor's claim example illustrates a situation where the CGL insurance carrier could argue that the claim should have been covered by the surveyor's PL policy. The PL carrier could have made the same argument against the CGL carrier.

In these situations, it is the insured that is caught in the middle while the insurance carriers debate over which policy should respond to the claim. A way to minimize the risk of this situation is to obtain both the CGL and PL policies from the same insurance carrier. The focus would then shift to resolving the claim instead of debating which policy should respond to the claim.

G. *Premises and Operations Liability*

There are several coverages included under BI/PD that provide protection for an insured against loss from legal liability arising out of the (1) operation, maintenance, or use of the premises and (2) liability arising out of business operations. This liability involves the insured's (1) legal responsibility for bodily injury or property damage to others on the insured's premises or (2) liability arising out of the insured's business operations or work performed by independent contractors on the named insured's behalf.

An example of a liability claim that would arise out of premises or business operations would be if a visitor walks into a design professional's office lobby, slips on the marble flooring after a rainstorm, and breaks her ankle. The injury occurred on the firm's premises and the injured woman would make a claim against the firm for her medical expenses resulting from her broken ankle.

This example illustrates the importance of maintaining safe building premises, proper warning signs for hazards, and prompt attention to hazardous situations that may increase the likelihood of an accident. It is also important to note that this policy does not provide BI/PD coverage for employee-related injuries. Such claims will likely be covered by workers' compensation insurance.

An example of a business operations claim would be the case of a manufacturer who accidentally flushes flammable liquids into a city's sewer system. An explosion in the sewer results and city streets are torn up, utility services are suspended, and dozens of businesses in the vicinity of the explosion must close because access to the neighborhood has been cut off. The businesses have sustained a loss of use of that property, even though the explosion did not damage the property of these businesses. The manufacturer's CGL coverage should respond to loss of use claims of this kind, just as it will to claims of direct physical injury.

H. *Contractual Liability Coverage*

The CGL policy provides broader contractual liability coverage than professional liability. This coverage is for specific contractual agreements. Under most PL policies, liability assumed by contract are excluded from coverage unless such obligations would have existed in the absence of the contract. This is commonly referred to as limited form contractual liability coverage.

The following contractual agreements are covered under the CGL policy:

- contract for a lease of premises;
- sidetrack agreement (related to railroad deliveries);
- easement or license agreement;
- an agreement, as required by ordinance, to indemnify a municipality; and
- elevator maintenance agreement.

I. *Personal and Advertising Injury*

Personal and advertising injury liability protects an insured against liability arising out of certain offenses, such as:

- libel;
- slander;
- false arrest;
- infringing on another's copyright;
- malicious prosecution;
- use of another's advertising idea; or
- wrongful eviction, entry, or invasion of privacy, which is committed during the policy period and within the coverage territory.

J. Libel and Slander

Libel is a written statement about someone that is personally injurious to that individual. An example of a libelous situation would be if a construction manager makes uncomplimentary comments in a letter to a newspaper about a construction contractor's performance or quality of work on a project.

Slander is similar to libel in that it is a spoken statement that is injurious to an individual. An example of slander would be if an architect verbally degrades the reputation or past work of another architect to a client in order to secure a project from that client. The act of speaking defamatory words by one architect to sway the client, whether justified or not, affects the other architect's means of livelihood, business, and reputation, which is a personal and advertising injury offense.

Design professionals need to be cautious about making comments, either written or verbal, about another professional or member of the project team. Truth is always a defense to libel and slander so statements should be based on facts or couched as a professional opinion based on the facts presented.

This is another area where there is potential conflict between CGL and PL coverage. This often arises in the context of a design professional commenting to the client about the qualifications of a contractor. When the alleged defamatory statements are made, the claim may be excluded under the CGL policy. Here again, having CGL and PL coverage with the same insurance carrier should minimize the risk of a claim being caught between two carriers as they debate which policy should respond.

K. Medical Payments

Coverage for medical payments includes payments for injuries sustained by members of the general public caused by an accident that takes place on an insured's premises or when exposed to the insured's business operations. Normally, injuries must be reported within one year of the accident. Medical payments coverage can be triggered without legal action from a third party. This provides for prompt settlement of smaller medical claims without litigation. It is included in the CGL policy and pays for all necessary and reasonable medical, surgical, ambulance, hospital, professional nursing, and funeral expenses for a person injured or killed in an accident arising out of the premises or business operations of the insured, regardless of negligence or liability. There is no defense or legal liability coverage as there is with BI/PD liability and personal and advertising liability since coverage is provided on a no-fault basis.

L. Additional Insured Status on the CGL Policy

Clients and contractors will often include design professionals as additional insureds on their builder's risk, wrap-up, or commercial general liability policies as members of the project team. An additional insured endorsement is simply another form of indemnification that gives the additional insured no greater coverage than that available to the named insured. The coverage provided by each endorsement specifically reflects the relationship between the named insured and the additional insured, including any limitations on coverage resulting from that relationship. For example, additional insured endorsements for design professionals explicitly exclude coverage for the additional insured's professional liability. Additional insured status on a professional liability policy is discussed later in this chapter but, generally speaking, is not available.

Whereas named insureds are automatically covered for any business activity that is not specifically excluded, additional insureds receive narrower coverage. Additional insureds are only covered to the extent of the named insured's operations on a particular project. Policy exclusions apply equally to named insureds and additional insureds unless specifically noted otherwise.

The Insurance Services Office (ISO) provides standard additional insured endorsement forms for all CGL insurance carriers. The ISO's standard additional insured endorsements for contractors, owners, and subcontractors have been specifically worded to balance the interests of all parties and avoid disputes. For this reason, ISO forms can be valuable, especially since general contractors are typically interested in acquiring the broadest coverage available while subcontractors are looking to provide the narrow coverage that results only from their sole negligence.

When discussing liability coverage, one must also consider owners' and contractors' protective (OCP) liability insurance as well. Generally speaking, these policies cover both the insured's direct liability for the supervision of contractors and the insured's vicarious liability for the activities of the contractor. Contractors are often required to purchase this insurance for the owner's benefit, with the owner as a named insured on the policy.

When this type of coverage is purchased, the owner is not as likely to be named an additional named insured on the contractor's CGL policy. The scope of an OCP policy is similar to that of a CGL policy; it covers the insured for bodily injury or property damage as well as costs incurred to defend a claim for covered damages. The OCP policy specifies two sources of liability for which coverage is provided: (1) operations performed by, or on behalf of, the designated contractor for the named insured at a specific location; and (2) the named insured's acts or omissions in connection with its general supervision of the work.

The most important consideration of the OCP policy is that its coverage benefits the named insured—typically, the owner—and not the insured contractor. Additionally, an OCP policy is written on an occurrence basis. As discussed earlier, the policy must be in effect when the bodily injury or property damage occurs to trigger policy coverage. While OCP and CGL coverages are very similar, two exclusions in the OCP policy distinguish them: the OEP policy excludes coverage for completed operations and the acts or omissions of the insured. The CGL policy does not. A knowledgeable insurance advisor can explain the risks associated with each policy.

M. *Construction Management: Agency vs. At-Risk*

In the ever-changing construction arena, many professionals are taking on the additional role of construction manager. The need for construction managers arose from a desire to improve project efficiencies and overall business management of construction projects. Both architects and contractors have sought to fill this role, as well as those who have developed specialized expertise as professional advisors.

There are generally two types of construction managers: agency and at-risk. Each type carries its own risks and rewards.

An agency construction manager is simply an extension of the owner. He or she acts as the owner's agent regarding contract matters by monitoring the design and construction phases of the project. This construction manager is generally not responsible for the construction means and methods and does not guarantee construction costs, time, or quality of the work. The agency construction manager usually contracts separately with the owner but may have administrative relationships with others on the project, including the architect, engineer, or general contractor.

Many design professionals have historically participated in the agency construction management as an extension of their design services and their traditional role as the owner's representative. As agency construction managers, design professionals can have a positive impact, especially by preventing or mitigating potentially serious cost overruns or delays, protecting the quality of the project and enhancing the owner's participation. However, when a design professional takes on this additional role, he or she must consider the scope of professional services carefully, including obtaining corresponding fees in proportion to the additional risk.

An at-risk construction manager typically offers advice and construction leadership during the planning and design phases. They may also manage contracts as well as supervise, coordinate, and control work during the construction phase. At-risk construction managers not only contract with the client but typically contract separately with the various contractors. As a result, they often control or are vicariously responsible for the construction means and methods, the safety of workers, and delivery of the completed work consistent with the owner's expectations.

Any design professional taking on this role needs to be aware of the risks, obligations, and insurance implications that arise from assuming this unique and challenging role. Where agency construction management is concerned, most professional liability policies for architects or engineers provide coverage for agency construction management. The associated duties and obligations are generally considered to fall within the scope of the intended coverage.

To obtain professional liability coverage for at-risk construction management, there must be a contract with a fee for construction management services that defines the construction management services provided. Such a policy will cover such acts as the failure to detect construction not in accordance with the construction documents. However, at-risk construction managers need to consider both the professional and general liability exposures posed by their role, all of which need to be addressed from an insurance perspective when participating in projects. The American Institute of

Architects (AIA), the Engineer's Joint Contracts Documents Committee (EJCDC), the Construction Management Association of America (CMAA), and the Associated General Contractors of America (AGC) have all issued standard forms that can be reviewed when negotiating risk allocation for the construction manager. Each of these contract forms is part of a document family with internally consistent and coordinated terms and conditions. However, each of the contracts reflects the perspective of its publisher. Consultation with an attorney is recommended when adapting standard contract forms to the circumstances of a particular project and to controlling law.

N. *Standard Exclusions*

Because professional liability coverage is so specific, it does not cover contractual obligations, except the obligation to carry out duties according to the professional standard of care. Most professional liability policies carry express exclusions for contractually assumed liability that would not otherwise exist in the absence of the contract.

How does the economic loss doctrine impact this contractual obligation and what interplay does this have with professional liability coverage? Although the economic loss doctrine has been discussed elsewhere in this publication, a brief overview will prove useful to our discussion here. The economic loss doctrine argues that a design professional has no liability for purely economic losses or damages to entities with which he or she has no contractual relationship. The rationale behind this doctrine is that the parties to a conflict should be able to resolve their disputes based upon the negotiated terms of the contract, to which all parties agreed.

The economic loss doctrine has had significant impact on the design and construction industry. Oftentimes, it is a contractor who argues for a design professional's negligence in designing a project, blaming delays on professional errors. However, since it is rare for a design professional to contract with a contractor, under the economic loss doctrine, the contractor's claim would be barred for lack of a contractual obligation, sometimes referred to as privity.

Whether or not the economic loss doctrine provides a valid defense for design professionals is a jurisdictional question that is specific to individual states. This answer has significant bearing on assessing the risks of design professionals. If the answer is yes—that the jurisdiction governing any given project recognizes the economic loss doctrine—then the design professional is afforded greater protection. If the answer to the question is no, then the design professional has to assess the direct exposure not only to the client but also to all others who could be impacted by the design professional's negligence.

The economic loss doctrine may also impact the coverage afforded under a professional liability policy. As discussed, the policy excludes any assumed contractual liability greater that what would be provided absent a contract. In those states in which the economic loss doctrine precludes a recovery by a third party, a design professional who agrees to provide deliverables for the benefit of any party with whom privity does not exist may find that commitment to be outside the scope of professional liability coverage. When held to a standard greater than the general standard of care, a design professional's risk and exposure increases. Moreover, this indirect exposure could greatly exceed any insurance protection provided by a professional liability policy.

On that cautionary note, it is worthwhile to indicate that, as of 2005 data, the CNA/Schinnerer program has not seen any significant difference in the number of or severity in professional liability claims in states where there is no economic loss doctrine versus states that have one.

O. *Express Warranties and Guarantees*

All professional liability policies also exclude express warranties and guarantees. An express warranty is an assurance by one party of the existence of a fact on which another party can rely. An express warranty is intended to relieve the relying party of any duty to ascertain the facts, which amounts to a promise to indemnify the relying party for any loss if the fact warranted proves to be untrue. While, in certain circumstances, such a promise will be presumed by operation of law, it is rare that such implied warranties are applicable to the services or instruments of service of a design professional. Design professionals can sometimes find themselves providing express warranties or agreeing to absolute responsibility, which are risky activities. They can also subject themselves to the similar problem of guaranteeing the actions of another.

Express warranties and guarantees are promises that are ancillary to the basic agreement between the design professional and a client. Therefore, an express warranty presents a different set of problems for design professionals from the normal contractual obligations of performance of professional services. For one, it is almost impossible for the design professional to have the power or control over all the factors that would lead to an express warranty. In addition, because an express warranty relieves the client of the burden to act in a responsible manner, the design professional no longer has the ability to rely on the rational participation of the client in the design process.

Accordingly, no professional liability insurance policy includes coverage for breach of express warranty claims. This includes a breach of contract action based on express warranty or guarantee provisions. Because such independent promises and their anticipated damages are impossible to assess, an insurance carrier has no way to ascertain the risk it would assume.

Can design professionals agree to absolute warranties and guarantees? Of course, but they need to be aware that such independent promises usually are within neither their capability nor their control—and often lie beyond the scope of professional liability insurance coverage.

P. *Limits of Liability*

Each professional liability policy carries a limit on coverage, called the limit of liability. This is simply the amount of money that the insurance company will pay on the insured's behalf for covered claims. The policy has two types of limits: a per-claim limit, which is the most that the insurance company will pay for any one claim, and the aggregate limit, which is the limit on the total amount of all of the claims during the policy period. These limits may be the same or, in some instances, the aggregate limit may be higher, depending on the needs of the insured and the frequency or severity of claims. Endorsements to a policy may adjust the limits available for various reasons, be it a specific project, client, or specific project type that allocate different limits available for coverage. These will be expressly stated and indicated on the endorsement to the policy.

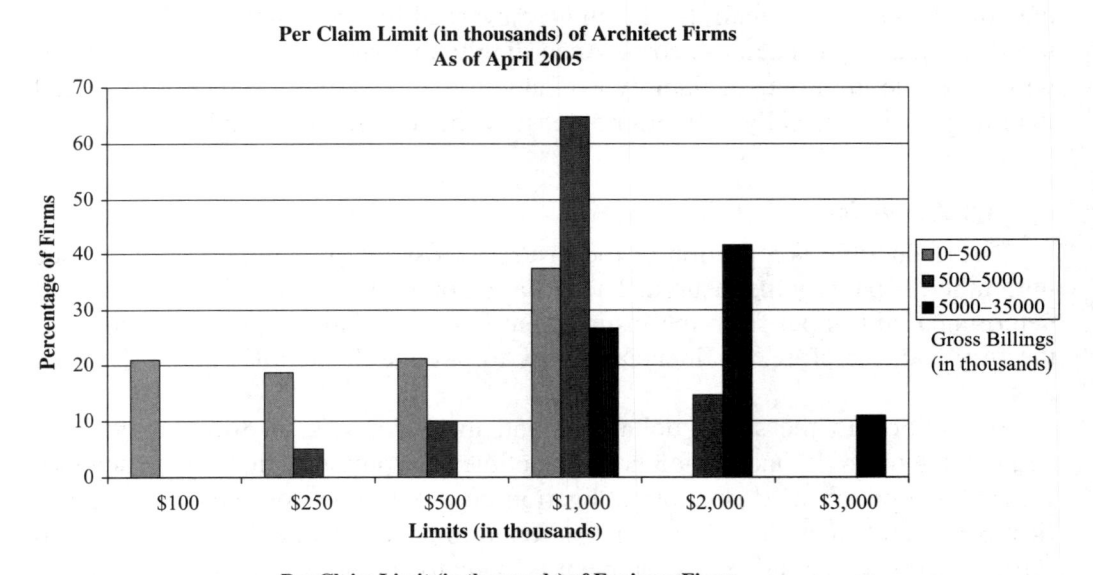

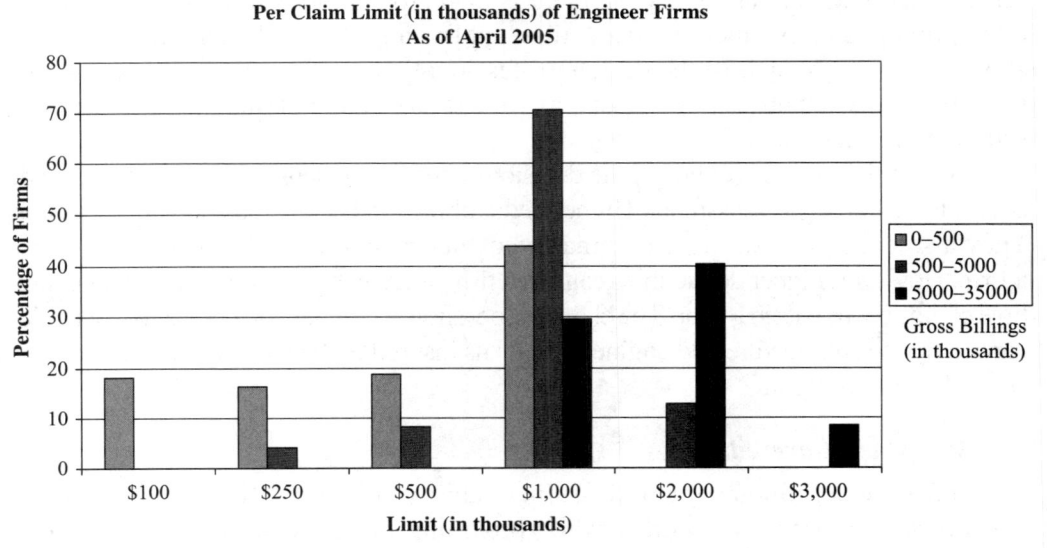

The question of what limits of liability a firm should purchase can be addressed only after a thorough assessment of their appetite for risk, frequency of claims, the financial strength of the company, and available limits in the marketplace. See charts noted above for a comparison of average per claim limits of liability purchased by engineering and architectural firms insured under the CNA/Schinnerer professional liability insurance program.

Some insurance policies may offer the additional feature of *defense outside the limits*, as opposed to *defense inside the limits*. Defense inside the limits means that the legal costs expended to defend an insured's claims can reduce and possibly exhaust the applicable limits of the policy, regardless of whether any indemnity payment is made. Defense outside the limits means that the insurance company will pay, on the insured's behalf, the indemnity awarded for the claims and, in addition to the limit, also pays defense costs. Accordingly, the payment of defense expenses will not erode the limits of liability available to pay indemnity. Most professional liability policies typically only offer defense within the limits of liability.

Q. *Deductibles*

The deductible is a portion of the insured's loss that is not paid by insurance but, rather, is paid by the insured. The dollar amount of the deductible is generally determined with input from the insured, but is usually based upon the billings of the firm and, therefore, its financial ability to pay the deductible should the need arise.

In addition to the stated dollar amount, there are several different types of deductibles offered. One choice is a deductible that provides first-dollar defense. Under this option, the deductible obligation does not apply to defense costs. The firm pays the deductible only when it is found to have been negligent or agrees to settle a claim and an indemnity payment is made to another party. Another deductible option is one in which the deductible applies any time a demand for money or services is brought against the design professional's firm. The deductible can apply to both defense expenses (legal costs) as well as any indemnity payments (damages, settlement, or judgments).

As with the limit of coverage, the decisions concerning deductibles should be made only after a thorough assessment. Higher deductibles can become cost-saving measures. They also encourage stronger risk management procedures and sensitivities within a firm. However, higher deductibles can put a firm at greater risk, particularly when multiple claims occur within a year. The charts below show a comparison between deductibles carried by the architecture and engineering firms insured in the CNA/Schinnerer professional liability program.

R. *Who's Covered*

Under the definitions of a professional liability policy, the firm and its employees are covered under the policy while providing professional services on behalf of the firm. Covered employees also include any partners past or present, officers, directors, members, and stockholders. In addition, any newly acquired subsidiaries are often included in coverage. Many insurance policies also cover retired individuals, as long as they are acting within the scope of their duties as a consultant for the firm, as well as spouses and heirs in the event of death.

Many clients, in a misguided effort to obtain better protection against third party claims, ask their design professionals to list them as additional named insureds on the design firm's professional liability policy. While additional insured status protection is afforded on some forms of insurance like CGL, it is neither an option nor a benefit under professional liability insurance programs; the design firm's client is not performing professional services and, therefore, does not have the kind of risk that the policy covers.

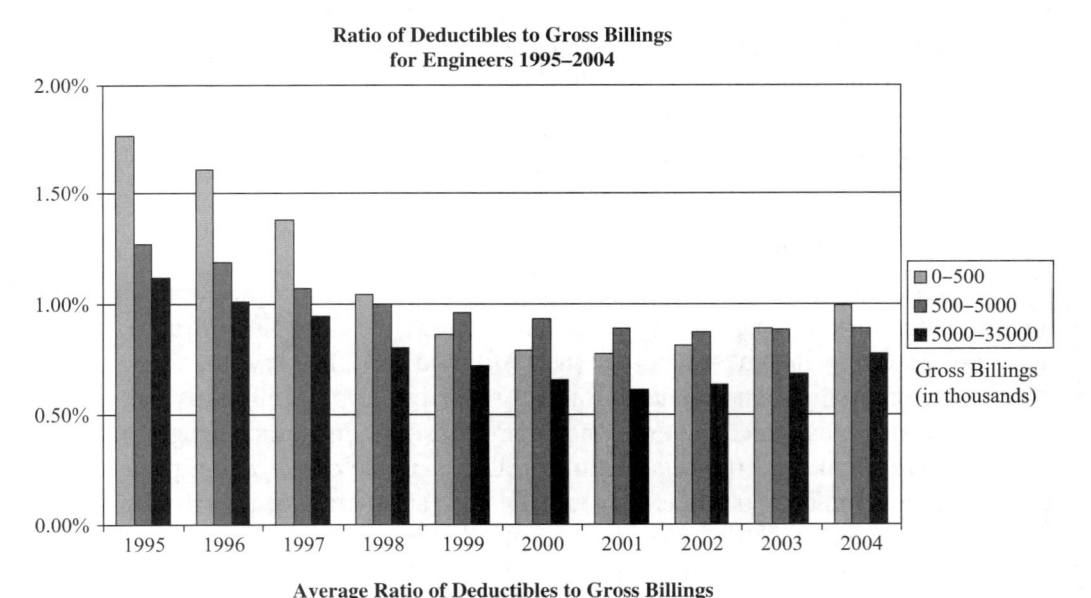

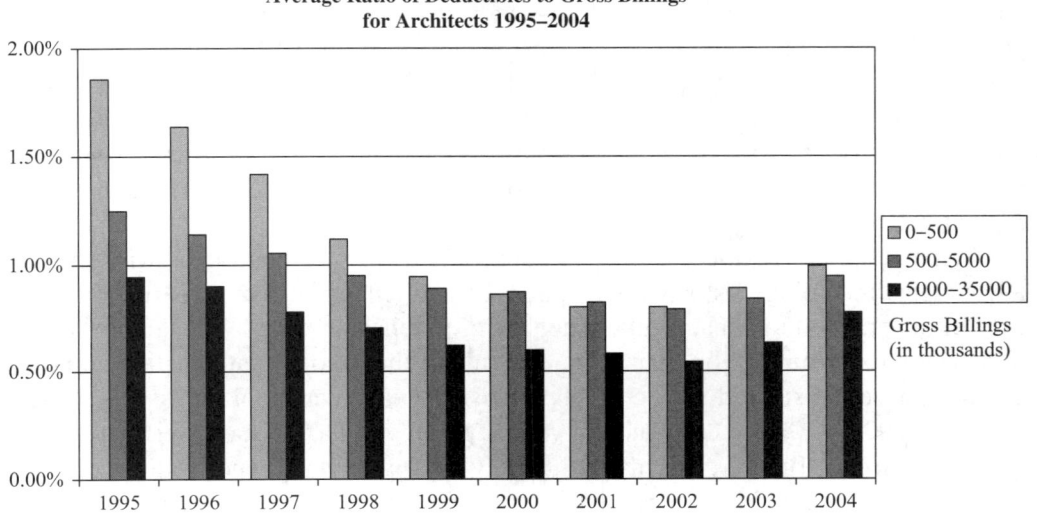

Ratio of Deductibles to Gross Billings for Engineers 1995–2004

Average Ratio of Deductibles to Gross Billings for Architects 1995–2004

Unlike other types of insurance policies, the professional liability insurance policy does not make payments to the named insured but rather pays on behalf of the named insured in the event that the named insured's negligence in rendering professional services causes damage or injury. If the design firm's client were named as an insured under the design firm's policy, the client would be unable to collect damages under the policy since typically an insured cannot make a claim against itself (including another insured) and recover.

An additional, though less common, problem occurs if the client has professional engineers, registered architects, or other design professionals on its staff. If that client were to be added to the design firm's professional liability policy, that coverage could apply to all of the professional activities of the client's in-house design professionals. It would be possible that any claim of professional negligence made against a professional member of the client's staff—whether that professional is involved with the design firm's project or not—would have to be defended and possibly even indemnified under the design firm's professional liability insurance policy. This presents an extremely precarious position; the design firm would be providing coverage for an entity or individual over whom the design firm has no control.

S. *Wrap-Up Project Policies*

When deciding on the participation in a specific project, it is important to be aware of the various types of project-specific insurance. The first, wrap-up insurance coverage, saw its development in the 1940s and early 1950s with the increasing number of very large construction projects. A wrap-up program wraps the various parties and their insurance coverage needs into a single insurance program. These programs are often used on large commercial and public construction projects and can be implemented on a single project basis or for a series of specified projects. Wrap-up coverages include both owner-controlled insurance programs (OCIP) and contractor-controlled insurance programs (CCIP). Although each wrap-up is unique and specific to its particular project, most wrap-up insurance programs insure against general liability, excess liability, workers' compensation, and employers' liability exposures. Builder's risk, professional liability insurance, contractor's pollution liability, and subcontractors default insurance may also be built into the policy. An OCIP generally does not cover vendors and suppliers, delivery persons, off-site fabrication or manufacturing, and work performed on other projects that are unrelated to the OCIP related project.

Wrap-ups can provide the program sponsor, owner, or contractor with distinct advantages, including cost savings. Contractors (often, an at-risk construction manager or general contractor) calculate the cost of purchasing wrap-up insurance versus purchasing the insurance on their own. The rule of thumb often cited is that an OCIP should not be considered as a cost-saving risk management tool unless the project exceeds $100,000,000 in construction values. However, OCIPs have been written on projects with construction costs of considerably less, particularly when rolling contracts (those utilized on continuous open-ended basis from project to project) are used.

Cost savings are not the only advantage of wrap-up policies. Traditional insurance approaches allow for the possibility that coverage can be canceled prior to a claim, exclusions on the policy may void coverage for specific types of claims, insurance limits may be eroded by other claims, or an insurer may become insolvent. Because the owner or contractor controls the insurance directly, a wrap-up avoids these issues.

In most instances, the owner will pay for the cost of the wrap-up insurance. However, the owner will usually ask the participating contractors to reduce their contract bids by the amount they would have otherwise paid for the insurance coverage provided by the OCIP. This calculation can become complex, since an OCIP covers all contracting parties to the construction project and hinges on the safety records of those parties. Poor safety records could push the cost much higher. Because the owner is given named insured status, he or she can select the insurer, types of coverage including enhancements, and intended limits.

T. *Considerations with Wrap-Ups*

When deciding whether to purchase an OCIP, a professional should consider the insurance market conditions, length of the project, complexity of the project, limits of liability to be purchased for the project, level of workers' compensation in a particular state, and the safety records of participants in the project, including what resources are allocated to project safety.

Tail coverage issues abound with OCIPs. Coverage typically ends at or near project completion, forcing a contractor to seek subsequent tail coverage elsewhere. This coverage can be difficult to procure, particularly since the insurer did not participate in the initial project yet is faced with assuming additional risk without the benefit of collecting earlier premiums. In determining whether or not to offer coverage, insurers will ask:

- How will completed operations coverage be handled?
- How will deductibles be handled and allocated?
- How will policy limits to be shared if a major catastrophe occurs?
- How will vehicle coverage be provided?
- What about coverage for supplier's delivery staff?
- How are the safety procedures to be developed?
- What impact do firms with poor claims experience have on the cost of the insurance?

Wrap-ups can also be labor intensive, especially where the owner is concerned. The owner must assume responsibility for purchasing the insurance coverage, reviewing and approving all program documents, facilitating the claims review process, and participating in underwriting meetings. Owners will often hire insurance professionals to help with this process.

The CCIP is also available to contractors, with primary control resting with the contractor instead of the owner. Many larger general contractors use this insurance, particularly because it relates to their subcontractor risks as well. In addition to the difference in control, OCIP and CCIP differ in the handling of completed operations coverage. Completed operations coverage typically extends three years beyond the construction period and often beyond the closing of the owner's construction loan. Even though coverage is in effect under both types of wrap-up policies to protect all insureds in the event of a completed operations claim, the owner remains responsible for the deductible for that three-year time frame under the OCIP. Under the CCIP, this deductible burden rests with the contractor.

U. *Attorneys' Fees and Litigation Costs*

Attorneys' fees and litigation costs are serious considerations in managing a claim. In the CNA/Schinnerer program, defending a claim costs an average of $50,000, a significant cost for any design firm. In an indemnification provision, the design professional is not required to pay for the defense, until the client's liability to the third party is attributed to the negligence of the design professional. The client will then seek reimbursement for damages including legal defense costs. It is also typical for a party seeking indemnification to recover reasonable attorneys' fees, but only those incurred in efforts defending the claim.

The overall cost of litigation is significant for any firm, regardless of its level of liability. The table below shows the percentage of claims resolved with both indemnity payments for damages, judgments, and settlements, as well as expense costs paid by the CNA/Schinnerer program.

Resolution of Claim	2001	2002	2003	2004
Closed without Any Payment by Insurer	43.3%	42.5%	43.1%	44.2%
Closed with Expense Payment Only	29.8%	30.4%	32.6%	32.5%
Closed with Indemnity Payment Made by Insurer	27.0%	27.1%	24.3%	23.3%
Total without Indemnity Payment	73.0%	72.9%	75.7%	76.7%

V. *Pre-Claims Assistance*

To assist with the rising cost of litigation, some professional liability insurance providers offer pre-claims assistance. Pre-claims assistance provides payment for investigating a potential claim. Successful pre-claims assistance can bring about an early resolution or successfully mitigate a nonpreventable dispute. By paying to employ experts in law, insurance, and construction in the early stages of the controversy, the insurer can prevent a dispute from maturing into expensive and lengthy litigation. Moreover, payments under pre-claims assistance are generally not subject to the deductible and are in addition to the limit of liability shown on the declarations page of the professional liability policy. Early reporting of a claim also allows the design professional to lock in the limit and deductible in force at the time the circumstance is reported. Approximately fifteen percent of claims filed in the CNA/Schinnerer program received the benefit of pre-claims assistance.

Library References

C.J.S. *Insurance* §§ 54, 392, 422, 429, 860–861, 934–950, 956–957, 970–973, 983, 1125–1127, 1129–1143.

West's Key No. Digests, Insurance <KEY>2263–2286, 2297–2316, 2359, 2391(1), 2391(7).

IX. Evidence of Insurance

In discussing insurance, it is important to address what constitutes evidence of compliance with the insurance requirements articulated in contracts. Owners, design professionals, construction managers, general contractors, and all parties to the contract for construction will find it in their best interest to make certain that all contracting parties obtain the required insurance coverage. This can prevent parties from being unprotected in the event of a loss. There are many claim scenarios where one party failed to maintain the required insurance and some other party (whether responsible or not responsible is a matter of debate) was made to pay in the event of the loss. Design professionals who are vicariously liable for the actions of their subconsultants find themselves particularly at risk when they fail to monitor the compliance of their subconsultants with the insurance provisions of the contract.

Two types of documents are typically mentioned when determining whether the required insurance has been procured: binders and certificates of insurance.

A. *Binders*

An insurance binder is the equivalent of temporary insurance until a formal policy is issued. Once the insurance policy is issued, the binder is no longer in effect. A binder provides an interim or temporary policy but it does not constitute part of the policy or create rights for the insured, other than its effective date, and remains subject to underwriting review of the risk intended to be covered. The binder can be canceled but unless and until it is, the insured continues to have coverage and is insured. Binders should include as many express terms and conditions of the intended insurance coverage as can be provided to avoid ambiguity in the event of a loss before issuance of the policy.

Problems typically arise for insureds when there are multiple layers of coverage on a policy (layers subject to varying underwriting constraints) and the applicable terms and conditions of each layer conflict or have different meanings, thus creating an issue of dispute between underwriters. The conflict is manifested when a loss occurs between the time the binder is issued and when the final policy language takes effect, which ultimately creates issues for litigation and takes substantial dollars to arrive at a coverage determination.

B. *Certificates of Insurance*

Certificates of insurance evidence that an insurance policy has been issued, the limits of liability in effect when the certificate was issued, the coverage period, and the type of insurance being addressed. There are standard certificates of insurance that provide limited information to a certificate holder. They do not actually create coverage, nor do they create any obligations on the insurance company that are not expressed in the policy. Some certificate holders seek to create legal obligations by requesting modification to the standard certificate language but most parties refuse to modify the certificates. Generally speaking, a certificate of insurance does not obligate an insurance company to the individual or organizations to which the certificate is issued. Design professionals, construction managers, and general contractors typically require evidence of insurance from those parties whose actions can give rise to vicarious liability. It is important to obtain certificates of insurance evidencing both professional liability and general liability coverage because, when either coverage is inadequate or unavailable, attorneys may assert legal theories in a claim that they hope will trigger coverage available under the other policy. Because insurance carriers' defense obligations are broader than their indemnity obligations, a considerable sum of money can be spent before these issues are sorted out. Remember that the defense obligations of an insurance carrier will trigger not only deductible obligations but also the cost of defending a claim under a professional liability policy that is within the policy limits. As a consequence, inadequate or nonexistent general liability coverage can erode the limits available under the professional liability policy.

Library References

C.J.S. *Insurance* §§ 266, 378–379, 386.
West's Key No. Digests, Insurance <KEY>1748, 1839.

X. Recent Claim Trends

Because the services of design professionals vary in complexity by discipline, firm size, geographic location, and many other factors, pinpointing emerging claims issues is not easy. However, several types of claims perennially arise.

A. *Meritless Claims*

One continuing trend that remains a critical issue is the significant number of meritless claims. Because of the nature of the United States judicial system and a misunderstanding of the services provided, design professionals are often subjected to defending themselves against unfounded allegations of negligence.

The CNA/Schinnerer program conducted a national study of claims from 1999 to 2003 and discovered that only 25.1 percent of all claims brought against design professionals in the program resulted in a payment. In the remaining 74.9 percent of claims, the design professionals were determined to have had no responsibility for damage or injury, as measured by having no indemnity payment made by the insurer on their behalf.

To combat this escalating problem, many states have enacted certificate of merit laws. These statutes require an affidavit from an expert to substantiate the factual basis of a claim. Meritorious lawsuits survive this process, while claims without merit are removed from the legal system.

B. Design–Build

The design–build project delivery method is another major source of claims for design professionals—one whose severity can be significant. The design–build project delivery method combines the designer's and builder's roles in a single point of responsibility, allowing a design professional to absorb a great deal of additional risk. When the lines between responsibilities of the contractor, subcontractor, and design professional begin to blur, a design professional's traditional protections erode on design–build projects. It is possible to mitigate some of this increased risk by taking care when assuming overly broad roles in design–build agreements and paying particular notice to indemnification provisions.

C. Construction Managers

The role of the construction manager has given rise to some unique and unanticipated third-party claims. One recent claim involved a construction manager hired by a lender to verify sufficient completion of construction in order for the lender to release funds to the project owner. Unfortunately, after the release of the funds, a new contractor had to be brought in to complete the project, at the price of significant cost overruns and delays. The owner filed a claim not only against the original contractor, but more interestingly, also against the construction manger, claiming that the construction manager's certification of completion of construction was negligent. Under this argument, the construction manager should have identified the significant construction problems existing at that time and notified the lender, who would have alerted the owner to the significant construction issues.

Library References

C.J.S. *Insurance* §§ 1525–1528, 1531–1533, 1642, 1725.
West's Key No. Digests, Insurance <KEY>3545.

XI. Further Resources for Understanding Risk Transfer and Sharing

Because of the complexities of insurance policies, the input of a qualified insurance professional is warranted when trying to determine the type of coverage needed.

Most design professionals who purchase professional liability insurance do so through independent insurance brokers. These brokers represent the interests of their clients, not those of the insurer. By contacting a broker experienced in professional liability insurance, a design firm can best be assured of finding insurance to suit its needs.

Some insurance companies are represented by agents who work for a single company in a predetermined territory. Because these insurance agents represent the interests of the particular insurance company, they may not have access to the entire insurance marketplace.

Regardless of whether a firm chooses an independent broker or exclusive agent, the firm will want to select its broker or agent in much the same way it selects its lawyer and accountant—with care and scrutiny of qualifications, services available, cost, as well as commitment and relationship skills.

Other risk management resources are available for design professionals through the insurance company that provides their professional liability insurance. Through the expertise of reviewing and insuring thousands of claims each year, each insurance company compiles resources that can inform the design professional's business decisions. Many even offer risk management information available regarding project types, contract negotiations, claims avoidance, and many other topics that can help the design professional navigate through the challenges of their business.

In today's world, design professionals should anticipate a claim in the course of their professional life. The use of a qualified legal advisor can help diminish these chances and, at a minimum, mitigate the potential damages. A qualified legal advisor can also help a firm proactively manage their risks, be it in negotiating favorable contract terms to handling the firm's risk management needs. Having the appropriate legal advisor can only benefit the quality of the work.

Design professionals may also find help in their professional associations, some of which have been mentioned in this chapter. They include The American Institute of Architects (AIA), National Society of Professional Engineers (NSPE), the American Congress of Surveying and Mapping (ACSM), the Construction Managers Association of America (CMAA), and the Associated General Contractors of America (AGC). These and various other organizations can provide insight into practice issues, supply standard contractual language, give guidance with regard to legislative impacts on the design professional, and provide general professional guidance.

XII. Conclusion

The complexities of indemnification and insurance requirements can sometimes appear to bog down a project. However, they can be meaningful risk-sharing tools when successfully utilized. Both can provide effective protection for design firms, allowing them to complete projects more smoothly and operate more competitively in the marketplace. Through careful thought and consideration of the issues presented in this chapter—*before* any professional services are rendered—design professionals can provide themselves the best protection against unwanted risk.

CHAPTER 15

Considerations for Subconsultant and Subcontract Agreements for Design Professionals and Construction Managers

L. TYRONE HOLT, ESQ.
CARRIE LYNN H. OKIZAKI, ESQ.

I. Introduction

This chapter will provide guidance and recommendations for negotiating subcontracts with the lead or prime design professional or construction manager (Prime Contractor) for construction projects in the United States. It will provide both theoretical and practical guidance to recommended practices and procedures for documenting these subcontractual relationships.[1]

Today's construction projects typically require the services of a team consisting of two or more professionals, including architects, engineers, cost estimators, schedulers, construction managers, and related construction industry personnel to deliver the full scope of professional services required for planning, design, and contract administration phases of a typical construction project, regardless of the nature of the project. The composition of a project design or construction management team may range from a relatively simple two-person group consisting of an individual architect or designer and one structural engineer who work together to produce a simple three- to five-sheet set of builder's plans for a single family, stick–built residence—to a twenty- or thirty-firm group of architects, engineers, costs estimators, construction managers, and specialty consultants who work together to produce a 250–400-sheet set of very complex drawings and a multivolume project manual for a large commercial, governmental, industrial, or multifamily project. One of the many characteristics that the single-family residence and governmental project

---

1. The principles in this chapter are applicable to the CM at-risk procurement method rather than the CM agency, in which the CM enters into prime contracts as the agent for the owner.

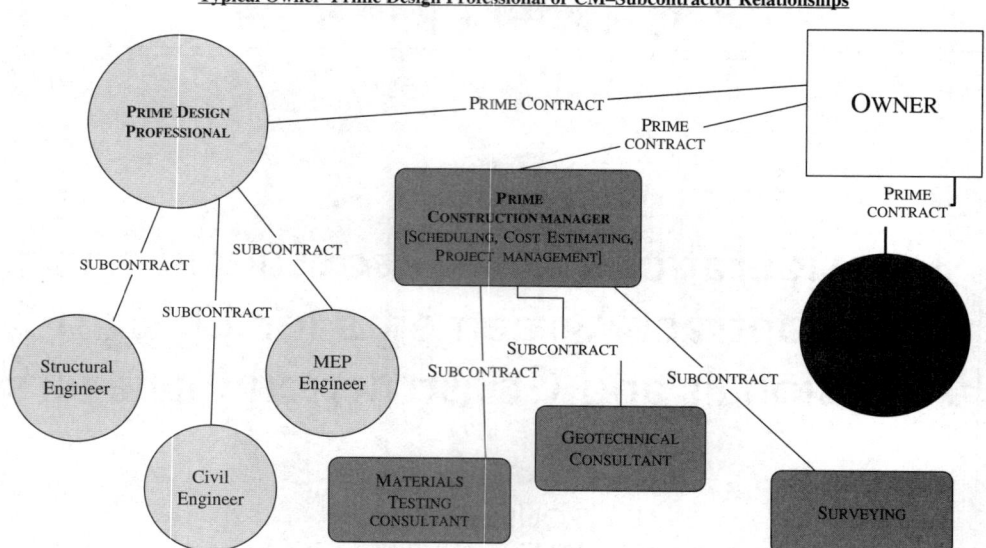

Typical Owner–Prime Design Professional or CM–Subcontractor Relationships

teams share is that each team will likely need one or more subconsultants or subcontractors[2] in order to complete its services for the project.

The above chart reflects typical contractual relationships for a construction project, including some typical Prime Contractor/subcontractor roles discussed in this chapter.

Often, negotiating and drafting the prime contract unintentionally receives the most attention and care. However, a construction project is made up of a complex web of contracts; each one is a small piece of the whole. Therefore, just as much thought and care should be taken in negotiating and drafting the subcontracts as with the major contracts. However, there are some legal considerations that are unique to the Prime Contractor/subcontractor relationship.

II. The Process of Choosing a Subcontractor

As a part of its efforts to obtain the prime contract for a construction project, the Prime Contractor typically solicits and reviews oral or written fee/scope proposals from a preselected group of potential subcontractors before it begins to negotiate the prime contract with the owner. This is necessary in order for the Prime Contractor to fully develop the scope and price for the services that it will provide to the owner.

Depending on the size and complexity of the project, the solicitation of the subcontractor's price and scope for its services for a particular project may be done in a

---

2. Design professionals typically refer to their subcontractors as *subconsultants*. Construction or project managers may use the term *subconsultant* or *subcontractor*. In this chapter, the terms *subcontractor* and *subconsultant* will be used interchangeably to refer to the lower-tier subcontractors of the Prime Contractor.

formal request for proposal (RFP) process, or simply by the Prime Contractor picking up the telephone to request an estimate from a trusted subcontractor it often uses.

Most public works or governmental projects have some sort of small and disadvantaged business and nondiscrimination policies to which both the Prime Contractor and its subcontractors must adhere. These policies will frequently require the Prime Contractor to solicit subcontractor proposals from businesses with whom it has little or no prior experience and/or will prompt subcontractors to submit proposals to the Prime Contractor on an unsolicited basis.

Potential subcontractors who are interested in working on the project with the Prime Contractor will typically provide specific information that enables the Prime Contractor to evaluate the price, background, experience, and scope of services required for the project. Invariably, this information is submitted in writing and is generally referred to as a subcontractor proposal.

As part of the preparation and submission of the proposals, each prospective subcontractor should become appropriately familiar with the project and all material aspects of the services on which it is proposing before submitting its proposal. There is usually a person (or persons) employed by the Prime Contractor who is the contact person from whom the potential subcontractor can either obtain the pertinent information or learn where such pertinent information can be obtained. After the potential subcontractor obtains basic information about the project, it should carefully review that information and direct any questions that are necessary to properly prepare the proposal for submission to the Prime Contractor in a timely manner. It is the subcontractor's responsibility to diligently review the requirements of the project and be absolutely certain that it is submitting an appropriate and fully responsive proposal. In the event the proposing potential subcontractor has any questions or any part of the information obtained lacks clarity, the potential subcontractor should seek all relevant information from the Prime Contractor or other appropriate sources.

Practice Tip

The Prime Contractor must read, understand, and negotiate the subcontractor's standard terms and conditions. Often, these are simply thrown in the file and not even attached to the contract. However, courts will enforce these terms and conditions like any other contract and, as such, the Prime Contractor may be bound by them, even in the event of a conflict with other terms of the subcontract.

A Prime Contractor may seek several competing proposals for a project, or it may elect to select a subcontractor with whom it has a long-term and successful relationship. In either case, each potential subcontractor should submit its proposal to the Prime Contractor in writing and in a timely manner for its review, comment, and use in preparing the Prime Contractor's proposal to the owner. The potential subcontractor should submit its price, scope of services, any exclusions from the scope of services, and all of the other *standard terms and conditions of service* in a timely manner. Typically, subcontractor proposals are signed and, as such, constitute the subcontractor's legally binding contractual offer to perform the services, subject to the terms and conditions in the proposal. It is a good practice for the potential subcontractor to contact the Prime Contractor. If the potential subcontractor has doubts about the contract requirements, it should make sure that it asks the Prime Contractor in a timely manner.

As it focuses on trying to be selected by the owner to provide the necessary design, construction, or project management services for the project, the Prime Contractor does not typically pay much attention to anything that it receives from potential subcontractors other than scope and fee information. It would be very unusual for a Prime Contractor to review and comment on any terms other than scope and fee from a potential subcontractor before being awarded the job. The failure to review and become generally familiar with all of the other terms and conditions submitted by the potential subcontractor is a very common mistake. This common mistake frequently leads to significant problems if the subcontractor who submitted a proposal is selected to perform services on the project but no formal written subcontract is negotiated and executed by the Prime Contractor and the subcontractor/proposer.

To minimize the likelihood of these problems occurring, the Prime Contractor should review the potential subcontractor proposals that it has received prior to its contract negotiations with the owner. At that time, it will normally identify the potential subcontractors it intends to use for the project. As part of this process, the Prime Contractor should carefully review and fully understand all of the terms and conditions of each potential subcontractor's proposal. It should identify and investigate any terms and conditions that are inconsistent with its requirements for the prime contract. If any of the potential subcontractors' proposals conflict with the prime contract, the Prime Contractor must address and resolve those conflicts before it relies upon that proposal for its submission to the owner. One should not just look at the scope and fee because other terms and conditions may have equally significant impacts on the project.

Once the Prime Contractor has been selected for the project by the owner and before it signs the prime contract, it should carefully review all of the potential subcontractors' proposals and confirm all of the materials, elements, and details of those proposals. It should confirm that the terms and conditions of those proposals are compatible with the terms of the prime contract. It should also confirm to each of the potential subcontractors that it intends to use them on the project and will be sending them a proposed subcontract agreement. Thereafter, once the prime contract is executed, the Prime Contractor should immediately turn its attention to negotiating the precise terms and conditions of its agreements with each of its sub-subcontractors.

Practice Tip

Once it has been awarded the project, and during the final negotiation phase with the Owner, the Prime Contractor should review both the prime contract and its subcontracts to make sure that the terms and conditions of each do not conflict, and that the Prime Contractor is fully aware of its liabilities and responsibilities to each of the parties. Additionally, the subcontractor should read the terms and conditions of the prime contract, especially if the subcontract contains a flow-down provision, discussed later in this chapter.

Upon the completion of the individual draft subcontracts for each potential subcontractor, the customized draft subcontract and a redacted copy of the prime contract should be sent to the subcontractor along with a request for a meeting to discuss the project, the potential subcontractor's proposal, and the draft proposed subcontract agreement. In its transmittal, the Prime Contractor should suggest that the parties and their counsel get together in person to discuss the project and the terms of the subcontract. While it is not necessary that all meetings be in person, it is advisable to hold at least one meeting in person with the subcontractor to discuss and review the subcontract agreement. If the Prime Contractor and the potential subcontractor have not worked together before, the first meeting or at the very least, one of the early meetings, should be at the potential subcontractor's office. This is an effective means by which the Prime Contractor can perform informal due diligence on the potential subcontractor's capabilities.

Library References

C.J.S. *Contracts* §§ 57, 59; C.J.S. *Counties* §§ 165–168; C.J.S. *Municipal Corporations* §§ 917–933; C.J.S. *Public Contracts* §§ 6, 14–24; C.J.S. *Schools and School Districts* §§ 409–414, 435, 761; C.J.S. *States* §§ 270, 280–288; C.J.S. *United States* § 113.

West's Key No. Digests, Contracts <KEY>17; Counties <KEY>115; Municipal Corporations <KEY>234; Public Contracts <KEY>5; Schools <KEY>80(2); States <KEY>98; United States <KEY>64.

III. Unique Legal Issues Applicable to the Prime Contractor/Subcontractor Relationship

Before turning to the mechanics of negotiating and drafting a subcontract agreement, it is important to consider legal issues unique to the Prime Contractor/subcontractor relationship. When a dispute arises between the Prime Contractor and the owner or the Prime Contractor and one or more of its subcontractors, it is important to know each party's rights and obligations to the project and to each other. The legal analysis relating to the liability exposure of each of the parties depends on the relationship between and among the design team, the construction team, the owner and/or the construction manager (CM), the likelihood of a particular jurisdiction to enforce common contractual clauses, the type of project, the type of services that are to be provided, and the applicable law relating to the issues of privity, economic loss, indemnity, and negligence. This section will focus on some of the more common liability issues that arise in Prime Contractor/subcontractor relationships.

In construction projects, the contractual relationships define and govern each of the parties' obligations. In most cases, the Prime Contractor will be liable for the acts and omissions of its subcontractor. Additionally, the Prime Contractor has certain obligations under its subcontract, most notably payment to the subcontractor, that are dependent on the owner's actions. As a result, many Prime Contractors utilize various risk-shifting contract clauses that limit their liability to both the owner and the subcontractor. Whether representing the Prime Contractor or the subcontractor, it is important to recognize and deal with these contract issues that may arise in the Prime Contractor/subcontractor context.

A. *Vicarious Liability, Indemnification, and Contribution*

When the Prime Contractor enters into the prime contract and various subcontracts, it is in contractual privity with both the owner and the subcontractors. As a result, the Prime Contractor is responsible to the owner for the acts and/or omissions of its subcontractors.[3] This, in its simplest terms, is the concept of vicarious liability.

Courts employ several legal theories when holding the Prime Contractor liable for any errors, acts, omissions, or breaches of contract caused by its subcontractors. Using agency law, a court may view the subcontractors as agents of the Prime Contractor, holding the Prime Contractor responsible for the subcontractors' conduct.[4] Additionally, the owner or the employer may have a breach of contract action against the Prime Contractor for errors caused by the subcontractors, depending on the terms of the contract.[5] For example, the contract may provide that the design professional is liable for the work of subcontractors they hire. The Prime Contractor may also be liable under tort law, which imposes on the design professional the duty to exercise a reasonable degree of care.[6]

---

3. *See, e.g.*, Scott v. Potomac Ins. Co., 217 Or. 323, 341 P.2d 1083 (1959) (architect held liable for failure of heating system designed by subconsultant engineer); Garver & Garver, P.A. v. Little Rock Sanitary Sewer Comm., 300 Ark. 620, 781 S.W.2d 24 (1989) (engineer held liable for supplier failing to meet specifications); Playskool, Inc. v. Elsa Benson, Inc., 147 Ill. Ct. App. 3d 292, 497 N.E.2d 1199 (1986) (architect held liable for failure of pre-cast concrete floor designed and fabricated by its subconsultants); Eastover Corp. v. Martin Builders, 543 So. 2d 1358 (La. Ct. App. 1989) (architect held liable for its mechanical consultant's failure to detect plumbing defect); R.D. Sepper Co. v. Venetian Iron Works, Inc., No. 89-C4174, 1990 U.S. Dist. LEXIS 17706 (N.D. Ill. Dec. 31, 1990) (architect held liable for its structural engineer consultant's failure to provide adequate specifications).

4. *See, e.g.*, Bayuk v. Edson, 236 Cal. Ct. App. 2d 309, 46 Cal. Rptr. 49 (1965).

5. *See* Getzchman v. Miller Chem. Co., 232 Neb. 885, 898, 443 N.W.2d 260, 270 (1989) ("If there is an express contract for architectural services, an architect's duties are determined by the contract for the architect's employment"); Kahn v. Terry, 628 So. 2d 390 (Ala. 1993) (quoting *Getzchman*).

6. *See, e.g.*, H. Elton Thompson & Assoc., P.C. v. Williams, 164 Ga. Ct. App. 571, 572, 298 S.E.2d 539, 540 (1982) ("The law imposes upon persons performing ... engineering[ ] and other professional and skilled services the obligation to exercise a reasonable degree of care, skill and

In an attempt to mitigate the liability exposure arising from agency law, some standard contracts describe subcontractors as independent contractors rather than agents of the design professional.[7] Moreover, independent of any contractual language, courts have held that the Prime Contractor is not liable for any acts by its subcontractor if the design subcontractor: (1) retained control over the manner and means of the work, or (2) was only responsible to the Prime Contractor.[8] Nevertheless, a court may still hold the subcontractor to be an agent of the prime professional depending on the amount of control exercised by the Prime Contractor.

Another method that Prime Contractors use to shift liability to their responsible subcontractors is by incorporating into the subcontract the legal concepts of indemnification and contribution. However, note that the statutes governing the application of indemnity and contribution vary from jurisdiction to jurisdiction. Therefore, it is important to determine how applicable statutes will enforce or imply indemnity and contribution principles as they relate to the Prime Contractor/subcontractor relationship. Often, these statutes dictate how such clauses must be drafted. For example, many jurisdictions have statutes that declare provisions that require one party to indemnify another party for the first party's own negligence as void because such broad indemnity provisions are against public policy. Additionally, whether a state has adopted joint and several liability or applies comparative fault principles can also affect the impact of such clauses.

Liability on a particular project may be shifted between a Prime Contractor and its subcontractors or among the subcontractors themselves through the incorporation of an indemnification clause in their contracts. An indemnification clause shifts all

---

ability, which generally is taken and considered to be such a degree of care and skill as, under similar conditions and like surrounding circumstances, is ordinarily employed by their respective professions.") (citations omitted); Milton J. Womack, Inc. v. House of Representatives, 509 So. 2d 62, 64 (La. Ct. App. 1987) ("The architect's duty is not to provide perfect plans but to exercise the degree of professional care and skill customarily employed by other architects in the same general area."); Nelson v. Commonwealth, 235 Va. 228, 236, 368 S.E.2d 239, 244 (1988) ("Absent a provision to the contrary, implicit in every contract of employment between an Owner and an architect is the duty of the architect to 'exercise the care of those ordinarily skilled in the business' ... This professional standard of care applies to the administration of project construction as well as to project design.") (citations omitted); *see also* Winsted Land Dev. v. Design Collaborative Architects, P.C., No. CV 960071571, 1999 Conn. Super. LEXIS 2180 (Conn. Super. Ct. Aug. 12, 1999); Riggins v. Bechtel Power Corp., 44 Wn. Ct. App. 244, 722 P.2d 819 (Wash. Ct. App. 1986); Laukkanen v. Jewel Tea Co., 222 N.E.2d 584 (Ill. Ct. App. 1966); Donnelly Constr. Co. v. Oberg/Hunt/Gilleland, 677 P.2d 1292 (Ariz. 1984); Vonasek v. Hirsch & Stevens, Inc., 221 N.W.2d 815 (Wis. 1974); Seiler v. Levitz Furniture Co., 367 A.2d 999 (Del. 1976).

7. *See, e.g.*, AIA Doc. C141-1987, Standard Form of Agreement Between Architect and Consultant, at 2.

8. *See* Schoedinger v. Hess, No. 99AP-1254, 2000 Ohio Ct. App. LEXIS 2409 (Ohio Ct. App. June 8, 2000).

or a portion of the liability from one party to another "in a manner that would not have occurred in a predictable manner under common law in the absence of the contract."[9] The scope of a contractual indemnity provision may range from a somewhat narrow obligation of requiring one party to reimburse the other for actual damages paid to a third party following an assessment of damages, to the much broader obligation of one party to provide legal counsel for the defense of another party based upon allegations which, if proved, would come within the scope of the indemnity provision.

Some states also recognize equitable indemnification as a part of their common law. Under this principle, if "a passively negligent tortfeasor is required solely through operation of law ... to pay for damages to a third person which have been primarily caused by the active negligence of another ... the active tortfeasor will be held to be the indemnifier of the passive tortfeasor."[10] Therefore, with equitable indemnification, one party may be required to indemnify the other as if there was a contractual indemnification clause in place *if* that tortfeasor actively caused the damages paid for by the passive tortfeasor.[11] Although the phrase "equitable indemnity" has been adopted, several states have adopted a hybrid of the concepts of indemnification and contribution.[12]

A second way in which liability may be shifted among multiple subcontractors or between the Prime Contractor and its subcontractor is through contribution. Contribution is a "[r]ight of one who has discharged a common liability to recover of another also liable, the aliquot portion which he ought to pay or bear."[13] Many states recognize the right to contribution following damages paid for negligent design or construction either under their common law or by statute.[14] Several states recognize

---

9. Robert L. Meyers & Debra A. Perlam, *Risk Allocation through Indemnity Obligations in Construction Contracts*, 40 S.C. L. REV. 898, 990 (1989); *see, e.g.*, Marino Constr. Co. v. Renner Architects, 214 Wis. 2d 589, 571 N.W.2d 923 (Wis. Ct. App. 1997); Kehoe v. Commonwealth Edison Co., 296 Ill. Ct. App. 3d 584, 694 N.E.2d 1119 (1998).

10. Green Constr. Co. v. United States, 506 F. Supp. 173, 178 (D. Mich. 1980).

11. *See, e.g.*, Woolard v. JLG Indus., Inc., 210 F.3d 1158 (10th Cir. 2000); Medallion Dev., Inc. v. Converse Consultants, 113 Nev. 27, 930 P.2d 115 (1997); District of Columbia v. Murtaugh, 728 A.2d 1237 (D.C. Cir. 1999); Beitzel v. City of Coeur d'Alene, 121 Idaho 709, 827 P.2d 1160 (1992); Winnsboro v. Wiedeman-Singleton, Inc., 307 S.C. 128, 414 S.E.2d 118 (1992); Berschauer/Phillips Constr. Co. v. Seattle Sch. Dist. No. 1, No. 38966-1-I, 1997 Wash. Ct. App. LEXIS 1801 (Wash. Ct. App. Oct. 27, 1997).

12. *See, e.g.*, Benner v. Wichman, 874 P.2d 949 (Alaska 1994) (recognizing right of "equitable apportionment" following state abolition of right to contribution); Far West Fin. Corp. v. D & S Co., 46 Cal. 3d 796, 760 P.2d 399 (Cal. 1988) (concept of comparative equitable indemnity covers whole range of apportionment of liability).

13. BLACK'S LAW DICTIONARY 328 (6th ed. 1990).

14. *See, e.g.*, Hartford Accident & Indem. Co. v. Scarlett Harbor Assoc., 109 Md. Ct. App. 217, 674 A.2d 106 (Md. Ct. Spec. Ct. App. 1996); Standhardt v. Flintkote Co., 84 N.M. 796, 508 P.2d 1283 (1973).

a statutory right to contribution, often stemming from comparative negligence principles rather than one grounded in the common law.[15]

B. *Exculpatory, Risk Allocation, and Limitation of Liability Clauses*

A subcontractor may also seek to limit its liability to the Prime Contractor through the use of a risk allocation or limitation of liability clause. Clauses limiting the liability of a construction professional are frequently found in construction contracts. For example, a structural engineer may limit its liability for performance under the subcontract to the fee earned on the project. These clauses, if clear and unambiguous, will normally be enforced by a court or arbitration panel. However, the laws of each jurisdiction should be reviewed to ensure that such clauses are not void as against public policy or that some unique legal formalities are required to assure their enforceability. A Prime Contractor or subcontractor that is bound by a risk allocation or limitation of liability clause is also well-advised to check with its insurance advisor to make sure that the clause is within the scope of its insurance coverage.

While the courts in some states have never directly addressed the enforceability of limitation of liability clauses contained in professional services contracts, they most likely have, however, considered the enforceability of exculpatory clauses in other contexts. Exculpatory clauses are very similar to those limiting liability; while a limitation of liability places a cap on damages recoverable by the other party to the contract, an exculpatory clause relieves one party of its liability entirely. In some cases, a court will uphold a limitation of liability clause where certain exculpatory or indemnity provisions would be unenforceable.[16] In other cases, courts have held limitation of liability clauses void as contrary to that jurisdiction's anti-indemnity statute.[17]

C. *Payment Clauses*

Timely payment for work performed on a construction project is one of the primary concerns for both Prime Contractors and subcontractors. Therefore, construction contracts frequently include payment provisions that attempt to condition

---

15. *See, e.g.,* Brochner v. Western Ins. Co., 724 P.2d 1293 (Colo. 1986); Rothberg v. Reichelt, 270 A.D.2d 760, 705 N.Y.S.2d 115 (2000); Clearwater v. L.M. Duncan & Sons, Inc., 466 So. 2d 1116 (Fla. Ct. App. 1985); Good v. Lemcon Dev., Inc., No. 12837, 1992 Ohio Ct. App. LEXIS 847 (Ohio Ct. App. Feb. 24, 1992).

16. *See, e.g.,* Valhal Corp. v. Sullivan Assocs., Inc., 44 F.3d 195 (3rd Cir. 1995) (upholding a limitation of liability provision under Pennsylvania law stating that "[l]imitation of liability clauses are not subject to the same stringent standards applied to exculpatory and indemnity clauses"); Marbro, Inc. v. Borough of Tinton Falls, 688 A.2d 159 (N.J. Super. 1996) (holding limitation of liability clause between municipality and engineer enforceable); Markborough California, Inc. v. Superior Court, 277 Cal. Rptr. 919 (Ct. App. 1991).

17. *See* City of Dillingham v. CH2M Hill Northwest, Inc., 873 P.2d 1271 (Alaska 1994) (Legislative history of Alaska's statute reveals that it was intended to prohibit not only indemnity clauses but also limitation of liability clauses.).

payment on receipt of payment from the upstream party and/or require payment within a certain time frame of the receipt of payment by the upstream party.

Even the largest architect/engineering (A/E) firms and general contractors cannot afford to routinely pay their subcontractors over an extended period without first receiving payment from the owner. Issues arise between the Prime Contractor and its subcontractor when payment from the owner or Prime Contractor's client is late or not forthcoming, or when a dispute arises regarding the services provided by the Prime Contractor that may be unrelated to services provided by all but one of its subcontractors, thereby causing the owner to withhold payment due everyone. These issues are so common they should be addressed as a fundamental provision of the subcontract. In more extreme situations, the owner may become insolvent or otherwise fail to make payment to the Prime Contractor, even though the Prime Contractor and its subcontractors have fulfilled their contractual obligations. Then, the challenge is determining which party can best bear the risk of the owner's nonpayment. Normally, without including risk-shifting provisions in its subcontract agreements, the Prime Contractor will bear the risk of making payments to its subcontractors for work performed despite the owner's nonpayment. However, two commonly used contract clauses may shift some or all of this risk to the subcontractor. These are known as the *pay-when-paid* and the *pay-if-paid* clauses.

In general, pay-when-paid clauses are construed merely as timing mechanisms, and do not operate to defer payment indefinitely. In other words, these clauses merely provide a general contractor with some reasonable period within which to make payment. The American Institute of Architects (AIA) A401 and the Associated General Contractors (AGC) 650 subcontract forms contain pay-when-paid terms.

From a subcontractor's perspective, these clauses are preferable to pay-if-paid clauses because they do not completely shift the risk of the owner's nonpayment to the subcontractor. With a pay-when-paid clause, as long as the cause for the owner's nonpayment to the general contractor is not related to the subcontractor's work, the general contractor will remain liable to the subcontractor immediately following the expiration of a reasonable time period. The time period is like a grace period for the general contractor, but if the general contractor does not receive payment from the owner through no fault of the subcontractor, the general contractor will still be liable to the subcontractor. An example of a simple pay-when-paid clause is: "Final payment to Subcontractor will occur within thirty days of payment to Contractor by owner."

On the other hand, pay-if-paid clauses will preclude a subcontractor from recovering payment from the Prime Contractor if the Prime Contractor is never paid by the owner. In other words, the risk of the owner's nonpayment is shifted to the subcontractor. An example of a pay-if-paid clause follows:

> The Contractor shall be under no obligation to make any payment to the Subcontractor except to the extent that the Contractor has received funds from the owner for the work performed by the Subcontractor. The Subcontractor agrees that payment by the owner is an express condition precedent

to contractor's obligation to pay subcontractor. Subcontractor will not be entitled to payment by the Contractor for work performed if, for any reason, including the owner's insolvency or lack of available funds, the owner fails to pay the Contractor in accordance with the General Contract.

When enforced, these clauses forfeit the subcontractor's right to payment under the contract when the Prime Contractor does not receive payment from the owner. However, courts disfavor these provisions and, therefore, require that they be clear and unambiguous. Some states, including New York and California, will not enforce pay-if-paid clauses at all, holding them to be void against public policy.[18] Additionally, North Carolina and Wisconsin have passed statutes voiding such clauses,[19] while other jurisdictions have statutes that do not void pay-if-paid clauses outright, but place limits on their enforceability.[20]

D. *Insurance Issues*

In allocating the risk among multiple parties, proper insurance coverage is essential. Many construction contracts, including the forms published by the Engineers Joint Contract Documents Committee (EJCDC), require that certain parties be listed as "additional insureds" on their Commercial General Liability (CGL) policy. Usually, additional insureds are parties who are upstream from the policyholder. For example, owners have often required general contractors to buy liability insurance that names the owners as additional insureds. In turn, general contractors often require subcontractors to do the same for the general contractor and possibly other upper-tier subcontractors.

Additional-insured endorsements allow upstream parties to make a claim directly to the potentially liable party's insurer. For example, if an owner finds a construction defect, he or she is most likely to sue the general contractor, even if the defect was caused by one of the general contractor's subcontractors. The endorsement allows the general contractor to make a claim directly on the responsible subcontractor's insurance.

In essence, parties named as additional insureds are entitled to much of the same benefits as the primary insured. Often, these types of arrangements exist even if the

---

18. Wm. R. Clarke Corp. v. Safeco Ins. Co., 15 Cal. 4th 882, 64 Cal. Rptr. 2d 578, 938 P.2d 372 (1997); West-Fair Elec. Contractors v. Aetna Cas. & Sur. Co., 87 N.Y.2d 148, 638 N.Y.S.2d 394, 661 N.E.2d 967 (1995).

19. *See* N.C. GEN. STAT. § 22C-2 (2006) and WIS. STAT ANN § 779.135 (2006).

20. *See, e.g.,* 770 ILL. COMP. STAT. ANN. 60/21 (2006); MD. CODE ANN., REAL PROP. § 9-113(b), (c) (2006); MO. REV. STAT. § 431.183 (2006). Still, other jurisdictions declined to adopt such legislation. *See* CONN. HB 5669, and RI 89-H5695 Substitute A.

additional insured has its own insurance policy. However, usually the policy against which an additional insured claims will be primary and noncontributory, meaning that the limits of that policy must be exhausted before the additional insured's own coverage is triggered. Additionally, the additional insured may be entitled to coverage for damages caused by its own negligence, something that would normally be barred if the additional insured was a third party claimant against the policy holder. Allowing coverage in such instances acts like a broad form indemnity provision in the contract, i.e., forcing one party to indemnify the other for the other's own negligence, which circumvents many states' anti-indemnity laws.

Subcontractors or Prime Contractors who agree to grant additional insured protection to others should be aware of some adverse consequences. These include the risk of exhausting the policyholder's insurance coverage and, for the cost of deductibles and higher premiums incurred, defending the indemnitee's negligence.

Alternatives to additional insured provisions include other available insurance products. For example, owners and contractors protective (OCP) policies, which are separate and distinct from the insured's CGL policy, provide vicarious liability coverage to the policyholder for damages arising from the insured's general supervision of the construction work.

On the design side, a project professional liability policy (Project Policy) is an alternative insurance product for large projects involving multiple design professionals. Under such a policy, coverage applies to design services provided to the project, rather than an individual firm employed on the project. Rather than having overlapping Project Policies for each design professional employed on the project, where gaps and conflicts in coverage may arise, a Project Policy covers the prime design professional as well as its subcontractors for their collective work on a project. Usually a Project Policy replaces the design professionals' practice policies with regard to services provided for that project, although members of the project design team may have the right to carry their practice policies as excess coverage. To assure that coverage will apply against any potential claims arising out of the project, an extended reporting period should be purchased so that coverage is maintained until the applicable statute of repose has expired.

Practice Tip

Make certain that terms used in the insurance sections of the contracts that are signed are fully understood, and address any related issues. For example, when a contract requires "additional insured" coverage or the parties agree to purchase a project policy with a significant deductible, consider drafting a deductible payment agreement to fairly allocate the insurance deductible among the various insureds should a claim arise. For design professionals, one straightforward way in which to divide the payment is to require each consultant to pay the same percentage of the deductible as the total design fees it will receive.

E. *Dispute Resolution*

If a dispute arises among the parties on a construction project and all efforts to resolve the conflict by informal negotiation fail, there are other more formal dispute resolution methods available to parties who wish to avoid litigation. These dispute resolution procedures, collectively termed Alternative Dispute Resolution or ADR, are usually less costly and more expeditious than litigation and they are widely used in construction disputes. In fact, some ADR techniques, such as arbitration and mediation, are so popular that calling them alternative is somewhat of a misnomer.

ADR procedures are especially popular in the construction industry in light of the fact that several of the standard contracts, including those from the AIA, incorporate mandatory mediation and/or arbitration clauses.[21] Many states have enacted statutes that regulate ADR procedures, including some version of the Uniform Arbitration Act.[22] There are a few issues that Prime Contractors and subcontractors should consider before agreeing to one form of dispute resolution over another.

The first issue is that many construction disputes involve multiple parties and a range of claims. However, parties cannot be forced to mediate or arbitrate a dispute except through an express contractual agreement. Therefore, if ADR is the preferred method of dispute resolution, it is in the best interests of the owner and the Prime Contractor(s) to ensure that all of the subcontract agreements contain similar ADR provisions. Otherwise, a party may be forced to litigate a dispute in two different forums with the risk of differing outcomes.

> *Example:* Assume that an owner sues the A/E for a design error caused by the A/E's structural engineer. However, the A/E's prime contract contains an arbitration provision with mandatory mediation, but the A/E's subcontract does not. Rather than being allowed to join the structural engineer as a third party to the arbitration, *passing through* the owner's claim to the structural engineer, the A/E must first defend the arbitration on behalf of its

---

21. *See, e.g.,* AIA Doc. A201-1997 General Conditions of the Contract for Construction. Section 4.5 contains both mandatory mediation and arbitration clauses. In fact, the mediation is made a condition precedent to arbitration.

22. *See, e.g.,* ALASKA STAT. § 9.43.010 to § 94.43.300 (2005); ARK. CODE § 16-108-201 to § 16-108-224 (2006); COLO. REV. STAT. § 13-22-201 to § 13-22-507 (2006); DEL. CODE ANN. § 10-5701 to § 10-5725 (2006); FLA. STAT. ch. 682.01 to § 682.22 (2006); HAW. REV. STAT. § 658A-1 to § 658A-29 (2006); IDAHO CODE § 7-901 to § 7-922 (2006); 710 ILL. COMP. STAT. § 5/1 to § 5/23 (2006); IND. CODE § 34-57-2-1 to § 34-57-2-22 (2006); KY. REV. STAT. § 417.045 to § 417.24 (2006); ME. REV. STAT. tit. 14, § 5927 to § 5949 (2006); MD. CODE ANN., CTS. & JUD. PROC. § 3-201 to § 3-234 (2006); MASS. GEN. LAWS ch. 251, § 1 to § 19 (2006); MINN. STAT. § 572.08 to § 572.30 (2006); MO. REV. STAT. § 435.350 to § 435.470 (2006); MONT. CODE ANN. § 27-5-101 to § 27-5-324 (2006); NEB. REV. STAT. § 25-2601 to § 25-2622 (2006); N.M. STAT. ANN. § 44-7A-1 to § 44-7A-32 (2006); N.C. GEN. STAT. § 1-567.1 to § 1-569.31 (2006); TENN. CODE ANN. § 29-5-301 to § 29-5-320 (2006); UTAH CODE ANN. § 78-31a-1 to § 78-31a-131 (2006); VT. STAT. ANN. tit. 12, § 5651 to § 5681 (2005).

engineer, and then sue the engineer in state court. If the engineer is found at fault in the arbitration, but not in the litigation, the result will leave the A/E in an unfortunate situation.

Practice Tip

The Prime Contractor should review all dispute resolution procedures in both the prime contract and each of its subcontracts to ensure that all parties are required to adhere to the same dispute resolution process. Additionally, if all parties agree to resolve the dispute through arbitration, there should be some language in the contract that allows joinder of all parties and claims.

F. *Flow-Down Clauses*

Flow-down clauses, sometimes referred to as *incorporation-by-reference clauses*, are contract provisions by which the parties incorporate the terms of the construction or design contract between the owner and the Prime Contractor into that of the lower-tier subcontractor. In other words, the subcontractor agrees to be bound by the obligations and responsibilities that the Prime Contractor has assumed toward the owner. Additionally, a well-crafted flow-down clause may allow the Prime Contractor to reduce its total exposure to its subcontractors by limiting its damages to only those recovered by the Prime Contractor from the owner and "provides a means to pass through to the responsible subcontractors the contractor's damage exposure to the owner."[23]

In addition to increasing the subcontractor's liability to the Prime Contractor, flow-down clauses may severely reduce the subcontractor's ability to recover damages on its contract, especially if the prime contract contains limitations of liability, limitations regarding rights to payment, or other enforceable exculpatory provisions such as no damages for delay or waiver of consequential damages.[24]

Often, it is difficult for subcontractors to remove these clauses from their contracts due to lack of bargaining power with the Prime Contractor. One commentator has suggested that one way to make such clauses more balanced is to insist on a

---

23. ANDREW HOWARD, *Flow-Down Clauses*, THE CONSTRUCTION CONTRACTS BOOK: HOW TO FIND COMMON GROUND IN NEGOTIATING DESIGN AND CONSTRUCTION CONTRACT CLAUSES 9-10 (DANIEL S. BRENNAN et al. eds., 2004).

24. L&B Constr. Co. v. Ragan Enters., Inc., 482 S.E.2d 279, 281-83 (Ga. 1997) (flow-down provision incorporating 'no damages for delay' clause from prime contract "remove[d] any conceivable doubt" that subcontractor's recovery was limited to time extension and was prevented from seeking monetary damages resulting from contractor's two-year delay on project).

flow-down clause that imposes reciprocal obligations: "For example, if the subcontractor assumes the Prime Contractor's obligations to the owner, then the Prime Contractor should be asked to assume the owner's obligations to the contractor, including the obligation of payment."[25]

However, not all flow-down clauses are necessarily detrimental to the rights of the subcontractor. There are instances where the prime contract actually affords the subcontractor more rights than the subcontract. These added benefits include access to "alternative dispute resolution,[26] payment terms, notice requirements, completion dates, delay and liquidated damages, indemnity obligations, and conditions of default."[27] To successfully evaluate whether a flow-down provision means added risks or benefits to the subcontractor, the subcontractor must take the time to review and examine the prime contract.

G. *Applicability of Economic Loss Rule*

While the contract is king on construction projects, tort principles regularly come into play. Most notable is the standard of care required of all design professionals, including consultants. While compliance with the standard of care is often made an express obligation under the contract or subcontract, its genesis is in negligence principles. In other words, if an A/E has breached the standard of care (acted in a way contrary to what a prudent A/E would do in that location under similar circumstances), the A/E has both been negligent and breached its contract. The economic loss rule, as discussed elsewhere in this handbook, is a judicially created doctrine that prevents a party from suing in tort to recover purely economic damages.[28] Under today's modern pleading rules, it is not uncommon for a plaintiff to allege a cause of action against the contractor or design professional for both breach of contract and negligence. However, in the absence of contractual language holding a particular party to a higher standard of care than that imposed by law, the standard of proof for breach of contract will, in most cases, be essentially the same as for negligent violation of a duty.[29] As a result, states in which the courts adhere to the economic loss doctrine will bar such actions in tort.[30]

---

25. HOWARD, *supra* note 23, at 9.

26. Turner Constr. Co. v. Midwest Curtainwalls, Inc., 543 N.E.2d 249, 252–53 (Ill. Ct. App. Ct. 1998) (subcontractor entitled to arbitrate its claims where prime contract contained an arbitration clause but subcontract did not).

27. HOWARD, *supra* note 23, at 10.

28. Economic loss is defined in the construction context as "the cost to repair or replace defective materials, damage to a structure, diminution in value of a damaged structure not repaired, loss of use or delay in utilizing property for its intended purposes and related lost profits, lost revenue, and costs." 6 BRUNER & O'CONNOR CONSTRUCTION LAW § 19:10 (2005) (citations omitted).

29. *See* Lee County v. Southern Water Contractors, Inc., 298 So. 2d 518 (Fla. Ct. App. 1974).

30. *See, e.g.*, National Steel Erection, Inc. v. J.A. Jones Constr. Co., 899 F. Supp. 268 (N.D. W. Va. 1995); World Trade Co. v. Westinghouse Elec. Corp., 256 A.D.2d 263, 682 N.Y.S.2d

With respect to subcontractors, the economic loss doctrine applies where there is no privity of contract. For example, an owner may claim that it was damaged by the negligence of one of the Prime Contractor's subcontractors and attempt to sue that subcontractor directly in tort. This situation also may arise where the general contractor or one of its subcontractors attempts to sue one of the A/E's subconsultants for negligence, or vice versa. Here the question is whether the economic loss doctrine will bar a direct action for damages from negligent conduct against a contractor or subcontractor not in privity with the plaintiff.[31]

As this principle is fully discussed in chapter 6, there is no need to reiterate this interesting debate here. However, the emerging trend by courts which are faced with this issue appears to favor expanding the applicability of the economic loss doctrine in cases even where no privity of contract exists between the parties.[32]

Library References

C.J.S. *Agency* §§ 17, 375, 377, 419–421, 423–445; C.J.S. *Contracts* §§ 271, 356, 442–445, 450, 508, 531, 592–593; C.J.S. *Contribution* §§ 5, 9–10, 12–13; C.J.S. *Insurance* §§ 54, 956–957, 970–973; C.J.S. *Negligence* § 58.

West's Key No. Digests, Alternative Dispute Resolution <KEY>112, 442; Contracts <KEY>114, 221(3), 319–321; Contribution <KEY>6; Indemnity <KEY>30(5), 33(5), 34, 67; Insurance <KEY>2359, 2391; Negligence <KEY>463; Principal and Agent <KEY>3(2), 159.

IV. Drafting Subcontract Agreements

This section provides practical drafting tips in addition to a check list of issues that should be considered in every subcontract. Due to the fact that the Prime

---

385, 39 U.C.C. Rep. Serv. 2d 1053 (1st Dep't 1998); Anderson Elec. v. Ledbetter Erection Corp., 115 Ill. 2d 146, 104 Ill. Dec. 689, 503 N.E.2d 246 (1986); Du Page County v. Graham, Anderson, Probst & White, Inc., 109 Ill. 2d 143, 92 Ill. Dec. 833, 485 N.E.2d 1076 (1985); Doran-Maine, Inc. v. American Eng'g & Testing, Inc., 608 F. Supp. 609 (D. Me. 1985); Berschauer/Phillips Constr. Co. v. Seattle Sch. Dist. No. 1, 124 Wash. 2d 816, 881 P.2d 986, 94 Ed. Law Rep. 610 (1994); American Towers Owners Ass'n v. CCI Mech., Inc., 930 P.2d 1182 (Utah 1996); Williams & Sons Erectors v. South Carolina Steel, 983 F.2d 1176 (2d Cir. 1993).

31. Philip L. Bruner et al., *2004 Construction Review*, CONSTRUCTION BRIEFINGS No. 2005-1 (2005).

32. Miller v. U.S. Steel, 902 F.2d 573 (7th Cir. 1990); *See also* Corporex Dev. & Constr. Mgmt., Inc. v. Shook, Inc., 106 Ohio St. 3d 412, 415, 835 N.E.2d 701, 705 (2005) (project owner could not evade the economic loss rule by suing a subcontractor in negligence; owner had to sue the general contractor for breach of contract and the general contractor could in turn sue the subcontractor); Digicorp, Inc., v. Ameritech Corp., 262 Wis. 2d 32, 662 N.W.2d 652, 666 (Wis. 2003) ("[E]ven in the absence of privity, the economic loss doctrine bars one party in the distributive chain from recovering economic losses in tort from another party in that chain."). Coastal Conduit & Ditching, Inc. v. Noram Energy Corp., 29 S.W.3d 282, 285–90 (Tex. Ct. App. 2000).

Contractor generally takes the lead in drafting the subcontracts, the check list is geared to the Prime Contractor's perspective. However, listing all of the issues will help subcontractors identify and negotiate important terms that affect their liability, obligations, and responsibilities under their subcontracts.

A. *Rule No. 1: Get It in Writing, and Get It Signed!*

Every lawyer who practices in this area has heard the typical client lament that "things are not as they used to be." This is usually in reference to the times when everyone in the construction industry knew everyone else and agreements were sealed by a handshake, not a written contract. Those were the days of complete honesty and perfect memories. Those were also the days when it was not unusual for someone to stay with the same company for his entire career. Without questioning exactly when those bygone days ended, it is safe to say they have long since passed.

The terms and conditions of every agreement between a Prime Contractor and its subcontractors should be reduced to writing and signed by duly authorized representatives of each party. Under some circumstances, it may be possible to have enforceable oral subcontracts. However, the statute of frauds, operable in every jurisdiction, will render most verbal construction subcontracts unenforceable. Notwithstanding this fact, unless the terms of the agreement are reduced to writing, disagreements over material and important issues are certain to occur. In general, it is not possible to predict how such disagreements will be resolved. Additionally, in many prime contracts, an obligation is imposed upon the Prime Contractor to pass down or impose certain provisions of the prime contract on the Prime Contractor's subcontractors; this obligation can only be accomplished in a written subcontract. Finally, documenting the terms and conditions of service before any dispute or disagreement arises will often prevent such disputes or disagreements from being unreasonably costly to resolve.

The number of times that a Prime Contractor and its subcontractor will spend weeks or even months negotiating a contract and then never bother executing it is baffling. While courts will usually enforce unsigned contracts if it conforms to the conduct of the parties, it saves both time and money simply to sign the final contract.

B. *Rule No. 2: Develop a "Form" Contract for All Subcontracts on the Project*

The prime contract with the owner will normally impose a duty and responsibility on the Prime Contractor to identify and select all subcontractors required for the project. The right of the Prime Contractor to select and employ these subcontractors will often be subject to the owner's approval. The owner may also have a conditional or contingent right of assignment for those subcontracts in the event of the Prime Contractor's termination or default.

> **Practice Tip**
>
> Before executing the prime contract, the Prime Contractor should deliver to the Owner a list of the potential subcontractors for the project. If the Owner is given the list and then requests a change in a subcontractor after the prime contract is executed, any additional costs related to the change in the subcontractor should be at the owner's expense.

1. *Form or Custom Subcontract Agreements*

There are many subcontract forms or templates available in the marketplace today. Among them are subcontract forms from the American Institute of Architects (AIA), the Associated General Contractors (AGC), the Engineers Joint Contract Documents Committee (EJCDC), the Design Build Institute of America (DBIA), and Construction Owners Association of America (COAA), to name a few of the most widely-known forms.

Many lawyers, Prime Contractors, and subcontractors prefer to use their own, customized subcontract agreements. It is beyond the scope of this chapter to evaluate and recommend one form or approach over another, but a typical customized subcontract agreement is attached as Appendix A to this chapter.

In general, the individual conditions of a particular project, coupled with the unique requirements of each contractor's subcontract terms and conditions, will require a great deal of unique language and terms or customization without regard to whether the initial template is a nationally recognized form or individually drafted subcontract agreement. In any event, it is essential to ensure that certain terms and conditions are the same for all subcontractors on the project, that all key subjects are addressed in the final, subcontract agreement, and that a properly signed subcontract is in place at the very beginning of the project.

2. *Developing a Standard or Template Base Subcontract Agreement*

As a starting point for negotiations with all of the subcontractors on the project, the Prime Contractor should prepare a form of subcontract agreement. Sometimes, this is done well in advance and it may be attached to the Request for Proposal (RFP) or Invitation for Bids (IFB). It does not matter whether the initial template is provided by the AIA, AGC, DBIA, EJCDC, other similar organization, or is a customized form that the Prime Contractor's lawyer has used on other projects or specifically drafted for the project at hand. Whatever the starting point, it is important to develop a template or form that contains all of the terms and conditions that apply to all subcontractors on the Prime Contractor's project team and their services required by the agreement. Remember that the prime contract may require that certain provisions or clauses be included in each subcontract agreement.

The next step is to prepare a draft of a customized subcontract for each of the subcontractors on the project using this template. Each individualized draft subcontract should include all relevant information, terms, and conditions, including information from the specific proposals previously submitted by the subcontractor. Once completed, the customized draft subcontract and a redacted copy of the prime contract should be sent to the subcontractor for its review and comment. The fee and reimbursable expenses information of the prime contract should be redacted. Any other confidential or proprietary information should also be redacted from the copy of the prime contract.

Some of the Prime Contractor's consultants will have significant professional responsibilities, and others will have relatively narrow scopes of work with short durations. For example, the structural engineer's scope of work for a project is typically more extensive and involves a greater fee than a subcontractor that is providing drafting or duplication services to the Prime Contractor for the project. To efficiently manage the development of a subcontract for these two different types of subcontractors, it is recommended that the Prime Contractor develop two versions of its subcontract template for the project to support efficient and properly focused contract negotiations with each of these subcontractors.

The Prime Contractor should develop a template that is comprehensive and detailed, or a long form of subcontract template, for the major subcontractors or subcontractors.[33] Major consultants are those whose expertise is typically required to produce the structure, facility, building, development, or project of the type at issue. The professionals who would be considered principal or major consultants vary depending on the type of project, the expertise possessed on the Prime Contractor's staff, and the deal made with the owner.

A more abbreviated short form subcontract agreement should be used for specialty subcontractors[34] with small fees and scopes of work and for limited or shared

---

33. On the design side, the Prime Contractor is an architect or engineer, and the major consultants will typically be the structural, civil, electrical, mechanical, and plumbing design professionals. The major subconsultants typically consist of licensed or certified professionals who have significant responsibility from the beginning of the design process through the production of final drawings and specifications, which must be stamped, sealed, signed, or certified for submission to the code review authority for approval before a building permit may be authorized for issuance. The principal or major subconsultants typically receive a significant portion of the total fee. Their services constitute significant portions of the total scope services that are to be provided by the Prime Contractor and/or may include significant potential risk or exposure for the Prime Contractor, if they are untimely or deficient.

34. The evolution of the standard of care that is a minimum performance standard for design professionals has caused some areas of consultancy—that thirty years ago were limited to only high-budget or unique projects—now to become as common and necessary as structural or civil engineering. Some examples include:
- lighting consultants on commercial, office, governmental, or other public facilities;
- security consultants;

risk subcontractors.[35] Specialty consultants are subcontractors to the Prime Contractor retained to provide specialized technical expertise for some particular system or portion of the project. On the design side, examples of specialty consultants include elevator or vertical transportation, interiors, kitchen, acoustical, scheduling, cost estimating, parking, surface transportation, graphic or signage, fire protection, communication, audio/visual, security, curtain wall, shoring, steel connections, and lighting.

Ideally, a representative of the Prime Contractor and each subcontractor should meet in person to discuss the draft subcontract and discuss their expectations of each other for their working relationship on the project. This provides an opportunity for each party to discuss the particular or unique characteristics and issues associated with the project. It also allows the parties to meet firsthand to address any particularly troublesome or difficult contract issues. After this first meeting, the remaining or unresolved subcontract issues can typically be addressed by exchanges of electronic redlines and clean drafts of the subcontract.

Master agreements are sometimes used between a Prime Contractor and a subcontractor who anticipates it will be working with the Prime Contractor on several projects together over the course of several years. Multiple tasks related to a single project or several unrelated projects may be subject to this master agreement. The master agreement contains the commercial terms and conditions that govern all

---

- acoustical consultants;
- communications consultants on governmental or headquarters office facilities; and
- elevator or vertical transportation consultants on office complexes, shopping centers, or multimodal transportation hubs.

35. In the modern world of design professional contract negotiation, drafting, evaluation, and review, a third type of subconsultant is frequently encountered. These subconsultants are technically subcontracted to the Prime Contractor but with some significant modification in the normally expected line of authority, reporting relationships, allocation of risk, or responsibility for performance. Unique contractual provisions that modify the normal level of liability of the Prime Contractor typify the agreements between these subconsultants and the Prime Contractors, which have corresponding provisions in the prime design contract. Some examples of these subconsultant disciplines include soils engineering, surveying, environmental, and hazardous materials or explosives removal. A significant characteristic of this grouping is that the discipline or specialty is not insured or insurable under the professional liability insurance of the Prime Contractor. For many years, the use of the project professional liability insurance policy reduced the need for a construction lawyer to have special agreements or provisions for this grouping. However, over the last several years and particularly since 9/11, we have observed project professional liability insurance become less available for certain types of projects, such as condominiums, due to the higher risk. Furthermore, the significant increase in premium costs since 9/11 has made project policies less attractive even when they are available. Accordingly, contractual risk limitation, sharing, or avoidance will make this grouping more common and special subcontracts more necessary.

projects involving the Prime Contractor and the subcontractor. Individual projects or tasks are then authorized in purchase orders or other simple scope documents identifying the specific work to be performed for that particular project, the price, and sometimes specific dates that must be met by the subcontractor.

Master agreements can streamline the contracting process. However, there is a danger that no two projects are alike and, therefore, the master agreement should be reviewed prior to commencing each project to determine whether special or supplemental terms and conditions are required to adequately reflect the assigned project and its scope of work. A master agreement can sit in the drawer for years, never reviewed by the contracting parties until a dispute arises. In the interim, laws may have changed, including the way in which the courts interpret or enforce various contract provisions. This means that the master agreement may produce unintended results if not regularly reviewed and updated.

C. *Rule No. 3: Identify Key Subcontract Terms and Conditions*

Certain issues need to be addressed in every Prime Contractor subcontract in order for it to be legally enforceable and in compliance with the terms of the prime contract. Other issues, subjects, terms, and conditions, based upon the common experience of many people in the construction industry, should always be considered for inclusion in Prime Contractor subcontracts. Good examples of these types of issues are discussed in Section III, above. Finally, there are other issues, subjects, terms, and conditions that may be necessary for a particular subcontract based upon the negotiation of the parties and/or the needs of the project.

Table 1 lists the terms and conditions that should be evaluated for inclusion in every Prime Contractor subcontract. This check list is based upon the customized Design Services Subcontract Agreement that is attached hereto as Appendix A.

Library References

C.J.S. *Contracts* §§ 67, 75.
West's Key No. Digests, Contracts <KEY>31, 35.

V. Subcontract Negotiations

The subcontract negotiations between the Prime Contractor and each of its subcontractors should be intended and conducted with the goal of developing a set of terms and conditions that fairly and accurately reflect the allocation of rights, responsibilities, obligations, and risks between the Prime Contractor and its subcontractors inherent in the project. The negotiations should not be intended to impose unreasonable burdens or liabilities upon either party.

During the process of the negotiations between the Prime Contractor and the owner, the Prime Contractor must be sensitive to the roles of its subcontractors and not agree to unusual or unduly burdensome responsibilities. To avoid this situation, the Prime Contractor should periodically advise its sub-subcontractors of the status of the negotiations.

Table 1
PROVISION CHECK LIST FOR DESIGN PROFESSIONAL/CONSTRUCTION MANAGER SUBCONTRACTS

Type/Subject Matter of Provision	Reason for, or Objective of, Provision(s)		Comments
Names/Identity of the Parties	The full and correct legal name of the parties must be properly reflected at the beginning of the document.	Essential	Drafting and reviewing counsel should check the Secretary of State for the jurisdiction in which the Parties are domiciled to make sure that they are in good standing. Use the full, legal name, including, if applicable, any fictitious name or "dba." In most jurisdictions, this task can be performed using the Internet.
Name or Title of Document	The name or title of the agreement should be reflected at the top of the first page.	Highly Desirable	
Project Description	Description of project, including address or location and "name"	Essential	The project description in the subcontract should be identical to the "project description" in the Prime Contract between the Owner and the Prime Contractor. Any differences can create significant problems if there are any claims, disputes, or litigation regarding the Project.
Effective Date of Agreement	Establishes the date on which the contractual relationship terms of the Agreement start to apply to their relationship	Essential	Frequently, the Prime Agreement establishes a date in the past as the date on which the design professionals, construction manager, and their consultants "started" to provide services.

Effective Date of Agreement *continued*			As a general rule, the "Effective Date of the Agreement" should be identical to the "Effective Date" in the Prime Contract between the Owner and the Prime Contractor. On a related topic, the parties may want to consider adding a date on which they agree that the statute of limitations begins to run for the purposes of disputes regarding the project.
Incorporation of Applicable Terms of Prime Contract	Requires that applicable provisions of the Prime Agreement that are required to be imposed on the subcontractors be incorporated into their subcontracts.	Essential	The prime contract will normally impose on the Prime Contractor an obligation to require that its subcontractors accept and agree to certain terms and conditions of the prime contract.
Notice/Contact/ Designated Representative Information	Identifies the person or persons who are responsible for receiving any official notice or information that is required by the terms of the subcontract agreement.	Highly Desirable	This provision sets forth the form or forms of notification that may be used for communicating information required or allowed by the subcontract. It also provides the specific contact information, i.e., name, title, address, e-mail, telephone, fax, and other pertinent details for communication between the parties.
Scope of Services	The scope or description of the services that are to be provided. This description should be very detailed and thorough.	Essential	

Table 1 *continued*

Type/Subject Matter of Provision	Reason for, or Objective of, Provision(s)		Comments
Standard of Performance			
Identification of Subconsultants		Highly Desirable	The Prime Contractor's duties and responsibilities under the provisions of the prime contract with the Owner will typically include the identification, selection, and management of the services and deliverables required of the Prime Contractor's subcontractors. In many prime design contracts, the Prime Contractor must obtain the Owner's approval of its subcontractors.
Schedule, Applicable to Provision of Services or Deliverables, if Any.		Highly Desirable	Ensure that the subcontractor's anticipated schedule—especially milestone dates—conform to the Prime Contract.
Description of Deliverables		Useful	
Budget or Financial Parameters Applicable to Prime Contractor's Services		Highly Desirable	
Coordination—Description of Relative Allocation of Duties and Responsibilities		Highly Desirable	
Compensation	Amount	Essential	
Terms of Payment	Payment application processes.	Highly Desirable	
	Pay-when-paid clause/Pay-if-paid clause.	Highly Desirable	In the event that the Owner does not pay Prime Contractor.

Additional Services	Sets forth the terms and conditions on which any out-of-scope or "additional services" may be authorized and the method of determining amount of compensation to be paid.		
Insurance	Sets forth the specific insurance types, coverages, and other parameters required of the subcontractor.	Highly Desirable	
Indemnification	Professional and contractual liability.	Highly Desirable	
	Intellectual property.	Highly Desirable	
	Mechanics' lien or claimant liability.	Highly Desirable	
	Damages incurred by Prime Contractor under the prime agreement due to subcontractor's acts or omissions The parties should also consider indemnity for exposure to hazardous materials, and violation of environmental laws.	Highly Desirable	This clause is especially important where the prime contract contains liquidated damages. Additionally, such provisions may allow the contractor to "pass through" the Owner's claim directly to the subcontractor, thereby incurring minimal expense in the event the Owner has a claim based upon the subcontractor's work.
Warranties	Warranties that run from the subcontractor to the Prime Contractor	Highly Desirable	It is important to make sure that the warranties in the subcontract and the prime contract are the same.
Dispute Resolution	Consider mediation and arbitration options.	Highly Desirable	Review dispute resolution procedures in both the prime contract and the subcontract to make sure they conform. Additionally, review and consider joinder issues.

Table 1 *continued*

Type/Subject Matter of Provision	Reason for, or Objective of, Provision(s)		Comments
Ownership and Use of Documents	Nonexclusive licenses.	Essential	In the event that the Owner terminates the Prime Contractor or the Prime Contractor terminates a subcontractor, consider whether the agreement allows an A/E to terminate its license, thereby possibly derailing the project, because the Owner or contractor no longer has authority to use the drawings.
	Indemnification for reuse.	Useful	
	Termination implications.		
Status of Subcontractor	Independent contractor, agency.		
Maintenance of Records		Optional	Usually, the records maintenance requirement will conform with the statutes of limitation and repose in the applicable jurisdiction.
Assignment/ Successor		Optional	Usually, the Prime Contractor will not allow assignment of the subcontract by the subcontractor. Otherwise, the Prime Contractor will not have the ability to effectively control subcontractors that it does not know or have a preexisting relationship with.
Authority to Enter into Agreement		Highly Desirable	

Integration Clause		Highly Desirable	
Counterpart Execution		Useful	
Third Party Beneficiary		Highly Desirable	The question to consider is whether the subcontractor is a third party beneficiary of the Owner/Prime Contractor contract (particularly the payment provisions) or the Owner is a third party beneficiary of the Prime Contractor/subcontractor contract.
Choice of Laws and Forum Selection		Useful	Usually laws of the state in which the project is located will apply.
Signatures		Essential	

VI. Subcontract Issues Unique to Offshore Subcontracting Relationships

Over the last twenty years, many architects, engineers, steel fabricators, and general contractors have used teams of engineers and technicians from other countries as subcontractors on their projects. Detailing and shop drawing for steel fabrication are just a few of the disciplines where personnel from Mexico, the Philippines, India, and other foreign locations continue to provide design or drafting services for projects in the United States.

The principal reason for the creation of these offshore relationships is usually the availability of less expensive services in these foreign locations or a lack of skilled personnel. These foreign personnel often do not hold any United States licenses or certifications. Their firms frequently do not possess workers' compensation, commercial general liability, or professional liability insurance. The entity providing the services is typically a lower-tier subcontractor and may not possess the background, experience, and training of comparable domestic personnel.

Before a Prime Contractor engages an offshore subcontractor, it should make sure that there are no formal or informal limitations in the contract that preclude such agreements or relationships. Frequently, governmental, public, or some large private projects limit or prohibit the use of such offshore subcontractors. Accordingly, at the outset, a determination must be made with respect to whether offshore subcontracting is allowed or limited.

If there are no limitations or prohibitions on the use of offshore agreements or personnel, when offshore contracting or subcontracting will be used, the following issues listed in Table 2 as well as the factors listed in Table 1, should be discussed and properly addressed.

Table 2
SUPPLEMENTAL CHECK LIST FOR OFFSHORE DESIGN PROFESSIONAL/CONSTRUCTION MANAGER SUBCONTRACTS

Type/Subject Matter of Provision	Reason for, or Objective of, Provision(s)		Comments
Form of Agreement	Will a written agreement be used? Does the written agreement have to comply with any provisions of the Prime Contract? Who will prepare it and in what language will it be prepared?	Essential	The costs of preparing such an agreement may erode any savings that were contemplated.
Communications	How will communications be handled? i.e., fax, e-mail, overnight delivery, Internet. What language will be used for the work product and for day-to-day communications?	Highly Desirable	How will questions and issues associated with understanding the scope of work be addressed?
Delivery Methods	Are there any limitations on the way the offshore subcontractor will obtain its labor or personnel?	Optional	
Payment	When, and in what currency, will payment for services be handled?	Essential	
Insurance	Will the offshore subcontractor carry any insurance? If so, are any requirements with respect to the size, "best rating," or location of the insurance carrier? How will claims, if any, be handled or adjusted?	Essential	Under the concept of vicarious liability, if the subcontractor does not carry insurance or it is not accessible for claims purposes, the next higher-tier subcontractor and the Prime Contractor are effectively providing insurance for it.

Disputes	Jurisdiction: who has it for purposes of asserting claims? What form of dispute resolution procedure will be used? Where will disputes be heard?	Highly Desirable	Frequently, arbitration is the only workable dispute resolution procedure.
Deliverables	What are the "deliverables" and where are they to be delivered?		

The use of offshore arrangements or subcontracts may initially appear to be more economical than they are in practice. As a result, their use should be very carefully considered.

VII. Conclusion

Every subcontract between a Prime Contractor and its subcontractor must be negotiated based upon the terms and conditions of the prime contract with the owner and include all of the terms and conditions of the subcontractor's service. The pertinent parts of the prime contract should be incorporated by reference into the subcontract, with a redacted copy of the prime contract attached to it. The subcontract and any amendments thereto must always be reduced to writing. Finally, it must be signed by the parties to it and copies must be distributed accordingly.

VIII. Practice Aids

A. *Standard Construction Manager and Subcontractor Contract Forms Available*

The most widely used standard forms are as follows:

- American Institute of Architects (AIA) Contract Documents
- Associated General Contractors of America (AGC) Contract Documents
- Engineers Joint Contract Document Committee (EJCDC) Contract Documents
- Design-Build Institute of America (DBIA) Contract Documents

American Institute of Architects
AIA Subcontract/Subconsultant Form Documents:

A401 Standard Form of Agreement Between Contractor and Subcontractor with Instructions (1997)

C105 Standard Form of Agreement Between Architect and Consulting Architect (2005)

C141 Standard Form of Agreement Between Architect and Consultant with Instructions (1997)
C142 Abbreviated Standard Form of Agreement Between Architect and Consultant with Instructions (1997)
C727 Standard Form of Agreement Between Architect and Consultant for Special Services with Instructions (1992)

Associated General Contractors of America
AGC Form Documents Regarding Subcontractors:

AGC 601 2004 Subcontract for Use on Federal Construction

AGC 602 2004 Standard Form Purchase Order

AGC 603 2000 Standard Short Form Agreement Between Contractor and Subcontractor (Where Contractor Assumes Risk of Owner Payment)

AGC 604 2000 Standard Short Form of Agreement Between Contractor and Subcontractor (Where Contractor and Subcontractor Share Risk of Owner Payment)

AGC 605 2004 Invitation to Bid/Sub-bid Proposal

AGC 606 2004 Subcontract Performance Bond

AGC 607 2004 Subcontract Payment Bond

AGC 610 1988 Subcontractor's Application for Payment

AGC 621 2004 Subcontractor Statement of Qualifications for a Specific Project

AGC 650 1998 Standard Form of Agreement Between Contractor and Subcontractor (Where the Contractor Assumes the Risk of Owner Payment)

AGC 650.1/655.1 2003 Standard Form Rider Between Contractor and Subcontractor for Storage of Materials at Subcontractor's Yard

AGC 655 1998 Standard Form of Agreement Between Contractor and Subcontractor (Where the Contractor and Subcontractor Share the Risk of Owner Payment)

The Engineers Joint Contract Documents Committee

Agreement Between Engineer and Geotechnical Engineer for Professional Services (Doc. No. 1910-27-B)

Amendment to Engineer-Consultant Agreement (Doc. No. E-571)

Standard Form of Agreement Between Engineer and Architect for Professional Services (Doc. No. 1910-10)

Standard Form of Agreement Between Engineer and Consultant for Professional Services (Doc. No. 1910-14),

Design-Build Institute of America

DBIA Document No. 501—Contract for Design-Build Consultant Services

DBIA Document No. 510—Design-Build Contracting Guide

DBIA Document No. 520—Standard Form of Preliminary Agreement Between Owner and Design-Builder

DBIA Document No. 525—Standard Form of Agreement Between Owner and Design-Builder—Lump Sum

DBIA Document No. 530—Standard Form of Agreement Between Owner and Design-Builder—Cost Plus Fee with an Option for a Guaranteed Maximum Price

DBIA Document No. 535—Standard Form of General Conditions of Contract Between Owner and Design-Builder

DBIA Document No. 540—Standard Form of Agreement Between Design-Builder and Designer

DBIA Document No. 550—Standard Form of Agreement Between Design-Builder and General Contractor—Cost Plus Fee with an Option for a Guaranteed Maximum Price

DBIA Document No. 555—Standard Form of Agreement Between Design-Builder and General Contractor—Lump Sum

DBIA Document No. 560—Standard Form of Agreement Between Design-Builder and Design-Build Subcontractor—Guaranteed Maximum Price

DBIA Document No. 565—Standard Form of Agreement Between Design-Builder and Design-Build Subcontractor—Lump Sum

IX. Appendix

A. Sample Subcontract Agreement
Consultant's Agreement to Furnish Services to
PRIME CONTRACTOR

PRIME CONTRACTOR Project No.

FROM:	TO:
Prime Contractor, A Professional Corporation ("**Consultant**") ("**PRIME CONTRACTOR**")	(Company or Firm name)
_____ _____	(Address) (City, State, Zip)
Telephone: _____	Telephone: _____
Facsimile: _____	Facsimile: _____
Attention: (Representative's Name)	Attention: (Representative's Name)
Title:	Title:
Direct Telephone:	Direct Telephone:
E-mail:	E-mail:

PROJECT DESCRIPTION AND LOCATION ("PROJECT"): *[Insert Project description and address]*

1. Consultant has fully acquainted itself with the available information relative to the Project and the services, which it will provide for the Project. Consultant has reviewed the Contract between PRIME CONTRACTOR and the Owner for the Project, which is attached hereto as Exhibit A. Consultant agrees that it will be bound by all of the pertinent provisions of Exhibit A (including any subsequent revisions, amendments, or modifications thereto), insofar as they apply to the scope, quality, character, processes, procedures (including without limitation those relating to procurement, affirmative action, nondiscrimination, and compliance with applicable codes, rules, regulations, statutes, laws, and ordinances) and manner of the services to be performed by the Consultant. In the event of any conflict or inconsistency between this Agreement and the requirements of Exhibit A regarding the services to be performed by the Consultant, the requirements of Exhibit A shall take precedent; otherwise this Agreement shall take precedence over all other documents and Exhibits hereto, and shall govern the relationship between PRIME CONTRACTOR and Consultant for this Project.

2. Consultant shall provide for PRIME CONTRACTOR, all *[describe the type of services, e.g., "food service, acoustical or vertical transportation"]* services, which PRIME CONTRACTOR is required to provide for the Project under the terms of Exhibit A *[or, as more specifically described in Exhibit B hereto]*.

   Consultant agrees to provide all of its services in a timely, competent, and professional manner, in accordance with applicable standards of care, for projects of similar quality and scope.

   Consultant shall provide all of its Deliverables to PRIME CONTRACTOR, in the form and format, by such means, i.e., by hand, telephonically, high speed data line, electronically or otherwise, and on such media, electronic, hardcopy or otherwise, as PRIME CONTRACTOR may specify. Additionally, PRIME CONTRACTOR may designate the software types, brands, formats, and versions, which will be used by Consultant for all computer and electronic media, telecommunications, and other transactions on the Project, e.g., AutoCAD © Release 2000 for drawings and MS Word © for text, etc. Consultant agrees that Prime Contractor is hereby licensed and authorized to copy, use and release the Consultant's Work Product (in all media and formats—electronic, hardcopy, telephonic or otherwise) developed or issued for this Project, including but not limited to all drawings and specifications in connection with Prime Contractor's services on this Project. Consultant further agrees that Prime Contractor may set reasonable terms and conditions on the licensing, use, and release of its Work Product, including but not limited to CAD releases for "backgrounds," shop drawings, Project Web site development and publication, or copyright licensing agreement, provided that all such licensing, copying and other uses are related to this Project.

SCHEDULE FOR CONSULTANT'S SERVICES:

3. _____

Requested Start Date: _____   Requested Completion Date: _____

COMPENSATION:

4. Compensation for the services shall be at the Consultant's standard billing rates in accordance with the attached *Exhibit ___, Hourly Billing Rates*. The maximum amount billable under this Agreement shall not exceed _____ _____ Dollars ($ _____).

OR — IF FIXED FEE AGREEMENT:
Compensation for Consultant's services shall be a fixed fee of _____ _____ Dollars ($ _____).

TERMS AND CONDITIONS OF SERVICE:

5. REIMBURSABLE EXPENSE: Reimbursable expenses shall be limited to those allowed in Exhibit A. Reimbursable expenses shall be billed at 1.0 times direct cost.

OR — IF EXPENSES ARE INCLUDED IN BASIC COMPENSATION, USE:

5. **REIMBURSABLE EXPENSE: Consultant's reimbursable expenses for this Project are included in the fixed fee set forth in Paragraph 4 above.**

6. Should PRIME CONTRACTOR request work not described and included in the above *Description of Services,* Consultant shall invoice PRIME CONTRACTOR for such services at an hourly rate in accordance with *Exhibit ___, Hourly Billing Rates.* No work shall commence without PRIME CONTRACTOR approval in writing.

7. a. Consultant's invoices with all appropriate supporting documentation shall be submitted monthly based on the amount of work complete (including other costs defined as Reimbursable Expenses). All invoices must be received by the 25th of the month in order to be processed with the next month's billing to Owner. Consultant shall submit invoices for this Project separately from all other projects. Consultant shall include PRIME CONTRACTOR's Project name, Project number, and Project manager's name on all invoices and shall direct all invoices to "Attention: Accounting." Upon the request of PRIME CONTRACTOR, Consultant shall submit its invoices and supporting documentation in a billing and/or invoice format established or approved by PRIME CONTRACTOR.

   b. Notwithstanding the percentage of completion of the Consultant's services, it is understood and agreed that Consultant's services shall be billed to Owner on the basis of the percentage of completion of PRIME CONTRACTOR's services for each Phase or the percentage of completion of the Consultant's services, whichever is the lesser, until all of the compensation due to Consultant has been billed to Owner. Consultant recognizes and agrees that Consultant's invoices will be presented by PRIME CONTRACTOR to the Owner on a regular basis and PRIME CONTRACTOR shall not be obligated to pay Consultant until PRIME CONTRACTOR has received payment for such services from Owner. PRIME CONTRACTOR will pay the Consultant the amounts due for services rendered and expenses incurred pursuant to the terms of this Agreement within thirty (30) days after PRIME CONTRACTOR receives payment for such services and expenses from Owner.

8. Consultant agrees to indemnify and hold PRIME CONTRACTOR harmless from any actions, causes of actions, proceedings, costs, losses, claims, demands, or expenses (including attorneys' fees, costs, and expenses of litigation) incurred by the Consultant as a result of any negligent errors, acts or omissions or breach of contract of Consultant, other consultants or subcontractors employed or retained by Consultant for the Project, *provided, however*, that if the scope of PRIME CONTRACTOR's indemnity to Owner under the terms of Exhibit A is broader than this Paragraph 8, the scope of Consultant's indemnity to PRIME CONTRACTOR with respect to Consultant's services shall be expanded to provide the same protection to PRIME CONTRACTOR which PRIME CONTRACTOR is obligated to provide to the Owner under the terms of Exhibit A.

9. PRIME CONTRACTOR and Consultant agree that from time to time, there may be conflicts, disputes, and/or disagreements between them, arising out of or relating the services of Consultant, the Project, or this Agreement (hereinafter collectively referred to as "Disputes"). It is the mutual desire, intent, and commitment of PRIME CONTRACTOR and Consultant that a sincere and good faith effort be made to resolve any and all such Disputes, promptly, in a mutually satisfactory manner, and without the need for incurring the time, expense, and delay occasioned by resort to litigation. Therefore, PRIME CONTRACTOR and Consultant agree that all Disputes that are not resolved as a result of the exchange of information and meetings shall be submitted, as a condition precedent to litigation or arbitration, to nonbinding mediation, in accordance with the Commercial Mediation Rules of the American Arbitration Association (AAA). The selection of the mediator, as well as, the date, time, and location of the mediation, shall be set by the mutual agreement of the parties or failing mutual agreement within a reasonable time, the selection of the mediator, as well as the date, time, and location of the mediation shall made by the AAA.

10. a. All Disputes that (i) are not resolved as contemplated in Paragraph 9 above shall be resolved by submission to binding arbitration in accordance with the Construction Industry Arbitration Rules of the AAA. No joinder in such arbitration shall be permitted of any other party without the prior written consent of Consultant, PRIME CONTRACTOR, and such party.

    b. This agreement to arbitrate shall be specifically enforceable under all applicable prevailing arbitration law. The award rendered by the arbitrator(s) shall be final and judgment may be entered upon it in accordance with applicable law in any court having jurisdiction thereof. Notice of the demand for arbitration shall be filed in writing with all parties to be joined in such arbitration and the AAA. The demand for arbitration shall be made within a reasonable time after the parties' failure to resolve the Disputes as contemplated in Paragraph 9 hereof.

11. a. Consultant shall purchase and maintain in a company or companies authorized to do business in the State of _____ _____ such insurance as will protect Consultant or anyone for whose acts it may be liable from claims set forth below which may arise out of or result from the performance of its obligations under this Agreement:

   i. Worker's Compensation: Statutory Limits; Employer's Liability: Limits—$100,000 Bodily Injury each accident, $100,000 Bodily Injury by Disease each employee, and $100,000 Policy Limit for disease. The worker's compensation insurance shall comply with the requirements of the statutes of the state in which the services are being performed and shall include an all-states endorsement and voluntary compensation.

   ii. Commercial general liability insurance, including coverage for bodily injury, property damage, personal injury (employee and contractual liability exclusions deleted), broad form property damage with limits of liability for each occurrence and in the aggregate, combined single limit for bodily injury, property damage, personal injury, and property damage for completed operations of not less than $_____, combined single limit and $_____, general aggregate.

   iii. Comprehensive Automobile Public Liability covering all owned, hired, or non-owned vehicles. Combined Single Limit—$1,000,000 per occurrence.

   iv. Limits of the Liability Insurance specified in Subparagraphs 11. a. ii. and iii. above, may be provided by any combination of primary insurance policies and excess liability insurance policies.

   v. Professional liability insurance policy in an amount not less than ___ _____ each claim; and _____ annual aggregate, and the policy shall contain contractual liability coverage in a form reasonably satisfactory to PRIME CONTRACTOR.

   b. Consultant agrees to maintain policies of insurance as required herein for as long as this Agreement is in force and effect, and as to the professional liability policy, if any, for a period of six (6) year(s) following the Date of Substantial Completion of the Project. Consultant's obligation to maintain the professional liability policy for six (6) years following the Date of Substantial Completion is subject to the general availability of such professional liability insurance policy in the marketplace, with no commercially unreasonable increase in premium therefor.

 c. With respect to all of the insurance required by this Agreement, Consultant shall deliver to PRIME CONTRACTOR, certificates of insurance, in a form and substance reasonably acceptable to PRIME CONTRACTOR, within ten (10) days after execution of this Agreement. With respect to the insurance coverages that are required to remain in force after the Date of Substantial Completion, a certificate of insurance evidencing continuation of such coverage shall be submitted along with the Consultant's invoice for its last payment under the terms of this Agreement, and at least annually thereafter where applicable.

The certificates of insurance, as well as insurance policies required by this Agreement, shall contain a provision that coverage will not be materially changed, materially altered, cancelled, or allowed to expire until at least thirty (30) days' prior written notice has been given to PRIME CONTRACTOR.

12. This Agreement shall constitute the entire Agreement between the parties hereto and shall supersede all prior contracts, proposals, representations, negotiations, and Letters of Intent, whether written or oral, pertaining to Services for the Project.

13. There are no third party beneficiaries to this Agreement.

14. Consultant may not assign this Agreement, nor may any of Consultant's duties hereunder be delegated, without the prior written consent of PRIME CONTRACTOR.

15. In the event of any litigation or arbitration between Prime Contractor and Consultant, arising out of or relating to the Project, Consultant's Services or this Agreement, the court or arbitrator shall awards attorneys' fees, costs, and expenses to the prevailing party in such litigation or arbitration.

16. The provision of services by Consultant and the interpretation of this Agreement shall be governed by the laws of the State of _____.

IN WITNESS WHEREOF, the Parties have caused this Agreement to be executed on the date first appearing above.

PRIME CONTRACTOR: **CONSULTANT:**

By _____   By _____

Title _____   Title _____

CHAPTER 16

Damages and Remedies

STEPHEN MILLER, ESQ.
STEPHEN A. HESS, ESQ.

I. Introduction

Previous chapters in this handbook have detailed the various tort and contract rights and obligations that design professionals and construction managers accept in the performance of their work. This chapter outlines the broad spectrum of remedies available to design professionals and construction managers when owners breach their contracts, as well as the remedies available against design professionals and construction managers who find themselves in breach of their own obligations to owners. Additionally, the chapter reviews a number of mechanisms by which parties attempt to limit the damages for which they might be responsible through their contracts.

Before a party can recover damages, it must establish its entitlement to relief by proving a claim against the tortfeasor or the party in breach of the contract. In legal parlance, proving a claim requires establishment of a discrete set of elements, and in the absence of proof of any element of a claim, recovery must be denied. Therefore, the chapter begins with a brief review of what a claimant must establish in order to recover on the types of claims most commonly brought in connection with construction disputes.

Two notes regarding the limitations of this chapter are in order. First, the function of this chapter is to provide an overview of remedies and not an exhaustive discussion of all the nuances of the different aspects of damages that might be recovered. Indeed, even full-length treatises cannot contain exhaustive discussions of all the aspects of construction damages. Second, the principles discussed below are general principles that are common to most jurisdictions. This chapter does not attempt to catalog the differences between the laws of all fifty states and the federal government. Identification of the laws of a particular state requires the services of a construction lawyer well-versed in the laws of a specific jurisdiction.

That much having been said, there is enough in common with respect to the general principles governing remedies and damages to make compact overview possible.

II. Causes of Action

When a party seeks to recover for a loss, it files suit (or an arbitration demand) seeking discovery for particular acts of malfeasance by the opposing party. There are well-established elements of recovery for different legal theories on which parties to a construction project may rely, and this section details (and comments briefly on) the legal theories most often encountered in construction litigation. When assessing the viability of a potential claim, the first task is to determine what elements a claimant must prove in order to recover, and then to measure the known facts against those elements.

A. *Breach of Contract and Related Claims*

The bulk of services performed by construction managers are governed by written or oral contracts, and relief for a breach takes the form of a claim for breach of contract. However, in some cases, the parties' negotiations may never reach the level of formation of a contract, or the contract may be unenforceable for a variety of reasons. The absence or unenforceability of a contract does not necessarily preclude recovery. However, the law may still provide a remedy under two different theories. First, under the theory of promissory estoppel, a party may be protected to the extent that it reasonably relies on another's promise, even if no contract is formed. Second, a party who provides services without a formal contract may be able to recover under a theory of unjust enrichment to preclude the other party from retaining the benefits of those services without some form of payment.

Accordingly, when an attorney pleads claims related to the provision of services on a construction contract, the complaint often includes alternative claims for breach of contract, promissory estoppel, and unjust enrichment. Although related, these claims have different elements and a party may recover on one where the others are unavailing.

1. *Breach of Contract*

The most common claim in construction litigation is for breach of contract. A breach of contract is nothing more complicated than a party's failure to perform express or implied obligations arising under its agreement with the opposing party. The scope and nature of such express and implied obligations for design professionals and construction managers are set forth in detail in prior chapters.

In order to recover for breach of contract, the claimant is ordinarily required to prove, by a preponderance of the evidence,[1] each of the following elements of the claim for breach of contract:

  a. the existence of an enforceable contract;
  b. breach of the agreement by the defending party;

---

[1] The general standard for all civil actions is *preponderance of the evidence*, which simply means a claimant must establish that it is more likely than not entitled to recover. This may be contrasted with the criminal standard under which a defendant may not be convicted unless his or her guilt is established *beyond a reasonable doubt*.

c. the claimant's performance of its own obligations (or a lawful excuse for its not having performed); and
d. damages proximately caused by the breach.[2]

Once established, the prevailing party is entitled to compensatory and other damages, the general calculation of which is spelled out below.

As may be gleaned from the chapters setting out the numerous express and implied obligations of design professionals and construction managers, there are many different types of claims that fall under the broad category of breach of contract claims. For example, a breach of warranty claim is a breach of contract claim at heart, where the warranty is simply that portion of the contract at issue. Breaches may be as simple as failing to pay an amount of money when due, or as complicated as providing plans that are not constructible in some material respect (thereby, breaching the implied warranty of suitability for construction).

The third element of a breach of contract claim embodies the doctrine of first material breach, and encompasses the principle that a party who is in material breach of a contract is not entitled to the benefit of the other party's return performance. Thus, an owner who has failed to pay a substantial sum due a design professional or construction manager can no longer insist on the return benefit of the contract, and the design professional or construction manager is excused from further performance (and may be entitled to terminate) as long as the owner's material breach remains uncured.

2. *Promissory Estoppel*

Many parties rely on the promises of another party and begin the performance of services or fashion their business relationships with other parties even before those promises are formalized into binding contracts. In many circumstances, the party relying on such a promise is entitled to protection notwithstanding the fact that a binding contract is never perfected. The doctrine of promissory estoppel may be summarized as follows:

Promise Reasonably Inducing Action or Forbearance
(1) A promise which the promisor should reasonably expect to induce action or forbearance on the part of the promisee or a third person and which does induce such action or forbearance is binding if injustice can be avoided only by enforcement of the promise. The remedy granted for breach may be limited as justice requires.[3]

---

2. *See, e.g.*, EUA Cogenex Corp. v. North Rockland Central Sch. Dist., 124 F. Supp. 2d 861 (S.D.N.Y. 2000).

3. RESTATEMENT (SECOND) OF CONTRACTS § 90(1) (1981).

In order to prevail on a claim for promissory estoppel, therefore, a claimant must establish that:

a. the opposing party made a promise intending to induce action or forbearance on the part of the promise;
b. the promisee reasonably relied on the promise to his or her detriment; and
c. injustice can be avoided only through enforcement of the promise.

It is important to note that the doctrine does not create a contract where none existed before, but it has much the same effect: namely, the promisor is prevented (estopped) from denying that he should answer for his failure to perform. Promissory estoppel differs somewhat from breach of contract in that the measure of damages available for a claim of promissory estoppel may be very different from the damages available for breach of a contract to perform the same promise.

By way of illustration, the doctrine of promissory estoppel is sometimes invoked to recover for professional services provided without an enforceable contract. Additionally, promissory estoppel serves as the legal basis under which some contractors attempt to recover from subcontractors who refuse to perform after the contractors win contracts after relying on the subcontractors' bids.

3. Unjust Enrichment

In some circumstances, a party may not be able to recover damages for breach of contract or for promissory estoppel. In such cases, a legal theory on which a party may attempt to seek recovery is unjust enrichment. As its name implies, the theory is designed to ensure that a party is not permitted to retain the benefit of another person's provision of services in circumstances in which it would be unjust to allow the uncompensated retention.

In order to prevail on an unjust enrichment claim, the party seeking relief must generally prove that it:

a. conferred a benefit on another party,
b. without intending to act gratuitously, and
c. the benefit was conferred in circumstances in which it would be unjust to allow the recipient of the benefit to retain the benefit without compensation to the party providing the service.

The most difficult element to apply is the third: what may be unjust is left largely to the conscience of the judge, jury, or arbitrator.

B. *Tort Claims*

Prior chapters set out the various obligations in tort that design professionals and construction managers assume in performing services. As noted therein, tort obligations are obligations imposed by the law apart from any contractual obligations the parties may assume expressly or implicitly. Because they are imposed by the law without regard to the parties' contractual obligations, these duties generally do not require express proof of the existence of a contract as an element of recovery.[4] Instead, a prima facie case for recovery under a tort theory requires that the victim establish:

a. the defendant owed a duty of care to the claimant;
b. the defendant breached that duty of care;
c. the victim suffered damages; and
d. the damages were proximately caused by the defendant's breach of duty.

The question of what duties are owed by a design professional or construction manager to their clients and third parties is addressed in more detail in several prior chapters. For purposes of this discussion, it is sufficient to note the general rule that "one who undertakes to render professional services is under a duty to the person for whom the service is to be performed to exercise such care, skill, and diligence as men in that profession ordinarily exercise under like circumstances."[5]

The failure to exercise reasonable care is typically branded as negligence, and may take many forms. As but two examples, contractors may maintain actions against design professionals to recover their losses under a variety of theories, including negligent supervision or negligent preparation of plans and specifications.[6] The last element, that of establishing that the damages claimed are *proximately caused* by the asserted negligence, is a judicially imposed limitation that precludes parties from recovering unpredictable damages. More precisely, proximate cause requires not only that the damages were *actually* caused by the negligence, but also that the damages are a natural and probable consequence of the negligence. For example, some jurisdictions have held that where it is foreseeable that a design professional's actions or inactions may create a risk of harm to a third-party contractor whose responsibility it is to implement the design work of the design professional, a cause of action lies against the design professional, even where no contract exists between the parties.[7]

---

4. In fact, under the economic loss rule, the existence of a contract may preclude recovery.
5. City of Eveleth v. Ruble, 225 N.W.2d 521, 524 (Minn. 1974).
6. Frank D. Wagner, Annotation, *Tort Liability of Project Architect for Economic Damages Suffered by Contractor*, 65 A.L.R.3D 249, § 2[a] (1975); *see also* Matrix v. Malow, 2006 WL 399762 (Mich. Ct. App. 2006) (unpublished opinion); Colonial Sur. Co. v. Lenard Eng'g, Inc., 2004 WL 2943640 (Sup. Ct. Conn. 2004) (unpublished opinion).
7. *Id.*

In outlining the elements of a claim based on the summary above, additional elements may be required for a particular claim as described further in the chapters discussing torts. Thus, for example, when a claim for negligent misrepresentation is brought against a design professional or construction manager, the claimant might not only be required to show a breach of the duty to represent information accurately, but also that the claimant reasonably relied on the misrepresentation at issue.[8]

Library References

C.J.S. *Architects* §§ 33–44; C.J.S. *Conflict of Laws* §§ 35–36, 39, 43–46, 87, 98; C.J.S. *Contracts* §§ 13, 15–18, 24, 26, 509–512, 600–612, 626, 630, 633–697, 701, 717–719, 726–736, 753, 756–760, 770–771, 777–778, 782–784, 795–804; C.J.S. *Estoppel and Waiver* § 90; C.J.S. *Fraud* §§ 10–17, 23–25, 29–30, 34–35, 59–60, 62–63, 67–75, 77–78, 80; C.J.S. *Implied and Constructive Contracts* § 4; C.J.S. *Negligence* §§ 163–164, 399, 595–596, 598–601, 618–619.

West's Key No. Digests, Contracts <KEY>324–355; Estoppel <KEY>85; Fraud <KEY>8; Implied and Constructive Contracts <KEY>3; Negligence <KEY>322, 1205(4)–1205(10).

III. General Measures of Damages

A measure of damages is simply a statement of how a court will calculate the amount of money a party may recover upon establishing a claim against the defendant. Although there are some differences in particular items of recovery among jurisdictions, the general rules regarding recovery are relatively consistent among jurisdictions.

The measures of damage are different for the various claims outlined above. However, there are some general principles common to all damage claims. First, the claimant bears the burden of proving the existence of damages with a reasonable degree of certainty. This requirement is established not so much by proving the exact amount of damages, but rather by establishing to the satisfaction of the fact-finder (judge, jury, or arbitrator) that the claimant has actually suffered some damage. Once the fact of damages has been established, courts are less exacting when it comes to quantifying damages. Courts do not require precise quantification, but instead only require a reasonable basis for calculating the damages claimed. Thus, a reasonable basis excludes damages that are merely speculative, as damages may not be recovered where their quantification is left to mere speculation without some basis in the record to support any determination of the amount of such damages.

A. *Breach of Contract and Related Claims*

1. *Breach of Contract*

Generally, "a party is entitled to have what he contracts for or its equivalent,"[9] which is another way of saying that a party is entitled to the "benefit of his bargain."

---

8. Presnell Constr. Managers, Inc. v. EH Constr., LLC, 134 S.W.3d 575 (Ky. 2004).
9. Charlotte v. Skidmore, Owings and Merrill, 407 S.E.2d 571, 580 (N.C. Ct. App. 1991).

When a breach by the opposing party deprives the claimant's benefit of that bargain, a court will endeavor to protect that benefit; in other terms, the goal in most breach of contract matters "is to award a sum that will put the non-breaching party in as good a position as he would have been had the contract been performed."[10] Generally, "rights of the parties with respect to a breach of contract are fixed at the time of breach and ... damages are measured as of that time."[11]

a. *Design Professional and Construction Manager Compensation*

A design professional's or construction manager's bargain is typically the compensation to which the design professional or construction manager is entitled under the terms of the contract with the owner. When the owner fails to pay the design professional or construction manager, the measure of damages is typically the difference between the contract compensation and the compensation actually paid. Of course, this measure is appropriate when the owner simply fails to pay. More often, disputes arise because the design professional or construction manager performs services that are arguably outside the scope of the original contract, and the measure of damages then becomes that additional compensation to which the design professional or construction manager is entitled for the extra work. Where the contract provides specified rates for such compensation, the damage calculation is readily performed. Where the contract does not specify how compensation is to be determined, courts will typically award *reasonable* compensation. What is reasonable is simply what a court (or arbitrator) will decide is reasonable after hearing evidence regarding the additional services and the value thereof. In some cases, the design professional or construction manager may also be entitled to recover interest or attorneys' fees permitted in the circumstances discussed below.

b. *Owner's Compensation*

The owner of a building or project typically contracts to get three elements related to its project. First, it wants the contract to be completed for the contract price. Second, it wants the contract to be completed within the contract time. Third, it wants the project to be of such quality as it has contracted for through the contract documents.

When the project is finished at a greater cost than promised, later than promised, or of lesser quality than promised, the owner may be entitled to recover. The general measure of damages is the same as that recited above: the owner is entitled to recover that amount of money that will put the owner in the same position in which it would have found itself but for the breach. In determining how these damages might be calculated, it is helpful to address the three separate dimensions of performance that the owner expects.

---

10. Hernandez v. Westoak Realty & Inv., Inc., 771 S.W.2d 876, 880 (Mo. Ct. App. 1989).
11. Fairway Builders, Inc. v. Malouf Towers Rental Co., 603 P.2d 513 (Ariz. Ct. App. 1979).

First, the owner is entitled to get the project completed for a specific price. If the contract is for a stipulated sum, this element is easier to measure. The owner is entitled to recover additional expenses when, for example, it must complete work that the contractor leaves unfinished or repair defective work. Similarly, when the contractor's breach forces the owner to spend more on design professional or construction manager fees, the owner is entitled to recover those additional fees so as to protect the benefit of its bargain. Of course, the owner is still required to pay the full contract price. If the owner has paid $9,500,000 of a $10,000,000 contract and then spends another $200,000 completing or repairing work, it is still required to pay the balance of $300,000 to the original contractor. It may not retain the difference as a penalty. Many construction contracts specifically detail the owner's obligation to pay the contractor the balance of the contract price once any damage for completing or repairing work is retained from the contract price.

Second, the owner is entitled to have the project completed on time. When a project is late, the owner may suffer such damages as extended design professional or construction management fees, lost profits or income from the delay in putting the project into service, and additional financing costs associated with the delay. Each of these may be recovered as part of the effort to put the owner in the same position it would have found itself but for the breach.[12]

Third, the owner is entitled to receive a project of the quality promised in the contract documents. When the construction is substandard, two ramifications may arise: the construction may need to be repaired or improved to the quality promised, and the value of the project may be diminished as a consequence of the substandard construction. When construction is substandard, the most common measure of damages to give the owner the benefit of its bargain is the cost of repair.[13]

However, where reasonable repairs may leave the project in a condition in which the fair market value is still diminished below that of the promised construction, the court may award an amount of money representing the diminution in value of the property after repair together with the costs of repair. For example, in *Northern Petrochemical Co. v. Thorsen & Thorshov, Inc.*,[14] an owner brought an action against its architect, structural engineer, and general contractor for the faulty construction and subsequent reconstruction of an industrial building. The Supreme Court of Minnesota held that "[t]he preferred measure of damages in a case such as this is to take either the cost of reconstruction in accordance with the contract, if this is possible without unreasonable economic waste, or the difference in the value of the building as contracted for and the value as actually built, if reconstruction would constitute unreasonable waste."[15]

---

12. As noted above, damages may not be speculative, and thus in some cases lost profits are not available because any award of lost profits would require too much by way of conjecture.

13. County Asphalt Paving Co. v. The 1861 Group, Ltd., 908 S.W.2d 184, 186 (Mo. Ct. App. 1995).

14. Northern Petrochemical Co. v. Thorsen & Thorshov, Inc., 211 N.W.2d 159, 165 (Minn. 1973).

15. *Id.*

c. *Contractor's Compensation*

From a general perspective, the contractor's ultimate goal is to protect its profit interest in the contract. When the contract as bid results in a loss to the contractor, of course, the contractor wants to limit its loss. Such a simplistic statement belies the complexity of determining damages from the contractor's perspective for breach by the owner.[16] In more complex cases, it is helpful to characterize the contractor's general expectation interest (that is, the *benefit of the bargain*) as the right to complete the contractual scope of work in accordance with plans and specifications, within the time permitted under the contract, and without interference from the design professional or construction manager. Ultimately, the frustration of any of these expectations affects the contractor's bottom line in a manner that ultimately is measured by reference to the original contract price. However, discussing each of these dimensions is instructive.

First, the contractor is entitled to complete the scope of work within the original plans and specifications. To the extent that the scope of work is expanded by the addition of other tasks not in the original contract, the contractor is entitled to recover its cost and reasonable overhead and profit associated with the additional work. When work is simply added to the contract, the claim is relatively straightforward. More complicated analysis is required when the plans and specifications are deficient in some respect, and the resolution of conflicts or deficiencies arguably adds to the work that the contractor is required to perform. However, from a conceptual standpoint, a contractor is required to perform the work set out in the original contract for the contract price.[17]

Second, the contractor is entitled to complete the contract within the time limits set out in the contract, at least to the extent the contractor's original schedule is reasonable. When the contractor's work is delayed as a consequence of problems attributable to others, the contractor will likely lose money through additional expenditures related to the delay. In protecting the benefit of the contractor's bargain, the court will allow recovery of sufficient damages to compensate the contractor for those additional expenses related to delay in completion of the work.

Finally, the contractor is entitled to complete the project without unreasonable interference by the owner or those acting on behalf of the owner; conversely, the contractor is entitled to cooperation and assistance from the owner in some respects. Many interference claims are inextricably intertwined with delay claims, but they are analytically distinct. Thus, for example, a contractor is entitled to reasonable access

---

16. When the breach is simply a payment default, damages are easy to calculate. This section presumes that the breach arises from interference by the owner, delay, bad plans and specifications, or other causes resulting in more typical damages described below.

17. This presumes that the contract is a lump sum contract or a cost-plus with a guaranteed maximum price. When the contract price is a cost-plus term, the battle will less often be fought over scope of work and instead will focus on the extent to which the contractor's charges for work are reasonable.

to the property for construction, reasonable response from the owner (or design professional) concerning owner-driven construction choices, reasonable responses to requests for information and submittals, and freedom from actual interference through the owner's, design professional's, or construction manager's unreasonable meddling. For example, when a portion of the property is not available for construction when promised, the contractor may suffer additional mobilization and demobilization costs even though the unavailable portion of the construction is not on the critical path. Of course, if access to the property is necessary to perform critical path work, delay damages may arise as well. Similarly, a contractor may have claims for inefficiency (discussed below) that do not necessarily delay ultimate completion of the contract, but that increase the contractor's cost of performance.

Ultimately, all of these elements of damages address different ways in which the benefit of the contractor's bargain—to earn a profit on the project—can be frustrated. Although they make computation of damages difficult in some cases, they do not change the general measure of damage that the court will impose for the owner's breach—awarding the contractor that amount of money that will put the contractor in the same financial position in which it would have found itself but for the owner's breach.

A number of these individual elements are discussed in more detail below, as they have their own nuances and rules with which the construction professional should be familiar.

2. *Promissory Estoppel*

The measure of damage for promissory estoppel is somewhat amorphous, as it requires that the court award the promisee an amount of money to protect its reliance on the promise. This *reliance interest* is typically calculated as the amount of money that the promisee lost in reliance on the promise, and in many cases does not include any element of profit for the promisee. In some cases, the reliance interest is very much like the calculation of damages as though a contract had been performed. Thus, for example, if a contractor justifiably relies on a steel erector's bid of $250,000 in bidding for the construction of a warehouse, and the steel erector refuses to perform, thereby forcing the contractor to find a replacement at $300,000, the reliance interest is $50,000—the difference between the bid and the actual cost. This is precisely the measure that would be used for simple breach of the steel erector's contract had it not shown up on the job.[18] However, that is not always the case, and the measure that protects the promisee's reliance should be distinguished from the remedy for breach of contract.

3. *Unjust Enrichment*

The value of the benefit conferred is the "reasonable value" of what it would have cost the party to obtain it from another person, or the extent to which the party's

---

18. Some courts analyze bid defaults under contract principles, others under principles of promissory estoppel.

property has been increased or its other interest advanced.[19] Most courts refer to this remedy as "restitution," as it forces the benefited party to disgorge the value of the benefit, where the value is measured from the perspective of *the enriched party*. Accordingly, unjust enrichment is distinct from both breach of contract and promissory estoppel, which look to the damage to the promisee. Unjust enrichment instead looks to the value conferred on the recipient and is not concerned with adequate compensation of the claiming party as much as it is concerned with making sure the recipient does not benefit unfairly. Thus, for example, if a tenant contracts for the improvement of a building and skips out before paying the bill, the contractor may be able to recover from the owner of the building. However, the measure of damages will not look to the contract price of the improvements, but rather will look to the value of the improvements to the building owner if the contractor can establish entitlement to recovery.

B. *Tort Claims*

The general measure of compensatory damages for negligence claims is similar to the measure for contract claims: it is that amount of money that will put the injured person in the same position as the injured person would have enjoyed but for the negligence of the defendant. In practical application to construction cases where the alleged negligence is founded on breach of the duty to provide suitable plans and specifications or not to design a project negligently, this measure results in calculations similar or identical to those that determine breach of contract damages.

However, there are two key differences to the measure of damages anchored on negligence rather than on breach of contract. First, negligence claims very often avoid many contractual limitations or conditions on damages. Second, contract damages are limited to those reasonably foreseen by the parties at the time of their contract as arising from any potential breach. Tort damages, however, are limited by proximate cause, which looks to what a reasonable person would have foreseen as a probable consequence of the commission of the tort at the time it was committed.

Some other tort claims have different measure of damages. For example, the measure of damages for negligent misrepresentation is the difference in value between what the victim would have enjoyed had the facts been as represented and the value of what the victim enjoyed, based upon the actual facts. Although this does not always result in a number that is different from the general calculation, it is a different measure in principle.

The scope of this chapter does not allow a thorough treatment of all the different measures of damage for every conceivable tort related to construction work, but most claims are subsumed within the category of general negligence.

---

19. RESTATEMENT (SECOND) OF CONTRACTS § 371 (1981).

Library References

C.J.S. *Damages* §§ 105–128, 141, 144–153, 163–164; C.J.S. *Parent and Child* § 344.

West's Key No. Digests, Damages <KEY>95–102, 117–126; Implied and Constructive Contracts <KEY>110–112.

IV. Calculation and Proof of Compensatory Damages

A. *General Damages*

The proof of damages that a party is entitled to collect is not a difficult task conceptually; rather, as the saying goes, the "devil is in the details." For example, when the owner contracts for the construction of a hospital for $40,000,000 (inclusive of design and contract administration fees) to be completed in the second year after the contract is let, and the project actually takes four years with substantial elements having been completed by the owner due to the contractor's default at a total cost of $48,000,000, the owner's damages are $8,000,000, plus interest and attorneys' fees when available, plus lost profits or carrying charges (or both). Determining the actual costs is never so simple as this example intimates, however. Similarly, when a contractor agrees to build the hospital for $40,000,000 over two years (with a profit of $3,000,000 built into the budget) but, because of delays by the owner and engineer, it costs the contractor $48,000,000 out of pocket and takes four years, the contractor's damages are $11,000,000 ($48,000,000 in costs plus $3,000,000 in profit, on top of the $40,000,000 contract price), plus home office overhead and consequential damages when recoverable. Of course, merely providing the numbers by way of example skips over the most difficult portion: figuring out what the recoverable costs really are.

The general measures of damages are set out above. This section endeavors to summarize some of the most important principles concerning the calculation and proof of determination of damages in construction cases.

B. *Direct Costs*

1. *In General*

Direct costs are typically those costs that can be identified specifically with and charged in whole or in part to a particular project and a particular cause within the contract. Where the contractor maintains a detailed and sophisticated cost accounting system, it may be possible to extract direct costs related to particular causes from the cost records. For example, if the engineer changes the specification of pumps mid-project, the additional compensation may be as simple as determining the difference in cost between the pumps specified and the pumps actually utilized.

Many claims are not as straightforward. If the substitution of pumps comes at a time when the pumps are on the project's critical path and there is insufficient lead time to order new pumps without delaying the project, the contractor may be entitled to recover not only the additional costs of the pump, but the results of the delay to the work as well. That delay, in turn, may be as simple as a couple of days of idle labor and equipment. More probable is that the calculation of the cost must accommodate the fact that otherwise idle crews or equipment were utilized to accelerate work on different parts of, perhaps diminishing the total delay or at the very least reducing the cost of, other work.

In short, the most accurate calculation of damages will always be to trace the direct consequences of a breach by one party, attribute and document a direct cost to every such consequence, and then award as compensation the sum of those direct costs. On discrete problems in the construction process, a reasonable approximation of direct costs may be had. The accuracy and reliably of direct cost calculations diminish as the ramifications of a particular breach widen.[20]

2. *Price Increases for Labor and Material*

It is not unusual for delay or disruption to a project to result in the performance of work during a later period of time at a higher cost than originally anticipated. The escalation in cost may result from scheduled increases in material costs or wage increases in a collective bargaining agreement. Escalated costs may also occur from unknown or unanticipated events (e.g., a natural disaster that increases construction material costs).

A schedule analysis may be necessary to establish that the uncertainty of the length of delay prevented the contractor from obtaining firm material, labor, or subcontract orders and to establish when such orders would have been placed but for the delay.[21] When changes in prices can be documented (both as to binding commitments for prices at the time work should have been performed and at the time work was actually performed), the difference in such prices may be recovered as part of the calculation of damages.

3. *Idle Equipment and Related Costs*

Where the owner causes delay to a project and a contractor has committed its construction equipment to the project but has been unable to use it, the contractor may be entitled to compensation for costs incurred while the equipment is idle. Equipment costs and delay claims are oftentimes the most significant costs and there is often substantial dispute regarding the rate at which the equipment should be charged.

There are different mechanisms for valuing idle equipment damages. For contractor-owned equipment, the measure is the contractor's actual cost to own the equipment. Ownership costs include the capital cost of the equipment, depreciation, major repairs, storage, taxes, and other direct costs and can be difficult to

---

20. Similarly, proof of direct costs diminishes with incompleteness, inaccuracy, or simplicity of the cost accounting system utilized by the contractor.

21. MICHAEL DODD & J. DUNCAN FINDLAY, STATE-BY-STATE GUIDE TO CONSTRUCTION CONTRACTS AND CLAIMS § 1.05[c][10] (2005).

prove. However, many contracts incorporate equipment schedules for the purpose of establishing equipment costs for change orders, and these schedules can be equally applied to establish damages for idle equipment. When the contractor rents equipment, the cost attributable to idle equipment is easily established by showing an invoice from a third party and payment thereof. If the contractor's record keeping is not sufficiently detailed to determine actual costs, then the contractor may resort to industry-published rates that set forth the average cost of ownership or third-party rental rates.

4. *Home Office Overhead*

When a contractor maintains a home office for the management of work over several projects, it expects to recover the cost of that home office through its general overhead and profit on the various jobs it performs. When one job is delayed and precludes the contractor from accepting other work, one effect is that the contractor's home office overhead is spread over fewer jobs, and thereby diminishes the contractor's ability to recover that overhead (in turn, lowering the contractor's overall profitability).

Accordingly, home office overhead is recoverable as an element of delay damages when the owner is responsible for the extended performance of a contract.[22] However, a claimant must establish a prima facie case for recovery of home office overhead, as the general description of the circumstances of home office maintenance are not always accurate. Thus, a contractor must establish that:

a. performance of the contract was delayed due to inexcusable actions or inaction of the owner;
b. the contractor suffered actual damages as a result of the delay;
c. the nature of the delay made it impractical for the contractor to undertake the performance of other work; or
d. the nature of the delay made it impractical to reduce home office overhead costs.[23]

Once established, the claimant must then determine the proper measure of calculating home office overhead damages. As a result of delay to a project, the contractor does not realize direct billings in the same time period as anticipated. Therefore, the home office overhead is unabsorbed or under-absorbed. Unabsorbed overhead occurs when there is no extra work associated with the extended time against which the contractor can bill its overhead. Under-absorbed overhead typically occurs with disruption or partial delay where the contractor is still realizing some direct billings, but the amount of work has been severely reduced.

---

22. Altmayer v. Johnson, 79 F.3d 1129 (Fed. Cir. 1996).
23. Mech-Con Corp. v. West, 61 F.3d 883 (Fed. Cir. 1995) (setting out first three elements). Many cases omit the fourth element as an express requirement.

There are several methods for calculating unabsorbed home office overhead. The most frequently used is the *Eichleay* formula. This formula is designed to calculate the daily home office overhead sustained in order to continue day-to-day operations during an owner-caused delay in a project.[24] In order to determine the amount of unabsorbed home office overhead, the *Eichleay* calculation requires three steps. First, the court must determine the overhead on the delayed contract by multiplying the contractor's total overhead by some ratio between the contractor's billings on the delayed contract and the contractor's total billings for the period of the delayed contract.[25] Next, a daily overhead rate for the delayed contract must be determined by dividing the contract overhead by the total days of contract performance.[26] Finally, the total recoverable overhead must be calculated by multiplying the daily overhead rate for the delayed contract by the total length in days of the delay.[27] Although the *Eichleay* formula is the preeminent method of calculating recoverable home office overhead, courts have developed other formulae that incorporate the same concepts.[28] For example, the *Hudson* formula, which is more prevalent in England and Canada, centers on the percentage of profits and overhead accounted for in the initial bid.[29]

In disputes where contractors have asserted *Eichleay* or similar formulae as a means of home office overhead recovery, owners have asserted a number of defenses that administrative agency decisions have upheld.[30] Among others, the list of defenses and mitigating factors include: (1) failure by the contractor to pursue substitute work;[31] (2) offset of the home office overhead claimed by the contractor for additional compensation paid by the owner during the delay period pursuant to change orders;[32] and (3) negation of the claim due to over-recovery by the contractor in active periods following the delay.[33] Although these and other defenses have prevailed in certain cases, in most situations where a contractor is able to show the existence of the aforementioned elements, *Eichleay* has become the bench mark means for recovery in owner-caused delay cases.[34]

---

24. John D. Darling, *Delay of Game: One of the Thorniest Issues in Delay and Disruption Litigation Is the Quantification of Actual Damages*, 27 L.A. LAW. 31 (2005).

25. Reginald M. Jones, *Recovering Extended Home Office Overhead: What Is the State of Eichleay?*, 40 PROCUREMENT LAWYER 8 (2004).

26. *Id.*

27. *Id.*

28. Patrick A. McGeehin & Carleton O. Strouss, *Learning from Eichleay: Unabsorbed Overhead Claims in State and Local Jurisdictions*, 25 PUB. CONT. L.J. 351, 358 (1996).

29. *Id.* at 359.

30. *Id.*

31. *Id.*

32. *Id.* at 362.

33. *Id.* at 364.

34. *Id.* at 359.

5. Lost Profits

Lost profits are subsumed within the damage calculations above when the contractor is working under a stipulated sum contract. When the contractor is working under a cost-plus contract, the contractor's lost profit is recoverable as a separate element when establishing the plus portion of the contract entitlement.

Design professionals generally do not seek recovery of lost profits for *additional* work, but instead seek recovery based on the contractual value of their services, which may include a component for profit. On the other hand, when an owner breaches a contract with a design professional by terminating the design professional, for example, the design professional is entitled to recover the value of payments promised less any costs the design professional avoided by not having to perform the work.[35]

Finally, an owner may be entitled to recover its lost profits for delay in completion of a project due to the fault of the general contractor or the design professional or construction manager. The recovery of lost profits for new business has never found much favor in the court inasmuch as lost future profit claims for new businesses involve substantial speculation at times. Indeed, the "new business rule" in many courts barred recovery of lost profits for new businesses.[36] The modern trend is not to prohibit the recovery of lost future profits for new businesses outright, but rather to treat the question of such recovery as a matter of evidentiary sufficiency under which the court must assess the extent to which a claim for lost future profits of a new business can survive challenges to the potentially speculative nature of such damage claims.[37] When a business owner can establish with reasonable certainty the fact of its damage, and provide sufficient proof as to what its future profits would have been had the building been completed on time, it may be entitled to recover damages for lost profits.

6. Delay/Disruption Damages

The delay of the completion of a contract increases both the owner's and the contractor's costs. In fact, in many cases damages arising from delay are substantially greater than other damages. By the same token, damages from delay are difficult to quantify, largely because the actual consequences of delay are not always easy to trace, as discussed above. Nonetheless, when a party can establish its additional costs attributable to delay (discussed further below), such costs are generally deemed to be recoverable as part of the overall scheme of compensatory damages. These are generally referred to as *compensable delays*.

There are some delays that excuse a contractor's tardy completion of work but for which the contractor is not entitled to additional compensation. For example, if

---

35. Some of the principles, such as the requirement that the design professional mitigate its damages, may alter the calculation in particular circumstances.

36. STATE-BY-STATE GUIDE TO CONSTRUCTION CONTRACTS AND CLAIMS § 1.05[c][13], *supra* note 21.

37. Drews Co. v. Ledwith-Wolfe Assocs., Inc., 296 S.C. 207, 371 S.E.2d 532 (S.C. 1988).

the project experiences unusually adverse weather, the contractor may be entitled to an extension of contract time without receiving additional compensation, even if the contractor experiences extra costs for idle equipment, etc. These are referred to as *excusable delays* to distinguish them from compensable delays.

7. *Loss of Efficiency*

While *delay* denotes an extension in the time necessary to complete all or part of a project, *inefficiency* generally denotes additional expense in the completion of work as a consequence of conditions in the field being different from those the contractor envisioned in accepting the work. At least sixteen types of inefficiency have been identified, ranging from the inefficiency that arises from working in adverse weather conditions to inefficiency related to dilution of supervision. Sometimes, it is impossible to segregate direct costs from costs incurred as a result of a loss of efficiency. However, the concept for calculating such damages can be simply stated: the actual costs with disruption less what it should have cost without the disruption equals the disruption damages. In practice, determining these two components is more difficult and is met with varying degrees of success. The method for determining damages for loss of efficiency often depends upon the nature of the work and the information available, and is discussed below regarding proof of additional costs.

8. *Liquidated Damages*

Liquidated damages are contractual instruments used to fix compensatory damages in the event of late completion of a defined scope of work.[38] The provision must represent a reasonable approximation of the damages to be incurred in the event of the stated event. If the provision is merely used to punish a party for delay or the provision is used as an incentive to prevent breach, the liquated damages provision is likely to be held void as an unenforceable penalty.[39]

A liquidated damage clause sets a specific amount of damages that one party may recover from another as a result of certain breaches of the contract. While contractors often view a liquidated damage clause with trepidation, such clauses may actually be advantageous to a breaching party. Although they relieve the injured party of the obligation to prove actual damage, such clauses may actually provide a significant limitation on the damage otherwise recoverable and provide the party with a known exposure. For instance, on a fast-track casino project, the owner will generally have pro forma projections well supported by historical evidence that can be used to establish large amounts of revenue and profit lost as a result of a late opening. A liquidated damage provision may limit the amount otherwise recoverable by the casino owner.

---

38. WILLIAM SCHWARZKOPF & JOHN J. MCNAMARA, CALCULATING CONSTRUCTION DAMAGES § 13.02, at 257 (2d ed. 2000).

39. *See, e.g.*, U.S. Fidelity & Guar. Co. v. Braspetro Oil Servs. Co., 369 F.3d 34 (2d Cir. 2004) (New York law).

Liquidated damages are appropriate only if genuine—that is, they represent the parties' estimate as to the damages that the injured party would suffer from late completion—and they cannot be a penalty in disguise. Simply referring to the provision as "liquidated damages" does not make it one; rather, the circumstances surrounding the contract and the objective intent of the parties determine the actual nature of the clause.[40] As a general rule, a penalty provision seeks to secure performance, whereas a liquidated damages provision seeks payment in lieu of performance.[41] Case law suggests there is often a fine line between the two concepts, and the distinction hinges primarily on reasonableness and, to a lesser extent, on party intent. A liquidated damages provision deemed unreasonable under the circumstances is void as a penalty; specifically, a provision purporting to award an unreasonably large amount of liquidated damages is void.[42]

Liquidated damages are valid only if actual damages are difficult to ascertain and the proposed sum is reasonable.[43] Liquidated damages are appropriate only where actual damages cannot be demonstrated with reasonable certainty. If damages for breach can be anticipated conclusively, a provision claiming to be liquidated damages is an unreasonable, inappropriate remedy and thus will likely be construed as a penalty.[44]

A provision is reasonable if: (1) the sum resembles the anticipated or actual harm from the breach; (2) there is difficulty proving the amount of loss; and (3) liquidated damages presents the most feasible avenue for adequate relief.[45] In some instances, the court equates "reasonable" with "conscionable," stating that a liquidated damages provision is unreasonable only if it "shocks the conscious" and violates public policy.[46] Regardless of how reasonable the sum for liquidated damages might appear, the provision must be considered a penalty if the drafting party makes no attempt to calculate the amount of actual damages that might be sustained in the event of breach.

The most common liquidated damage clauses concern late completion and are expressed as a specific amount for each specific day that the project completion is late. However, liquidated damage clauses need not be limited merely to delays to project completion.[47]

---

40. United Tunneling Enters., Inc. v. Havens Constr. Co., 35 F. Supp. 2d 789 (D. Kan. 1998).

41. Unified Sch. Dist. No. 315 v. DeWerff, 6 Kan. Ct. App. 77, 79 (1981).

42. KAN. STAT. ANN. § 84-2718.

43. *Unified Sch. Dist. No. 315*, 6 Kan. Ct. App. at 82.

44. Kvassay v. Murray, 15 Kan. Ct. App. 2d 426 (1991).

45. KAN. STAT. ANN. § 84-2718.

46. TMG Life Ins. Co. v. Ashner, 21 Kan. Ct. App. 2d 234 (1995).

47. STATE-BY-STATE GUIDE TO CONSTRUCTION CONTRACTS AND CLAIMS, *supra* note 21, § 1.02(e), at 13.

9. *Interest*

Interest on a claim may be provided by contract or by statute. If provided by contract, the percentage of interest is either specified in the contract or paid to a bench mark, such as one percent above a specified prime rate as published in the *Wall Street Journal*.

If no rate of interest is provided in the contract, most states will have a default rate of interest that will apply. Some jurisdictions also have a Prompt Pay Statute that provides a specified rate of interest. Oftentimes, the interest rate will be the default rate of interest. However, if payment was withheld without just cause or excuse or in bad faith, then an enhanced rate of interest may apply.[48]

Pre-judgment interest on the principal amount is generally recoverable provided that the amount of the principal is liquidated and the date the principal amount became due is known or readily ascertainable. Whether or not it is liquidated depends upon whether it is undisputed or, although disputed, is capable of determination by mathematical computation.

Post-judgment interest is generally recoverable from the date of the judgment. Post-judgment is interest generally fixed by statute or may be tied to some economic standard, such as the U.S. Treasury rate. In such instances, the applicable rate at the time of the judgment will become the post-judgment rate of interest and will apply throughout collection of the judgment.

10. *Attorneys' Fees*

Generally, under U.S. jurisprudence, a prevailing party is not entitled to recover its attorneys' fees from any other party. This system, in which each party is responsible for its own attorneys' fees, is known as the *American Rule*. However, there are several widespread exceptions to the American Rule, and parties may be entitled to recover attorneys' fees when:

a. there is a contractual provision which provides for the recovery of such fees;
b. a statute permits recovery of such fees; or
c. the opposing party has acted either fraudulently or in bad faith in bringing or maintaining the lawsuit.[49]

Under the first exception, many contracts contain a prevailing party attorneys' fees clause under which a party that prevails in litigation, under contract, may be awarded its attorneys' fees. The second exception for statutory attorneys' fees is found most often when a statute is designed not only to protect parties but also when the claim is of a nature that the legislature wants to ensure that claimants are

---

48. Mo. Rev. Stat. § 34.057(2006).
49. Calculating Construction Damages, *supra* note 38, § 10.01, at 213.

not deterred from potentially meritorious claims because of the relative expense of litigation in light of the size of the typical claim. The third category covers frivolous or groundless suits. Although some parties take comfort in the fact that attorneys' fees may be awarded for the defense against a frivolous or groundless claim in most states, the comfort should be tempered by a realistic understanding that such fee awards are very difficult, as a practical matter, to obtain.

C. *Alternative Methods for Calculating Damages*

As noted above, the most satisfactory and most reliable method for proving damages is to tie particular costs to particular causes. When this method is possible, courts expect proof that is anchored in direct demonstration of the resulting costs of another's breach of duty. At the same time, courts recognize that tying individual consequences and costs to particular causes is not always an easy task, and if such demanding proof were required in every case, many parties with meritorious claims would be left without any remedy for want of their ability to prove exact damages through the direct cost method. As a consequence, and with the goal of not denying relief where the fact of damages is certain but the quantification is not susceptible to precise proof, courts have accepted a number of alternatives to proof of direct costs.

1. *Total Cost*

The simplest but also the most controversial and least accepted method is the total cost method. The total cost method calculates damages as the difference between the total actual costs incurred, including overhead and profit, and the bid amount.[50] While the simplicity of the method appeals to claimants, courts have criticized the method due to the assumption that all cost overruns are the responsibility of the project owner.[51] As such, courts disfavor and, in some cases, reject use of this method to calculate damages.[52]

The courts have been critical of the total cost method because it assumes that the contractor's original bid was reasonable and the work could have been completed for the bid amount absent breach of the contract. It further assumes that all of the cost overruns were caused by the conduct of the breaching party and none of them were caused by the contractor's own problems or inefficiencies. To protect against the inaccuracies inherent in the total cost method, courts have developed four elements that a claimant must establish before proceeding with the total cost method.[53] The claimant seeking damages through the total cost method must prove that:

a. the nature of the particular losses makes it impossible or highly impracticable to determine them with a reasonable degree of accuracy;

---

50. *Id.* § 1.03[C], at 14.
51. *Id.* § 1.03[C], at 14.
52. *Id.* § 1.03[C], at 14.
53. *Id.* § 1.03[C], at 15.

b. the contractor's bid or estimate was realistic;
c. the contractor's actual costs were reasonable; and
d. the contractor was not responsible for the cost overrun.[54]

If the claimant can establish each of the four premises, then the claimant may proceed with the calculation of damages under the total cost method.

2. Modified Total Cost

The modified total cost method was developed to address some of the inherent flaws in the total cost method.[55] Essentially, the modified total cost method reduces the amount calculated under the total cost method by any errors in the bid, costs attributable to the contractor, and costs attributable to parties other than the owner.[56] The resulting damage is presumed to more accurately reflect the damage directly suffered by the claimant.[57]

In a modified total cost, the contractor carefully scrutinizes its bid and takes responsibility for any errors. The contractor also analyzes its own performance on the project and assumes responsibility for any of its own inefficiencies. This is not to say that the contractor must reduce its claim by all inefficiencies that the contractor experienced, as some normal inefficiencies would be assumed in the contractor's bid. The contractor need only assume responsibility for those problems or inefficiencies outside the norm.

3. Measured Mile

In a measured mile approach, one compares the productivity achieved on an unaffected portion of the project with the productivity achieved on an affected portion.[58] Productivity is typically based on hours or dollars per unit of work achieved.[59] It is one of the preferred measures of inefficiency or loss of productivity because it uses actual historical data from the project.[60] It is critical that one prove that the affected and unaffected areas compared represent comparable work, where the only material variable is the disruption caused by the breaching party.[61] In addition, the unaffected work must be large enough to be material and to be representative of the type of work on the project as a whole. If the non-breaching party cannot prove these criteria, the other party will attack the base line productivity calculations for being faulty.[62]

---

54. *Id.* § 1.03[C], at 15.
55. *Id.* § 1.03[D], at 21.
56. *Id.* § 1.03[D], at 21.
57. *Id.* § 1.03[D], at 21.
58. John D. Darling, *Delay of Game: One of the Thorniest Issues in Delay and Disruption Litigation Is the Quantification of Actual Damages*, 27 L.A. LAW. 31, 36 (2005).
59. *Id.*
60. *See id.*
61. *Id.*
62. *Id.*

By its very nature, the measured mile approach requires that some portion of the project be unaffected by the delay or disruption. As a result, this method may not be available where such delay or disruption has continued throughout the project.

4. *Should-Cost Estimates*

For whatever reason, when the measured mile approach is not available, an estimate of what the cost should have been without disruption may be a reliable substitute. Where the measured mile approach compares performance on the same project, performance on other projects for substantially similar work may form the basis for reasonably estimating what performance on a particular project should reasonably have cost had there been no disruption.[63]

5. *Industry Standards and Studies*

In certain instances where the measured mile approach and the should-cost estimate may not be available (the entire project was affected and the contractor had not previously performed comparable work), it may be necessary to rely upon industry standards or studies. Such standards and studies have been developed in the industry through the MCAA, Business Round Table, and R.S. Means.

6. *Time-and-Motion Studies*

Where a contractor cannot provide an actual, unaffected performance period for the current job that would allow a measured mile approach or other contemporaneous jobs that would allow a should-cost estimate, the contractor may consider using a time-and-motion study. These studies are generally done by experts, and they attempt to measure the loss in both labor and equipment efficiency by approximating an unaffected sequence and comparing it to an affected sequence.[64]

7. *Jury Verdict Method*

Where damages cannot be easily determined from a contractor's evidence, a court may allow a jury to consider the evidence and make a reasonable judgment. This is generally expressed by the jury in a lump sum. The court may then still have the option of determining the percentage of damage attributable to the owner's own inefficiency.[65]

The jury verdict method allows the jury to determine the amount of damages suffered based on equitable considerations.[66] The trier of fact is responsible for determining what it considers a fair and reasonable amount as damages.[67] Oftentimes, the

---

63. STATE-BY-STATE GUIDE TO CONSTRUCTION CONTRACTS AND CLAIMS, *supra* note 21, § 1.05[c][6].
64. *Id.* § 1.05[c][8].
65. *Id.* § 1.05[c][9].
66. CALCULATING CONSTRUCTION DAMAGES, *supra* note 38, § 1.04[A], at 29.
67. *Id.* § 1.04[A], at 29.

total cost method or the modified total cost method is used by a party to establish a basis for the jury or fact-finder to make its decision.[68]

D. *Consequential Damages*

Consequential damages or "special damages" are damages that are caused by a breach of duty, but whose existence or impact is less readily or directly traceable to the breach than compensatory damages.[69] These damages may include lost rents, lost profits other than profits anticipated directly from the work, extended financing costs, loss of business, and loss of bonding capacity.[70] Consequential damages may be recoverable in actions based in contract if reasonably foreseeable. They are generally not recoverable in actions based on unjust enrichment. Consequential damages may be recoverable in tort, depending upon the nature of the tort, where the issue is proximate causation and not reasonable foreseeability.

E. *Punitive Damages*

Punitive damages are intended to punish a malicious wrongdoer and to deter the wrongdoer and others from behaving in a similar fashion in the future. Punitive damages in the construction industry are rare, and many jurisdictions do not allow the recovery of punitive damages. Because punitive damages are generally not available in breach of contract actions in general, in order to recover there must be an independent tort involving outrageous or malicious conduct.[71]

Library References

C.J.S. *Admiralty* § 229; C.J.S. *Costs* § 125; C.J.S. *Damages* §§ 21–33, 38, 42–45, 53–74, 80–91, 154–162, 175–201, 208–217, 380–382; C.J.S. *Interest and Usury* [**TO AUTHOR: INSERT SECTIONS**]; C.J.S. *Consumer Credit* §§ 34, 37, 41; C.J.S. *Parent and Child* § 344; C.J.S. *Torts* § 26.

West's Key No. Digests, Costs <KEY>194.16; Damages <KEY>15–17, 21–30, 35–46, 66, 74–94, 226; Interest <KEY>29, 31, 38.

V. Defenses to and Mitigation of Damage Claims

A. *Economic Waste*

While the cost of repair is the basic measure of damage for many breaches of construction contracts, this measure may not be applied where the repair cost is exorbitant, taking into account the nature of the damage, the value of the work at issue, the value of the project in general, and the diminution in value of the product

---

68. *Id.* § 1.04[A], at 29.
69. *Id.* § 1.03, at 9.
70. *Id.* § 1.03, at 9 note 10.
71. STATE-BY-STATE GUIDE TO CONSTRUCTION CONTRACTS AND CLAIMS § 1.05[f].

in light of the damage. In other terms, where the cost of repair is manifestly excessive in light of the other elements, a court may not award the cost of repair as the measure of damages.[72] "Where such a case occurs and it is also a case where completion or repair will be inordinately expensive, the policy against economic waste suggests use of the value formula rather than the cost formula."[73] As applied, the value formula consists of the diminution in value of the structure as opposed to the cost actually required to repair the structure.[74]

As an example of economic waste, in the *County of Maricopa v. Walsh Oberg Architects, Inc.*, the county obtained a judgment against its architect for the negligent preparation of plans and specifications for an underground parking garage that was designed to be waterproof but failed to prevent water from dripping onto cars below. The county alleged that it would cost approximately $498,169 to remove landscaping and cover the top of the underground structure with a waterproof membrane. The Arizona Court of Appeals was persuaded by the fact that it cost the county far less to perform an alternate remedy that would protect the vehicles from water damage and prevent further deterioration. As such, the court of appeals recognized that it would be economic waste to award the county more than one-third of a million dollars to merely remove landscaping when an alternate remedy existed.[75]

B. *Failure to Mitigate Damages*

Even though victimized by the conduct of another party, an injured party has an affirmative obligation to mitigate its damages. That is, the injured party must take reasonable steps to minimize the damages it will suffer. For instance, where a contractor has been prevented from working on one area of the project, the contractor may have an affirmative obligation to make its best effort to continue to work by moving to another unaffected area of the project.

When a party does not undertake reasonable steps to mitigate its damages, it does so at the risk of losing part of its right to recover damages. In particular, a party generally may not recover as part of its damages any money that represents damages that the claiming party could have avoided through more prudent conduct.

C. *Betterment*

The concept of betterment is widely recognized in the construction industry.[76] As a general principle, when there is a defect in the work or an omission in the plans,

---

72. Charlotte v. Skidmore, Owings and Merrill, 407 S.E.2d 571, 581 (N.C. Ct. App. 1991).

73. *Id.* (citing D. DOBBS, HANDBOOK OF THE LAW OF REMEDIES § 12.21 (1973)).

74. *Id.* at 580.

75. County of Maricopa v. Walsh and Oberg Architects, Inc., 494 P.2d 44 (Ariz. Ct. App. 1972); *see also* Stovall v. Reliance Ins. Co., 107 P.3d 1219 (Kan. 2005).

76. *See* Jerome V. Bales, Shamus O'Meara & Mark R. Azman, *The "Betterment" or Added Benefit Defense*, 26 CONSTRUCTION LAWYER 14 (2006).

the owner should not be placed in a better position than if the error had not occurred. This concept is known as betterment, added value, or added first benefit.

The courts have generally held that if the design is incomplete, the owner should pay what the omitted component would have cost if it had been included in the original design.[77] Logically, the designer should pay any additional costs to the owner, such as any premium associated with the change order adding the component to the project.

When a design error is not discovered until during or after construction, the owner may be entitled to recover the cost of any retrofitting expense, waste, or an intervening increase in the cost of the labor or materials necessary to correct the error.[78]

Although there are no reported decisions on betterment involving construction managers in an advisor role, the concept should apply to the same extent that a designer could raise the betterment or added value defense.

Contractors are obligated to build a project in accordance with the plans and specifications. However, in some circumstances contractors and construction managers at-risk may also rely upon the defense of betterment. For example, if the owner must repair or replace defective work, the courts have generally held that it should not recover the cost of any enhancement exceeding the quality evidenced in the contract documents.[79] When the repairs do not involve any enhancement, the owner's damages may still be reduced to the extent of any extended useful life of the building component.[80]

Library References

C.J.S. *Contracts* §§ 377, 565–566, 575–576; C.J.S. *Costs* §§ 125, 127; C.J.S. *Counties* §§ 159, 169; C.J.S. *Damages* §§ 46–52, 165–174; C.J.S. *Mechanics' Liens* § 31; C.J.S. *Municipal Corporations* §§ 944–946; C.J.S. *Principal and Surety* §§ 78, 82, 120; C.J.S. *Public Contracts* § 28; C.J.S. *Schools and School Districts* §§ 437–440, 446–450; C.J.S. *States* §§ 291–294; C.J.S. *United States* §§ 105–106, 127–130, 137, 139–148.

West's Key No. Digests, Contracts <KEY>284(1), 308; Costs <KEY>194.32; Counties <KEY>128, 129; Damages <KEY>58–65; Mechanics' Liens <KEY>36; Municipal Corporations <KEY>253, 254; Principal and Surety <KEY>75, 77; Public Contracts <KEY>22; Schools <KEY>85; States <KEY>107; United States <KEY>67, 73.

---

77. *See, e.g.,* Lochrane Eng'g, Inc. v. Willingham Realgrowth Inv. Fund Ltd., 552 So.2d 228 (Fla. Ct. App. 1989), *rev. denied,* 563 So. 2d 631 (Fla. 1990).

78. *See, e.g.,* Grossman v. Sea Air Towers, Ltd., 513 So. 2d 686 (Fla. Ct. App. 1987), *rev. denied,* 520 So.2d 584 (Fla. 1988).

79. *See, e.g.,* St. Joseph Hosp. v. Corbetta Constr. Co., Inc., 316 N.E.2d 51 (Ill. Ct. App. 1st Dist. 1974).

80. *See, e.g.,* Allied Chem. Corp. v. Van Buren Sch. Dist. No. 42, 575 S.W.2d 445 (Ark. 1979).

VI. Protection for Performance and Payment Rights

A. *In General*

A party need not rely on its ability to obtain and collect a court judgment to protect its rights to payment for performance of a construction contract. Rather, there are several other protections afforded the parties, as described below.

B. *Retention and Limitations Thereon*

The first protection that an owner usually has against a contractor's failure to complete work is retention of a part of the contract price until final payments. This nominally ensures that the owner has a source of funds from which it can pay for completion of any work that the contractor does not finish. It also provides some motivation to the contractor not to walk off the job as soon as all but punch list items are done.

At the same time, there are often statutory limits on the amount of money an owner may hold pending completion of work. These statutes specify the amount of money, by percentage, that an owner may withhold from the contractor pending final payment. In general, the percentages are fixed (e.g., five percent or ten percent) and provide that the public owner may continue to retain funds until certain prescribed events, normally substantial completion, in order to ensure performance and payment by the contractor. The statutes alter the amount and timing of the payments that would otherwise be due to a contractor. Such a statute may also govern the relationship between contractors and subcontractors by providing that retention held from subcontractors may not exceed that withheld from the general contractor.

Retention laws are sometimes found within the context of Prompt Pay Statutes. At other times, they enjoy their own statutory scheme. In general, retention laws do not apply to design professionals. Very typically, retention laws are not applicable on private projects but retention may be, and often is, required by private contract.

Both the percentage of permissible retention and the time period for which one may hold retainage vary from jurisdiction to jurisdiction. On large projects, the amount of retainage may be substantial and the length of time in which it may be retained may be long. This can cause a substantial cash flow drain or, at a minimum, a loss of use of funds. As a result, some jurisdictions have allowed a payee to recover interest on its retainage.[81] Whether a person is entitled to interest on retainage varies from jurisdiction to jurisdiction. However, in almost all jurisdictions, retainage will accrue interest at the default statutory rate if not paid by the required date.

Some jurisdictions have addressed the competing interest between an owner's need for security and the contractor's requirements for cash flow by permitting a contractor to substitute a bond in lieu of retainage[82] or by allowing a contractor to

---

81. *See, e.g.,* OR. REV. STAT. § 701.420(2) (2005).
82. *See, e.g.,* MO. REV. STAT. § 436.312 (2006).

pledge collateral, such as marketable securities.[83] An issue may sometimes arise as to whether or not limitations on retainage or entitlement to interest can be waived by contract. Again, as in other instances where an attempt is made by contract to waive statutory protections, such waivers are sometimes found to be void because they violate statute or are found to be unenforceable on public policy grounds.

C. *Labor and Material Payment Bonds*

Payment bonds, in and of themselves, are merely private contracts that guarantee the payment of certain claimants when specific conditions are met. Generally, these bonds serve to guarantee that materialmen are paid for the materials, equipment, and supplies services they provide.84 In many cases, these bonds are required by either state or federal law in situations where subcontractors and materialmen cannot avail themselves of the protections of the state's lien statutes.

D. *Performance Bonds*

Unlike payment bonds, performance bonds are designed to protect owners from a contractor's failure to complete its contractual obligations.[85] Performance bonds are three-party agreements between the owner, also referred to as the *obligee*, the contractor, also referred to as the *principal*, and the surety. Under a performance bond, if a contractor/principal, fails to fulfill its obligations under the contract, the surety has an obligation to complete the project. The scope of the surety's obligations depend largely on the terms of the bond itself.[86]

E. *Mechanics' Liens*

While all states have enacted mechanics' lien laws, these "laws present an extraordinarily varied approach, in substance, and in language, to the issues involved in mechanics' lien legislation. In fact, variation among the states may be greater in this area than in any other statutory area."[87] However, while an individual state's laws must be referenced with regard to any specific question, general observations can be made.

Most states originally enacted mechanics' lien laws to protect construction laborers and contractors.[88] However, in recent years, many states have amended these

---

83. *See, e.g.,* Or. Rev. Stat. § 279C.560.

84. D. Robert Beaumont, *Assuring Performance in International Construction Contracts*, 19 Cons. Law. 5, 6 (1999).

85. 2 Construction Law Handbook § 35.01, at 1276 (Robert Cushman & John Meyers, Eds. 1999).

86. *Id.*

87. Phil Bruner and Patrick J. O'Connor, Bruner & O'Connor on Construction Law, *Payment Process* § 8:124 (2002) (citing Unif. Constr. Lien Act, *Prefatory Note, reprinted in* 7 U.L.A., pt. II, at 382).

88. *Id.* § 8:128.

laws to include the liens of design professionals. Yet, like the underlying lien laws themselves, the states have varied approaches to the protection. For example, Pennsylvania requires that the design professional contract directly with the owner to claim a lien.[89] Additionally, under Pennsylvania law, the design professional's lien is limited to administration services performed under the contract, as opposed to fees associated with preparation of the plans and specifications.[90] Typically, the critical consideration will be whether the services of a design professional or construction manager contributed to the improvement of real property.

F. *Miller Act*

The Miller Act was enacted to protect persons who supply labor and materials for construction of federal buildings.[91] The Act requires a performance and payment bond be furnished before any contract is awarded for the construction or repair of any U.S. public buildings or public work projects for contracts in excess of $100,000.[92] "The statute limits valid claims under the Act to persons who have a contractual relationship with the prime contractor, called a 'first tier' claimant, or with a subcontractor, a 'second tier' claimant."[93] Nevertheless, "a person who contracts with an entity holding a management contract for a public facility is also generally considered by the courts to be a Miller Act prime contractor."[94] The Act is highly remedial in nature and is to be construed liberally.[95] Because of this, courts tend to look at the overall relationship between the parties and not to the contract alone when determining if a party is a proper claimant.[96]

Although there is no express statutory requirement for an obligee to give notice to the contractor or surety that it is bringing an action against the performance bond, the bond itself could require notice. Claimants under the payment bond who are not in privity of contract with the contractor typically must give the contractor written notice of the work performed within ninety days from the date of last providing labor and material for which a claim is made.[97]

State statute of limitations governs the time frame for filing a performance bond lawsuit. In contrast, actions on a payment bond must be brought in the district court

---

89. *Id.* § 8:128.
90. *Id.* § 8:128.
91. 17 AM. JUR. 2D Contractor's Bonds § 134 (2004).
92. 40 U.S.C. §§ 270a–270f (2005).
93. CONSTRUCTION LAW HANDBOOK, *supra* note 85, § 36.03[A], at 1318.
94. *Id.* § 36.03[A], at 1318.
95. F.D. Rich Co. v. United States *ex rel.* Industrial Lumber Co., 417 U.S. 116 (1974).
96. Reagan v. Tri-County Excavating, Inc. 62 F.3d 501 (3rd Cir. 1995); United States *ex rel.* K&M Corp. v. A&M Gregos, Inc., 607 F.2d 44 (3rd Cir. 1979); United States *ex rel.* Greenwald Industrial Products Co. v. Barlows Commercial Constr. Co., 567 F. Supp. 464 (D.D.C. 1983).
97. 40 U.S.C. § 270b.

for the district in which the contract was to be performed within one year of last providing labor or materials.[98]

The Act does not contain a specific prevision allowing the prevailing party in an action against a bond to recovery attorneys' fees. Even when state law would allow the recovery of attorneys' fees against a surety, the general rule is that attorneys' fees are not recoverable under a Miller Act bond.[99]

G. *State Public Works Claims (Little Miller Acts)*

Most states' public works acts limit claims against payment bonds.[100] While a potential claimant must examine the laws of the particular state, many states judicially define a claimant by examining the following factors:

i. whether an entity constructs a definite or substantial part of the work called for in the prime contract;
ii. whether the work is performed according to the plans and specifications of the prime contract; and
iii. whether the plans and specifications of the prime contract call for a unique product not readily available on the open market.[101]

However, "design professionals and providers of services typically are not eligible as state public works bond claimants unless they perform services for the contractor or a subcontractor."[102]

H. *Prompt Payment Statutes*

Many jurisdictions have enacted legislation that provides enhanced remedies to contractors and subcontractors in the event that payment does not occur within a specified amount of time. Such legislation is commonly referred to as *prompt pay* legislation. These statutes may be found within the context of public works or lien statutes, or they may have their own legislative framework. Generally, they apply to public contracts, but some jurisdictions have expanded the protection to private contracts as well.

The time allowed for payment may vary by jurisdiction. The time period is generally fairly short and rarely exceeds thirty days.

Prompt pay statutes will specify the interest a party may recover on a late payment. In some jurisdictions, an unpaid party may recover interest at a rate in excess of the statutory default rate plus attorneys' fees, if the unpaid party can demonstrate

---

98. 40 U.S.C. § 270b.
99. F.D. Rich Co. v. United States *ex rel.* Industrial Lumber Co., 417 U.S. 116 (1974); United States *ex rel.* Leno v. Summit Constr. Co., 892 F.2d 788 (9th Cir. 1989).
100. CONSTRUCTION LAW HANDBOOK, *supra* note 85, § 36.03[B], at 1320.
101. *Id.* § 36.03[B], at 1321.
102. *Id.* § 36.03[B], at 1321.

bad faith or a lack of a good faith basis on behalf of the debtor for not paying.[103] An issue sometimes arises concerning whether parties to a construction contract may alter or waive the protections of a prompt pay statute. For instance, if the prompt pay statute requires payment within thirty days, will the parties that specify forty-five days for payment alter the parties' respective rights and obligations? Even more dramatically, may the parties expressly provide that they are waiving the protections of the Prompt Pay Act and the remedies available thereunder will not apply? While parties are, in general, free to contract, as with other contract clauses that attempt to waive statutorily granted rights, courts sometimes declare such terms void and unenforceable on public policy grounds or because they violate the express terms of the statute itself. In general, the courts have been very reticent to find a waiver of Prompt Pay Act rights since such waivers are often obtained as a result of an unequal bargaining position or inability to negotiate, and the legislative intent is frustrated if waivers are permitted.

Library References

C.J.S. *Architects* §§ 4, 16–17, 19, 21–24, 32; C.J.S. *Contracts* §§ 11, 341, 343, 347, 356, 359, 442–445, 450; C.J.S. *Damages* § 113; C.J.S. *Mechanics' Liens* §§ 252–254; C.J.S. *Negligence* §§ 381–385, 388–393, 398; C.J.S. *Workers' Compensation* §§ 1587–1591, 1656–1662.

West's Key No. Digests, Contracts <KEY>196–199, 205, 221(3); Damages <KEY>118, 122; Indemnity <KEY>33(2), 33(5); Mechanics' Liens <KEY>208; Negligence <KEY>1260; Workers' Compensation <KEY>2084, 2142.25.

VII. Contractual and Other Limitations on Liability

A. *Exculpatory Clauses*

Exculpatory clauses attempt to limit a party's liability by stating specific activities the design professional is to perform and excluding the specific activities that the design professional is not to perform. These clauses can be extremely helpful in limiting the liability of the design professional.[104] Oftentimes, design professionals will include exculpatory language in their contracts with owners in order to eliminate the possibility of third party liability; for example, the American Institute of Architects's *Standard Form of Agreement Between Owner and Architect with Standard Form of Architect's Services*. Exculpatory clauses are generally enforceable to the extent that they are not the products of oppression, although there are exceptions where the clause seeks to immunize a party from liability for personal injuries.

---

103. *See, e.g.*, Mo. Rev. Stat. § 34.057 (2006) (interest at 1½% per month if payment was not withheld in good faith).

104. These clauses are not uniformly practiced. See James Acret, Architects and Engineers § 9.31 (3d ed. 1993).

B. *Waivers of Consequential/Incidental Damages*

Consequential damages are those damages that indirectly flow from the consequences of an act.[105] Contract clauses that attempt to limit these types of damages are intended to shield design professionals from the sometimes far-reaching risks associated with the design professional's errors.[106] In fact, recognizing the unpredictability associated with consequential damages, American Institute of Architects (AIA) Document B141 provides: "[t]he Architect and Owner waive consequential damages for claims, disputes or other matters in question ... ."[107] As such, the AIA's mutual waiver clause attempts to promote predictability by relieving participants from liability for abnormally large losses because of unanticipated events occurring after formation of the contract.[108] Waivers of consequential damages are generally enforceable.

C. *Indemnification Clauses*

Clauses that require one party (indemnitor) to defend and indemnify another party (indemnitee) from claims arising from activities on the project are a common way to shift risk and, hence, the damages for which a party may otherwise be liable. Even though they are almost always included in the same clause, the litigation and defense obligation is distinct from the obligation of indemnity.[109] The obligation to defend requires the indemnitor to either undertake directly the defense of the indemnitee or pay for the cost of defense. The obligation to indemnify requires the indemnitor to either satisfy any perfected claim against the indemnitee or reimburse the indemnitee for costs incurred.

In theory, the fairest way to allocate responsibility is for each party to indemnify the other from damage caused by the indemnitor. That is, each party would bear its share of damages in proportion to its fault. Notwithstanding issues of fundamental fairness, contracts, including industry standard contracts, often allocate risk other than on a comparative fault basis, thereby fundamentally altering the damages and remedies available to parties.

Such risk-shifting clauses are generally enforceable but courts may void them as unconscionable, against public policy or a violation of specific statutes. Therefore, in determining rights and remedies available to a design professional and construction manager, recourse must be made not only to the specific contractual term that may shift normal responsibility, but also to statutes and case law that may void those very indemnity clauses. Indemnity clauses that create exposure that would not otherwise exist absent such a term may have significant risk for design professionals. It may be possible to obtain commercial general liability insurance to cover such contractual liability, but it is generally not possible to obtain professional liability

---

105. *Id.* § 17.6, at 414.
106. *Id.*
107. CONSTRUCTION LAW HANDBOOK, *supra* note 85, § 31.05, at 1158.
108. *Id.* § 31.05, at 1158.
109. STATE-BY-STATE GUIDE TO CONSTRUCTION CONTRACTS AND CLAIMS, *supra* note 21, § 1.02(b).

insurance because it covers only damages that would have been imposed absent the contract.[110]

Well-drafted indemnification clauses can greatly protect owners and architects and expand the liability of general contractors.[111] For example, an indemnification clause commonly used by owners states that:

> To the fullest extent permitted by law, the contractor shall indemnify, defend, and hold harmless the owner, the architect, and the architect's consultants, and their agents and employees, from and against any and all claims, suits, causes of actions, damages, losses, and expenses, including but not limited to attorneys' fees, arising out of, related to, or resulting from the performance of the general contractor's work in any fashion to the extent that the loss is caused in whole or in part by acts and/or omissions of the general contractor or any person or entity directly or indirectly employed by the general contractor.[112]

D. *Disclaimer of Warranties*

As described in prior chapters, most construction contracts include express warranties and the law imposes implied warranty for many such contracts. However, most states permit the disclaimer of warranties related to service contracts as long as the disclaimer at issue is conspicuous in the contract in which the disclaimer is contained.

E. *Waiver of Lien Rights*

All states provide some protection to contractors, subcontractors, suppliers, laborers, and design professionals by providing a claim against the property and/or improvements thereon for amounts unpaid for labor, equipment, materials, or services furnished for the benefit of the real property and/or improvements. Terms are sometimes inserted in contracts requiring a party to waive its lien rights. Given the strong public policy in favor of protecting those who improve real estate, broad form advance waivers of statutory lien rights (prior to payment) are sometimes rendered null and void or unenforceable by the courts on public policy grounds or by a statute specifically prohibiting such waivers.[113]

F. *Limitations of Liability*

Oftentimes, a party will attempt to limit its liability by placing a cap on its potential exposure. This may be expressed in a specific dollar amount or a percentage of the design professional's or construction manager's fee. In such instances, the damages are not liquidated; actual damages must still be proven. Such caps do not reduce

---

110. *Id.* § 1.02(b), at 10.
111. CALCULATING CONSTRUCTION DAMAGES, *supra* note 38, § 14.02[B], at 282.
112. *Id.* § 14.02[B], at 282.
113. STATE GUIDE TO CONSTRUCTION CONTRACTS AND CLAIMS, *supra* note 21, § 1.02(c), at 12.

insurance coverage. They also do not provide indemnification; they merely limit a party's exposure.

Some states find such clauses unenforceable as against public policy or contrary to a specific statute.

G. *No Damages for Delay*

While owners quite commonly will insert liquidated damage clauses providing specific compensation for contractor-caused delays, they will also frequently include a clause precluding recovery of damages for delay caused by the owner. Such clauses will normally provide that a contractor's sole remedy for delay, regardless of the cause, is for additional time and that no additional compensation will be paid.

No damages for delay clauses are generally enforceable. However, courts will narrowly construe them and scrutinize them under equitable principles. This has resulted in the development of several exceptions regarding the enforceability of no damage for delay clauses. Such clauses are generally not enforceable if the cause of the delay is:

   i. the bad faith of the party seeking to enforce the claim;
   ii. so long in duration that it amounts to abandonment of the contract;
   iii. beyond the contemplation of the parties at the time of contract formation; or
   iv. the result of active interference of the parties seeking to enforce the clause.[114]

In addition, these provisions may also be precluded by state statutes.

H. *Pay-if-Paid/Pay-When-Paid Clauses*

It is not uncommon for construction contracts to contain a clause providing that payment by the owner is a condition precedent to the contractor's obligation to pay its subcontractor. Such terms are generally referred to as a *pay-if-paid* clause. If enforceable, the contractor does not have to pay the subcontractor, prior to receiving payment from the owner.

Such clauses may be enforceable, but are generally not favored by courts. Therefore, they have been narrowly construed. Many courts have found that unless a pay-if-paid clause expressly and unambiguously states that payment from the owner is a *condition precedent* to payment to the subcontractor by the contractor, then the clause is ambiguous and will be treated as a pay-when-paid clause, requiring the contractor to pay the subcontractor within a *reasonable time*, regardless of whether the contractor has been paid by the owner. Other jurisdictions have struck a middle ground of finding that such clauses may be enforced against a subcontractor unless it can be shown that nonpayment by the owner was caused by some failure of performance on behalf of the contractor for which the subcontractor is not responsible.

---

114. John E. Green Plumbing & Heating Co. v. Turner Constr. Co., 500 F. Supp. 910 (E.D. Mich. 1980), *aff'd*, 742 F.2d 965 (7th Cir. 2004).

The *pay-when-paid* clause is a less harsh limitation on a subcontractor's ability to recover. Under a pay-when-paid clause, there is not an absolute bar to recovery, as may occur with a pay-if-paid clause. Pay-when-paid clauses generally lack clear and unambiguous language establishing that payment by the owner is a risk assumed by the contractor and is a condition precedent to the contractor's obligation to pay the subcontractor. Examples of a pay-when-paid clause would include:

"The contractor will pay the subcontractor within thirty days of receipt of payment by the contractor from the owner;" or
"The contractor will pay the subcontractor when the contractor receives payment from the owner."

Under such clauses, it is common for the courts to find that the subcontractor need wait only a reasonable time for payment. However, the contractor must ultimately pay the subcontractor regardless of payment from the owner. The courts generally prefer such an interpretation, finding that the contractor was in the best position to determine the financial stability and creditworthiness of the owner with whom it contracted, while the subcontractor looked primarily to the financial capabilities of the contractor.

I. *Immunity under Workers' Compensation Laws*

All jurisdictions have some form of workers' compensation laws which will protect an employer from claims for bodily injury to, or the property damage of, its employees by providing immunity for such claims to the employer. The immunity exists whether or not the claims arise from site safety or some other cause. The sole remedy against the employer is generally limited to benefits specified by statute (in monopolistic jurisdictions) or the employer's workers' compensation carrier (in non-monopolistic jurisdictions).

In general, an employer will enjoy this immunity if it follows the statutory requirements and its conduct is not intentional, criminal, or reckless. This immunity can be compromised if the employer has agreed to indemnify another party for bodily injury to the employer's own employees. Under workers' compensation laws, claims against parties are not precluded. Therefore, the contractor may have the situation where it has agreed to defend, indemnify, and hold harmless an owner for all claims arising from the contractor's negligence. When an employee of the contractor is killed on the project as a result of the contractor's negligence, the worker's estate sues the project owner, and the project owner, in turn, sues the contractor, thus allowing a potential *end run* around the statute.

Although most contract rights can be waived, contract provisions that would result in an express or effective waiver of statutorily granted right or immunity may not be enforceable. This needs to be examined on a state-by-state basis.

J. Immunity from Site Safety Claims

Traditionally, courts have held that design professionals have no legal duty to protect the construction workers from unsafe job site conditions. The courts have reasoned that there is a lack of privity between design professionals and the workers and that the general contractor is in the best position to control the job site and protect the worker. However, this traditional protection has been eroded in recent years in many jurisdictions. Courts have tended to examine the issue based on facts and circumstances and have examined a number of factors in determining whether a legal duty exists. These factors include:

  i. retention of any right to supervise and control;
  ii. actual supervision and control of the work;
  iii. participation in the activities at the construction site;
  iv. supervision and coordination of the subcontractors;
  v. responsibility for safety practices;
  vi. authority to issue change orders; and
  vii. knowledge of unsafe conditions.

In response to the erosion of common law protections, some jurisdictions have enacted legislation giving immunity to design professionals from liability for site safety claims with certain well-defined exceptions (e.g., contractual responsibility for site safety, actual control over a portion of the site where the worker was injured, or an injury as a result of the design professional's negligent preparation of plans and specifications).

K. One-Call Statutes and Damage to Underground Structures

Many states have enacted *one-call notification* or utility locate statutes that require contractors to call prior to commencing any excavation. This notice is intended to allow public and private underground utility owners to mark the location of the utilities. The contractor that complies with the statutory requirements is generally immune from responsibility for damage to the utilities.

On the other hand, the failure to comply with such statutes may subject a contractor not only to liability for damages to the utility, but also to consequential damages for interruption of service provided by the utility, as well as civil penalties.

CHAPTER 17

Alternative Dispute Resolution and the Construction Dispute

DAVID W. RUDY, ESQ.

I. Introduction: Litigation and Alternative Dispute Resolution

Litigation has always been a time-consuming, expensive, and draining mechanism for parties to resolve disputes. Increasingly crowded court dockets, courts, and juries' lack of knowledge regarding the construction process and the widely held perception that court (and jury) decisions are often little better than coin tosses have led parties to consider alternatives to judicial processes. This, in turn, has led to a collection of nonjudicial dispute resolution mechanisms that fall under the broad heading Alternative Dispute Resolution (ADR). Individual instances of ADR often have only one common thread: they are alternative to the traditional resolution of a legal dispute through a court or jury trial. For example, arbitration is a quasi-judicial decisional process. It actually has more in common with a court trial than it does with mediation. Yet, arbitration and mediation are both commonly understood as ADR mechanisms simply because they are alternatives to traditional judicial resolution.

Although ADR procedures are usually less formal than judicial proceedings, that does not suggest that they do not invoke their own strategic considerations. Just as an educated use of the judicial process is necessary to obtain the greatest benefit from court proceedings, the practitioner and professional benefit from an understanding and measured use of ADR. This chapter summarizes the most popular forms of ADR (Section II), and then focuses most of its attention on discussing the effective use of the two forms of ADR most common to construction disputes: mediation (Section III) and arbitration (Section IV).

II. Overview of Types of ADR

A. *Arbitration*

Arbitration is a decisional process in which the disputants submit their conflict to one or more independent, neutral persons for decision. The decision may be binding[1] or nonbinding.[2] Although in most contexts, discussions of arbitration refer to binding arbitration in which parties are legally bound by the decision in the same manner in which they would be bound by a court decision.

Arbitration generally requires the parties' consent to be effective; this consideration arises from the nature of judicial power. In order for any tribunal to impose an enforceable decision against the will of any party, the tribunal must be vested with jurisdiction over the dispute and over the parties. Courts derive jurisdiction from the constitutional and statutory laws of the United States or individual states. On the other hand, arbitrators have no jurisdiction except as the law[3] or the agreement of the parties may provide. Of course, even a voluntary arbitration agreement would be of little effect if the parties were free to ignore the decision imposed by an arbitrator. Accordingly, federal and state laws give specific legislative blessing to arbitration agreements such that (with some exceptions) voluntary agreements to arbitrate are enforceable, and arbitrators' decisions are enforced through confirmation by courts of record.

There is no mystery as to how parties vest an arbitrator with authority to resolve their dispute: they usually include a binding arbitration provision in their contract before any dispute arises. For example, design professionals most often contract with an owner to provide design and related services using contract forms issued by the American Institute of Architects (AIA). These forms (B141-1997 CA) currently contain provisions for binding arbitration. If the parties execute these form agreements without deleting the arbitration provisions, any dispute between them will be resolved through binding arbitration and not through the ordinary judicial process. The fact that binding arbitration provisions find their way into many of the construction contract forms utilized in the construction industry is a sign of the general favor that binding arbitration has found in the industry. Binding arbitration clauses can be used in any agreement, and many parties who draft their own contracts, rather than use industry standard forms, frequently include such provisions.

---

1. Binding means that the decision of the arbitrator(s) will typically be converted into a final judgment, as discussed further below concerning confirmation of arbitration awards.

2. Sometimes called *advisory*, the purpose of such a procedure is to present the parties with the opinion of a respected neutral as to the validity of their claims and defenses. At the conclusion of a nonbinding arbitration, the parties are juridically in the same position as they were before the procedure commenced. Thus, it is little used in resolving the construction dispute.

3. Many states have statutory arbitration provisions which confer jurisdiction on arbitrators who are appointed pursuant to court order. Typically, such arbitrations are nonbinding and would not be used in construction litigation.

B. *Mediation*

Mediation bears little resemblance to arbitration, even though both are *alternative* dispute resolution mechanisms. Mediation is a voluntary process[4] through which the parties invoke the good offices of an independent neutral to assist them in *negotiating* a resolution of the conflict. The chief attribute that distinguishes mediation from arbitration is that mediation does not produce a decision that can be given effect over the will of one of the parties.

Although persons unfamiliar with the process will sometimes ask, "Is mediation binding?" the question betrays a basic (albeit common) misunderstanding of mediation. There is no decision or ruling made by a mediator that is susceptible to enforcement. Rather, a successful mediation results in an agreement between the disputants to resolve the dispute in accordance with the terms and conditions of the settlement agreement. If drafted properly, such an agreement will be enforceable and, thus, *binding*. However, the binding nature of the resolution comes not as a result of the process of mediation, but rather from the nature of the contractual agreement into which the parties enter at the conclusion of the process.

Mediation may also be called *mediated negotiation*. The latter name is perhaps most expressive of the essence of the mediation process. When resolution is achieved in mediation, it is the result of the agreement of the parties through negotiation with the assistance of the mediator.

C. *Partnering*

In the last fifteen years or so, the concept of partnering has emerged in the construction arena. Partnering is a contractual arrangement between the parties that redefines the construction enterprise in a nonadversarial way. The principal actors in construction, from design professionals to construction managers to contractors, begin by viewing their relationships and interaction as collaborative instead of oppositional. Although partnering approaches can radically affect all aspects of a construction project and the relationships that underlie it, the feature of partnering that is relevant here is its impact on dispute resolution.

By substituting a collaborative approach for an adversarial one toward problems and issues in the construction project, partnering can anticipate and avoid disputes before they take root. It is possible from the outset to fashion dispute resolution procedures and techniques creatively into the contractual relationships between the parties. As but one example of how partnering can be implemented into the parties' relationship at the outset of a construction project, the construction documents might contain integrated provisions with the following attributes:

---

4. Under some statutes and court rules, mediation may be ordered by a court in an appropriate case. Mandatory mediation, as it is sometimes called, is really an oxymoron, since the process itself is only effective based on the voluntary participation of the parties. One can force a party to attend a mediation but one cannot force a resolution of the dispute through mediation.

- A project manager, construction manager, and architect representative are required to meet weekly at the job site, and require a disgruntled party to raise and to discuss any claim or dispute at the meeting before the dispute can be escalated to the next level of dispute resolution.
- An independent mediator/facilitator is designated in the contract documents to assist in the resolution of any disputes that might arise. If the weekly meetings among representatives of the primary parties do not produce resolution, the parties must invoke the services of the designated mediator/facilitator, who may be required to attend a job site meeting with the disputants on as little as two or three days' notice.
- In the event that the facilitated dispute resolution meeting is unsuccessful, the parties must attend mediation on a short timeline to attempt to resolve the dispute before the dispute produces major damages due to delay, disruption, penalties, etc.
- In the event that the mediation does not produce resolution, the issue must be arbitrated by a pre-approved project arbitrator within twenty-one days of the first job site meeting at which the dispute is raised.

Partnering with a dispute resolution process as described above brings with it both advantages and disadvantages. The process requires appointment of a facilitator/ mediator and arbitrator even at the contract stage. If these professionals require a retainer, there is expense involved, even though disputes may not arise on the project. In addition, valuable project time may be lost while the disputants are involved in meetings, mediations, and arbitrations during ongoing construction. The representatives of the disputants are usually the managers and supervisors whose field presence is essential to the functioning of the construction project.[5]

On the other hand, resolving the issues promptly when they arise saves valuable project time and minimizes the ever present risks of: (1) historical revisionism in the post-project litigation phase, as resolution of a dispute contemporaneous with the key events generally ensures a more complete and accurate record on which the dispute can be resolved; (2) a tradeoff of a modest loss of management time when the dispute is resolved through early mediation as opposed to the significant loss of time that could accompany subsequent dispute resolution procedures; and (3) the aggravation of damages through protraction of a dispute. In an appropriate project, working collaboratively, anticipating disputes, and building in mechanisms designed to stop disputes before they gather momentum can be a valuable and well-advised approach.

D. *Judicial or Quasi-Judicial Procedures*

In addition to the mechanisms summarized above, there are a pair of quasi-judicial procedures that parties may invoke through the court system.

---

5. It does not help to appoint lower-level employees who do not have the power to bind the company to resolutions achieved through negotiation.

1. Early Neutral Evaluation

Courts are as mindful of their own docket congestion as litigants, and many courts have become proactive in managing their dockets through imposition of ADR requirements. These may include such simple obligations as requiring that the parties undertake some independent ADR before setting a trial date, or they may include intervention of judicial officers such as magistrate judges or senior attorneys annexed to the court for purposes of mediation.

One such program is early neutral evaluation. Pioneered by federal district courts, early neutral evaluation is a process that utilizes the skills of knowledgeable and respected practitioners to give the parties, counsel, and insurers an independent valuation of the strengths and weaknesses of their respective cases based on whatever information the parties provide to the neutral. Essentially, the decision-makers, counsel, and insurance representatives meet with the evaluator and present their cases.

There is great flexibility in presentation, which may be through argument of counsel, actual calling of witnesses (although without rules of evidence, direct or cross-examination), documentary or electronic presentations, etc. The evaluator encourages the parties to narrow or focus the dispute as appropriate by suggesting areas of agreement, disagreement, possible terminal motions, assisting with discovery plans, or the like. All of these are designed to make the litigation more efficient and focus the areas of dispute to be resolved. Following the hearing, the evaluator retires and writes a reasoned assessment of the strengths and weaknesses of the parties' positions, the likelihood of recovery, and estimated ranges of value. The parties may elect to receive the evaluation or to postpone its delivery and pursue settlement discussions, focused discovery, or dispositive motions and then reconvene the early neutral evaluation. The process is used by many federal government agencies and as a form of ADR associated with particular federal and state courts.[6]

2. Mini-trial

In a mini-trial, the parties seek to obtain a rough guess of how a real trial would turn out through a nonbinding, abbreviated process. A real jury is impaneled by the court through an abbreviated jury selection procedure. The lawyers then present opening statements and evidence in the form of representations by the attorneys made in good faith consistent with what they believe the actual evidence would show. The lawyers are free to use demonstrative exhibits, documentary evidence, and the like. In an appropriate case, the parties can stipulate that key witnesses actually testify before the jury. The attorneys then deliver closing arguments as they would in a real trial, and the court instructs the jury, who then retires to deliberate,

---

6. A good example of the process can be seen in the United States District Court, Northern District of California's Local Rules on alternative dispute resolution, available on the Internet at www.adr.cand.uscourts.gov.

ultimately rendering an advisory verdict. Then, the parties proceed to settlement discussions with the benefit of the jury's advisory verdict. If the settlement discussions fail, the case proceeds to trial as it otherwise would. Somewhat more popular several decades ago, the mini-trial is still a procedure which can be used in unique cases.

Of all the settlement-oriented procedures available to parties in litigation, the mini-trial consumes the most court resources. In a sense, it is the ultimate mock jury proceeding, since the actual adversaries present their real cases in front of a real jury. Use of a mini-trial as a settlement mechanism is a high-stakes venture. Although neither side is compelled to accept the results of the verdict; as a practical matter, the result may be devastating to one side. The settlement position of the prevailing party in the mini-trial may well harden as a result of the verdict, and the defeated party may well lose traction in subsequent settlement negotiations. Even in the face of this high-risk aspect (and in some cases, because of it), mini-trials may be the only way to bring reality to an attorney or party with unrealistic expectations of the case. Parties who believe that their own witnesses are angelic and the opposing side's witnesses are incorrigible liars are often given a proverbial wake-up call by an adverse verdict.

Finally, careful practitioners will consider that one side is usually favored in an abbreviated presentation, even though the opposing side might ultimately have a stronger case in a fully developed presentation. Because of the impact of the mini-trial result—especially on the prevailing party—this increases the high-risk aspect of the mini-trial as a settlement option. For these reasons, the mini-trial should be (and has been) used sparingly and only after careful consideration of the potential consequences.

Library References

West's Key No. Digests, Alternative Dispute Resolution <KEY>111, 112, 178, 229, 353, 382, 441, 500.

III. Strategies in Arbitration

A. *Overview*

Many ADR mechanisms anticipate resolution through settlement by agreement of the parties after a process that includes the use of intermediaries, who assist the parties, but have no authority to bind the parties to a determination. Arbitration, on the other hand, constitutes the submission of a dispute to a neutral third party whose determination results in the imposition of a binding judgment or award, the enforcement of which does not require the consent of the affected party (except to the extent that the party appeared in arbitration voluntarily). In this sense, arbitration is far closer to trial than mediation or the other ADR procedures reviewed above that rely on the negotiation of settlements.

However, arbitration is dissimilar to trial in other important respects. It is essentially intended to be faster and more efficient than a full evidentiary trial by court or jury to conclusion. Arbitration also makes litigation more efficient by eliminating most grounds for an appeal by a party that is disgruntled with the result.[7]

Because arbitration is a process that results in a binding determination of the parties' respective rights and responsibilities, it is a procedure that should not be taken lightly, however informal it may appear at times. Indeed, because arbitration comes in part at the cost of sacrificing any meaningful appellate rights in most cases, it is especially important for the parties to treat the prospect with due respect for the impact of the decision it will yield.

This section discusses some of the attributes of arbitration that counsel special attention, from the initial contract compelling the parties to arbitrate, through the culmination of the arbitration process in the issuance of a final, binding decision.

B. *Law Governing*

Court proceedings are governed by relatively sophisticated procedural rules that dictate the manner in which claims are made, the process by which parties disclose evidence to the other side, obtain evidence through formal discovery means, and the manner in which pre-trial proceedings and trials themselves are conducted. In addition to the procedural rules, sophisticated evidentiary rules restrict the evidence that can be presented in formal litigation.

Although arbitrations result in decisions whose financial impact is every bit as great as judicial decisions, the rules are far less specific, and even evidentiary rules are often treated more as advisory than binding in arbitration hearings. Accordingly, an attorney or design professional does well to understand the potential sources of rules that might apply in arbitration proceedings and the processes by which such rules may be modified.

---

7. There are two noteworthy exceptions. First, the parties may agree to have a binding arbitration, but with limited rights of appeal to a second panel of arbitrators who review the award issued in much the same as the appellate court would review the decision of the trial court following a judgment rendered after trial in the lower tribunal (for example, *see* JAMS's *Optional Arbitration Appeal Procedure*, at www.jamsadr.com/rules/optional.asp). Second, the arbitration may be nonbinding. In this case, the arbitrator(s) hear(s) the evidence and issues an award, but the parties are not bound to accept the award. Typically, the objecting party can proceed to trial, but there may be some consequence to the party rejecting the award if he or she proceeds to trial and obtains a less favorable result than the arbitration award. Nonbinding arbitration most often results from court-mandated arbitration programs. Such programs apply only to litigation pending in the court system. Due to the cost and complexity of a full evidentiary presentation in construction litigation, it is not a favored procedure for construction litigants. In most court-mandated programs, the assigned judge has the discretion to waive the required nonbinding arbitration. Often, the judge will be receptive to use of mediation instead of arbitration in a construction case.

1. *The Contract Provision*

With rare exceptions, parties are compelled to submit a dispute to binding arbitration only when their contract provides for such resolution. However, a provision that simply states that the parties agree that any dispute arising out of or concerning the performance of the contract shall be resolved by binding arbitration may well be given effect to compel binding arbitration, but it does nothing to establish the ground rules for such an arbitration.

In the absence of such procedural ground rules, the parties must rely on the Federal Arbitration Act or their applicable state statutes to determine what rules apply. Even then, the rules are only sketchy guidelines that grant substantial discretion to an arbitrator to mold the proceeding. Accordingly, parties do well to specify in their contracts not only that their disputes will be resolved by binding arbitration, but also the important procedural rules that govern the arbitration. These rules may include such elements as: (1) the number of arbitrators; (2) the method of selecting arbitrators, including provisions for resolving deadlocks if the parties are given the charge of agreeing to an arbitrator in the first instance; (3) the number of depositions that will be permitted—if any—before arbitration; (4) the parties' rights and obligations to produce and to discover documentary evidence or expert opinions from the other party; (5) timing requirements for the submission and adjudication of claims; (6) the jurisdiction whose law governs the resolution of the dispute; (7) the location of any arbitration hearing; and (8) whether the arbitration will be administered through a third-party administrator such as JAMS, the National Arbitration Forum, or the American Arbitration Association (AAA).

Of course, many parties do not have the time or inclination to go to such lengths in specifying arbitration procedures in their initial contracts. Rather than abdicating all specification of such procedures, the parties can incorporate the rules of an established third-party arbitration administrator.[8] In the construction industry, the most

---

8. For example, AIA Document B181 is a commonly used form of contract between architect and owner. In the 1984 edition, Article 7 provides:

7.1.1 Claims, disputes or other matters in question between the parties to this Agreement, arising out of or relating to this Agreement or the breach thereof, shall be subject to and decided by arbitration in accordance with the Construction Industry Arbitration Rules of the American Arbitration Association currently in effect, unless the parties mutually agree otherwise. Demand for arbitration shall be filed in writing with the other party to this Agreement and with the American Arbitration Association. A demand for arbitration shall be made within a reasonable time after the claim, dispute or other matter in question has arisen.

7.1.2 No arbitration arising out of or relating to this Agreement shall include, by consolidation, joinder or in any other manner, an additional person or entity not a party to this Agreement, except by written consent containing a specific reference to this Agreement signed by the Owner, Architect, and any other person or entity sought to be joined. Consent to arbitration involving an additional person or entity shall not constitute consent to arbitration of

widely adopted rules are the AAA[9] Construction Industry Arbitration Rules and Mediation Procedures (Including Procedures for Large, Complex Construction Disputes),[10] although other ADR providers have generated equally reputable rules.[11]

In cases where there is no provision in the contract requiring arbitration, it is possible for construction litigation to be initiated in courts of competent jurisdiction, and subsequently be transferred to arbitration. For example, the assigned judge may, under many mandatory arbitration state law programs, send the matter to a nonbinding arbitration at some point in the judicial process. More commonly, the parties may agree to join in an arbitration already pending involving different parties but the same construction project,[12] or to initiate a new arbitration proceeding. This arbitration by stipulation will (with the permission of the court) remove the matter from the pending trial calendar and put the judicial proceeding in suspense until such time as the arbitration award becomes final.[13]

---

any claim, dispute or other matter in question not described in the written consent or with a person or entity not named or described therein. This agreement to arbitrate and other agreements to arbitrate with an additional person or entity duly consented to by the parties to this Agreement shall be specifically enforceable in accordance with applicable law in any court having jurisdiction thereof.

7.1.3 In no event shall the demand for arbitration be made after the date when institution of legal or equitable proceedings based on such claim, dispute or other matter in question would be barred by the applicable statute of limitations.

7.1.4 The award rendered by the arbitrator or arbitrators shall be final, and judgment may be entered upon it in accordance with applicable law in any court having jurisdiction thereof.

AIA forms, however, are currently under revision, and recent forms (Design Build) have removed binding arbitration language.

9. The American Arbitration Association (AAA) is the oldest provider of neutral services in the United States. It publishes both rules of evidence and procedure for different types of disputes. These rules are of long standing, and provide the practitioner with the advantage of numerous court decisions interpreting and enforcing them. Use of the AAA rules comes with a body of decisional law that makes both the process and the outcome more predictable than in a case where the parties choose to use untested or unlitigated rules.

10. The rules are available to the public at no charge on the AAA's Web site, http://www.adr.org/sp.asp?id=22004.

11. It should be noted that parties may choose to use the AAA rules without using AAA to administer the arbitration. The practitioner should consider the advantages and disadvantages of using the AAA as administrator (discussed below).

12. One scenario that can quickly become a nightmare involves multiple parties pursuing multiple forms of relief in multiple fora in relation to a particular construction project.

13. Of course, a party cannot be forced to binding arbitration without consenting to it either at the time or beforehand, for such would constitute an involuntary waiver of basic rights to due process, jury trial, etc.

Finally, arbitration agreements do not necessarily need to be found in the underlying contract documents. Parties can always agree to submit their dispute to arbitration at any point in the dispute resolution process, even without a prior agreement to do so. Indeed, many parties spend little time thinking about dispute resolution until the need arises, and at that point the prospect of expensive and time-consuming litigation goes far toward convincing the parties to accept arbitration.

2. *Contractual Preconditions*

Often, construction contracts provide for mediation or other dispute resolution procedures to be completed as a precondition to initiation of either arbitration or litigation. Such clauses are enforceable and must be complied with.[14] The practitioner should begin with a careful examination of the contract provisions. For example, if there is a provision requiring completion of mediation before initiation of arbitration, it must be complied with before arbitration.

Similarly, it is not uncommon for counsel to agree that arbitration may be initiated and progress made in selecting arbitrator(s) and even calendaring the hearing(s) before initiation of the mediation process, provided that the mediation itself is completed before commencement of the arbitration.

C. *Multiple Party Disputes and Fashioning Adequate Relief*

One of the most intractable problems of arbitration is that it does not always facilitate the resolution of a complex dispute in one forum. The typical construction dispute involves more than two parties. In any modestly complicated construction dispute, large numbers of parties may be involved. The owner may have a dispute with both the design professional and the general contractor. The design professional may in turn contend that if there is any legal responsibility, it is with the design subconsultants. The contractor may contend that although he may be responsible to the owner for construction deficiencies, it is really a subcontractor who caused the problem. The subcontractor may in turn contend that it was his subcontractor or materialman (a sub-sub) who actually caused the problem. The possibility of collateral insurance coverage disputes increases the number of potential parties even more.

Inevitably, claims for partial or complete indemnity or contribution suggest that the most efficient manner of handling the dispute is for all parties to participate in arbitration together (or mediation for that matter). However, for a number of reasons, some parties may not want to participate in a single arbitration proceeding and instead may prefer that their claims and defenses be resolved in separate proceedings

---

14. For example, The Federal Arbitration Act specifically states, "A written provision in any... contract... to settle by arbitration a controversy thereafter arising out of such contract or transaction, or the refusal to perform the whole or any part thereof, or an agreement in writing to submit to arbitration an existing controversy arising out of such a contract, transaction, or refusal, shall be valid, irrevocable, and enforceable, save upon such grounds as exist at law or in equity for the revocation of any contract." 9 U.S.C. § 2 (2005).

(or to the same effect, they may prefer that a claimant be required to engage in piecemeal litigation of its claims so as to discourage pursuit of such claims).

Consider the following hypothetical example:

Owner and general contractor have a written construction contract that provides for binding arbitration of all disputes between them. Owner has a written contract with architect that provides for binding arbitration of all disputes between them. Contractor has a written contract with primary subcontractor, which is silent as to arbitration. Primary subcontractor has a written contract with Secondary Subcontractor, which requires arbitration. Architect has a written contract with civil engineer that is silent as to arbitration. Owner has a written contract with construction manager that requires arbitration.
Owner notes about a dozen deficiencies after substantial completion of the project. Owner believes that some are design-related, while some represent construction defects; some are in owner's view both design- and construction-related.
Contractor denies responsibility for all deficiencies, contending that architect and construction manager are responsible. In any event, contractor argues that if there is a construction deficiency, primary subcontractor is ultimately responsible.
Architect denies responsibility for all deficiencies, contending that contractor and manager are responsible. In any event, architect argues that if there is a design deficiency, civil engineer is ultimately responsible.

The resolution of all these disputes presents a procedural quagmire. Since the owner contracted separately with the architect and the contractor, he must file separate claims against each of them. The claims overlap, as to both deficiency and amount of damages sought. As a result, he faces double-filing fees at AAA, based on two times the amount of his full actual damages. Worse, the owner must pursue two separate arbitrations which could produce inconsistent and unproductive results.

Example: Contractor arbitration results in award for 40 percent of damages, finding that 60 percent is design deficiencies. Architect arbitration results in award for 20 percent of damages, finding that 80 percent is construction deficiencies. Each arbitrator further reduces the award for fault of construction manager: Contractor Arbitrator by 20 percent and Architect Arbitrator by 10 percent.

If the owner wishes to pursue the manager, the owner must initiate a third arbitration proceeding. In order to proceed against the primary subcontractor, the contractor files a lawsuit in state court. The architect also files a lawsuit, but in federal court, against the civil engineer, who is a resident of a neighboring state. Finally, the primary subcontractor initiates arbitration against secondary subcontractor.

The expense and unproductivity of the procedures contemplated above are apparent. The possibility (even probability, on these facts) of inconsistent results is manifest. The owner is engaged in three separate arbitrations. There are two lawsuits and two other arbitrations. Seven proceedings in all are dealing with what is essentially one factual dispute. Soon, the efficiency and timesavings that a binding arbitration clause promised vanish, and even full-blown litigation among all the parties begins to look preferable!

What options are available to any of these parties to resolve the multifaceted dispute in reasonable fashion? First, all of the parties could agree to participate in mediation (discussed below). If successful, mediation can resolve all issues among all parties in a cost-effective manner that obviates any further proceedings.[15]

Second, the parties without an arbitration provision in their contract can nonetheless agree to arbitrate their claims and defenses, and further agree to a consolidated arbitration involving all parties and all issues in the case. Although the threat of duplicate proceedings may suggest that this is a reliable solution, pragmatic considerations often stand in the way. For example, the primary subcontractor may not want to suffer the extra expense and time of sitting through the entire design portion of the arbitration. Further, the secondary subcontractor, for similar and other tactical reasons, is required to submit to arbitration, but may easily refuse to submit to the combined arbitration in mind. If he does so, the primary subcontractor may then also refuse.

Without proceeding further, it is apparent that unnecessarily complex, unproductive, and inefficient litigation is the result. Even worse, there is a strong possibility of multiple inconsistent decisions.

There are no easy solutions to the problem described. As long as different parties entertain different dispute resolution procedures in their respective contracts, the problems above will continue to haunt parties to construction disputes. Still, the thoughtful practitioner can take some steps to try to ameliorate the problem.

First, when advising one of the parties at the time of contract, counsel can gather and examine the other written contracts to see whether arbitration or litigation is in store for the client's portion of the project in the event of a subsequent dispute. By including an arbitration clause in the contract, counsel can potentially minimize the likelihood of multiple modalities of litigation in the future, especially if the clause contemplates and requires joinder of other parties to obtain a single forum for resolution of the disputes.

Second, when confronted with a situation such as that presented above, counsel can endeavor to promote a group commitment to mediation before any party embarks on a litigation or arbitration course. If the matter can be resolved in mediation with

---

15. Since mediation does not impose a solution by a third party, but generates a solution through negotiation between the parties, all parties who agree to the solution can live with the outcome by definition.

all parties participating, great expense can be saved and the other problems avoided. Mediation may not succeed too early, if the decision-makers in settlement discussions are not aware of all of the facts and cannot fully evaluate risk and exposure. But with a commitment to mediation, pre-litigation discovery can occur, experts can be retained, and the parties can gather the requisite information to make informed decisions and comfortably adopt realistic settlement positions.

Third, counsel representing parties who may be entitled to indemnity from others, or parties who may be required to indemnify others, should seriously consider deferring those disputes until the primary disputes are resolved. For example, the secondary subcontractor will never be required to pay more than the primary subcontractor. If primary can successfully defend against or minimize his liability, then secondary will have little or nothing to absorb. But primary's liability is only determined after his dispute is resolved. By proceeding sequentially as opposed to simultaneously in arbitrating and litigating the various claims, a number of parties may well be able to settle their claims without litigation or arbitration at all.[16]

Fourth, counsel should seriously consider recommending joinder in existing arbitration or litigation to curtail the number of proceedings. While there may be a short-term benefit in being the odd man out and refusing to join the *party*, it usually is a short-term benefit. Over the long term, forcing multiple and inconsistent proceedings is a high-risk strategy that can easily backfire on the one proposing it. Making the resolution of the dispute manageable is ultimately in the best interest of every party, regardless of their positioning in a particular case.

D. *Selection of the Forum and Procedural Considerations*

As noted above, in many construction contracts, arbitration is required by the terms of the contract. In such cases, absent a stipulation to *remove* the dispute from arbitration (virtually unheard of), the parties will be proceeding to arbitration. The forum, in that sense, is preselected. Even in those cases, several questions still remain concerning selection of the forum.

First, who will administer the arbitration? Administering the arbitration includes everything from facilitating selection of arbitrators to coordinating communications among the parties and the arbitrator(s), as well as participating in scheduling matters related to hearings, preparing orders, resolving procedural issues, accounting for all costs and fees, providing hearing rooms, finalizing awards, and the like.

Second, how many arbitrators will there be? If the contract is silent as to the number, but provides that the AAA Construction Industry Arbitration Rules (hereafter

---

16. A *caveat* is in order. In some states, statutes of limitation for indemnity claims may expire before conclusion of the underlying action. Counsel must be intimately familiar with statutes of limitation and may have to obtain tolling agreements as necessary to keep the indemnity issues from being litigated too soon.

CIAR) will apply, there is one arbitrator unless the parties agree otherwise or the AAA decides otherwise in its discretion.[17] The parties may agree to any number of arbitrators.[18] Typically, the choice is whether there should be one or three arbitrators. Cost is a significant factor. Arbitrator fees vary geographically and by arbitrator, but generally are between $200 and $1,200 per hour. With a single arbitrator charging over $1,000 per day, the cost of multiple arbitrators can quickly become an issue in many disputes. However, with three arbitrators, each party has the assurance of multiple opinions and reactions to their presentation of evidence, as well as the dynamic of arbitrators being able to talk to and sometimes convince each other almost independent of the presentation of evidence. There is also some hedge against an aberrant ruling from a single arbitrator who may surprise the parties by going in an unexpected direction for one reason or another.

If there are to be three arbitrators on the panel, counsel must decide whether they are three independent or neutral arbitrators, or whether two of them are party arbitrators. In essence, party arbitrators are chosen solely by one party and are charged to a greater or lesser extent with persuading the other arbitrators to agree with the position of the party that retained them. Although there is a range of roles they may perform, they are to a greater or lesser extent advocates for their party. It is even common (and expected) that party arbitrators will communicate with counsel for their party *ex parte*. On the other hand, the neutral arbitrator is beholden to no one in the dispute. This arbitrator may often become an umpire between the two party arbitrators. The neutral arbitrator may perceive himself or herself as, in reality, the sole arbitrator.

Many counsel are uncomfortable with the concept of party arbitrators, and elect instead to have three neutral arbitrators. This creates in effect a small jury where different arbitrators perceive evidence and witnesses slightly differently, and work together collaboratively to produce a theoretically better-informed and better-reasoned result.

There is another matter critical to the selection of arbitrators. It is that the parties are entitled to full disclosure when they accept or reject an arbitrator as to that arbitrator's relationship to a party, counsel, or witness, and to other information that may bear on bias, prejudice, or the appearance of them. This is one of the most litigated issues involving arbitration, and improper disclosure is one of the few bases on which an arbitration award may be subsequently vacated by the courts.

---

17. CIAR Rule R-16 states, "If the arbitration agreement does not specify the number of arbitrators, the dispute shall be heard and determined by one arbitrator, unless the AAA, in its discretion, directs that three arbitrators be appointed. A party may request three arbitrators in the demand or answer, which request the AAA will consider in exercising its discretion regarding the number of arbitrators appointed to the dispute." Typically, the AAA will require three arbitrators if a party requests and the dispute involves more than $1,000,000 on either side.

18. The author has been selected, with some trepidation, as one of a two-member arbitration panel.

Finally, the parties must agree on the rules under which the arbitration is to be conducted. For example, the parties could stipulate to an individual named arbitrator and also agree that the arbitration is to be conducted under the CIAR, or the Federal Rules of Evidence plus the Federal Arbitration Act, Uniform Arbitration Act, or provisions of applicable statutes in a particular state.

If the parties agree to arbitration or agree to modify the contract provisions relating to arbitration, it is essential that the agreement be reduced to writing in the form of a Stipulation to Arbitrate, or a stipulated Grant of Authority to the arbitrator. The stipulation should contain, at least, the following provisions:

- careful description of the matter submitted to arbitration and the authority of the arbitrator to rule on the matter;
- all parties who will participate and by whom they are to be represented;
- designation of the arbitrator(s) by name;
- declaration that the arbitrator's award shall be final and binding and provision for enforcement of the award (typically, that the prevailing party may file in a particular court to confirm the award and turn it into a final, executable judgment);
- type of award (reasoned or summary);
- provisions for payment of arbitrator fees and expenses;
- agreements (if any) regarding discovery (in the absence of agreement, discovery is typically at the discretion of the arbitrator, in accord with whatever governing law is to be applied);
- governing law of rules for the arbitration, including both rules of evidence and rules of procedures;
- provision for discovery and similar matters to be handled only by the chair of the panel instead of all three arbitrators;
- provision for abbreviated motion filing, timing, and telephonic hearings on all pre-arbitration hearing matters; and
- deadline, if any, by which the arbitration must be completed.

In addition to the stipulation, it is prudent for the parties to enter into a retention agreement with the arbitrator. This agreement should cover the arbitrator's fees and expenses, billing mechanism, required deposits, and cancellation fees. It may cover disclosures (conflicts) and disclosure issues, if any, and include a provision protecting the arbitrator from subpoenas to testify or produce documents in a subsequent proceeding.

E. *Selection of the Arbitrator(s): Additional Considerations*

One of the chief benefits of binding arbitration is that parties can avoid resolution of their dispute by a trial judge who has only a passing acquaintance with construction law and techniques, or a panel of jurors who cannot tell soffits from fascia. At the same time, if parties are not careful in the selection of an arbitrator, they find that their rights and obligations are decided by an individual who has plenty of construction experience, but no practical experience in adjudicating disputes. Accordingly, in order to reap the full benefit of arbitration, parties should choose their arbitrators carefully. The factors that may affect the desirability of a particular proposed arbitrator may include:

- whether she is a full-time neutral (that is, mediator or arbitrator) or maintains an active professional practice as an architect, engineer, contractor, or attorney;
- her level of sophistication in construction matters or in specific construction issues involved;
- her track record as an arbitrator—his or her attentiveness to the evidence, willingness to make difficult decisions, tendency to split the baby, etc.;
- her previous experience with parties or lawyers;
- her personality and demeanor including such factors as:
  - how she will conduct the hearings;
  - her flexibility with respect to procedure;
  - the extent of her ability and commitment to follow the law;
  - her inclination to do justice as she perceives it (perhaps, at the expense of following strict legal rules);
  - her likely reaction to problem evidence or personalities;
  - her approach to contract interpretation and performance and, in particular, whether she tends to enforce contracts strictly in accordance with their literal language; and
  - whether you perceive that your client will be satisfied with his or her day in court at the end of the process, based on her handling of the case.

Selection of arbitrators is sometimes a difficult and demanding process. Ultimately, counsel may have only limited say in the selection. For example, when a provider furnishes a list of names to all counsel, each of whom can remove a certain number of names from the list, the provider will select from among the remainder. Counsel should not hesitate in communicating with an administrator[19] to note that a case is extremely complicated and calls for an experienced arbitrator, or that this case is volatile with respect to relationships and calls for a firm arbitrator.

Finally, the best method for discerning a particular arbitrator's nuances is to talk to as many parties as possible who have utilized the arbitrator. In the context of construction disputes, it is usually not difficult to obtain details about an arbitrator's prior experiences just through word of mouth.

---

19. Of course, it is generally improper to communicate with a proposed *arbitrator* directly, and a party who attempts to communicate with a proposed arbitrator outside the presence of the adversary may be subject to sanctions. However, arbitration *administrators* are usually sensitive to the particular needs of the parties. After all, they are in the business of providing ADR services and they will find themselves without clients if they ignore the needs of their clients.

F. *Pre-Hearing Issues*

1. *Discovery*

Discovery—the formal process through which parties apprise themselves of the other party's case—is one of the most expensive aspects of litigation, and a key reason why some parties prefer arbitration. By the same token, nobody likes surprises and this tension between saving money and being prepared makes discovery a complex and sometimes divisive issue in arbitration. In some fora, discovery proceeds as if the case were in litigation, with the arbitrator ruling[20] on motions to compel responses to interrogatories and the like. Typically, however, most arbitration rule sets and statutes provide for no discovery, except as stipulated by the parties or ordered in the sole discretion of the arbitrator.

Arbitration of a construction dispute without some measure of discovery is often foolhardy. Construction litigation is usually very document intensive. It is essential that both sides have access to all contemporaneous documentation. It is also essential that both sides have access to all expert reports and documents on which experts rely. Often, the arbitration will be more fair and more efficient if the parties have, at least, complete documentation and expert reports and are allowed to take depositions of experts. In many cases, a small number of additional depositions may be allowed where the need is demonstrated. Counsel must bear in mind that the responsibility of the arbitrator includes making arbitration a (relatively) speedy and cost-effective alternative to litigation. Experienced arbitrators are aware that in addition to congestion in the courts, one major cause of the expense and inefficiency of litigation is virtually unlimited discovery, both as to amount and scope. The arbitrator must balance the legitimate need of the disputants for essential information on central issues with the need to make the arbitral process expeditious and cost-effective.

Counsel should endeavor to stipulate, as far as possible, that basic discovery be undertaken in a given case and, only those limited matters on which agreement cannot be achieved, presented to the arbitrator.

2. *Expert Retention and Preparation*

In construction arbitration as well as in litigation, experts have become the norm. Experts are necessary both with respect to liability issues (e.g., standard of care is always an issue in a case against architects and other professionals on the design team) and damage issues. The entire outcome can and often does turn on an expert's credibility and the articulated bases of his opinion.

Therefore, in terms of presenting the case at arbitration, proper expert selection and preparation are essential. The focal point of expert testimony in arbitration is the

---

20. In a three-member arbitration panel, prudent counsel will consider stipulating with their opponent that discovery and other procedural pre-hearing decisions will be rendered by the chair of the panel, who is authorized to consult as he or she sees fit with the other arbitrators. This saves the time and expense of having three arbitrators hear each discovery motion.

expert's ability to persuade the arbitrator. The length of the expert's resume will not save an unpersuasive expert, although it may add credence to an the opinions of an expert who is otherwise persuasive. Obviously, an expert who is inarticulate, does not think well on his feet, or has an abrasive personality should normally be rejected by counsel.[21]

Assuming counsel has identified an articulate expert who is well-qualified in the precise area at hand,[22] the next issue is whether the expert will be persuasive. The biggest single factor in persuasiveness is credibility. The biggest single factor in credibility is a mastery of the relevant facts coupled with a theory that fits all of them. Simply put, the best expert *performance* will be given by the best-prepared expert, all other things being equal. Preparation of the expert is a function of both the lawyer and the expert. Some experts will work in an organized and thorough fashion to obtain mastery of the facts. Some will not, preferring to wing their performance to some extent. Use of the latter makes litigation haphazard, at best.

An expert who has mastered the facts and thought through the implications of his opinions with respect to the facts is very difficult to successfully impeach on cross-examination. On the other hand, an expert who falls short of such mastery and thoughtfulness is easy prey in the hands of a skillful cross-examiner.

A final caution: timeliness of expert disclosures, last-minute consideration of new data, and modification of opinions seem to plague arbitrators as they do judges in construction cases. While an arbitrator is not constrained to a mechanical application of the procedural law regarding exclusion of experts who do not disclose their opinions or the basis thereof in a timely manner, a good arbitrator has some desire to achieve justice at the end of the day. It does not sit well with most experienced arbitrators that counsel or experts have been playing games (or appear to have done so) during the pre-hearing phase. Indeed, such an impression can predispose an arbitrator against a particular expert. Therefore, counsel should pay close attention to deadlines as they occur in the pre-hearing process and respect them as much as possible. In the event that deadlines cannot be precisely met, counsel should seek relief or extensions wherever possible to eliminate the impression of gamesmanship.

3. *Pre-Arbitration Hearings*

The need arises from time to time for the parties to present contested issues to the arbitrator for resolution before the arbitration hearing. Such pre-arbitration hearings may involve discovery issues or problems, relief from deadlines, or any other

---

21. However, note that experts whose logic is plain and simple may be preferable to those who appear slick. Some experts with abrasive personalities are also extremely persuasive despite their attitude.

22. An expert will be more persuasive in direct proportion to the proximity of his experience and knowledge to the precise issue or sub-issue at hand. Expert battles frequently can be won by having an expert whose specialty is closer to the precise issue than that of the opposing expert.

matter affecting progress toward the arbitration hearing. Experienced counsel and arbitrators will anticipate such matters in the first preliminary hearing. It is a good idea to have a procedure in place for making motions or applications to the arbitrator. Can it be done electronically (e-mail)? How much time needs to elapse between the motion and the response? How soon can a hearing be scheduled after the filing of the motion? Can every pre-arbitration hearing be telephonic? If there is to be a stenographic record of the arbitration, must there be a stenographic record of each such hearing?

These and similar questions can be addressed with a pre-hearing protocol put into effect as soon as the arbitrator is selected. Some arbitrators prefer more formality than others. Counsel needs to communicate with the arbitrator (either at the time of retention or in the first preliminary hearing) in order to determine these matters (which may well bear on the conduct of the arbitration hearing as well).

Dispositive motions (e.g., motions for summary judgment) are rapidly becoming a fact of life in the arbitration process, as at the courthouse. Although there is nothing in most rule sets (including AAA's CIAR) prohibiting the making or granting of a summary judgment motion, most arbitrators (like most trial courts) are reluctant to preclude a party's day in court by way of summary judgment, and such a motion is not likely to end the parties' entire dispute.[23] Nonetheless, as in litigation, counsel may desire to alert the arbitrator to a weakness or defect in her opponent's case. Summary judgment (or other terminal motion) may be an appropriate and tactically astute way to accomplish that.

4. Briefing

As part of the arbitration scheduling order, there should be a briefing schedule for opening briefs as well as the exchange and service of pre-marked exhibits. Briefing schedules may follow the opening-response-reply pattern in which the party with the burden of proof is allowed opening and closing briefs, or the simultaneous exchange of opening and response briefs (favored by many arbitrators) in which each party is allowed two briefs but each submission is simultaneous. Ultimately, briefing is a matter of substantial discretion with the arbitrator.

---

23. Exceptions would include pure matters of law, where the arbitrator's ruling on a matter of law will be determinative whenever it is made, and inability to meet a legal threshold such as the statute of limitations. These matters seem to lend themselves to summary judgment in a special way. Arbitrators do have inherent jurisdiction to control the arbitral process. Bifurcation may be appropriate in these cases, such that instead of a summary procedure, each party is afforded the full evidentiary hearing, but limited in scope to certain issues whose determination may affect the remainder. Thus, it would be natural where the statute of limitations (or other affirmative defense) is a major issue to have a phase one arbitration hearing limited to evidence bearing on the statute of limitations. If the ruling is for respondent, the arbitration would terminate at that point with an award in favor of respondent. If the ruling is for petitioner, the arbitration would proceed to phase two, an evidentiary hearing on the merits of petitioner's claim.

Accordingly, counsel are encouraged to ask the arbitrator what his or her preferences are with respect to briefs, advance consideration of evidence, and detailed discussion of evidence before its presentation (in the presence of the opponent, of course). In any case, it is suggested that counsel consider the possibility of a relatively short opening brief, designed to set the stage, review the relief sought or opposed, give an overview of why the evidence supports that party's contentions, and address other issues that are sure to come up during the hearing. Then, this brief can be augmented by short issue-specific briefs on known evidentiary questions and substantive law issues likely to arise during the hearing. Arbitrators, just as trial judges, will often request briefing during the arbitration on selected points. Indeed, it is impressive when an advocate is prepared on such issues ahead of time and submits a one- or two-page brief that the arbitrator can read on the spot and make a ruling as the hearing progresses.

As in a court proceeding, briefs should contain not only argument but liberal reference to appropriate case and statutory law. If presented during the hearing, the brief should be accompanied by a copy of the one or two most significant legal authorities—again, so that the arbitrator can consider the matter fully on the spot and rule without further research or delay.

G. *The Arbitration Hearing*

1. *Scheduling*

When should the matter proceed to a hearing? Also, what intermediate steps need to be calendared along the way? Thinking through the arbitration process at the beginning of the dispute helps experienced arbitration counsel formulate a comprehensive, consistent, and thorough game plan.

The arbitrator(s) have the responsibility to try to keep the arbitration process cost-effective and speedy. This means that lengthy intervals without explanation will often be discouraged or even prevented by the arbitrator(s). Thus, as a general rule, the hearing should proceed as soon as discovery is complete, experts are retained and exchanged, and all schedules permit. In some cases, arbitration can occur three to six months following the submission of the original demand for arbitration. In others, the hearing may be one year to two years after the process is initiated. It is and should be a rare case when an arbitration hearing is as far removed from inception of the dispute as a formal trial would be. In addition, consideration should be given to less obvious issues such as bifurcation of the hearing for greater efficiency. The arbitrator has great flexibility in hearing logistics. Sometimes, bifurcation is in the best interest of all parties.[24]

---

24. Bifurcation is not limited to liability-damages phases. In multi-cases, claims and defenses can be heard in a reasonable order to maximize both efficiency and clarity. Sometimes, particular issues, claims, or defenses can be separated out of the whole dispute. After a ruling on those, the parties may be able to resolve the remaining issues.

2. Special Issues in the Arbitration of Construction Cases

In one sense, arbitrating a construction case is no different from arbitrating any other type of dispute. In another sense, construction cases have unique issues, party relationships, and problems that separate them from other cases. We consider several substantive and procedural issues that arise with some frequency in both construction claim and defect arbitrations.

a. *Historical Revisionism*

While the charge is not atypical in litigation, construction cases seem especially prone to claims that the litigation picture bears little resemblance to the project as it actually proceeded. Arbitrators are often curious as to whether particular claims or defenses are the product of creative legal thought after the fact, or whether they appear to have arisen on the job site, contemporaneous with the construction project.

The problem is exacerbated by the frequent unavailability or pronounced bias of the most knowledgeable witnesses. It is further compounded by the lack or rarity of contemporaneous documentation, almost always on at least one side and sometimes on all sides of the dispute. It is often the case that between the time a dispute arises on the job site and the time that litigation (arbitration) is initiated, project managers, superintendents, responsible professionals, and others closest to the action at the time are no longer employed by their previous employer. Thus, it falls to middle and sometimes senior management to carry the ball forward in an arbitration setting. While these overseers may have had some connection with the underlying facts (varying from minor to major), they were not the principal company representatives charged with direct project management at the time of the construction. Much of their testimony will inevitably be based on documentation and a general personal acquaintance with the underlying facts, as opposed to an intimate and detailed daily experience with the job itself.

As a result, use of individuals with direct knowledge of the dispute is often a strategic advantage. Of course, the simple fact that an individual is no longer employed by the party does not mean that she is unavailable as a witness. On the contrary, the problem may be that the witness has been laid off or fired, or even quit, and has a bias against her former employer.

Another defense to the threat of historical revisionism is the presence of a well-documented contemporaneous record. Although the general contractor, subcontractor, and construction manager are all charged with (usually) maintaining a daily record of job activity, the documentation is frequently either absent or sketchy at best. Alternatively, one party may have considerably better documentation than the other(s). Other things equal, the party with the better documentation usually has a distinct advantage. After all, if a major goal of the arbitrator is to get to what really happened as opposed to legal spin, contemporaneous documentation would seem to head toward the heart of the matter, and it appears essentially more objective than the testimony of witnesses months or probably years after the events at issue.

b. *Delay Claims*

Parties often assert damages caused by work delay or disruption in both construction defect and claim disputes. Although some parties feel that delays and their impacts are intuitively obvious, as a practical matter, the establishment of critical path delays and the quantification of resulting damages are complex and require expert testimony. Parties should not underestimate the difficulty of establishing such claims and should prepare accordingly.

c. *Formality of the Process*

By the time the hearing commences, the advocate should have a clear understanding of how the arbitrator will preside over the process. Arbitration is almost universally less procedure-bound and formal than trial. In the larger and more complex construction case, formality may be required simply to preserve order and keep the process more efficient. In a two-sided arbitration, the arbitrator has at least the option of being extremely informal in manner and procedure. Some arbitrators allow counsel great latitude in scheduling witnesses, in setting starting and concluding times, and in the manner in which the hearing is conducted. Others may be somewhat rigid and inflexible on the same subjects.

In cases where one of the parties has opted for a stenographic record of the proceedings, the process inevitably becomes more formal. Experienced counsel and arbitrators all speak and act more self-consciously where there is a record.

Well-prepared counsel knows in advance what to expect of the arbitrator's preferences and process, and tailors his presentation and conduct of the hearing to accommodate the arbitrator. Disputes with the arbitrator are almost never productive for the advocate or for his client.

d. *Discontinuity in the Hearing*

Fortunately, lawyers are not paid based on the accuracy of their time estimates. In fairness, it is extremely difficult, especially months or years before discovery is completed, for lawyers to accurately predict how long a hearing will take. Arbitrators (who are typically lawyers as well) do not seem to fare much better in this respect. Therefore, it is not uncommon for a hearing to take longer than originally scheduled. What course is best to follow in such a situation, given the acute frustration that parties experience with any delays in a process that is chosen in part for its efficiency?

First, there are at least two ways of precluding the problem. Anticipating the length of the hearing, the parties may reserve extra time with the arbitrator. Even if it is necessary to pay for the arbitrator's time, it is preferable to have more time than is necessary, rather than to suspend and continue the hearing. During the hearing, there is another opportunity to correct for an underestimate. Counsel can shorten witness presentations (often, with little sacrifice in presenting the essentials). If the arbitrator is willing, hearing days can start earlier, end later, and lunch time and other recesses can be reduced. If it appears inevitable that the hearing will have to be recessed and reconvened, counsel should again consider a reasonable bifurcation of issues. It is preferable to sever the issues neatly than to merely insert a break into the middle of what would otherwise be a cohesive presentation.

Second, if a continuance is unavoidable and neat division impossible, the advocate should understand the damage caused thereby and strive to minimize it. As the effect of the discontinuity increases over time, every effort should be made to reconvene the hearing at the earliest possible time. Counsel should also consider creative ways to assist the arbitrator in remembering what transpired in the first part of the hearing at the next setting. For example, the advocate could submit a brief or written argument before the second hearing which recapped the evidence presented at the first and tied that to what is expected to occur at the next portion of the hearing. Knowing that the hearing is going to run longer than the time set aside, counsel can also reserve an important lay or expert witness until the end of the first part of the hearing. Instead of rushing to finish that witness before the first hearing concludes, counsel could end in the middle of that testimony. At the next setting, counsel can resume with the same witness and use that both to establish continuity and to briefly recap and reintroduce the subject matter of the earlier testimony.

Finally, counsel should keep in mind that, as troublesome as some continuances can be, the failure to grant a justified continuance is one of the few grounds that most jurisdictions recognize as a basis for invalidating an arbitration award.

e. *Use of Technology in Proceedings*

Since construction cases are document intensive, the most efficient way of handling documentary evidence may be through electronic means. Modern technology adds certain pyrotechnics to a presentation that can have a substantial impact on the case.

There are several considerations in determining whether and how to implement such technology as part of an evidentiary presentation. First is the question of cost. It is apparent that digitizing, editing, and categorizing videotaped depositions is not inexpensive. It can be accomplished through technology consulting and support firms, legal assistants, associates, or principal trial attorneys. Although the technology may be effectively employed, it may be too expensive in terms of impact and results in smaller cases. However, in larger damage construction matters, technology may add a level of interest to the presentation and greatly enhance the persuasiveness of evidence.

Second, counsel must consider the reaction of the arbitrator. While some arbitrators may be intrigued by and interested in a technological dimension to the presentation, others may find it an unamusing distraction to the business at hand. Since the purpose of every aspect of the advocate's presentation is persuasion, any technique or device that hinders persuasion should be eliminated.

Accordingly, counsel should raise the technology issues as early as possible with the arbitrator. In addition, counsel should make provision to accommodate the opponent in every way possible, in order to minimize the appearance of raw overpowering and unfairness (e.g., offering to the opponent to allow the use of the professional and the database to project any exhibit that the opponent may desire). Additionally, counsel should make provision to review specially edited exhibits with the opponent before the hearing, to allow for appropriate objections. If counsel manages the utilization of technology so as to enhance the presentation in a fair way while making the arbitrator's job easier and not more difficult, it can be an extremely effective way of augmenting the presentation of evidence.[25]

f. Nature of the Award

In construction cases, it is important for a variety of reasons to have not only the decision but the thought process of the arbitrator. Consequently, parties usually should request a *reasoned award* rather than a simple decision indicating how much one party owes the other party. For example, CIAR Rule R-43(b) provides for a reasoned decision only if all the parties request it in writing or the arbitrator exercises his discretion to issue such a decision. Under other rules, a reasoned award may or may not be required. When a reasoned award is required, it is possible that an award that is not reasoned will be reversed on that basis.[26] It is the responsibility of counsel to attempt to provide for such an award under whatever rules are applicable.

H. Post-Hearing Procedures

1. Requesting Correction or Modification of the Award

Once an award is entered, it is very difficult to seek reconsideration or amendment of the arbitrator's decision. For example, CIAR Rule R-47 provides that "[w]ithin twenty calendar days after the transmittal of an award, the arbitrator on his or her initiative, or any party, upon notice to the other parties, may request that the arbitrator correct any clerical, typographical, technical or computational errors in the award. The arbitrator is not empowered to redetermine the merits of any claim already decided." Consequently, the function of this rule is only to permit correction of evident errors in the decision and not to provide a launching pad for attacks on the substance of the award itself.[27]

2. Confirming the Award

An arbitration award finds its power in the fact that it can be confirmed as a judgment of a court in the jurisdiction in which the award is entered. Once the award

---

25. All the ramifications of the use of electronic technology to present the construction case are beyond the scope of this article. It should be sufficient to note that many consultants earn decent livings assisting attorneys with the presentation of complex construction cases to judges, juries, and arbitrators.

26. *See, e.g.*, Vold v. Broin & Assocs., Inc., 699 N.W.2d 482 (S.D. 2005).

27. Arbitrators are frequently allowed such broad discretion that the rules—including this one—may be observed in the breach. The author is aware of a Colorado case in which the arbitrator utilized this rule to essentially re-decide the case such that the prevailing party changed, as did who paid and received money as a result of the award. The courts refused to disturb what was essentially a dramatically reversed decision of the arbitrator, justified as correction of "computational error."

is so confirmed, the prevailing party has the power to enforce the award through compulsory means such as garnishment, foreclosure of the judgment lien, and other such means.

The precise manner and procedural requisites for confirming an arbitration award are spelled out in the Federal Arbitration Act[28] and in state arbitration statutes. In general, the prevailing party is entitled to commence a formal suit for confirmation of the award, although the confirmation process is abbreviated to provide expeditious resolution of any dispute to confirmation of the award. It would give the parties to an arbitration little comfort if the process of confirming an arbitration award were as cumbersome as regular litigation, and indeed it is not.

3. Vacating the Award

One attribute of arbitration that both attracts and annoys litigants is that arbitration awards are truly final in the sense that they are extremely difficult to attack on substantive grounds. The process through which an award is confirmed by a court does not invite (nor does it allow) the disgruntled party to revisit the arbitrator's decision on the basis that the arbitrator's decision was erroneous or against the manifest weight of the evidence. Instead, an arbitration award is generally subject to attack only on narrow grounds specified by the applicable arbitration law—either the Federal Arbitration Act or the applicable state statute.

By way of example, the Federal Arbitration Act permits vacation of an award (or refusal to confirm the award) only on one of the following grounds:

(1) where the award was procured by corruption, fraud, or undue means;
(2) where there was evident partiality or corruption in the arbitrators, or either of them;
(3) where the arbitrators were guilty of misconduct in refusing to postpone the hearing, upon sufficient cause shown, or in refusing to hear evidence pertinent and material to the controversy; or of any other misbehavior by which the rights of any party have been prejudiced; or
(4) where the arbitrators exceeded their powers, or so imperfectly executed them that a mutual, final, and definite award upon the subject matter submitted was not made.[29]

Although one might read into the sufficiently broad language of these four exceptions on which a substantive challenge might be based, these conditions for vacating an award have regularly been construed narrowly to prevent substantive

---

28. 9 U.S.C. § 9 (2005).
29. 9 U.S.C. § 10(a) (2005). The Uniform Arbitration Act includes similar provisions. Many states permit vacatur of an award on the additional express grounds that there was no valid agreement to arbitrate.

attacks on the pretense of raising one of these procedural challenges. As a result, parties to arbitrations should understand from the outset that the arbitrator's decision will be final in the absence of some very fundamental defect in the proceedings.

Library References

West's Key No. Digests, Alternative Dispute Resolution <KEY>111, 112, 125, 152, 153, 220–222, 239, 250–252, 255, 256–267, 307, 316–335, 343, 344, 454, 353, 391, 316–335.

IV. Strategies in Mediation

A. *Selection of the Process*

1. *Overview of the Mediation Process*

Although some experience with mediation is presumed, it may be helpful to briefly review the steps in the mediation process. After any pre-mediation communications, submission and consideration of briefs, the mediation typically begins with all participants in one room and an opening presentation by the mediator. This session may last for as little as fifteen minutes to thirty minutes, or continue for as long as several days.

The opening presentation by the mediator is usually brief. It is designed to acquaint the parties with the process used by the mediator, to relax the participants, and to cover certain essentials of the process, including the mediator's style and approach to the process, any prior knowledge of the mediator related to the case, an overview of what the participants can expect in terms of timing and process, how confidential communications will be handled by the mediator, and basic and critical issues of logistics, breaks, stopping time, etc. At the same time, the mediator may remind the parties that the alternative to their resolution of the dispute in the mediation session is prolonged and expensive litigation.

The opening statement of the mediator is followed by opening statements and sometimes presentations by the advocates and/or experts for the parties. Information exchanged in the opening session is the beginning of direct negotiation between the participants and is usually encouraged by the mediator, but the decision on whether and how to participate is typically left to the advocates. The facilitative mediator may ask questions and validate that he or she has heard what the advocate or expert has said, but almost never challenges a participant in front of the others. The evaluative mediator may be considerably more aggressive in participation.

An important part of the opening session in a construction mediation, especially if the parties have not met with the mediator ahead of time to define the process, is setting forth the parameters of the process—what we might call the logistics of the mediation. This comprises everything from identifying the issues to be discussed, and the order in which they are to be discussed, to a time estimate[30] for when specific parties can expect to meet with the mediator.[31] Careful management of these logistics can help to diffuse tension, frustration, and impatience, although they invariably cannot eliminate it for all participants.

A series of caucuses (private meetings with the mediator) follow the conclusion of the opening joint session (usually defined by the facilitative mediator as the point at which the parties believe that further conversation in the group is likely to be less productive than private conversations with the mediator). Each caucus may consist of a meeting with one party or with a group of parties (the mediator may meet initially with all of the defendants, all of the subcontractors, all of the design team, etc.). The mediator then works sequentially through the parties and groups until returning to the first party or group.[32] The caucuses continue until the parties decide to reassemble in a larger group, individual parties have resolved their disputes and are dismissed, or the process comes to an end through irresolvable impasse.[33]

2. Selection of a Specific Process

Unlike arbitration, which is still largely bound by formal rules and procedures, mediation is generally unrestricted beyond loose parameters described above. Although there are some rules "around the edges" of the mediation process,[34] the

---

30. Recall that mediators are mostly lawyers too, and do not necessarily have a better track record than trial advocates at time estimates.

31. Because of the number of participants and the disparity in their likely substantive participation, the question of when and how long the mediator spends with various participants is a complex and sometimes provocative one. For example, while the mediator may wish for the venting sub-subcontractor to have appropriate time to do so, the logistics may make that a difficult option. Others who have a much more significant stake and role in the process may become irritated and impatient with the mediator who spends what is perceived as too much time with a minor subcontractor. On the other hand, the advocates themselves may expect too much of the mediator if they expect to see him anytime soon with six or more other teams to talk to on the way around the table.

32. Alternatively, as appropriate, the mediator may move back and forth, speaking multiple times with certain groups or parties before completing the circuit. There are an infinite variety of possibilities.

33. False impasses frequently occur in mediation. It is the mediator's responsibility both to head off the building of impasse walls and to try to break through them without declaring the mediation at an impasse.

34. Two examples are case management orders or rules of court requiring mediation at certain times and under certain circumstances or as ordered by the court, and post-mediation issues of enforceability of mediation awards, discussed below.

conduct of the hearing[35] is unspecified by any law or rule.[36] There are certain models of mediation followed to a greater or lesser extent by individual mediators, and there are styles of individual mediators that define to some extent how the hearing will proceed.

Because the very process of mediation changes depending upon the style (for lack of a better word) of the mediator, selection of the mediator is critical.[37] This section, *Selection of the Process*, is placed before the next, *Selection of the Mediator*, to draw the reader's attention to the importance of the nature of the *process* as an essential part of selection of the mediator.

It is outside the scope of this work to discuss the theory of the mediation process in great detail.[38] It is nonetheless appropriate to discuss some mediation process theory in order to understand how mediators work and what mediator might be a better or worse selection in a particular case.

Rather than discuss all of the available mediation models and their variations, let us consider the two extremes of the mediation continuum, sometimes called *facilitative* and *evaluative*. The facilitative mediator is essentially committed to allowing the parties to negotiate with each other to achieve their own resolution, through the facility of the mediator. In pure facilitation, the mediator does not express her opinion, does not challenge the positions taken, and concentrates on facilitating the negotiating dialogue between the parties. In short, the facilitative mediator does not use power at all, at least in a visible way.

---

35. Some mediators would even bristle at the word *hearing*. The word connotes a formal proceeding of some sort, with someone in authority. Mediation is an informal process in which the mediator is in authority only to the extent necessary to keep things moving in the right direction, make administrative decisions, and the like. Put another way, hearing implies that the judge or hearing officer has vested, institutional authority. Mediators have only such authority as the parties give them by consent through the process.

36. There are certain mediation processes that are more highly regulated. Examples would include labor-management mediation and conciliation and, to some extent, domestic relations mediation. This chapter does not address those processes that are not relevant to mediation of construction disputes.

37. Selection of the mediator is as critical to the mediation process as selection of the arbitrator is to arbitration, but for very different reasons. Arbitrator selection is critical because the individual chosen will dictate the outcome of the dispute. That outcome produces at least one loser, who is bound to be displeased. Mediation can be a frustrating or rewarding process, depending upon how it is conducted. Since mediators use vastly different processes, selection of the individual determines the process that will be followed. Mediation may be unproductive simply because of dissatisfaction of parties or lawyers with the process itself.

38. For those interested in a detailed examination of the process, see CHRISTOPHER MOORE, THE MEDIATION PROCESS: PRACTICAL STRATEGIES FOR RESOLVING CONFLICT (2003).

Some lawyers are cynical about facilitative mediation. Expressions like "win-win"[39] do not seem to assuage the cynicism. In fact, the facilitative mediator performs a service of great value in negotiation. Keeping the parties at the table is often a major achievement in itself. A negotiation that would derail early and often without a trained neutral will frequently persist to the end through the efforts of a capable mediator. From allowing venting of parties (and lawyers) to go on in a nondestructive way (the other side never hears most of the venting), to building confidence that the dispute is indeed capable of resolution, to convincing the parties that progress is being made even though there is no appearance of progress, to killing messages that would be terminal to the negotiation, the facilitative mediator enhances the negotiation without ever taking a position which directly confronts the positions of others, or wielding obvious power.

The facilitative construction mediator also needs to be sensitive to and experienced in maintaining the delicate balance of process satisfaction versus outcome satisfaction, and to focusing the principals on the differences between goals and real interests. Parties may be quite willing to pay more or receive less than their predetermined goal once they understand that there is more to the resolution than simply counting dollars. Appreciation of the risk of receiving an adverse outcome at trial may introduce flexibility which the party is quite willing to accommodate. The belief that the party's position has been heard, understood, and communicated effectively to the opponent[40] goes a long way towards creating process satisfaction, which can be traded off to some extent with outcome satisfaction in the party's eventual perception. Similarly, the satisfaction that, in some tangible way, justice is being achieved even though the result does not conform to the party's preference may produce an outcome that is quite acceptable to a party who initially wanted to pay less or receive more consideration as the outcome of the negotiation. In sum, if the party perceives the process as both fair and valuable, he is much more likely to consider outcomes that are different than his original negotiating goals. It is the author's uniform experience that almost every case will settle where the parties are satisfied with the process.

The mediator must also be alert to potential process dissatisfaction caused by the parties themselves. Parties will often translate frustration with each other's negotiation styles, long waiting times, and the like, into dissatisfaction with the process and/or the

---

39. *See* ROGER FISHER, WILLIAM L. URY & BRUCE PATTON, GETTING TO YES: NEGOTIATING AGREEMENT WITHOUT GIVING IN (1991). Many mediators begin the process by telling the parties that a good settlement is one in which nobody leaves happy—a proposition disputed by the present author. Understanding the delicate balance between *outcome satisfaction* and *process satisfaction* is critical to understanding and defining success in any negotiation

40. In effect, this is the mediation party's equivalent of a *day in court*.

mediator personally. The frustration may even be the result of dissatisfaction by one team member with another member of the same team.[41]

The mediator must also be careful to work with parties to help them make informed risk-benefit decisions. Parties (and lawyers) often take positions with respect to the issues in controversy. The positions may be motivated by a wide range of factors, including the perception of right and wrong, the belief that a certain position is necessary to achieve a certain monetary result in negotiation, and an opinion of the merits of one's own case or the other party's case. However, parties often take positions that are inconsistent with their real interests. For example, certainty of outcome, precluding further expenditure of time, financial, and personnel resources, may be of far greater import than the monetary amount of settlement consideration. However, parties often focus on the amount of consideration and ignore all these other factors. Positional bargaining tends to produce linear negotiation, in which only dollars are traded. Interest-based bargaining is multifaceted and often produces resolution where linear negotiation would fail.[42]

At the other end of the spectrum, the evaluative mediator comes to his or her own opinion as to the merits of different arguments made and positions taken, confronts parties directly in response and, in a visible way, takes substantial responsibility for fashioning the settlement or outcome of the process. An evaluative mediator often uses institutional power as the basis for engaging in a highly directive process. For example, the evaluative mediator may be a settlement judge or magistrate, a court-appointed special master or referee, or a retired judge[43] who still has the appearance of the judicial cloak.

In a construction case, pure facilitative mediation is rarely the model used by successful mediators. The logistics of a construction case make facilitative mediation almost impossible. Even in cases where the arbitration or lawsuit may involve only two parties, the mediation often involves many more parties. The mediator has

---

41. The author recalls a mediation, where after spending about an hour in a joint opening session and one and a half hours with the plaintiff (the length of the session caused by a particularly talkative party plaintiff), plaintiff's counsel commented, "We've already been here for two and a half hours, and I don't have an offer yet!"

42. Much more could be said—indeed many volumes have been written—about positional versus interest-based bargaining. A detailed discussion is once again beyond the purview of this chapter. The brief mention of the subject here is to remind the practitioner of the importance of the mediation process and how it is managed to successful outcomes.

43. It should be noted that many retired judges are also skilled and successful facilitative mediators. There are also well-known non-judge mediators (especially in the construction arena) who are much more evaluative than facilitative in their approach.

44. This is often an extremely complex task. Mediators do not necessarily know who the most important parties are nor who or what the major apparent obstacles to settlement are until well into the negotiation process. Construction cases sometimes begin with a lack of clarity as to who will be paying and who receiving money when the dust settles.

to fashion the process in a visible way so as to check in with all parties with some frequency and focus on the most important persons to the resolution of the matter.[44] The construction mediator must necessarily wield at least some visible power from the outset of the process.[45]

The perceptions of the parties and advocates are also extremely important. Discouragement easily turns into disgust with the process and the belief that the whole affair is a waste of time. Mediators must keep the parties at the table until resolution is achieved. Thus even the facilitative mediator must show and exercise power in the management and administration of the process which is often invisible in a two-party negotiation.

Virtually every construction mediator is somewhere between facilitative and evaluative on the continuum, but each is anchored more or less to one of the extremes, and tends to follow what could be characterized as a modified facilitative or modified evaluative approach.

In selecting the process desired in a particular case, counsel should keep a number of factors in mind, including the following:

- preference of the advocate;
- preference of the client;
- candid assessment of both your and opponent's negotiating style and type of process which will work best;
- candid assessment of both your and opponent's negotiating skills and how they are best augmented;[46]
- unrealistic expectations of your own client and how they will be best controlled and revised; and
- level of overt hostility between the parties or counsel.

B. *Selection of the Mediator*

It is apparent that selection of the process, discussed immediately above, is the largest factor in selection of the mediator. In construction cases, it is rarely productive to select a mediator who is a practicing lawyer unskilled[47] as a neutral but

---

45. This is not to say that the parties have no say in defining the process. The best construction mediators consult with the principal counsel before commencement of the mediation and spend some time defining a process that is acceptable to them. One example of a frequent issue in complex construction mediation is whether subcontractors' time with the mediator precedes or follows sessions with the principal parties (owner, architect, general contractor, etc.). The counsel always know more than the mediator about the underlying relationships and issues; some of them may not always know more (and frequently know less) about successful negotiation as it applies to those relationships and issues.

46. The best negotiators typically prefer a more facilitative mediator, since she allows them much greater control over the negotiation and ability to use their skills more effectively.

47. Note the difference between skill and training. Although training of neutrals is important, much of the skill of a mediator is intuitive. It is possible to be an effective mediator with little or no formal training. One can also be quite ineffective as a mediator despite substantial training.

extremely knowledgeable in the field central to the dispute. It is the author's opinion that it is uniformly preferable to have a skilled mediator ignorant of construction than an expert in construction who is inept as a mediator.

Selecting experienced practicing lawyers who are skilled mediators is appropriate in many cases.[48] Counsel should take care in the selection, since practicing construction lawyers tend to identify with one group of principals (owner, architect, general contractor, subcontractor) more than others over time. An ideological bias can accompany the repeated representation of certain interests.[49] Experience with the proposed mediator *as a mediator* is critical to the selection.[50]

Over the past decade or so, a cadre of skilled professional mediators has grown up who are knowledgeable and experienced in construction litigation issues. Some of these are former practicing construction attorneys or judges and some have picked up what they know through experience. They make excellent candidates as mediators in construction cases. Due to the availability of both professional mediators who specialize or are knowledgeable in construction matters and practicing lawyers who are skilled mediators, it is usually unnecessary to consider professional mediators who are inexperienced in construction.

The success rate, reputation, and personality of a mediator ought also to be considered in the selection process. When counsel have questions, or lack firsthand experience with a proposed mediator, they should not hesitate to contact the mediator and discuss his training, experience, statistical success, and process he follows in some depth.[51] Counsel should also ask for references and incorporate those into the decision.

---

48. As always, it helps to know the mediator in her role as mediator as much as possible.

49. Equally, an individual with an ideological bias may bend over backwards to be receptive to those who are usually his opponents.

50. The author is aware of a number of practicing construction lawyers who are skilled and highly effective as mediators. Mediators who fit this description should not be rejected simply because they continue to practice law.

51. It seems that counsel are so conditioned about ex parte communication that they are often reluctant or unwilling to contact the mediator in advance of the mediation. They should resist this inbred caution when dealing with the mediator. After all, once the mediation begins, it will not be long before most of the communication between the counsel and mediator is ex parte. There is no good reason to avoid it before inception of the formal hearing. It is far more important to address questions directly to the mediator candidate than to discover friction and discontent with the mediator's style, experience, or personality during the mediation. The mediator, for his part, has the responsibility to ensure that the process is symmetrical and shows the appearance as well as the substance of impartiality. For that reason, if there is extensive communication with any counsel before the hearing, the mediator should ensure that there is communication with all (significant) counsel, insofar as practicable.

C. Planning for and Attending the Mediation

1. How Much Expert Work before Mediation?

One of the great attractions of mediation is that it is cost-effective, i.e., it saves the parties money compared to arbitration or a trial.[52] One issue that comes up in every construction case (in every case for that matter) is the timing of mediation. A mediation should be able to be successfully conducted shortly before trial, after all discovery is completed and all experts have completed their reports, testing, etc. However, mediation is not nearly as cost-effective at that stage as one conducted before the retention of experts, for example. On the other hand, mediation before the retention of experts may be unlikely to succeed and therefore not be as cost-effective as it may appear.[53]

Counsel should determine early in the case when the appropriate time for mediation will arrive. In a construction case, it is not unwise and frequently fruitful for the parties to select a mediator early in the case and meet for half a day or one day to acquaint the mediator with the case. Such a session can be remarkably productive in identifying required discovery and expert disclosures and sequencing those activities so that the case can proceed to a successful mediation or resolution on a future occasion.[54] Typically, it is possible in a construction case to target specific critical depositions and appoint experts, but avoid full expert workup, depositions, etc., before mediation.

These considerations ultimately resolve to questions of persuasion. With an evaluative mediator, what is the minimum necessary to persuade the mediator so that he or she, in turn, may prevail upon your opponent to consider your position more or less meritorious? With the assistance of the facilitative mediator, what is the minimum necessary to persuade your opponent that your position, viewed objectively, presents a degree of risk that has significant value to the opponent? We might call the threshold

---

52. Note that the cost-effectiveness of mediation can involve complex calculation. For example, mediating a construction case in which one of the principal parties has highly unreasonable expectations, which his lawyer has tried without success to manage and get under control, can take a long time. It still may be preferable (economically, as well as in other ways) to a trial or arbitration in which the client's expectations are revised in a more blunt and forceful way.

53. The author has witnessed the resolution of even condominium construction defect litigation before filing of the lawsuit or arbitration, but only in a handful of cases.

54. Early mediation can also open the possibility of creative use of off-the-record discovery to allow for settlement discussions and preserve the parties' rights. For example, an expert can be retained to do a preliminary analysis and issue a preliminary, informal letter opinion on the stipulation that the letter is not admissible in evidence, nor is the fact that the expert expressed that opinion. Experts can be allowed to speak with each other, without fear that each can use the other's statements as admissions at a later stage of the proceedings. Many other creative uses of informal discovery can advance the cause of getting the case to productive mediation earlier than would otherwise be possible.

"credible threat."[55] Attending a mediation before the retention of a delay expert and expecting the opponent to accept your delay position on the basis that if the case does not settle, you will retain a qualified delay expert who will prove up your side of the case does not rise to the level of a "credible threat." On the other hand, retaining a delay expert who is respected or feared by your opponent and having only preliminary, as opposed to completed opinions on hand, may well accomplish the purpose. In all cases, the threshold threat level is determined not by *your* view of the evidence, but by your opponent's (or the evaluative mediator's). What is the quantum of proof that is sufficiently risk-raising—as perceived by the other side—that the negotiation will be productive from your perspective? The assistance of the (facilitative) mediator can be invaluable in the presentation of such proof to the other side.[56] What no mediator can accomplish is the creation of the perception of a credible threat in the absence of the party meeting the opponent's perceptual threshold before the mediation.

2. *Use of Technology, Live Experts at the Mediation*

Technology has been discussed at length above as an arbitration consideration. What of its application in the mediation process? In those cases where electronic document management and/or digitized video selections are intended to be used at trial or arbitration, it makes good sense to use a judicious selection of them at mediation. Distraction of the advocate is not nearly as dangerous in the mediation context, because the use of technology can easily and effectively be limited to a pre-prepared presentation that follows a script. At the same time, effective use of technology at mediation, in an appropriate case, can devastate the opponent who is exposed to a thorough presentation, as the opposing parties and principals can see a real threat in three dimensions in front of them.

Effective use of such technology at mediation may persuade the opponent of any or all of the following:

- your mastery of the extensive documentation, witnesses, and ability to try the case in a penetrating and precise manner;
- the persuasive effect of "instantaneous impeachment;"[57]

---

55. It makes sense to talk in these terms since the experienced negotiator ultimately makes and evaluates settlement proposals based on a risk-benefit analysis. A credible threat is one that raises risk in the estimation of the opposing negotiator.

56. There is the *magic* of what we may call the *neutral mouth*. Many statements made by one advocate will be shrugged off by the other. When the mediator makes the same statement, it may well be heard and received by the opposing counsel, party, and/or principal.

57. Combining a witness' videotaped testimony with a facially contradictory or impeaching document, or two contradictory documents with each other, and projecting the two simultaneously produces a dramatic effect, almost impossible to equal with traditional cross-examination. The latter relies on memory and attention span, since the cross-examiner first lays a foundation and then proceeds with impeachment. The former does not, since the foundation and impeachment occur simultaneously.

- the graphic impact of visuals as opposed to conversation or argument about witnesses and evidence;
- the virtual inability to respond to the technological presentation except by a counter-technological presentation (extremely rare);
- in short, the high risk that accompanies the organized and powerful presentation of the weaknesses of the opponent's case.

Negative considerations to using technology at the mediation include:

- Distraction of the presenting counsel (usually remedied by having the technology professional at the mediation).[58]
- Cost of the preparation and presentation.
- Some mediators and/or opponents have an aversion to the technology and may react with anger rather than fear. This can be counterproductive.
- By definition, the revelation of impeachment before trial or arbitration gives the opponent a preview which allows her to prepare to meet your presentation at the decisional hearing. Many lawyers are quite reluctant to inform their adversaries and eliminate what little surprise remains under current discovery rules.

If counsel elects to use a technological presentation at the mediation, keep it as short as possible (the idea is to give a preview, not to thoroughly present the entire case). A brilliant presentation is worthless to the negotiator unless it persuades his opponent to evaluate or reevaluate risk in going forward to arbitration or trial.

On a lesser scale, it has become relatively commonplace for advocates or their experts to use PowerPoint (or other presentation software) presentations at the mediation. These presentations can range from relatively simple lists of relevant points to graphic depictions and even animations where appropriate. The utilization of such demonstrative aids to a mediation presentation harmonizes with the decision of when to retain an expert and how much to invest in the expert before negotiations.[59]

3. Developing a Mediation Strategy

a. Pre-Mediation Negotiation or Positioning

Parties, principals, and advocates each insist on some occasions on setting pre-mediation negotiating bench marks before mediation. Often, such positioning takes

---

58. It is also possible for someone other than lead counsel—assistant counsel, for example—to present the technological case, although it is frankly more impressive when lead counsel is the advocate.

59. There are a number of ways to utilize experts in mediation other than formal presentation by the experts. One side's expert(s) can be present in order to listen to and make note of opposing experts' presentations. By her presence at the mediation, the expert convinces the opposing party that she is engaged and ready to proceed. The expert can, in private caucuses during the mediation, provide invaluable assistance to the negotiator in responding to claims from the opponent, and putting the best *spin* on various technical issues within the expert's purview.

the form of required monetary offers or demands before mediation. Sometimes, the negotiation is over who pays what share of the mediator's fees. Most mediators (if asked) will strongly discourage such negotiations. They come with a high price tag. If the parties are already displeased with each other from a negotiation perspective at the beginning of mediation, the mediation (and the mediator) starts in a hole. It takes some time to recover momentum broken by pre-mediation positioning. Additionally, the mediation is the best forum within which to evaluate the chances of success. Pre-mediation attempts to establish the good faith of the opponent are likely to produce false negative predictions about the opponent's bottom line negotiating position. They also assume (implicitly) that there will be no flexibility or change in the opponent's bottom line during the mediation process. If that were true, there would be a much lower percentage of cases that settled during mediation. Changing perspective, changing expectations, and revised risk-benefit calculations under the tutelage of the mediator are frequently required in order to resolve cases. These are the cases in which mediators earn their keep.

b. *Proactive Participation in Defining the Process*

Especially when dealing with a facilitative mediator, counsel should not hesitate to take a proactive role in defining the process and dealing with special issues. As in the selection process, counsel should resist a conditioned reluctance to speak to the mediator ex parte concerning the process and issues likely to arise. Mediation is largely an ex parte process in the sense that, unlike trials and arbitrations, it is expected that parties will communicate directly with the mediator outside the presence of the other parties and even share confidential information with the mediator. There is nothing to prevent ex parte communications with the mediator after the mediator is retained and the hearing confirmed but before its commencement. In many cases, such communications assist the mediator and help to prepare him to anticipate problems, issues, and logistical concerns.

In this phase as well, the mediator must be mindful of symmetry or fairness in the process.

When counsel refuse or fail to initiate such pre-mediation communications, the mediator should do so, especially in a construction case, which usually presents logistical and sequencing issues. Alternatively, the parties can discuss mediation process and protocol with each other before the hearing and agree on the process to be utilized. Such pre-mediation sessions can (and usually should) include the mediator as well.

The point of these pre-mediation communications is to actively help design the process instead of leaving such decisions to the mediator during the hearing without advice or assistance from counsel. Often, a party has preferences about the order in which the parties talk to the mediator, or about how issues should be scheduled in the mediation itself. Sometimes, one advocate defines the issues differently than her opponent, and this difference itself needs to be resolved before substantive negotiation can occur. Sometimes, one side believes that required attendees are absent from an opposing side (e.g., insurance adjusters, project managers) and that such absence will be fatal to negotiation progress. In such cases, it is far better to discuss these matters with the mediator in advance of the hearing and define, with the mediator's assistance and ultimately the concurrence of the other party(ies), the structure of the process, pre-conditions, attendance of proper persons, etc.

c. *Making Offers vs. Discussing Positions*

Two questions that inevitably arise in the construction mediation are (1) the level of evidentiary detail at which discussions should be had, and (2) the point in the process at which discussions should turn from positions to dollars. If the mediation turns into a full evidentiary presentation, it ceases to be cost-effective since, absent a settlement, the evidentiary presentation will have to be repeated at the decisional hearing. On the other hand, attempts to negotiate dollars without discussing issues are rarely productive for several reasons:

- Once the discussion proceeds to dollars, it is virtually impossible to go back and discuss evidence, witness credibility, or other matters that actually are the foundation for risk-benefit analysis.[60] On the other hand, if such matters are discussed initially, it is possible to revisit them even during financial negotiation.
- Such premature financial negotiation precludes the possibility of a party *changing* its financial bottom line based on hearing something new or re-appraisal of previously known facts and circumstances.
- Dollar discussions are essentially linear, single-topic discussions. Such linear negotiations have a much higher risk of derailing or terminating conversation than do issue and fact-driven discussions.
- Parties simply do not change expectations or reevaluate risk and benefit as a result of their opponent's financial position. Risk may be expressed in dollars, but it is not determined by dollars.

It is frequently necessary to discuss construction cases issue-by-issue and party-by-party. Some level of evidentiary detail is usually essential to the discussion. With thought beforehand, one team can arrive at a realistic expectation of how much detail and preparation is required, and how long the process ought to take, leaving room for some flexibility in that appraisal during the mediation process itself.

It is also frequently the case that some of the participants at the mediation are less prepared than others. This may be caused by a party's lesser role and unwillingness to spend the dollars required to get up to speed on all of the essential issues. It may involve laziness, inattention, unavailability of sufficient time to prepare, last-minute substitution of counsel, principals, or party representatives, or even the proverbial "workup by the associate, negotiated by the lead counsel" problem. This problem

---

60. In other words, negotiating financially at the inception is likely to produce agreement only in those cases where the parties' expectations essentially overlap, the decision-makers are using reasonable risk-benefit analysis, and the parties are in possession of essentially the same material facts and knowledge of the case.

can almost always be solved through mediation, but it requires time and patience on the part of others, who may be justifiably upset at the lack of ability to proceed on the part of the opponent, when they have done their homework and are quite ready to proceed.[61] The sooner the participants resolve to go to a certain level of detail, the more efficient the mediation process will be. Arguing about the level of detail consumes time which may be better spent in negotiating in greater detail.

Often, the problem is not that *all* of the parties are unwilling or unprepared for the necessary level of detail, but rather that *some* of them are in such a posture.[62] This introduces additional complications to the mediation process. Differences in expectation, whether as to process or eventual result, are a major impediment to resolution, especially in complex cases (like construction cases). Again, this is a strong argument for careful consideration and design of the mediation process ahead of time, in consultation with the mediator and with the agreement of the parties.

d. *Briefing*

Briefs are normally provided to the mediator in most mediations, and certainly are the norm in construction cases. Briefs may be short or long and may go into extraordinary detail or virtually none. The best mediators do not force parties with respect to whether briefs are filed, a list of required contents, their length, or level of detail. Rather, they rely on the advocates to make reasonable decisions on these matters. Successful mediators are usually quick studies. They should be able to pick up significant amounts of information in short periods of time, verbally as well as in writing. Parties and advocates are always free to present additional material as they deem appropriate during the hearing.

Briefs certainly assist the mediator to understand the relationships of the parties (which can be quite complex in a construction case), the nature of the dispute, the status of the legal proceeding (trial or arbitration date, extent of discovery completed, etc.), an overview of the issues to be discussed, and the like.

---

61. In essence, this presents a typical example of how the (facilitative) mediation process works. The mediator identifies such problems (on a good day, before they erupt into flashpoints in the negotiation) and works at resolving them, while allowing other participants to vent, and keeps them engaged in the dialogue while the problem is addressed. Interim frustration of advocates and parties with each other does *not* justify the conclusion that, at the end of the day, the advocates and parties will be unable to find a framework of agreement in which to resolve the dispute. It does produce anger and impatience, with which a skilled mediator ought to be able to deal, and most significantly it threatens to derail the negotiation, which results in either a postponement or preclusion of settlement.

62. This is another problem that can often (but not invariably) be resolved with forethought and the help of the mediator before the mediation. Resolution of the problem before the mediation will lessen stress and frustration on the part of some or all of the participants. However, if not resolved beforehand, the problem can almost always be resolved in mediation. Again, it simply requires time and patience.

Briefs may either be exchanged with the other party(ies) or given confidentially to the mediator. The latter should be reserved for true secrets which will assist the mediator in managing the process.[63] Remember that with a facilitative mediator, the negotiation will be conducted with the opponent through the mediator, as opposed to negotiation conducted with the mediator who will then negotiate with the opponent (evaluative model). Since the parties have to negotiate with each other in some fashion in order for the case to settle, it is advisable during briefing to exchange as much information as possible with the opponent. This applies to joint sessions during the mediation as well.

e. Client and Advocate Expectations of Process and Result

People negotiate differently. Styles of negotiation often reflect underlying personality differences in the negotiators. Differences are also produced by training (or, more often, untrained assumptions)[64] about better and worse ways to negotiate to maximum advantage. Sometimes, the negotiation style of the lawyer is produced by pressure from the client. In a surprising number of cases, one team member has a negotiation style that is unacceptable to other(s) on the same team. In such cases, the mediator must often engage in intra-team mediation (which can be quite subtle at times). It is also not unusual for some team member(s) to have unrealistic expectations as compared to other member(s) of the same team, which produces frustration and friction within that team. All of these factors, and a number of others, bear on the relationship of participants to the mediation process.

(1) Pace

It seems inevitable that in any given negotiation, one of the negotiators is relatively impatient, desiring to proceed immediately to the bottom line. Sometimes, that negotiator is well-prepared. At other times, he or she may be quite unprepared. The opponent, on the other hand, wants to proceed carefully and slowly to address various issues. The former expects the mediation to be concluded within half a day. The latter expects the mediation to take two to three days.[65] The mediator can provide valuable assistance by buffering the frustration produced by the resulting friction and making the parties understand the reason for the frustration. The slow negotiator may be able to speed up to some extent, but the impatient negotiator will always have to be more patient than she wants to be in order to resolve this problem.

---

63. Surprisingly, counsel sometimes submit *confidential* briefs which contain very little, if any, confidential information. This way indicates that counsel expects an evaluative versus facilitative mediation process.

64. It is shocking that lawyers receive so little training in negotiation. Until quite recently, it was difficult to find even one negotiation course at many law schools. Yet, negotiation is a skill that is in frequent demand for the practicing construction lawyer.

65. This is another problem that can be addressed and often resolved by pre-mediation attention.

The problem is compounded when the problem is intra-team, or even worse, both inter- and intra-team. This is the kind of problem that makes negotiation fall apart without the presence of a mediator and can threaten an otherwise potentially successful negotiation even with a first-rate mediator.

(2) *Negotiation Style*

Negotiation styles have been discussed above to some extent. Suffice it to say that like pace, there are other friction points created by disparity in negotiating style. The question of style could be discussed in some detail outside the scope of the present chapter. Suffice it to say that all questions of disparity of approach to mediation fall into the same general category. They constitute potential irritants which can derail the process and which require the patience of advocates and parties and the close attention of the mediator.

(3) *Risk-Benefit Decision-Making*

One additional and special issue concerns the manner in which the decision-makers determine their intermediate and bottom line negotiating positions. This is the methodology by which decisions are made. A good settlement from the perspective of a party is that which advances his true interest more than the alternative.[66]

Decision-makers often make bottom line decisions using fuzzy-headed logic or for entirely wrong reasons. The expectations of the decision-makers as to eventual outcomes may be utterly unreasonable in light of the case specifics, although they may not appear to be unreasonable on the surface.

The same problem attends many if not most decision-makers, even corporate executives and experienced party representatives. Sometimes, they become emotional about the dispute, and emotion clouds the evaluation of the alternatives. At times, the evidence changes as the case proceeds and an expectation that was reasonable several months ago has become unreasonable with new discovery, the presence or absence of certain witnesses, etc.

Adjusting or correcting unreasonable expectations is a natural and predictable result of the mediation process. Mediators are trained to accomplish this task. The process works, in one situation after another, with decision-makers who are experienced, inexperienced, well-informed, badly informed, emotional, and unemotional. However, it requires time and patience.

A working relationship (usually unspoken) is helpful between the mediator and other member(s) of the team that contains the member with unrealistic expectations. Many lawyers are unwilling to confront unrealistic clients. They realize that counseling an unrealistic client can easily lead to feelings or charges of disloyalty

---

66. Facilitative mediators work towards helping parties to identify and understand what has been called their BATNA (best alternative to negotiated agreement) and to identify and understand their real interests, as opposed to negotiating positions.

("Whose side are you on, anyway?"). Sometimes the lawyer's role as counselor conflicts with the lawyer's role as advocate in precisely this way. Skilled mediators recognize the expectations problem when they see it, and are willing and able to work with the advocate to assist in revising unrealistic expectations in a productive manner.[67] However, the mediator must be trustworthy not to undermine counsel in the process. Unfortunately, not all mediators are as sensitive as they should be to guarding the attorney-client relationship while resolving expectation problems.

Adjusting unreasonable expectations and assisting decision-makers to refine their decisional grid or methodology are bread-and-butter issues for experienced mediators. These issues are not unique to construction cases. Rather, they are common to virtually every case in mediation. When added to the other complexities in mediating construction cases, they make construction cases even more challenging and rewarding for the mediator and the participants as well.

4. Wrapping Up the Successful Mediation

Before commencement of the mediation, counsel should address the issue of settlement terms, mechanics, and written agreements. At an early stage, counsel for the principal parties should consider who should draft any settlement agreement. It is prudent to bring specimen settlement documents (possibly, finalized prior agreements from similar cases) to the (final) mediation session. Two or three hours spent negotiating the final written documentation can save days of time after the fact.[68]

Counsel should pay special attention to enforceability of settlement agreements entered into at mediation and any special rules that may apply in the relevant jurisdiction. Counsel should also have thought through and be prepared to address issues relating to partial settlements, pursuit of non-settling parties, etc. As with basic written settlement agreements, smaller dollars expended earlier achieve larger savings of time and expense after the mediation is over.

Even more, settlement agreements perfected at the time of the settlement prevent all parties from a change of heart or mind after the mediation, which are followed by attempts to back out of the agreement. Even when parties do not attempt to disavow a settlement agreed to in mediation, they (or their advocates) may attempt to improve their position in the settlement documentation negotiation sure to follow. In cases as complex as construction disputes are almost certain to be, negotiating through the settlement terms in detail at the mediation and reducing the agreement to final documentation have tremendous advantages and no known disadvantages. Even in cases where there are disagreements over documentation, it is preferable to have those disagreements in the presence of the mediator, where they are more easily resolved and

---

67. A somewhat more difficult but equally soluble problem occurs when the unrealistic expectation belongs to the lawyer and not the client. This problem manifests itself with surprising frequency.

68. Well-prepared counsel often either make logistical arrangements or bring laptop, printer, and paper with them to the mediation.

momentum has been established for resolution, than to have them by telephone, fax, or letter after the fact and without a mediator to help avoid and resolve impasses.

Finally, counsel should consider the mediator for a potential post-settlement role. With some frequency, the parties agree that the mediator whom all respect should be appointed by stipulation as a referee, special master, or binding arbitrator as to all issues of terminology and documentation or enforcement of settlement terms. In this way, the parties have provided for the most knowledgeable neutral concerning the case and the settlement to have authority to enforce it. This is not always acceptable to all parties and, sometimes, is unacceptable to the mediator. In fact, if the mediator agrees, he usually does so as an accommodation to the parties. Mediators, unlike arbitrators, are used to solutions which are agreeable to all parties. When the mediator becomes an arbitrator or judge, he runs the risk of forcing a decision on the parties and wielding decisional power directly and bluntly. One or more of the parties may well be unhappy with a decision imposed on them from outside. For the mediator, it is a complete role reversal, contrary to the voluntary nature of the whole process up to that point. Nonetheless, if the mediator and the parties do consent, this step can save expense, time, and heartburn further down the road.

Library References

C.J.S. *Compromise and Settlement* §§ 1, 6–7, 9–13, 19–21.

West's Key No. Digests, Alternative Dispute Resolution <KEY>472, 481; Compromise and Settlement <KEY>5.

CHAPTER 18

Special Topics in Litigation of Disputes Involving Design Professionals or Construction Managers

KENNETH STRONG, ESQ.
ERNEST ISOLA, ESQ.

I. Introduction

Litigation is a large enough topic that an entire Section of the American Bar Association is devoted to litigation, and one of its Committees is dedicated to the topic of construction litigation. Any attempt to summarize all aspects of litigation in a handbook like this would either be so cursory as to be unhelpful or so long as to be impractical. In lieu of providing an overview of the litigation process (there are many perfectly reputable overviews that are available to the professional), this chapter discusses in some detail certain aspects of litigation as they are especially important to the litigation of disputes involving design professionals and construction managers.

II. Time Limits for Filing Suit

A. *Statutes of Limitation and Related Laws*

All jurisdictions have statutes of limitations that govern the time period within which a party must file suit after the cause of action accrues. Most jurisdictions utilize the discovery rule under which a cause of action accrues and the deadline to file a suit commences, running on the date that the claimant discovered (or reasonably should have discovered) the existence of a cause of action. Where the discovery rule does not apply, the statute commences running at some date that does not depend on the claimant's discovery of the cause of action. For example, the statute of limitations for a breach of contract claim in some states commences at the time of the breach, regardless of whether the claimant knew of the breach at that point.

It is important to be aware of shorter time limitations that may apply either by contract or by code. Virtually all governmental entities are protected by some form of requirement that a claim must be filed against the entity before filing suit. Normally, such statutes provide that a claim cannot be filed until some pre-litigation notice and demand is filed. These statutes further provide that suits be filed against the public entity within a relatively short time, often six months to one year, following accrual of the claim or it is barred.

Many states also empower public entities to prescribe, by written contract, procedures for resolution of claims that are different from those otherwise provided by statute. These procedures may vary or shorten the time to submit claims. In addition, private entities may agree by contract to shorter periods for filing claims, which effectively imposes a *private statute of limitations*. Contracts may also specify that all claims of a particular type are deemed to accrue as of the date of substantial completion.[1]

Therefore, it behooves the practitioner to obtain and review all relevant contracts for contractual time limitations on claims to see if such clauses provide a defense or impose a barrier to otherwise viable claims. Where governmental entities are parties or potential parties, strict compliance with contractual or statutory deadlines is essential.

B. *Statutes of Repose*

In states that utilize the discovery rule to date the commencement of the running of the statute of limitations, there is a danger that a claimant might not discover the cause of action until several years after construction is completed, thereby extending the construction professional's exposure almost indefinitely. For example, if the state utilizes a two-year statute of limitations for claims involving a design professional's negligence, but the owner of a building does not know about the design professional's asserted negligence until fourteen years after a structure is completed, the owner would have the ability to file suit—and the design professional would be exposed to liability—sixteen years after completion of the improvements.

This potential exposure unlimited by any temporal constraints presents numerous practical difficulties; states have enacted *statutes of repose* to avoid these difficulties. A statute of repose limits the time period when certain types of claims may be brought against parties involved in construction and is keyed to some definite time that is not fixed to anyone's discovery of a claim. Thus, for example, a statute of repose may provide that any claim against a construction professional related to any deficiency in the improvement of property must be brought no more than six years after substantial completion or it will forever be barred. Forty-nine of the fifty states have some form of statute of repose.[2]

---

1. *See, e.g.*, American Institute of Architects A201-1997.
2. Kentucky had a statute of repose but it was declared unconstitutional as conflicting with Kentucky's constitutional grant of open access to courts. Tabler v. Wallace, Ky., 704 S.W.2d 179 (1985), *cert. denied*, 479 U.S. 822 (1986). Kentucky has not stricken its statutes of limitation as unconstitutional.

When a state uses the discovery rule for its statute of limitations and has a coexisting statute of repose, the shorter of the time periods will govern the time before which suit must be filed. Suppose a state uses a two-year statute of limitations in which the period commences running upon discovery of the cause of action and also imposes a six-year statute of repose. If the owner of the building discovers that it has a cause of action against the contractor exactly one year after the building is completed, the owner must file suit by the third anniversary of completion of the improvements. The six-year statute of repose does not come into play because the shorter time period imposed by the statute of limitations applies.

On the other hand, if the owner discovers the cause of action five years after substantial completion, the statute of limitations would permit filing of a claim until the seventh anniversary of the completion of improvements. However, the statute of repose bars any claim not filed within six years after the completion of the improvements and, thus, would serve as the outside limit for the filing of a claim. In that case, the statute of limitations is of no practical effect, as the shorter statute of repose governs the deadline for filing suit. One consequence of the operation of a statute of repose is that a claimant might actually lose the right to bring a claim before discovering the existence of the cause of action.

Different states take different approaches as to what claims are barred and the time frame in which the statute of repose applies. Tennessee is the shortest at four years,[3] while Iowa's statute is fifteen.[4] Nine states use a six-year period.[5] Twenty-three states use ten years as the time frame.[6] Some states split the time frame depending on the nature of the defect. For example, California provides a four-year statute for patent defects[7] and a ten-year statute for latent defects.[8] Other states take a similar bifurcated approach.[9] California's statutes do not apply to claims for personal injuries and most state statutes do not apply to claims by public entities. While most state statutes create a complete bar to claims brought after the time frame set forth in the statute, a few state statutes, such as Kentucky's, create a rebuttable presumption that there was no design defect after five years.[10]

---

3. TENN. CODE ANN. § 28-3-202 (2005).

4. IOWA CODE § 614.1 (2005).

5. Colorado, Delaware, Idaho, Massachusetts, Mississippi, North Carolina, Utah, Vermont, and Washington.

6. Alaska, District of Columbia, Hawaii, Indiana, Kansas, Kentucky, Maine, Maryland, Michigan, Minnesota, Missouri, Montana, New Jersey, New Mexico, North Dakota, Oklahoma, Oregon, Rhode Island, South Dakota, Texas, West Virginia, Wisconsin, Wyoming.

7. CAL. CIV. PROC. CODE § 337.1 (2006).

8. CAL. CIV. PROC. CODE §§ 337.15 (2006).

9. Florida, Illinois, Nebraska, and Nevada have different time frames for latent and patent defects.

10. KY. REV. STAT. § 198b (2006).

Finally, states vary as to when the statute of repose begins running. Many states use substantial completion of the improvement at issue, but a state may use the owner's occupancy of the improvement[11] or the completion of the design to trigger the commencement of the running of the statute of repose.

C. *Using Tolling or Standstill Agreements to Defer Litigation*

There may be strategic reasons why parties may want to defer litigation beyond the time permitted under an applicable statute of limitations or statute of repose. For example, a contractor may assert a claim for delay and disruption against an owner. In those jurisdictions where it is permitted, the contractor may also sue the architect or the construction manager claiming they are responsible for a portion of its damages. Whether the design professional is sued directly by the contractor or is subject to cross-claim by the owner, the owner and architect or construction manager may want to avoid direct claims against each other during the pendency of the contractor's lawsuit. For example, while they may ultimately have claims against each other, they may want to ensure that the trial does not require them to indict each other's work and, instead, may prefer to present a united defense.

At the same time, the owner and architect recognize that they ultimately may not prevail against the contractor and a victory by the contractor would breathe life back into their claims against each other. However, by the time the contractor's litigation is completed, the statute of limitations or statute of repose may bar those claims.

As a means of avoiding this time problem, the architect and owner can enter into a waiver and tolling agreement under the terms of which the time for either of them to file suit against each other (under a statute of limitations or a statute of repose) will not re-commence running immediately. Instead, the time periods will either be "tolled" indefinitely[12] to be commenced upon notice by the other party, or will commence running at a certain later time such as upon entry of a judgment, settlement in the contractor's litigation, or the expiration of a particular product warranty.

Library References

C.J.S. *Employer to Employee Relationship* § 87; C.J.S. *Limitations of Actions* §§ 87, 131–142, 144, 146, 148, 159, 164–183, 198–205; C.J.S. *Physicians, Surgeons, and Other Health-Care Providers* § 107; C.J.S. *RICO* (Racketeer Influenced and Corrupt Organizations) § 16.

West's Key No. Digests, Limitation of Actions <KEY>46, 55, 95, 104.5.

---

11. FLA. STAT. ANN. § 95-11(3)(c) (2006).

12. In some jurisdictions, contracts seeking to extend the statute of limitations for an indefinite period of time are unenforceable. T&N PLC v. Fred S. James & Co., 29 F.3d 57 (1994) (based on N.Y. GEN. OBL. LAW § 17-103 (1993)).

III. Certificates of Merit

Design professionals such as architects, engineers, and construction managers provide critical services to construction projects. The role of these consultants typically begins as early as the conceptual design phase and continues through completion of the project. Architects, engineers, and construction managers have always faced significant exposure for breach of contract to their clients. However, in recent years, the number and severity of claims asserted against design professionals and construction managers have increased.

One reason for this increase is that as the barriers of privity and the economic loss doctrine have eroded, consultants are increasingly facing claims from contractors and other third parties. In many jurisdictions, homeowners can sue construction professionals directly for design defects. Developers have taken advantage of the law to form limited liability companies or have simply gone out of business after project completion. Latent defects may not become apparent until many years after completion of construction. By that point, the professionals may be the only viable entity to sue. An additional reason design professionals may be sued is that professional liability insurance often covers economic damages. Where an owner suffers lost profits due to a delay in completion, it may be easier to recover those damages under an errors and omissions policy as compared to pursuing a contractor's assets.

Although these reasons explain some of the increase in litigation against design professionals, that is not to suggest that these suits are all meritorious. Indeed, over the past two decades, there has been a perception that the number of frivolous suits against licensed professionals—including professionals involved in the construction process—has increased. As a means of trying to curb the number of frivolous suits, several states have passed statutes that require an attorney obtain a *certificate of merit* or similar assurance in connection with litigation against a licensed professional.[13] Although initially targeting the medical field to protect against non-meritorious malpractice claims, these protections were broadened to encompass other professions, including architects and engineers.

As of 1995, only five jurisdictions had enacted certificate of merit legislation. Over the past ten years, a growing number of states have enacted their own certificate of merit legislation. As of 2005, 11 states had enacted certificate of merit statutes that protect architects and engineers.[14] In addition, several other states have proposed legislation in furtherance of a certificate of merit statute. These states include Hawaii, Missouri, New York, North Carolina, and South Carolina. Although the specific provisions of each statute vary, the underlying purpose of each of these

---

13. Not every state uses the same terminology, but these will be referred to as certificates of merit in this discussion for the reader's convenience.

14. These states include Arizona, California, Colorado, Georgia, Maryland, Minnesota, Nevada, New Jersey, Oregon, Pennsylvania, and Texas.

statutes is to protect professionals from meritless lawsuits in the early stages. In some form, they require that before an attorney can pursue a claim for professional negligence against a design professional, there must be some form of consultation with or review by one or more design professionals to determine whether, in their opinion, the claim against the design professional has merit.

A. *Who Is Covered?*

In virtually all jurisdictions that have enacted certificate of merit legislation, the statutes cover licensed professionals who work in the construction field, including licensed architects, professional engineers, and land surveyors. In some jurisdictions, the language is more general—applying to *design professionals* and may include interior designers. The authors have not located any certificate of merit statutes that explicitly protect construction managers.

B. *What Are the Requirements?*

While the specific requirements in each jurisdiction are provided by statute and vary from state to state, there are generalizations that can be made. The statutes take one of three forms. Those state statutes that are the least restrictive from a procedural standpoint simply require an attorney affidavit. States following this pattern include California, Colorado, Oregon, and Pennsylvania. Generally, an attorney is required to submit an affidavit to the court either at the same time as the complaint is filed or shortly thereafter. A certificate of merit requires the plaintiff's attorney to consult with a third party design professional prior to filing suit. Generally, the attorney must review the facts of the case, consult a design professional licensed and practicing the same discipline as the defendant, and receive an opinion regarding the professional negligence of the defendant. Only after the consultant's opinion has been given can the attorney file the certificate with the court concluding that there is reasonable and meritorious cause to file the lawsuit.

However, such provisions do not pose much of a deterrence to meritless claims. First, most states do not require that the consultant's opinion be in writing and, therefore, there is no ready way to confirm whether the consultant opined that the professional being sued was negligent.[15] In most jurisdictions, the attorney is not even required to disclose the identity of the consultant in the certificate. Indeed, there is no realistic way to confirm whether the attorney, in fact, consulted with a licensed professional in the same field as the consultant being sued. California provides a mechanism for testing whether or not the attorney did in fact consult with a licensed professional before filing suit. However, since this mechanism can be invoked only after the conclusion of the litigation, it is worthless for all practical purposes.

---

15. Pennsylvania does require that the expert's statement be in writing and has provisions whereby the defendant can request a copy of the written statement, if the case is dismissed. Sanctions can be awarded against the attorney filing the certification if the court finds there is no written statement meeting the code requirements. See Penn. R.C.P. §§ 1042.3, 1042.7 (2006).

Some statutes allow an attorney to waive the certificate requirement if she is unable to obtain a consultant after three separate good faith attempts with three separate professionals. These provisions were enacted out of fear that professionals would not testify against fellow professionals and, therefore, attorneys would not be able to obtain certificates. In practice, finding design professionals willing to testify against fellow professionals has not been a problem.

Failure to file a certificate results in a demurrer or motion to strike. Unlike other states, California does not require dismissal with prejudice for failure to comply with its statute. Instead, a demurrer or motion to strike permits leave to amend, which is routinely and liberally granted to allow the plaintiff to cure the alleged defect. By granting leave to file an amended complaint, the court can give the plaintiff an opportunity to comply fully with the statutory requirements for filing a certificate of merit.

A more stringent variation of the certificate of merit process is followed in Hawaii and Kansas. In those states, the attorney must file a request for a review panel of professionals that issues a nonbinding opinion concerning whether or not the standard of care was met. The cost of the panel either is shared or is the responsibility of the losing party before the panel.

A second approach requires that an expert's affidavit be filed with the initial pleading or shortly thereafter. Arizona, Georgia, Maryland, Nevada, and New Jersey have statutes that follow this approach. Under Georgia and New Jersey law, the failure to file such an affidavit is grounds for dismissal with prejudice of the complaint.[16]

Nevada is an example of this approach. Nevada requires that an expert report be attached to the attorney affidavit. Under the Nevada statute, the attorney must file an affidavit with the first pleading stating that she reviewed the facts of the case, consulted with an expert knowledgeable in the defendant's discipline, and concluded that the action has merit. In addition, a report prepared by the expert consulted must be submitted that includes the expert's resume, a statement that the expert is experienced in all disciplines in the report, and the conclusions of the expert on the merits of the action. The attorney must also include a copy of each non-privileged document used by the expert in preparing his report, including documents deemed relevant by the expert to the allegations of negligent conduct. Failure to submit either the attorney's affidavit or the expert's report results in dismissal of the action. There is no provision for an extension to file the expert report. However, at the time of filing the attorney affidavit, if the attorney files an additional affidavit stating that she made reasonable efforts to obtain the required non-privileged documents but was unable to obtain them, she will not be required to submit those documents or the expert's conclusion in the report. However, this *escape valve* does require that the expert's report then must be amended to include such documents and information as soon as available.[17]

---

16. Stamps v. Johnson, 535 S.E.2d 1, 2–3 (Ga. 2000); Burt v. West Jersey Health Sys., 339 N.J. Super. 296, 771 A.2d 683 (2001).

17. Nev. Rev. Stat. 40.6884 (2006).

Statutes like Nevada's provide the design professional an early indication of the precise area in which his professional services are contended to have fallen below the standard of care and caused the plaintiff's damage. This is particularly beneficial in multiparty cases, such as subdivision defect claims, in which architects ordinarily would have no means of knowing which of the often numerous categories of defects their work is alleged to be deficient. Before suing a design professional, it forces the plaintiff to investigate the claim and at least theoretically provide the expert's details on the claim. In practice, there often is not enough information available at the time a suit is filed to provide the level of detail envisioned in the statute. That is where Minnesota's approach may be more effective.

Minnesota requires an attorney affidavit be served on the defendants with the pleadings. The affidavit must state that the attorney has reviewed the facts with an expert who, based on that review, is of the opinion that the design professional breached the standard of care. This satisfies the desire that attorneys at least consult with an expert about the merits of a claim for professional negligence before filing suit. Minnesota goes on to require that, within 180 days of filing suit, the attorney file a second affidavit identifying the experts who will testify, stating the facts and opinions about which the expert is expected to testify, and summarizing the grounds for each opinion. Failure to file the required affidavits in a timely fashion results in mandatory dismissal with prejudice of the causes of action for professional negligence.[18]

The Minnesota approach allows a party a reasonable time after commencing a suit to investigate and develop the evidence to support a claim for professional negligence. However, it places time limits in order to prevent frivolous or unsupported claims from being dragged out. Where a court finds an affidavit was not made in good faith, it has the power to issue sanctions including attorneys' fees.

Library References

C.J.S. *Negligence* §§ 650–652, 654–655, 728.
West's Key No. Digests, Negligence <KEY>1506.

IV. Management of Litigation through the Use of Special Masters

Complex construction litigation often involves many different parties litigating a multitude of legal and factual issues. For example, a typical construction defect claim may be initiated by homeowners, or homeowners' associations on behalf of an entire development or subdivision, against developers, general contractors, construction managers, and design professionals, which in turn file cross-complaints against subcontractors of every tier, engineers, subconsultants, material suppliers, and other participants in the construction project.

---

18. MINN. STAT. ANN. § 544.42 (2006).

Therefore, construction claims can involve twenty to thirty or more parties litigating literally hundreds of issues. Each party retains counsel and experts to assist in prosecuting and defending against the various design and construction defects alleged, resulting in months of percipient and expert depositions and hundreds of thousands of documents to review, organize, analyze, and prepare for depositions and trial. This process culminates in a jury trial which can span months, sometimes seasons.

Trial judges are typically averse to tying up their courtrooms for five months and subjecting jurors to complex, expert-intensive testimony over the course of a long construction trial. Therefore, judges generally will do everything in their power to avoid sending a complex construction case out to trial. Consequently, mediation and other alternative dispute resolution procedures are generally exhausted before the litigants in a complex construction case are provided a courtroom and trial is commenced.

A growing trend in alternative dispute resolution that has gained popularity in California and other western states is the use of special masters to coordinate discovery and settlement in multiparty construction litigation. Here, the parties to the litigation will generally stipulate or file a motion for the appointment of a special master to oversee pleadings, certain motion work, discovery, investigations, information exchange, and settlement. The special master will be responsible for coordinating most aspects of the case with an eye toward ultimately positioning the case in its most favorable settlement posture.

State laws allowing an order of reference are patterned after Federal Rules of Civil Procedure Rule 53(b), which allows a special master to be appointed in jury cases "only when the cases are complicated."[19] In cases to be tried without a jury, the appointment is made only "upon a showing that some exceptional condition requires it."[20] The process by which an order of reference to a special master is obtained varies from state to state. Most states have promulgated specific procedures that must be followed in order to appoint a special master.

For example, California allows for the appointment of a special master by stipulation of all parties,[21] by motion of a party to the case, or on the court's own motion.[22] California courts have the authority to order that a special master act as discovery referee and mediator and that the parties participate in the mediation.[23]

After the parties stipulate in writing to an order of reference to a special master, the fully executed stipulation is then sent to the judge for approval. The approved order of reference is then filed with the appropriate court, and the special master is

---

19. FED. R. CIV. P. 53(b).
20. Id.
21. CAL. CIV. PROC. CODE § 638 (2004).
22. CAL. CIV. PROC. CODE § 639 (2004).
23. Lu v. Superior Court, 55 Cal. App. 4th 1264, 64 Cal. Rptr. 2d 561 (1997).

granted the powers specified in the order of reference. Then, the special master will coordinate most aspects of the litigation and periodically report back to the judge as to the progress of the case. The actions taken by the parties during the special master process are generally afforded full confidentiality, as this allows the parties to conduct objective investigations and evaluate the case without fear of evidentiary fallout later at trial.

Library References

C.J.S. *References* §§ 3, 5–7, 27, 29–31.
West's Key No. Digests, Reference <KEY>3, 24, 27.

V. Use of Experts

In order to establish a claim for professional negligence against an architect, engineer, or construction manager, it is necessary in most cases to offer expert opinion that the professional failed to meet the applicable standard of care. For that reason, the preparation and testimony of retained experts is critical. Three recent decisions by the U.S. Supreme Court have set new criteria for evaluating the experts offered in litigation and the testimony they propose to give in federal court. Many jurisdictions that model their procedures on the federal rules have adopted these standards.

A. *The Trial Judge Has a Gatekeeping Obligation Pursuant to* Daubert, Joiner, *and* Kumho Tire

In light of *Daubert, Joiner,* and *Kumho Tire*,[24] lawyers who practice construction law have begun to challenge expert testimony with greater frequency. This trio of cases establishes and defines the trial court's gatekeeping function in making preliminary assessments of proposed expert testimony to ensure that the subjects of expert testimony are truly susceptible to such treatment. *Daubert, Joiner,* and *Kumho Tire* do not limit the trial judge's gatekeeping obligation regarding expert testimony to testimony involving scientific evidence. For example, in *Kumho Tire*, the Court held that the trial judge's gatekeeping obligation, as described in *Daubert*, applies not only to scientific expert testimony, but to all expert testimony, including that based on technical and other specialized knowledge. Indeed, the *Daubert* gatekeeping obligation requires the trial judge to make an inquiry into both the relevance and reliability of the expert's testimony to exclude "expertise that is *fausse* and science that is junky."[25]

---

24. Daubert v. Merrell Dow Pharms., 509 U.S. 579 (1993); General Elec. Co. v. Joiner, 522 U.S. 136 (1997); Kumho Tire Co. v. Carmichael, 526 U.S. 137 (1999).

25. *Kumho Tire*, 526 U.S. at 159.

It is worthwhile to note that the court's gatekeeping function is not "to make ultimate conclusions as to the persuasiveness of the proffered evidence."[26] The sole purpose of a *Daubert* hearing is to determine admissibility through a preliminary assessment of the methodologies underlying the opinion."

An attorney who seeks to rely on an expert's testimony must understand how to lay the groundwork to ensure admission of the testimony. The skillful litigator begins to lay the groundwork on the day he or she retains the scientific or technical expert.

B. Daubert/Kumho Tire *Extends to Both an Expert's Testimony and the Conclusions Reached by the Expert*

Federal Rule of Evidence 702 was amended in 2000 and now encapsulates the key rulings from the Supreme Court's trio of cases:

> If scientific, technical, or other specialized knowledge will assist the trier of fact to understand the evidence or to determine a fact in issue, a witness qualified as an expert by knowledge, skill, experience, training, or education, may testify thereto in the form of an opinion or otherwise, if (1) the testimony is based upon sufficient facts or data, (2) the testimony is the product of reliable principles and methods, and (3) the witness has applied the principles and methods reliably to the facts of the case.

The trial court may apply the specific factors of evidentiary reliability set forth in *Daubert* to the testimony of engineers and other experts who are not scientists. While the courts have not specifically stated that *Daubert/Kumho Tire* apply to other technical but not scientific experts like design professionals and construction managers, the *Daubert* and *Kumho Tire* decisions likely encompass such experts, because the Court based its opinions on technical or specialized knowledge. Indeed, the federal courts tend to have a more liberal standard compared to the state courts when applying evidentiary rules. There are a number of cases applying *Daubert/Kumho Tire* outside the arena of scientific knowledge, such as expert testimony on damage issues in patent infringement cases as well as to delay damages. Therefore, *Daubert/Kumho Tire* would seem to clearly apply to design professional and construction management experts. While a design professional may be more of a technical expert than a construction management expert, lawyers throughout the country typically name both design professional experts and construction management experts in construction cases under appropriate situations.

The trial courts will screen such testimony by applying the specific factors of evidentiary reliability set forth in *Daubert* "where they are reasonable measures of the reliability of expert testimony."[27] *Kumho Tire* cleared any ambiguity regarding whether the *Daubert* gatekeeping obligation applies to all expert testimony,

---

26. Quiet Tech. DC-8, Inc. v. Hurel-Dubois U.K. Ltd., 326 F.3d 1333, 1341 (11th Cir. 2003).
27. *Kumho Tire*, 526 U.S. at 150.

including testimony involving technical or specialized knowledge. The *Daubert* factors apply to many areas of expertise, including construction scheduling experts who are frequently engineers and typically testify based on their experience, technical, and other specialized knowledge. The *Daubert* gatekeeping obligation ensures that experts and their opinions in construction cases do not go unchecked by the court.

The following table summarizes the use of the *Daubert* trilogy throughout the United States:

States Adopting *Daubert, Joiner,* and *Kumho Tire*	States Adopting *Daubert/Kumho Tire,* Not *Joiner*	States Adopting *Daubert,* Not *Kumho Tire/Joiner*	States Holding That *Daubert* Is Instructive	Non-Frye[28] States That Reject *Daubert*
Arkansas	Kentucky	Alabama	Colorado	Idaho
Delaware	New Hampshire	Alaska	Hawaii	New Jersey
Illinois	North Carolina	Connecticut	Indiana	Nevada
Louisiana	Ohio	Montana	Iowa	North Dakota
Massachusetts	Rhode Island	New Mexico	Maine	South Carolina
Mississippi	South Dakota	Oregon		Utah
Nebraska		Vermont		Virginia
Oklahoma		West Virginia		Wisconsin
Texas				
Wyoming				

C. Daubert/Kumho Tire *Applies in Construction or Closely Related Cases: It Must Be Reasonable for an Expert to Rely on That Which Forms the Basis of His or Her Opinion*

A number of courts have used *Kumho Tire* to judge the reliability of engineers who offer testimony on a variety of issues, including causation and standard of care. Several of these cases have involved engineers who provide testimony in construction or closely related cases.

In *Ballard v. Buckley Powder Co.*,[29] the court applied *Kumho Tire* to the testimony of an architectural engineer. Applying *Daubert/Kumho Tire*, the court determined that the engineer's technique was vulnerable to significant sources of unreliability. Specifically, the architectural engineer could not rule out the possibility that ordinary settling or other environmental forces may have caused the damage in question, because his inspection occurred ten months after the defendant's blasting activities. The architectural engineer also could not reliably differentiate between damage caused by ordinary settling and damage caused by blasting. Moreover, the engineer's inspection method had never been subjected to peer review and the architectural engineer offered

---

28. *See* Frye v. United States, 293 F. 1013 (D.C. Cir. 1923).
29. 60 F. Supp.2d 1180 (D. Kan 1999).

no proof that other members of the engineering field generally accepted his method. Based upon this analysis, the court excluded the expert's testimony because it failed to meet the standards for reliability required by *Daubert/Kumho Tire*.

Similarly, in *Carbotrade, S.P.A. v. Veritas*,[30] the court applied *Kumho Tire* to discredit a naval architect's testimony. The expert's method did not satisfy the court that it could establish, even through calculations, the reduction in hull strength by corrosion and the loading conditions, in terms of bending moment, sustained by a sinking ship. By comparing measurements of the ship's as-built deflection with measurements of its actual deflection before its last voyage, the naval architect in *Carbotrade* calculated that the hull's strength had decreased by about seventeen percent to twenty percent. After applying the *Daubert/Kumho Tire* factors, the court determined that no treatise or scientific literature supported the expert's method of computing corrosion and hull-strength deterioration from compared deflection measurements. The court determined that the method's accuracy had gone untested empirically and questioned the expert's method for his failure to consider the margin of error inherent in his method of measuring deflection. The court also found that two other available computer programs appeared to produce more reliable bending moment values than the one on which the expert relied and noted flaws in the charterer's weather expert's data on which the expert based some calculations.

Courts have also applied *Kumho Tire* to experts offering testimony as to an applicable standard of care. These courts stress that experts should look for published industry guidelines or industry regulations to define the proper standard of care. If published guidelines do not exist, then the courts will scrutinize the expert's alternative bases for determining the standard of care under *Daubert/Kumho Tire*.

Kumho Tire has increased the duty of trial judges to scrutinize the reliability of engineer testimony to an unprecedented level. Litigants who fail to offer evidence satisfying the *Daubert/Kumho Tire* standards run a significant risk that the court will exclude their engineer's testimony.

Library References

C.J.S. *Evidence* §§ 597–598, 601–609, 611–614, 619, 634, 636–652, 662–665, 681–682, 686, 713; C.J.S. *Negligence* §§ 650–652, 654–655, 728, 782, 811.

West's Key No. Digests, Evidence <KEY>538, 555; Negligence <KEY>1506, 1633, 1662.

VI. Discovery, Management, and Use of Evidence in Complex Construction Cases

A. *Document Management and Use of Databases*

Complex construction litigation often involves large construction projects. Each party to the case, including developers, general contractors, design professionals, subcontractors, material suppliers, etc., can possess extensive project documentation that must be produced to all other parties. These documents must then be organized, reviewed, and

---

30. 1999 WL 714126 (S.D.N.Y. 1999).

analyzed by each party during the course of the litigation. The larger the project, the more documentation that each party must sort through in order to prepare the case for mediation or trial. Eventually, after depositions are completed and all document work is done, a smaller, but still substantial, subset of documents is ultimately introduced into evidence at trial. The process whereby trial counsel begins literally reviewing rooms full of project documents and hard drives full of electronic information, and methodically whittles it down to relevant categories of documents and ultimately trial exhibits is daunting, to say the least. Fortunately, technology has streamlined the process somewhat and aided the practitioner in managing this seemingly unmanageable sea of information.

The ability to effectively organize voluminous documentation produced during the course of litigation will greatly assist the construction attorney in advocating on behalf of his client. One must take the time at the outset of the litigation to review and organize the documents in a way that subjectively makes sense, because in many cases a huge number of potentially relevant documents must be reviewed quickly and efficiently before a deposition. Some organize relevant documents by issue and prepare issue binders that are put in chronological order. Others prefer to segregate the documents by category, e.g., contracts, change orders, requests for information, meeting minutes, correspondence, etc., and then organize them within that category chronologically or by number. Cross-separation of documents by witness may also be helpful.

Depending on the issues and claims raised, the damages claimed, and the manner in which the original project documentation was kept, an attorney should utilize some combination of these approaches that intrinsically makes sense within the framework of the claims at issue. For example, in a delay case, it may be helpful to first segregate the documents by task, whether it be critical path, concurrent delay, etc. Then, within each task, the attorney can continue to organize documents chronologically or separate them by issue, trade, document type, etc., and this will assist the attorney and expert in analyzing whether a critical path delay took place. On the other hand, in a defect case involving a discrete number of issues, it may be helpful to first organize the documents by issue, and then order them chronologically or by document type within each issue category. Again, the final determination is a subjective one that depends on the peculiar circumstances of the project and the litigation.

In recent years, construction litigators have come to rely on document databases to assist them in organizing and accessing the voluminous records that are routinely produced in complex construction disputes. Technological advances have made document databases more intuitive and user-friendly, enabling the construction attorney to electronically search through countless documents at any time. A document database is an electronic file comprised of documents that have been scanned and imaged onto an electronic medium. Construction attorneys will typically hire an outside document imaging vendor to scan the client's project files and store the electronic image in a retrievable document database. The attorney can then access all imaged documents through her computer, as opposed to manually leafing through boxes of paper documents.

Document imaging vendors are also offering practitioners advanced features that allow varying levels of text searches within documents that have been scanned and imaged. For an additional cost, each page that is scanned can be run through optical character recognition (OCR) software, which effectively transforms the image of a document into a searchable database of all text on that document. For example, if a letter is scanned and run through OCR, word searches can be run through the full text of the document or multiple documents. This ability to run keyword searches through an entire database of documents allows parties to efficiently identify potentially relevant documents through a few mere key strokes, as opposed to hundreds of hours of intensive document review.

The cost of running large numbers of documents through OCR can be expensive, usually around a quarter to a third of the cost of scanning and imaging the page. In addition, the OCR technology has not progressed to the point where it is completely accurate, so one runs the risk that the OCR will not accurately pick up every word on a given document. This imperfection in the technology means that practitioners should not rely on OCR as a complete substitute for conducting comprehensive document reviews where warranted.

Whereas OCR is an automatic encoding method, attorneys may choose to manually encode the imaged documents using an objective encoding system. Here, the attorney can select a number of fields, e.g., author, recipient, carbon copies, type of document, reference subject of the document, etc. Then, as the imaging vendor is scanning the documents into the database, he is encoding any applicable objective requirements within the document. Later, as the attorney is manipulating the imaged documents, she will then be able to run searches and quickly identify documents because of the encoding work performed at the outset. Once the database is created, documents can easily be accessed and retrieved by the attorney. In addition, as the attorney is perusing the database of imaged documents, electronic notes and comments can be kept by the attorney and encoded into the database along with any document or series of documents. As with OCR, document encoding can be cost prohibitive, so attorneys should solicit price quotes from imaging vendors, obtain the consent of the client, and undertake a rigorous cost-benefit analysis before purchasing these litigation tools.

B. *Electronic Discovery*

Parties to litigation have expansive discovery obligations. They are generally obligated to respond to other parties' discovery requests by producing all information or documentation that is likely to lead to the discovery of admissible evidence. Many states have patterned their rules governing the scope of discovery after the Federal Rules of Civil Procedure Rule 26, which defines the scope of discovery in federal lawsuits and provides that all documents and tangible things that are relevant to a claim or defense of any party must be produced at the outset of the litigation.[31] The standard is not admissibility; it is relevance to the subject matter of the litigation.[32] As a consequence of the liberal rules governing the scope of discovery, it is the most costly and time-consuming aspect of most construction cases.

---

31. FED. R. CIV. P. 26(b)(1).
32. *See, e.g.*, CAL. CIV. PROC. CODE § 2071(a) (2005) ("[A]ny party may obtain discovery regarding any matter, not privileged, that is relevant to the subject matter involved in the pending action,... if the matter either is itself admissible in evidence or appears reasonably calculated to lead to the discovery of admissible evidence.").

Federal Rules of Civil Procedure Rule 34 broadly defines the term "documents" to include "information in any tangible format."[33] It has been firmly established since 1985 that the definition of documents in Rule 34 includes electronically stored information, including e-mails and data compilations.[34] Thus, parties to construction litigation are obligated to produce electronic information that meets the discovery relevance standard just as they are with paper documentation.

Advances in technology have drastically changed the way participants in construction projects communicate with each other, and this has materially affected discovery obligations in construction litigation. It was not long ago that one could trace the material communications in a given construction project through the paper documentation generated and stored by the various parties in their project files. If something needed to be documented, a letter would be mailed and/or faxed, and the sender and receiver would each retain copies of the document in their paper files. If litigation ensued, with limited exception, an attorney representing a particular party would sort through the papers in a discrete project file in order to comply with discovery demands made by other parties during the course of litigation.

The rise of electronic media has affected the construction industry just as it has the rest of corporate America. Construction participants now communicate and store information electronically by e-mail, through spreadsheets, and comprehensive project management software, in scheduling and estimating programs, in AutoCAD and *pdf* files, to name a few. Developers, contractors, design professionals, and construction managers now pride themselves on paperless projects, in which all communications are sent electronically, stored on computer hard drives, and backed up on storage tapes. Advances in technology relating to the backup and archiving of electronic information have resulted in less printouts of project documents and more reliance upon electronic storage methodologies.

The effect of this trend upon the discovery obligations of parties to construction litigation has been significant, and the concept of electronic discovery has surfaced and begun to evolve. Electronic discovery refers to the discovery of information that is stored electronically, on hard drives, backup tapes, CDs, DVDs, floppy disks, and other electronic storage media. Litigants must preserve, protect, and ultimately produce its electronic files, in order to comply with discovery requests propounded by the other parties to litigation.

A series of decisions arising out of one case in New York have defined many of the standards governing electronic discovery and served as a model for other

---

33. FED. R. CIV. P. 34.
34. Bills v. Kennecott Corp., 108 F.R.D. 459, 461 (D. Utah 1985); Rowe Entm't, Inc. v. William Morris Agency, Inc., 205 F.R.D. 421, 428 (S.D.N.Y. 2002).

jurisdictions. *Zubulake v. UBS Warburg LLC*[35] involved an employment discrimination claim by Laura Zubulake, an equities trader, against her employer, UBS. In August of 2001, after Zubulake had filed her initial complaint with the Equal Employment Opportunity Commission, UBS's in-house counsel instructed certain key personnel within the company to retain all electronic documentation pertaining to her claims and segregate it from other materials. This request was repeated both verbally and in writing on several occasions over the course of the litigation.[36] As it turned out, a number of key employees deleted relevant electronic data despite the many instructions given, and discovery disputes arose between the parties as a result.[37] The court found that UBS failed to preserve relevant electronic data notwithstanding the early warnings by its counsel.[38] Sanctions were awarded against UBS, and the jury was given instructions that permitted the inference that the electronic data not produced would have been unfavorable to UBS.[39]

Several decisions ensued that have largely defined the scope of electronic discovery obligations. The *Zubulake* decision can be boiled down to the following general rules:

- First, companies should be prepared to produce all information that is relevant to the subject matter of the dispute, even if the information is in electronic form and exists only on backup tapes or media.[40]
- Second, the obligation to preserve electronic information requires reasonable and good faith efforts to retain such information that may be relevant to pending or threatened litigation.[41]
- Third, the obligation to preserve arises whenever the company "should reasonably know that the evidence may be relevant to anticipated litigation," at which point key players in the company should be ordered to preserve all such information.[42]

---

35. There are seven published decisions that came out of the *Zubulake v. UBS Warburg, LLC* case, and they are numbered sequentially: *Zubulake I*, 217 F.R.D. 309 (S.D.N.Y. 2003); *Zubulake II*, 230 F.R.D. 290 (S.D.N.Y. 2003); *Zubulake III*, 216 F.R.D. 280 (S.D.N.Y. 2003); *Zubulake IV*, 220 F.R.D. 212 (S.D.N.Y. 2003); *Zubulake V*, 229 F.R.D. 422 (S.D.N.Y. 2004); *Zubulake VI*, 231 F.R.D. 159 (S.D.N.Y. 2005); *Zubulake VII*, 382 F. Supp. 2d 536 (S.D.N.Y. 2005).
36. *Zubulake IV*, 217 F.R.D. at 215; *Zubulake V*, 229 F.R.D. at 423.
37. *Zubulake V*, 229 F.R.D. at 423–24.
38. *Zubulake V*, 229 F.R.D. at 436
39. *Zubulake V*, 229 F.R.D. at 437, 439–40.
40. *Zubulake I*, 217 F.R.D. at 317.
41. *Id.*
42. *Zubulake IV*, 220 F.R.D. at 218; *see also* Unigard Security Ins. Co. v. Lakewood Mfg. Corp., 982 F.2d 363 (9th Cir. 1992) (court found that the plaintiff in control of both the timing of litigation and the preservation of evidence was obligated to preserve evidence two years before he filed suit).

- Fourth, both inside and outside counsel are obligated to oversee the retention and production of relevant electronic information and communicate directly with key personnel in control of such information.[43]
- Fifth, courts will engage in a seven-factor balancing test to determine which party or parties should bear the costs of retrieving and producing electronic information that has been stored or is otherwise costly and time-consuming to retrieve,[44] but the company storing/producing the information should be prepared to pay the majority of the costs.[45]

Other courts have followed the *Zubulake* lead and imposed significant sanctions for failure to preserve electronic documents after notice of litigation. In *Coleman v. Morgan Stanley*,[46] Morgan Stanley was sanctioned, subject to partial default judgment, and adverse facts were deemed *conclusive* for its intentional failure to produce all electronic information. Additionally, companies that do not follow their internal document retention policies may be subjected to monetary sanctions as well as issue/inference sanctions (as in *Zubulake*).[47] In *U.S. v. Philip Morris USA, Inc.*,[48] high-ranking employees of defendant Philip Morris lost or destroyed e-mails and did not follow the company's internal procedures for retaining electronic documentation.[49] The company was sanctioned $2.75 million and precluded from calling those high-ranking employees to the stand at trial.[50]

Since many construction participants have not perfected their policies for electronic document retention, organization, and storage, attorneys and their clients often find it difficult to comply with electronic discovery requests. First, many e-mails

---

43. *Zubulake V*, 229 F.R.D. at 432–34.

44. *Zubulake I*, 217 F.R.D. at 322 (The factors are: (1) the extent to which the request is specifically tailored to discover the relevant information; (2) the availability of such information from other sources; (3) the total cost of production, compared to the amount in controversy; (4) the total cost of production, compared to the resources available to each party; (5) the relative ability of each party to control costs and its incentive to do so; (6) the importance of the issues at stake in the litigation; and (7) the relative benefits to the parties of obtaining the information.).

45. *Zubulake III*, 216 F.R.D. at 289–91.

46. Coleman (Parent) Holdings, Inc. v. Morgan Stanley & Co., 2005 WL 679071 (Fla. Cir. Ct. Mar. 01, 2005); Coleman (Parent) Holdings, Inc. v. Morgan Stanley & Co., 2005 WL 674885 (Fla. Cir. Ct. Mar. 23, 2005).

47. *Cf.* Arumburu v. Boeing Co., 112 F.3d 1398 (10th Cir. 1997) (where a party engages in willful or bad faith spoliation of evidence, the other party is entitled to a rebuttable inference in its favor concerning the evidence), *with* Residential Funding Corp. v. DeGeorge Fin. Corp., 306 F.3d 99 (2d Cir. 2002) (negligence can support a rebuttable inference, and it is not necessary to show bad faith).

48. U.S. v. Philip Morris USA, Inc., 327 F. Supp. 2d 21 (D.D.C. 2004).

49. *Id.*

50. *Id.*

and other electronic data are not segregated by project. One can imagine the burden of going through each employee's general inbox, outbox, deleted e-mails, etc., in order to comply with discovery requests and produce all communications that pertain to the project in litigation. Under *Zubulake*, the majority of the costs associated with sifting through these electronic data will be borne by the party that should have organized it in the first place, and that company risks the imposition of monetary and issue sanctions against it for failing to comply with these rules.

Many companies have simply failed to promulgate or enforce any electronic document policies that would govern the retention, archiving, and disposal of documents after a certain period of time. As explained earlier in this chapter, the statute of repose for bringing claims for latent defects in twenty-three states is ten years, and this can be extended by tolling or through claims for indemnity. Accordingly, a design professional or construction manager in these jurisdictions should not only retain its project files for at least ten years, but also organize and store the electronic files in a way that it is not unduly burdensome to access them at a later date. For example, many companies back up their electronic data by compressing mass amounts of data that have not been identified to any particular project. Later, when litigation is instituted against that company, its representatives and attorney must decompress and sort through an enormous amount of undifferentiated files in order to comply with discovery obligations. The failure or inability to do so may result in the sanctions described above.

Additionally, project Web sites have become more commonplace in large construction projects. Here, the construction participants will log onto a Web site and be able to communicate with others and access documentation and project management information electronically. Different individuals typically have different levels of access to the information stored on the Web site. Project Web sites can create interesting discovery and document production issues, especially for construction managers and design professionals with extensive construction administration duties. For example, a construction manager with access to many or all areas in the project Web site may be obligated to produce such information in litigation, and it can be costly and time-consuming to retrieve such information.

C. *Demonstrative Evidence*

During the course of a trial, attorneys for the parties will attempt to introduce testimony, documentary evidence, and tangible evidence to prove or defend against a given claim. The substantive evidence to be admitted into evidence during a construction trial will consist primarily of project documents, and attorneys must comply with the rules of evidence for that particular jurisdiction in order to successfully introduce the documents into evidence. For example, tangible evidence can consist of construction components that were removed and analyzed during the course of investigating a particular claim. Once the judge admits a document or tangible thing into evidence, the jury may take these items into the jury room when deliberations begin.

Construction trials tend to be very technical and expert oriented. Jurors are susceptible to getting lost in the technical aspects of the case. Therefore, counsel for construction participants should strongly consider using appropriate visual aids to assist the jurors in comprehending the flood of information that passes their way during the course of a complex construction trial. The adage that *a picture tells a thousand words* is very appropriate in trial work, especially in construction trials.

Demonstrative, or illustrative, evidence is therefore commonly relied on by construction attorneys to make a point or explain complicated or technical information to jurors. It can include charts, maps, graphs, slides, models, lists of information, enlarged evidence, video recordings, timelines, computer-generated simulations or animation, or basically anything that provides a visual representation of the evidence. Demonstrative evidence is not actually admitted into evidence. During the course of a trial it may be used by counsel for various purposes, including illustrating or amplifying a point, synthesizing, visually portraying, or explaining admissible testimony or evidence.

Construction litigators typically use demonstrative evidence in opening statements to convey a particular theme to the jury and preview the evidence. Attorneys also routinely rely on demonstrative evidence in closing arguments to synthesize key points and arguments and provide the jury with a summary and visual representation of the oral testimony and voluminous project documentation. Jurors are most attentive to and impressed by computer-generated evidence that is projected onto a large screen, but blown-up poster boards can also be effective on a smaller budget to convey the visual information that jurors may need in order to better understand the issues.

The Federal Rules of Evidence allow attorneys to employ demonstrative evidence as long as certain requirements are met. Under Rule 102, the trial judge has the authority to fashion evidentiary procedures not otherwise covered in the Rules, and demonstrative evidence is one category of information that falls within this discretionary watchdog power.[51] Demonstrative evidence "may be excluded if its probative value is substantially outweighed by the danger of unfair prejudice, confusion of the issues, or misleading the jury, or by considerations of undue delay, waste of time, or needless presentation of duplicative evidence."[52] Most states have similar rules governing the allowance of demonstrative evidence during a trial.

---

51. FED. R. EVID. 102; United States v. Bibbs, 564 F.2d 1165 (5th Cir. 1977).

52. FED. R. EVID. 403; *see, e.g.*, Kehm v. Procter & Gamble Mfg. Co., 724 F.2d 613 (8th Cir. 1983) (trial court has the discretion to admit demonstrations where relevant, as long as they do not create a serious misimpression as to how an event might have occurred); Finchum v. Ford Motor Co., 57 F.3d 526 (7th Cir. 1995) (demonstrations purporting to illustrate that one side's version of the facts are subject to exclusion if there is a substantial risk that the jury, despite instructions, may misuse the evidence as a recreation of a real-life event, i.e., where the demonstration is not a close enough approximation of the real-life event).

In construction cases, computer-generated simulations and graphs are a very effective way to visually portray the complex issues that were framed by expert testimony during the presentation of evidence. For example, attorneys representing design professionals and construction managers in delay cases should strongly consider offering a computer-generated presentation during closing argument that is projected on a large screen and which synthesizes the mass of confusing and technical information that was presented during the trial. The presentation may contain comparative schedules which outline differences between baseline and as-built schedules, charts depicting the project's critical path and containing information as to how design issues did not negatively impact the critical path, visual aids to assist the jury in understanding home office overhead calculations under the *Eichleay* formula, bar charts characterizing job cost reports, and total cost information, etc.

A litigator may therefore put on a strong case in defense of her design professional or construction manager client, but the points typically need to be illustrated with as many visual aids as possible in order to persuade jurors who simply are not familiar with these types of issues.

As another example, a computer presentation program may be used as demonstrative evidence in a defect claim to underscore the impressive design and resulting structure that is the subject of the litigation. Too often, a defect trial focuses on the negative, i.e., the defects. However, the defects at issue usually constitute a very small percentage of the project; the remainder of the project may have progressed smoothly and without significant problems. If the jury walks into the deliberation room with only the problems and defects in mind, then the plaintiff has prevailed in limiting the jury's frame of reference to only the negative issues.

Where appropriate during closing argument, counsel for a design professional should therefore consider offering the jury a presentation which diverts jurors' attention from the two percent of alleged negligence and illustrates the ninety-eight percent of design professional competence and resulting attractiveness of the finished product. The presentation may contain renderings, photographs depicting aesthetically pleasing design elements, CAD designs, models, computer-simulated views of the project from various vantage points, etc. This will place the defect issues in their proper perspective and help temper jurors' anger if the defects caused the plaintiff to incur significant damages.

Library References

C.J.S. *Discovery* §§ 71, 86–87, 89–91; C.J.S. *Evidence* §§ 791–794, 813–815, 904–915, 917–931, 934, 940–956, 958–981, 984–999, 1002, 1023–1025, 1065.

West's Key No. Digests, Evidence <KEY>194, 195, 350–359; Pretrial Procedure <KEY>331, 352, 371.

VII. Using Joint Defense Agreements to Avoid Finger-Pointing and Manage the Dispute

A complex construction project is a collaborative process. When problems arise, it is often unclear who is at fault. The natural reaction for litigants in construction dispute may be to point the finger at the other project participants trying to shift the blame. When this occurs, an owner may be able to sit back and watch the members of the design team undercut each other and, from a practical viewpoint, establish the owner's claims that deficiencies in the design in construction administration resulted in defects or increased project costs.

Rather than accepting that litigation will be fought on such terms, parties who are naturally aligned, with respect to particular defense strategies, may engage in joint defense agreement, under the terms of which the aligned parties cooperate in presenting common defenses, even if they reserve the right to pursue claims against each other at some later point. The parties with common interests may include separate design professionals, who are defending against claims related to asserted deficiencies in plans and specifications, or they may include an owner and its design professionals in fighting delay claims by a contractor.

The benefits of establishing a joint defense arrangement may be particularly significant in a complex case. These advantages go beyond the presentation of an united front, or a mechanism by which codefendants pool resources to address a common legal threat. A joint defense agreement can provide an effective framework for evaluating alternative theories and developing a mutual strategy. The joint arrangement can expedite the discovery and sharing of information at a time before parties have committed to positions in written discovery or deposition. The information can also be shared at significantly less cost than formal discovery. Indeed, this sharing of information and resources is the strength of a joint defense arrangement. A properly conceived and executed formal joint defense structure can establish an effective litigation framework for the rest of the dispute, including an equitable structure for apportioning litigation costs for expert witnesses, court reporters, and graphic consultants to implement a shared theory of the case. It may even allow them to share counsel in limited circumstances.

There are risks associated with a joint defense agreement. These risks can be broken into three areas: (1) a concern regarding disclosure of confidential attorney-client communications and attorney work product to potentially adverse parties and future adverse use of such information; (2) the potential for waiver of these privileges through disclosures; and (3) the uncertainty as to how any remaining disputes will be resolved.

With respect to the primary concern of a joint defense agreement, the parties must be careful to structure their relationship to ensure that they do not waive the benefit of the attorney-client privilege. Generally speaking, discussions between attorneys and clients are protected from disclosure to outsiders—most importantly, these communications cannot be discovered by the opposing party. However, the client must treat these communications as confidential in order to maintain the benefit of the privilege. If a client discloses the communication to an outsider (or an outsider is present), the privilege may be lost. If an architect and her attorney meet with the owner and its attorney to discuss defending claims against a contractor, all of their discussions fall outside the attorney-client privilege and may be discovered by the contractor in the absence of a joint defense agreement.

A joint defense agreement provides protection against this problem. The joint defense privilege or *common interest rule* is viewed by most jurisdictions as an *extension* of the attorney-client privilege and attorney-work-product doctrine.[53] It is designed to protect communications between individual defendants and their attorneys who, in concert with other defendants, are acting as "part of an on-going and joint effort to set up a common defense strategy."[54] The privilege embodies the fact that the disclosures to third parties were intended to be insulated from exposure beyond the confines of the group by demonstrating that the disclosures would not have been made but for the sake of securing or supplying the legal representation.

The joint defense privilege is not equal to the attorney-client privilege but is an exception to it. In fact, the rationale behind the attorney-client privilege and attorney-work-product doctrine seemingly runs counter to the purpose of the joint defense arrangement, whose purpose is to divulge confidential attorney-client information to other attorneys representing potentially adverse parties. Case authority recognizes that just as the attorney-client privilege is required to promote the free flowing exchange of information between attorney and client in order to improve the quality of the representation, the joint defense privilege encourages attorneys and their clients to communicate to third parties in recognition of their common interests, which will likely result in a more effective and efficient representation of each client.

Case authority recognizes a variety of approaches to the exchange of confidential information among codefendants. For example, there is specific case authority in some jurisdictions establishing a *per se* joint defense privilege,[55] whereas other jurisdictions acknowledge the rationale for protecting confidential information but

---

53. United States v. Stepney, 246 F. Supp. 2d. 1069 (N.D. Cal. 2003).

54. United States v. Bay State Ambulance & Hosp. Rental Serv., 874 F.2d 20, 28 (1st Cir. 1989). *See also* United States v. Moss, 9 F.3d 543, 550 (6th Cir. 1993); In re Imperial Corp. of Am., 167 F.R.D. 447, 455 (S.D. Cal. 1995); In re Megan-Racine Ass'n, 189 B.R. 562, 570 n.4 (Bankr. N.D.N.Y. 1995) ("Most commentators and courts view it as an extension of the attorney-client privilege or work-product doctrine.").

55. Metro Wastewater Reclamation Dist. v. Continental Cas. Co., 142 F.R.D. 471, 478 (D. Colo. 1992) ("widely accepted by courts throughout the United States"); *see also* People v. Pennachio, 637 N.Y.S.2d 633, 634 (Sup. Ct., Kings County 1995) (stating that the privilege exists in Arkansas, Florida, Hawaii, Louisiana, Minnesota, Nevada, Oregon, South Dakota, Texas, Virginia, and Wisconsin); State v. Maxwell, 691 P.2d 1316 (Kan. Ct. App. 1984) (holding privilege exists in Kansas).

refuse to recognize an explicit independent privilege. Those jurisdictions view the joint defense privilege as being an exception to or extension of the attorney-client privilege and attorney-work-product doctrine, with limitations on their application to specific factual scenarios.[56] The privilege is not recognized in the Federal Rules of Evidence[57] although many jurisdictions have codified the common law attorney-client privilege.[58]

Even in jurisdictions where there is no *per se* joint defense privilege, the case authority follows the same basic analysis to determine whether the confidentiality of communications among and between codefendants has been waived. California courts have used such a waiver analysis since the legislature has limited the court's ability to expand evidentiary privileges. Instead, the parties claiming confidentiality based on the common interest doctrine must show first that the confidential information is protected under traditional attorney-client or work product privileges and then establish that disclosing the information beyond the attorney-client relationship does not result in a waiver because the communication is reasonably necessary to further the interests of the litigant.[59]

Consistent with the other privileges, the burden of establishing the joint defense privilege is on the party asserting it.[60] To assert the joint defense privilege, one must first establish the existence of an attorney-client privilege applying to the communications among attorneys and their clients for the common defense. If the attorney-client or attorney work product umbrellas do not protect the underlying communication, there is no privileged status under which the joint defense protection will apply.[61]

The party seeking to establish the privilege must show that the communications were confidential and made in the course of a joint defense effort.[62] This effort can

---

56. Raytheon Co. v. Superior Court, 256 Cal. Rptr. 425, 429 (Cal. Ct. App. 1989) (stating that "there is no 'joint defense privilege' as such in California").

57. FED. R. EVID. 501.

58. United States v. Moss, 9 F.3d 543, 550 (6th Cir. 1993); United States v. Schwimmer, 892 F.2d 237, 244 (2d Cir. 1989), *cert. denied*, 502 U.S. 810 (1991).

59. Oxy Resources California LLC v. Superior Court, 9 Cal. Rptr. 3d 621 (2004).

60. United States v. Moss, 9 F.3d 543, 550 (6th Cir. 1993); United States v. Schwimmer, 892 F.2d 237, 244 (2d Cir. 1989), *cert. denied*, 502 U.S. 810 (1991).

61. Metro Wastewater Reclamation Dist. v. Continental Cas. Co., 142 F.R.D. 471 (D. Colo. 1992); Sackman v. Liggett Group, Inc., 167 F.R.D. 6, 19 (E.D.N.Y. 1996); In re Megan-Racine Ass'n, 189 B.R. 562, 571 (Bankr. N.D.N.Y. 1995); *see also* United States v. McPartlin, 595 F.2d 1321 (7th Cir.), *cert. denied*, 444 U.S. 833, 100 S. Ct. 65, 62 L. Ed. 2d 43 (1979); Hunydee v. United States, 355 F.2d 183 (9th Cir. 1965); Continental Oil Co. v. United States, 330 F.2d 347 (9th Cir. 1964); In the Matter of Grand Jury Subpoena, 406 F. Supp. 381 (S.D.N.Y. 1975).

62. United States v. Bay State Ambulance and Hosp. Rental Serv., 874 F.2d 20, 28 (1st Cir. 1989); In re Megan-Racine Ass'n, 189 B.R. 562, 571 (Bankr. N.D.N.Y. 1995) ("The joint defense privilege is only applicable where the party asserting it can demonstrate an agreement between the parties privy to the communication that such communication will be kept confidential.").

be defined as the parties' protection of common interests which need not be identical but may be aligned on some issues and adverse on others.[63] However, communications about matters of conflicting interest or those that do not further the common interests of the codefendants are not deemed protected by the privilege.[64] The privilege will not attach where the parties' interests are adverse and the information is provided with the expectation that the relationship is not confidential.[65]

Of course, the fact that a joint defense agreement will permit the parties to avoid disclosure of attorney-client communications and their joint work product does not mean that a joint defense is always appropriate. Instead, the parties must first determine whether a joint defense is more beneficial than the potential harm that arises from the mutual disclosure of the joint parties' litigation strategies and evidence. This decision must be based on several factors. Before entering into the joint defense agreement, an attorney should ask:

1. What is your current evaluation of the potential exposure of your client?
2. What is the status of discovery?
3. Have all necessary parties been joined?
4. What is the relative exposure of other defendants and their ability to respond to an adverse judgment?
5. What is the estimated expense of litigation?
6. Are they viable joint defense theories that will limit the collective exposure of the participants?

Once the determination is made to pursue a joint defense, the parties can undertake the task of crafting their relationship through an appropriate agreement.

A. *Practical Considerations for a Joint Defense*

There are a number of practical and strategic issues that must be addressed, with respect to both the pending action and how the parties to the joint defense agreement will ultimately allocate any adverse judgment or settlement.

The first question determines who will participate in the joint defense arrangement. The nature of the claims and the contractual relationships between the defendants will often control whether the agreement can or should include all of the named defendants or a subset thereof. For example, if the case is one of clear liability on the part of the defendants but the damages are highly suspect, all defendants may be able to admit liability to the plaintiff and then defend the case jointly on the damage issues.

---

63. United States v. McPartlin, 595 F.2d 1321, 1335–1336 (7th Cir.), *cert. denied*, 444 U.S. 833 (1979); Griffith v. Davis, 161 F.R.D. 687, 692 n.6 (C.D. Cal. 1995); In re Megan-Racine Ass'n, 189 B.R. 562, 572 (Bankr. N.D.N.Y. 1995); Visual Scene Inc. v. Pilkington Bros., 508 So. 2d 437, 440 (Fla. Dist. Ct. App. 1987).

64. People v. Pennachio, 637 N.Y.S.2d 633, 634 (Sup. Ct., Kings County 1995).

65. Eisenberg v. Gagnon, 766 F.2d 770, 788, *cert. denied*, 474 U.S. 946 (1985).

Alternatively, if the dispute involves conflicting theories of liability, such as whether the damages were the result of a faulty design or poor construction, the design team might decide to adopt a joint defense which does not include the contractors. It is not necessary that all of the participants in the joint defense have identical positions. Rather, they only need to have a sufficient commonality of interests to warrant sharing resources and information to achieve a mutually advantageous result.

One possible advantage of a joint defense agreement is that it may permit substantial cost savings through the sharing of experts or even counsel. Without a joint defense agreement, discussions with the joint experts may not be privileged. The agreement will also facilitate the collaboration between individually retained experts, since the communications would fall within the joint defense privilege.

Where the terms of the joint defense agreement ensure a unity of interests at least through trial, the parties may even consider agreeing to representation by one law firm to minimize defense costs. However, without express waivers the law firm would not be able to represent any of the individual parties in the event that lingering disputes between the parties to the agreement must be resolved through subsequent alternative dispute resolution (ADR).

The joint defense agreement contemplates that none of the participants will do anything to prejudice the interests of the other parties to the agreement or to help the plaintiff. Ideally, the parties to the joint defense arrangement will agree at the time of forming the arrangement to share responsibility in equal percentages or on some other agreed allocation. Unfortunately, it is a rare case where this is possible. In most circumstances, the parties view their respective liability differently. Without having certainty as to the exposure, the parties leave for a later determination the ultimate allocation of any judgment or settlement.

In most circumstances, the participants will have agreed not to direct formal discovery towards each other and may agree not to cross-examine the other participants' witnesses during deposition. However, there may be significant strategic and cost impacts from this agreement depending on how the participants intend to allocate an adverse judgment or settlement. For instance, information that is both relevant and necessary to a dispute over allocation of an adverse judgment may not be adequately developed or refuted in the case with plaintiff. This can and will have a potentially serious impact if third party witnesses are not available in the allocation proceeding. Those third party witnesses may also be uncooperative if they have already been deposed in the initial dispute and do not want to be deposed a second time by many of the same parties.

One of the more contentious issues in drafting a joint defense agreement is whether parties can or should be permitted to settle separately with the plaintiff(s).[66]

---

66. Settlement strategy is just one of the ethical pitfalls that counsel representing multiple parties under a joint defense agreement need to consider. These ethical issues are discussed in detail in Bruce King & Carol Patterson, *Representation of Multiple Parties in the Construction Arena: Ethical Issues*, 25 Fall CONSTRUCTION LAWYER 5 (Fall 2005).

It is fundamental to most joint defense agreements that none of the participants may settle separately. This is particularly true in jurisdictions where there are barriers to pursuing indemnity claims against settling parties. Otherwise, a participant whose relative liability was not exposed by the other participants could escape with a relatively small share of responsibility. In most agreements, the parties must either settle as a group or try the case as a group. This can be a particularly difficult issue where one of the participants has low limits and/or a wasting insurance policy.

An alternative arrangement may permit the parties to settle separately with the plaintiff but still require them to participate in a subsequent method to resolve the allocation issue. Any settlements are aggregated with any judgment and reallocated through an ADR process. A risk of this approach is that the non-settling parties may perceive they have little to lose by attending trial since any unexpected verdict will be subject to reallocation through the subsequent process. Alternatively, a participant may agree to an unreasonably large separate settlement early to save itself litigation costs, knowing that much of the settlement will be reallocated to other participants.

The cleanest way to avoid disputes and maximize the utility of a joint defense is to include in the joint defense agreement a term allocating specific percentages of responsibility for an adverse judgment or settlement to each participant. If the parties can agree up-front to share responsibility in equal percentages or on some other agreed allocation, most potential areas of disagreement disappear. All parties will have an equal interest in minimizing the judgment or settlement and will have certainty as to their financial contribution. Of course, it is important to consider the other participants' financial resources since any agreement reallocating a verdict effectively makes the participants joint tortfeasors. This scheme may not make sense where some of the participants have limited insurance or other assets to satisfy an adverse judgment. In most cases, the parties to a joint defense agreement will decide to resolve any allocation issues through ADR. This allows the participants to preselect the means and method for dispute resolution and to structure the process in a way that makes economic sense. The parties can agree on an expedited process involving only written position papers, a multistep ADR involving mediation and/or a mini-trial, or use traditional binding arbitration under rules similar to the American Arbitration Association's Construction Industry Rules.

B. *The Necessity of a Writing*

The potential for conflict and dispute over the scope and obligations of a joint defense means that any joint defense agreement should be in writing. Moreover, while an oral joint defense agreement may be respected for purposes of fending off unwanted inquiries into the parties' attorney-client communications and work product, reliance on an oral agreement is haphazard and raises evidentiary concerns that can be avoided easily. As one court explained, "Obviously, a written agreement is the most effective method of establishing the existence of a joint defense agreement, although an oral agreement whose existence, terms and scope are proved by the party asserting it, may be enforceable as well."[67]

In the most pragmatic terms, a joint defense agreement can be crafted in far less time than a brief responding to an opponent's motion to compel discovery of attorney-client communications shared with another party.

Library References

C.J.S. *Contracts* § 67; C.J.S. *Witnesses* §§ 316–340, 374–376.
West's Key No. Digests, Contracts <KEY>32; Witnesses <KEY>199(2), 222.

---

67. Minebea Co., Ltd. v. Papst, 228 F.R.D. 13, 16 (D.D.C. 2005).

INDEX

Note: Page numbers with *t* indicate material found in tables.

A

accreditation, construction managers 239
ADA, design-build and liability regarding 132
additional insured status on the CGL policy 476
additional services, design professional contracts 76
adjacent site investigation, construction management and 289
agencies, governmental agencies and design professionals' role 61
agency CM 259
AIA Ethical Standards and Rules of Professional Conduct
  enforcement 31
  generally 30
AIA, National Ethics Council 31
allocation of risks, contracts and design professionals 74
alternative dispute resolution (ADR)
  generally 566
  judicial or quasi-judicial procedures 568
American Council for Construction Education, construction managers and 239
approvals, governmental agencies and design professionals' roles 61
arbitration 566
  award 588
  briefing 583
  confirming the award 588
  correction or modification of the award 588
  delay claims 586
  discovery 581
  experts and 581
  forum selection 577
  hearing 584
  law governing 570
  multiple party disputes 574
  post-hearing procedures 588
  pre-arbitration hearings 582
  pre-hearing issues 581
  scheduling 584
  selection of arbitrators 579
  strategies 570
  technology 587
  vacating the award 589
Architectural Works Copyright Protection Act 194
ASCE Code of Ethics and Rules of Professional Conduct 32
associated schools of construction, construction managers and 239
attorneys' fees
  damages 547
  insurance 486
audit, construction management 372

B

basic services, design professional contracts 76
betterment 552
binders, insurance 487
Board of Ethical Review, NSPE 35
bodily insurance injury 473
bonds
  construction management and 365
  claim limitations in design-build 138
  design-build contracts and 132, 141
  labor and material payment bonds 555
  performance 555
breach of contract
  claims 530
  damages 534
breach of fiduciary duty, design professionals and 151
bridging, design-build 142
budget estimates, design professionals and 68, 80
builder's risk insurance, construction management and 309

building codes, construction manager liability and failure to satisfy 384
building permits, design professionals and 62

C

certificates of insurance 487
certificates of merit, professional liability claims and 611
certification, construction managers 238
changes
    construction management and 306
    construction manager and design-build 424
claims
    breach of contract 530
    breach of fiduciary duty by design professional 151
    causation in design professional cases 149
    copyrights 197–203
    design professionals 145–54
        and claims by owner 108
        and contractor claims 104
        and contractor/subcontractor claims 110
        and evaluation of contactor claims 121
    environmental liability 159
    False Claims Act 113
    fraud by design professional 153
    gross negligence and design professionals 149
    joint defense agreements 628
    negligence and design professionals 146
    negligent misrepresentation by design professional 154
    OSHA 113
    personal injury and design professionals 111
    preparation of plans and specifications by design professional 158
    promissory estoppel 531
    standard of care for design professionals 147
    standing to bring against design professionals 160
    strict liability and design professionals 145
    third party claims and design professionals 109
    tort claims generally 531
    unjust enrichment 531
close-out, design professionals and documentation of 121
CMAA, Standards of Practice 254
co-authorship, copyrights and 186
code of ethics
    American Institute of Architects 32
    generally 27
    historical development 28
    NSPE 34
collusion with contractors, construction manager liability 398
commissioning and turnover, construction management and 370
commissioning, use of agents in design-build by construction manager 446
common law copyrights 191
compensation
    construction management 270, 304, 421
    damages 535
conflicts of interest
    construction manager liability 398
    construction managers and design-build 417
consequential damages 550
    construction management and waiver of 313
    waiver 559
constructability
    construction management and 299
    design-build and the construction manager 441
construction cost estimate, design-build and 440
construction documents, design professional services and construction management 287
construction management
    adjacent site investigations 289
    agency construction management 259
    at-risk 260
    audit or cost verification 372
    claims by construction manager 311
    commissioning and turnover 369
    compensation 270
    constructability 299
    contractual risk allocation 311–15
    coordination 373
        of changes and modifications 354
        of trades 354
    cost control 355
    cost estimates 294, 304

customer service and customer relations 367
design professions distinguished 256
design services 290–303
engaging design professional and 287
environmental impact statements and 284–86
funding 287
green buildings 301
guarantees and 371
historical evolution 248
insurable risks 364
insurance for construction manager 308
LEED 301
management of
  on-site work 354
  trade contractor changes 357
owner's representative services 262
permits and approvals 303
pre-construction duties 303
pre-design duties 283
procurement of trade contractors 351
program management services 261
project progress scheduling 297
project scope 283
punch list 368
quality control and quality assurance 361
recognition as profession 253
risk management and 363
safety and 366
schedule control 358
site logistics 300
site surveys 286
stand alone pre-construction agreement 316
standard form agreements 263
standard form contracts and documents 375
surety bonds and performance risk 365
suspension or termination 310
trade contractor bidding and selection 353
trade contractor qualification 352
types 258
utility service 290
value engineering 298
  and substitutions 358
  services 262
warranty management 371
work scope allocation 352
construction manager liability
  defenses 410–13
  third parties 400

construction manager licensing, matrix of state laws 241
construction manager regulation 205–44
construction means and methods, design professionals and 101
construction phase, design professionals and 96, 107
consultants, design professionals
  and design consultants 60
  and project consultants 60
contact documents, interpretation by design professionals 104
contacts, design professionals and negotiation of 68
contingencies, construction management and 355
contingent fees, construction managers and 232
contract clauses
  additional services for design professionals 76
  basic services 76
  consequential damages 84
  construction means and methods 101
  flow-down 137
  indemnification 75
  indemnification in design-build 137
  limitations of liability 85
  "murder clauses" in design-build 135–37
  pay-if-paid 136
  payment to contractor 104
  pay-when-paid 136
  quality assurance 102
  rejection of work 103
  safety 102
  site visits by design professional 98
  submittal review by design professional 100
  warranties in design-build 136
contract documents, ownership and design-build projects 435
contracts
  amendments 87
  construction managers and 233
  custom 73
  design professional duties 55–92
  extra-contractual roles of design professionals 56
  generally 71
  interpretation by design professionals 104

pre-design phase 56
standard form 71
contractual liability coverage 475
contribution
   construction manager 412
   subcontracting 496
contributory/comparative negligence, construction manager defenses to 411
coordination of trades, construction management 354, 360
copyrights
   Architectural Works Copyright Protection Act 193
   attorneys' fees 202
   common law 191
   conveyance of 189
   defenses to infringement 199
   derivative works 194
   design professional contract and 81
   distinction from patents 184
   fair use defense 200
   generic features 195
   history of 192
   idea/expression dichotomy 195
   independent creation defense 199
   infringement 197
   injunctive relief 202
   licenses 187–89
   ownership of 185, 188, 197
   protections afforded by 192, 194–97
   registration 191
   reverse licenses 187
   transfer of 190
   useful article exception 192
   works made for hire 185
   works protected 184
corporations, design-build 140
cost estimates
   construction management and 294, 304, 393
   design professionals and 66
cost verification, construction management and 373
custom contracts, 73

D

damages
   attorneys' fees 547
   betterment 552
   breach of contract 534
   consequential damages 84, 550
   contractor's compensation 537
   copyright infringement 201
   defenses to and mitigation of damage claims 551
   delay/disruption 544
   design professional and construction manager compensation 535
   direct costs 540
   disclaimer of warranties 560
   economic waste 551
   exculpatory clauses 558
   failure to mitigate 552
   general damages 540
   home office overhead 542
   idle equipment and related costs 541
   indemnification clauses 559
   industry standards and studies 550
   interest 547
   jury verdict method 550
   limitations of liability 85, 560
   liquidated damages 545
   loss of efficiency 545
   lost profits 544
   measured mile 549
   modified total cost method 549
   no damages for delay clause 561
   one-call statutes 563
   owner's compensation 535
   price increases for labor and material 541
   promissory estoppel 538
   proof of compensatory damages 540
   punitive damages 550
   should-cost estimates 550
   time-and-motion studies 550
   tort claims 539
   total cost method 548
   unjust enrichment 538
   waiver of consequential damages 84, 559
   waiver of lien rights 560
Daubert hearings, expert testimony and 616–19
deductibles, insurance 482
defamation, construction manager liability to third persons 405
defective design/design errors, construction manager liability 389
delay claims, arbitration 586
delay/disruption, damages 544

design development, design professional
services and 89
design professional
advisor role 57
and construction manager compensation 535
ethics, generally 24
licensure 39
professional organizations 51*t*
state licensure boards 52*t*
design professional practice
firm practice 11
interjurisdictional checklist 17
sample interstate commerce letter 20
design professional regulatory boards
NCARB 19
NCEES 19
design services
construction documents 89
construction management and 290–93
design development 89
generally 87
schematic design 88
design-bid-build
generally 94
traditional roles of project participants 94
design-build
ADA liability 132
bonding 132
bridging 126
commissioning 446
construction cost estimate 440
construction managers and
conflicts of interest 417
contingencies 426–29
control of changes during design 424
management of design team 425
management of total project costs 418
predicting permits and fees 426
protection of the work 429
vicarious liability for design 417
contractor-led 135
costs and *force majeure* 420
design professional
as leader 123, 128
as subcontractor 125
generally 123–144
roles 123
fast-track construction 126

insurance 132
joint ventures
less complete design 127
liability 12834
licensing and 134
review of payment applications by construction manager 443
schedule and construction manager 443
strict liability 131
value engineering with construction manager 431–33
direct costs 540
discipline, design professional, standard of review 46
disclaimer of warranties 560
discovery
arbitration 581
complex construction cases 619
electronic 621
dispute resolution
design professional contracts and 82
subcontracting and 503
Document management, complex construction cases and 619

E

early neutral evaluation 569
Economic Loss Doctrine
design-build and 134
exceptions 165
generally 160–166
negligent misrepresentation exception 165
subcontracting and 505
sudden/calamitous event exception 166
economic waste 551
entitlements, Design Professionals and 61
environmental impact statements, construction management and 284–86
environmental liability, design professionals and 159
ethics
AIA National Ethics Council 31
American Institute of Architects 30
code of ethics 26
construction managers
and codes of ethics 240
and owners 230
design professionals generally 24

dilemmas 38
indemnification 34
law and 36
limitations of liability 34
PLUS decision-making guidelines 39
public projects and construction managers 231
rules of professional conduct 26
evidence of insurance 487
evidence
  complex construction cases and 619
  demonstrative evidence in complex construction cases 625
exclusive dealing agreements, design-build and 142
exclusivity of remedies, defense to claims against construction manager 412
exculpatory clauses
  damages 558
  subcontracting and 499
experts
  litigation and 616–19
  necessity in design professional claims 148

F

failure to mitigate damages 552
fair use, copyrights and defense of 200
False Claims Act 113
fast-track construction, design-build and 126
faulty workmanship, construction manager liability 391
fee structures, construction management 270
fiduciary duty, construction manager liability for breach 385
final completion, design professionals and 119
float, construction management and schedules 361
flow-down clauses 137, 504
fraud, design professionals 151
funding, construction management and 287

G

general liability overage 472
geotechnical investigations, construction management and 286

governmental agencies
  building permits and 62
  design professionals
    and implied obligations 62
    roles dealing with 61
    roles obtaining 61
  entitlements 61
  inspections and the design professional 62
green buildings, construction management and 301
gross negligence, design professionals and 149
guarantees, insurance 480

H

hazardous waste sites 180
home office overhead, damages 542
Hyatt Regency sky walkway collapse 47

I

idle equipment and related costs 541
immunity, site safety and design professionals 176
implied obligations, design professionals and design 90
  government agencies 62
improper certification of progress payments, construction manager liability 395
improper inspection/supervision, construction manager 391
improper processing of submittals, construction manager liability 397
improper schedule preparation, monitoring and coordination, CM liability 396
indemnification clauses 559
indemnification
  broad form 464
  comparative (limited) form 465
  construction management 315
  construction manager 411
  construction manager and design-build 449
  design professional contracts 75
  design-build 137
  ethics 34
  generally 464
  insurance impacts 468
  intermediate form 465
  restrictions 469
  site safety and design professionals 171
  subcontracting 496

independent creation, copyrights and defense of 199
industry standards and studies, damages 550
industry standards, construction manager liability and 384
inefficiency, damages 545
initial proposals, design professionals and potential liabilities 63
injunctive relief, copyrights and 202
inspections, design professionals and 62
insurance
   additional insured status on the CGL policy 476
   attorneys' fees 486
   binders 487
   bodily injury 473
   certificates 487
   construction management and 307
      agency vs. at-risk 478
   construction managers 489
   contractual liability coverage 475
   deductibles 482
   design professional contracts and 83
   design-build 132, 140, 489
   evidence of insurance 487
   express warranties and guarantees 480
   general liability coverage 472
   generally 46987
   libel and slander 476
   limitations of liability 480
   medical payments 476
   meritless claims 488
   personal and advertising injury 475
   persons covered 482
   pre-claims assistance 486
   premises and operations liability 473
   prior acts coverage 471
   professional liability coverage 471
   property damage 473
   standard exclusions 479
   standard policies 469
   subcontracting and 501
   tail coverage—extended reporting periods 472
   wrap-ups 484
intellectual property, design professionals and 183–203
intentional/negligent misrepresentation, construction manager liability 386
interest, damages, 547
interference with contract, construction manager liability to third persons 404
interjurisdictional practice by design professionals, checklist 17
international building code, sealing of design documents 12
interpretation, design professionals and contract documents 104

J

joint defense agreements, 628–33
joint ventures
   design-build 139
   design professionals in design-build 125
judicial review, design professional discipline 45
jury verdict method, damages 550

L

land acquisition, construction management and 287
leadership in energy and environmental design, construction management and 301
LEED, construction management and 301
less complete design, design-build 127
licenses
   copyrights 189
   design professionals 39, 40
      revocation and suspension, 43
      disciplinary powers, 42
      enforcement 41
      investigative powers, 42
      powers 40
      Web sites 52t
licensing
   construction manager liability of failure to comply 381
   construction managers and state regulation 205
   contractors 227
   design professional discipline, judicial review 45
   design professionals 2, 9, 39, 226
   design professionals, reciprocity 44
   design-build 134
   illegality of unlicensed design-builder 138

matrix of construction manager licensing laws 241
state requirements for construction managers and design build projects 418
limitations of authority. defense to claims based on construction manager's conduct 412
limitations of liability
   construction management and 314
   damages 560
   design professional claims 151
   design professional contracts 85
   ethics 34
   insurance and 480
   subcontracting and 499
limited liability companies, design-build 139
limited liability partnerships, design-build 140
liquidated damages 545
loss of efficiency, damages 545
lost profits, damages 544

M

measured mile, damages 549
mechanic's liens 555, 560
mediation
   developing a mediation strategy 599
   experts and 597
   generally, 590–606
   selection of the mediator 595
   selection of the process 590
   technology and 598
Miller Act, state claims (Little Miller Act) 557
mini-trial 569
modified total cost method, damages 549
murder clauses, design-build 135–38

N

National Ethics Council, AIA 31
negligence
   claims against design professionals 146
   construction manager liability to third persons 400
   construction managers 378
negligent misrepresentation
   construction manager liability to third persons 402
   design professionals 154
negotiation, design professionals and contracts 68

no damages for delay clause 561
noncompete clauses, design-build and 141
NSPE Code of Ethics, generally 34
NSPE, Board of Ethical Review 35

O

offshoring, subcontracting and 517
one-call statutes, damages and 563
OSHA
   design professionals and 177
   design professionals and claims under 113
   fines 177
owner's representative services, construction management 262
owners, initial design concept 59

P

partnering 567
partnerships, Design-build 140
pay-if-paid clauses 136, 561
payment clauses, subcontracting 499
payment
   construction management fees 270
   contractor claims and design professionals 104
   design professional 78
   Miller Act 556
   payment statutes 557
pay-when-paid clauses 136
permits and approvals, construction management 303
personal and advertising injury insurance 475
personal injury claims, design professionals and 111
PLUS decision-making guidelines 39
pre-design phase, design professional roles 56
premises and operations liability 473
price increases for labor and material, damages 541
prior acts coverage 471
privity, economic loss doctrine and 160
procurement codes, construction manager liability for failure to comply 384
professional liability coverage 471
professional organizations, design professionals 51t
program management services, construction management 261

project managements institute, construction managers and 239
project parameters 64
project progress scheduling, construction management and 297
project scope, construction management and 283
project size, owners and initial design concept 59
project team, design professional and formation of 59
project use, owners and the initial design concept 59
promissory estoppel
    claims 531
    damages 538
prompt payment statutes 557
proof of compensatory damages 540
property Damage insurance 473
punch lists
    construction management and 368
    design professionals and 118
punitive damages 550

Q

quality assurance, design professionals and 102
quality control/quality assurance 361

R

reciprocity, design professional licenses 44
registration interstate commerce letter, sample 20
registration of copyrights 190
regulation of construction managers
    express regulation as contractors 211
    express regulation as design professionals 219
    generally, 20544
    implicit regulation as contractors 216
    state procurement laws 221
regulatory boards 19
rejection of work, design professionals and 103
retention 554
reverse licenses 187
revocation, design professional licenses 43
risk allocation
    design-bid-build 95
    subcontracting and 499

risk management, construction management and 363
risks, allocation of in design professional contracts 74
rules of professional conduct 26

S

safety
    construction management and 366
    design professionals and 102
schedule estimates, design professionals and 67, 81
schedules
    construction management and 359
    design-build and the construction manager 443
schematic design, design professional and design services 88
scope creep 157
scope of services, design professional contracts 76
sealing of design documents
    by whom 11
    documents required to be sealed 14
    general 11
    international building code 12
    manner of sealing 14
    timing 14
self-dealing, construction manager liability 398
should-cost estimates, damages 550
site logistics, construction management and 300
site safety
    claims against design professionals 172
    contractor obligations 168
    design professionals and 167
    design professionals and safety of employees 179
    immunity and design professionals 176
    owner obligations 168
site surveys, construction management and 286
site visits, design professionals and 98
special masters, litigation and 614
standard form agreements, construction management 263
standard of care
    assumption of higher standard by design professional 155

construction managers 255, 379
  contractual modifications by design professionals 157
  design professionals 147
standard of review, design professional discipline 46
standing, claims against design professionals 160
state public works claims (Little Miller Acts) 557
statutes of limitations
  copyrights 201
  generally 607
  tolling agreements 610
statutes of repose 608
strict liability
  claims against design professionals 145
  design-build and 131
subcontracting for design services
  choosing subcontractors 492
  contribution 496
  dispute resolution 503
  drafting agreements 506
  economic loss rule 505
  exculpatory clauses 499
  flow-down clauses and 504
  indemnification 496
  insurance issues 501
  legal issues 495
  limitation of liability clauses 499
  negotiations for 511
  offshoring 517
  payment clauses 499
  risk allocation clauses 499
  vicarious liability 496
submittals, review by DP 100
substantial completion, design professionals and 115
suspension
  construction management and 310
  design professional licenses 43

T

tail coverage 472
teaming agreements, design-build 142

termination
  construction management 310
  design professional contract 78
  design professionals and termination of contractor 106
third persons, owner's liability to based on construction manager misconduct 409
time-and-motion studies, damages 550
tolling agreements, statute of limitations 610
torts
  construction managers generally 377–413
  damages 539
  design professionals generally 145–181
total cost method, damages 548
trade Secrets, design professionals and 203

U

unfair trade practices, construction manager liability 387
United States Green Building Council 301
unjust enrichment
  claims 531
  damages 538
USGBC 301
utility service, construction management and 290

V

value engineering, construction management 262, 298
value engineering, construction manager and design-build 431–33
vicarious liability, subcontracting 496

W

warranties
  construction management and administration of 371
  design professionals and administration of 120
  design-build 128, 136
  disclaimers by design professionals 156
  insurance 480
worker's compensation, immunity under 562
wrap-up project policies 484

AMERICAN BAR ASSOCIATION APPLICATION FOR FORUM MEMBERSHIP

YES, I want to join the **Forum on the Construction Industry** for $40.00. I am currently an ABA member and belong to at least one dues-paying Section which is a prerequisite to enrollment. Dues for members of the ABA Law Student Division are $5.00.

(Please print or type only)

NAME /__/
FIRM/ORG /__/
DEPT/STE/APT /__/__/__/__/__/__/__/__/__/__/__/__/__/__/__/__/__/__/__/
STREET /__/
CITY /__/
STATE/ZIP /__/__/   /__/__/__/__/__/ - /__/__/__/__/
BUS. PHONE /__/__/__/ - /__/__/__/ - /__/__/__/__/
   EXT. /__/__/__/__/
E-MAIL ADDRESS _____ @ _____

PLEASE INDICATE YOUR PREFERRED METHOD OF PAYMENT:

__Check (made payable to ABA)   __Visa   __MasterCard   __American Express

Card Number
/__/__/__/__/__/__/__/__/__/__/__/__/__/__/__/__/

Exp Date /___/___/___/___/ Signature _____

(Dues for membership in the American Bar Association and the Forum are not deductible as charitable contributions for federal income tax purposes. However, such dues may be deductible as a business expense.) You will begin receiving publications within 90 days of returning this application.

COMPLETE AND RETURN THIS FORM TO:
   American Bar Association
   Forums Department MS 18.2
   321 N. Clark Street
   Chicago, IL 60610

OR FAX TO: 312/988-5677 (credit card only)
OR CALL THE ABA SERVICE CENTER AT 800/285-2221